THE CASE FOR MARS V

EDITOR
Dr. Penelope J. Boston — Complex Systems, Incorporated

SERIES EDITOR
Robert H. Jacobs — Univelt, Incorporated

ASSOCIATE EDITOR
Kelly R. McMillen — Boulder Center for Science & Policy/ University of Colorado, Boulder

Front Cover Illustration:

From Mars, With Love: NASA/JPL/Caltech/Sojourner (TM), (Image # PIA01342; 06-17-99).
First, Mars seemed to be greeting Mars Global Surveyor (MGS) with a Happy Face. Now, it seems as if the planet is sending its love with the latest picture from MGS's Mars Orbiter Camera (MOC). This valentine from Mars is actually a pit formed by collapse within a straight-walled trough known in geological terms as a graben. Graben are formed along fault lines by expansion of the bedrock terrain. The graben in which the pit formed can be seen in this full MOC image. The heart-shaped pit is about 2.3 kilometers (1.4 miles) at its widest. The image was targeted by the MOC team in order to examine the relationship between a lava flow and the graben and pits that disrupted and cut across the flow. The graben, pit, and lava flow are located on the east flank of the Alba Patera volcano in northern Tharsis. The MOC images are illuminated from the left. Malin Space Science Systems and the California Institute of Technology built the MOC using spare hardware from the Mars Observer mission. MSSS operates the camera from its facilities in San Diego, CA. The Jet Propulsion Laboratory's Mars Surveyor Operations Project operates the Mars Global Surveyor spacecraft with its industrial partner, Lockheed Martin Astronautics, from facilities in Pasadena, CA and Denver, CO. Photo Credit: NASA/JPL/Malin Space Science Systems.

American Astronautical Society

THE CASE FOR MARS V

Edited by
Penelope J. Boston

Volume 97
SCIENCE AND TECHNOLOGY SERIES

A Supplement to Advances in the Astronautical Sciences

Proceedings of the fifth Case for Mars Conference held May 26-29, 1993, at the University of Colorado, Boulder, Colorado.

Published for the American Astronautical Society by Univelt, Incorporated, P.O. Box 28130, San Diego, California 92198

AAS Publications Office
P.O. Box 28130
San Diego, California 92198

Affiliated with the American Association for the Advancement of Science
Member of the International Astronautical Federation

First Printing 2000

ISSN 0278-4017

ISBN 0-87703-459-1 (Hard Cover)
ISBN 0-87703-460-5 (Soft Cover)

Published for the American Astronautical Society
by Univelt, Incorporated, P.O. Box 28130, San Diego, California 92198

Printed and Bound in the U.S.A.

CONTENTS

FOREWORD

The Case for Mars was started as an informal and open forum for the discussion of human exploration to Mars. The first conference in March 1981 brought together a diverse group of scientists, engineers, social scientists and interested laypersons. Formal presentations were complimented by workshops and informal discussions. The openness and diversity of the Case for Mars conferences has been maintained through the subsequent conferences and The Case for Mars V reflects these qualities.

When the first Case for Mars was in the planning stages official interest in Mars exploration was at a historic low. The concept of human exploration was distant at best. The Case for Mars conferences helped define a new paradigm in Mars exploration. Rather than large self-contained missions with limited scope and duration, the Case for Mars scenario envisaged a continued human presence on Mars with missions that built an ever expanding capability. Central to this new way of exploring Mars was the use of Martian resources to produce basic life support consumables, building materials, rocket fuel, and even for the growth of food in greenhouses. Mars was no longer like the Moon, a place just to visit and explore. Instead, Mars was a destination for future long term human activity. The main science goals for the Case for Mars was life: past, present and future. Past life meant the search for fossil or biochemical evidence of life that may have existed on Mars several billion years ago. Present life meant the maintenance of human life on Mars and the greenhouses necessary to provide food for the human explorers. Future life meant the possibility that in the long term Mars could be terraformed and made once again into a planet with a biosphere.

From the Case for Mars I, interest in Mars grew steadily reaching a climax in the Space Exploration Initiative announced by then president Bush in July 1989. As planning for Case for Mars V began in early 1993 it was clear that the Space Exploration Initiative would not be implemented. Interest in Mars exploration was waning fast. During both the periods of growing interest in Mars as well as during periods of waning interest, the Case for Mars conferences provided a vehicle for maintaining a continuous dialogue amongst people in the technical community who were exploring problems and issues related to human exploration of Mars.

The proceedings from the conference provide the concrete embodiment of this continuity. These proceedings contain an important set of papers that deal with broad concepts for how to motivate the human exploration of Mars. Of particular significance is the paper by the late Dr. Thomas Paine. Tom was truly a pioneer that led the way in planning out a bold but practical plan for space exploration. Other papers deal with the technical aspects of robotic and human exploration of Mars.

Finally, the concept of terraforming Mars is considered. The concepts and visions of making Mars a second home for life provide a powerful motivation; here is a task worthy of every human endeavor and contribution of intellect. Let the adventure begin.

Christopher P. McKay
NASA Ames Research Center

CHAPTER 1:
PLANNING FOR MISSIONS TO MARS

MARS COLONIZATION: TECHNICALLY FEASIBLE, AFFORDABLE, AND A UNIVERSAL HUMAN DRIVE*

Thomas O. Paine†

In the 21st century men and women from Planet Earth will establish a new human society on the resource-rich Red Planet. Several years ago I presented a 100 year "Timeline for Martian Pioneers" at the Case for Mars II Conference at the University of Colorado [Paine, 1984]. Revisiting the rationale and schedule for Mars colonization, I find prospects are now brighter than ever for international investment in the settlement and economic development of the Inner Solar System. This universal human drive will spur continuing technological advance, global cooperation and economic growth. The settlement of Mars on the proposed schedule can eliminate Malthusian limits to the aspirations of humanity, and open our way to the stars.

INTRODUCTION

The future of mankind will be decided by the race between two competing human drives, one unleashing military power to compete for Planet Earth's finite resources, the other organizing international cooperation to provide access to unlimited extraterrestrial resources. The outcome of this fateful competition will be decided within our lifetimes. I believe that rational statesmen will choose peace and the extension of humanity to other worlds. To establish the required transportation and communication network linking bases throughout the inner solar system we must:

- Avoid major wars through increased international cooperation and armament reductions;
- Reduce global tensions by raising living standards on all continents; and
- Apply a small fraction of the vast capital, technological, and educated human resources thereby liberated to advance spaceflight technologies and learn to live off the land in self-sustaining Martian communities.

This paper looks ahead ten decades to consider briefly: (1) The prospects for international cooperation to achieve peace and prosperity that will make Mars settlement technically feasible and affordable; (2) The rationale that will make Mars settlement a universal human drive; and (3) A potential decade-by-decade timeline for creating an independent, high-technology Martian society.

* This paper was first published in *Strategies for Mars: A Guide to Human Exploration*, C. R. Stoker and C. Emmart, eds., Vol. 86, AAS *Science and Technology Series*, pp. 579-591, 1996. The paper is being republished in this volume in honor of Dr. Paine.

† Administrator - NASA, 1968-70; Chairman - U.S. National Commission on Space, 1984-86; Member - Advisory Committee for the Future of the U.S. Space Program, 1989-91. Dr. Thomas Paine is now deceased.

INTERNATIONAL COOPERATION FOR PEACE AND PROSPERITY

The Stone Age may return on the gleaming wings of science.

— Winston Churchill

This short paper cannot rigorously address geopolitical issues of war and peace, yet the avoidance of major wars and catastrophic nuclear exchange is an essential precondition for the settlement of Mars, as is global economic and technical advance to promote stability and prosperity on Earth. These conditions will activate Martian settlement in the 21st century. Brief comments therefore follow on the chances of achieving these prerequisites.

Avoiding Major Wars and Nuclear Armageddon

This is clearly imperative! Since World War II nuclear weapons have increased in numbers and destructive power by more than a factor of a thousand, and are already in the possession of at least seven nations. Their awesome power effectively blocked their use in the sporadic conflicts of the post-colonial era, and deterred the nuclear superpowers from escalating warfare. In the dark shadow of the hydrogen bomb responsible leaders of the United States and the Soviet Union avoided direct armed conflict despite Cold War tensions, while well-disciplined military controls prevented an accidental firing which could trigger Armageddon. The threat of nuclear holocaust is diminishing in the new geopolitical environment of the 1990s. With good luck and responsible statesmanship, there appears to be a good chance that non-proliferation treaties and international inspection can avert nuclear war. Diplomacy based upon enlightened self interest must minimize armed conflicts while a vigorous Martian civilization becomes self-sustaining. This will represent a great victory of life over death.

Enough fissionable material exists in the world's nuclear warheads to provide fuel to power Martian civilization for centuries. Beating terrestrial nuclear swords into extraterrestrial plowshares provides statesmen with a compelling rationale for mutual contribution of nuclear arms to provide power for Mars, thus removing these destructive weapons from Earth. The conquest of Mars, unlike terrestrial military invasions, provides a unique unifying force around the Earth that reaches out to all people. A vigorous international program to settle Mars can provide an alternative to destructive wars that could decimate high-tech civilization on Earth, and humanity's chance to reach the stars.

Prospects for Global Technical and Economic Advance

If major conflicts can be avoided through wise diplomacy and effective international institutions, can we achieve the positive prerequisite of global technical and economic progress on a scale that makes sustained Martian investment affordable? There are many alternative scenarios for future world economic development, from neo-malthusian doom and gloom to technocratic optimism. In his classic *The Two Cultures*, Lord C. P. Snow [1959] observed that non-technical writers often project a pessimistic view of the future, based on their fears of debased human nature manipulated by despotic bureaucracy (e.g. *Brave New World* [Huxley, 1932] and *1984* [Orwell, 1949]) while technically-oriented writers often paint euphoric pictures of advancing science and technology rationally applied for human benefit in global democratic enterprise (e.g. *2001, A Space Odyssey* [Clarke, 1968], and *The World, the Flesh and the Devil* [Bernal, 1929]). This paper assumes that basic human nature changes very little, but that exponential technological advances initiate revolutionary social change. For the purposes of this paper I have selected as most likely my late friend Herman Kahn's "Guarded Optimism" scenario, which sees the steady economic growth of the last two hundred years continuing as the industrial revolution continues to spread. His provocative book, *The Next 200 Years* [Kahn, 1976], is based in large part on an analysis of the technical, economic, and social changes that have taken place in industrializing countries since the start of the Industrial Revolution in late 18th century Europe, and the implications for future advances.

One quarter of Earth's population now lives in the 80 percent urbanized and industrialized nations of Europe, North America, Australasia, and Japan. This group enjoys a per capita Gross Domestic Product exceeding $10,000 and a life expectancy 50 percent greater than that prevailing in the least developed nations. This uneven progress of economic development is reflected in a roughly 50:1 disparity in productivity and wealth between modern industrialized and traditional agrarian economies. In spite of this gap, the impact of science and technology is felt in every nation. Witness, for example, the unprecedented global medical campaign that completely eradicated smallpox from our planet, and the proliferating "Green Revolution" that in the space of two decades made possible a grain surplus for a third of the human race in Asia.

Comparing different eras and economies involves complex problems of concept and methodology, so results can be ambiguous. With this limitation in mind, Table 1, from *Pioneering the Space Frontier* [National Commission on Space, 1986], summarizes two-centuries of steady economic growth in the United States.

Table 1
U.S. REAL ECONOMIC GROWTH - 1800-1985

Mid-year	Population (1000s)	Total GDP (1985 $ billions)	GDP per person (1985 $ billions)
1800	5,297	4.9	923
1855	27,386	48.3	1,762
1885	56,658	156.4	2,761
1900	76,094	277.7	3,650
1933	125,579	517.3	4,120
1950	152,271	1,245.7	8,181
1965	194,303	2,157.6	11,104
1985	238,631	3,769.4	15,796

In the past 185 years, America's population has multiplied about 45 times, Gross Domestic Product about 770 times, and per capita Gross Domestic Product about 17 times. This experience is paralleled throughout the industrialized world.

In 1800, Planet Earth's estimated total population was about 900 million people, 10 percent urbanized, producing a Gross World Product (GWP) of about $350 billion, or $390 per capita (all monetary units are in 1985 U.S. dollars). Today's world population is estimated to be 5,320 million people, about 40 percent urbanized, producing a GWP of about $20,000 billion, or $3,750 per capita (about the U.S. per capita output of 1900). In other words, the first 200 years of the Industrial Revolution saw the population of the world multiply about 5.9 times, Gross World Product about 57 times, and GWP per capita about 10 times.

This momentum appears likely to continue through the next 100 years, but with uneven progress in differently managed economies. In the 21st century an increasing percentage of humanity will possess advanced technologies, growing capital resources, trained scientists and engineers, and future-oriented leaders. This combination will encourage many nations to participate in pioneering the space frontier.

Because most of the 100 new post-colonial nations are of subcritical size, free world trade is essential for their economic growth. Less developed nations need ready access to the industrialized world's technology, capital and markets. With good management, sustained growth can then be achieved, as has been demonstrated by Japan and the four neoconfucian economies of East Asia: South Korea, Taiwan, Hong Kong and Singapore. Between 1974 and 1983 the percent growth in GNP (in constant dollars) of these economies was: U.S. 22%, Japan 47%, South Korea 89%, Taiwan 104%, Hong Kong 108%, and Singapore 124%. The South Korean farm family who received on average $747 in 1970 earned $4,800 in 1983; the factory worker who was paid $0.10 an hour in 1970 got $1.69 in 1983. Beijing's emphasis on economic development for world markets has impelled China to follow the path of its East Asian neighbors, improving prospects for global economic growth by bringing another quarter of the human race into productive global trading and investment relationships. These successes are consistent with the observations condensed in Table 2, a one-page commentary on the organization of productive economic institutions at seven levels of aggregation based upon my personal observations in international business. Others may disagree, but I believe that this pragmatic view is basically correct, and that similar considerations will apply to super-high-productivity industry on Mars.

The point here is that the development of advanced industrial organizational skills and the attainment of much higher productivity and wealth on every continent of Earth appear well within reach in the next 100 years. New technologies, like biological engineering, optical computer networks, robotics and artificial intelligence, will accelerate this trend. There are many uncertainties, but I believe that productivity and living standards will continue to advance in much of the world. These demographic and economic trends could lead to a 2090 world population on the order of 20 billion people producing a Gross World Product around $100 trillion, for a per capita GWP of $5,000. This is half of the $10,000 per capita GNPs of developed nations today, but will require a challenging eightfold increase in GWP.

Table 2

ONE-PAGE COMMENTARY ON ORGANIZING FOR PRODUCTIVITY

The most productive economic institutions at seven levels of human organizations are:	
Level I	A trained, well-equipped team of healthy, motivated workers; within
Level II	A well-managed, internationally competitive industrial enterprise that balances the interests of customers, investors, and workers; within
Level III	A decentralized modern urban complex with excellent financial, educational, and cultural institutions linked to the world through efficient communication and transportation; within
Level IV	A 99% literate, 90% urbanized electrified nation with an effective legal system, equal opportunity, minimal gap between rich and poor, and globally-oriented, incorruptible leaders; within
Level V	An actively cooperating region with ready access to world markets, capital, technology and resources; within
Level VI	A freely trading world at peace; within
Level VII	A solar system rich in energy and materials under cooperative human exploration and development.

(Note: The world's *least* productive worker is an illiterate, unhealthy, low cast peasant working without incentives for a bureaucratic, xenophobic, military dictatorship in a corrupt, war-ravaged, agrarian nation.)

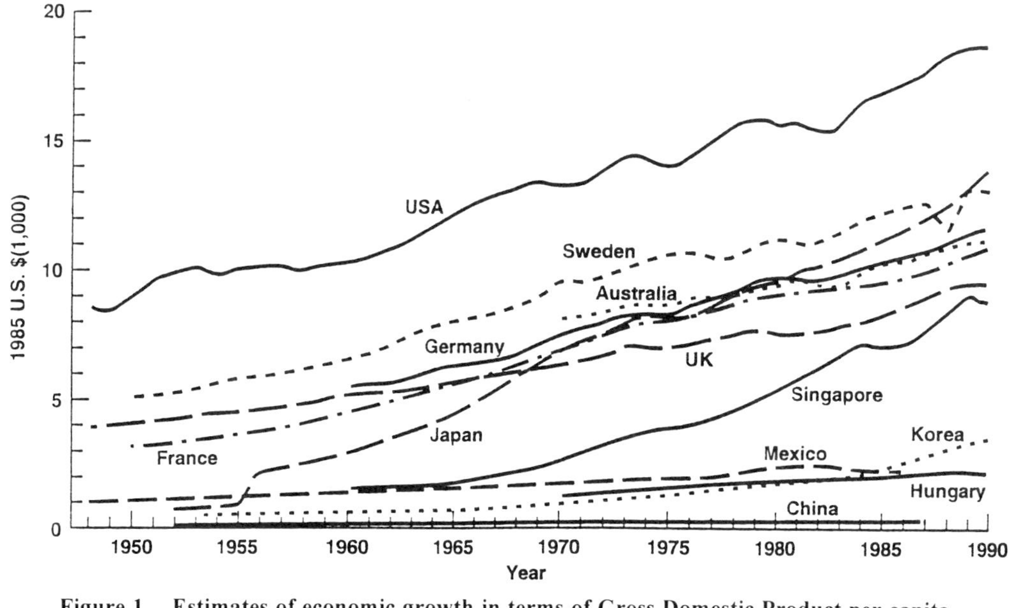

Figure 1 Estimates of economic growth in terms of Gross Domestic Product per capita.

Figure 1 provides a sense of historical perspective by depicting estimates of comparative real economic growth, in 1985 dollars and exchange rates, for 12 nations across the development spectrum in terms of Gross Domestic Product per capita. As the figure shows, some countries, such as Germany, Japan, and Singapore, are achieving relatively rapid rates of economic growth, surpassing the per capita incomes of other countries with long histories of higher ranking positions. Virtually no one looks on such statistics as ends in themselves, but they do correlate with other social measures to indicate nations whose citizens enjoy greater opportunities and higher standards of living.

Adjusting to technological change will bring many social problems as the broadening industrial revolution continues to transform our planet. Strengthened international trade and environmental institutions will be required to guide wise economic growth, but five fundamental factors suggest favorable prospects for international cooperation in the 21st century:

- 45 years of Cold War ending without a nuclear Armageddon;
- The prospect of declining military expenditures freeing capital to update the world's industrial, agricultural, communication, and transportation infrastructure (without economic distortions due to superpower competition);
- Accelerating technical progress across a broad spectrum that will improve public health, education, and spur further widespread productivity gains;
- Demonstrated rapid growth of national economies in well-managed, global-market-oriented nations; and
- Increasing capabilities of effective international organizations like Intelsat to bring nations together in challenging high-tech enterprises.

It therefore appears likely that the substantial economic, human and technical resources required for the settlement of Mars in the 21st century will be broadly available on an increasingly affluent Earth.

This brings us to our basic question: Will Mars settlement be affordable within the probable 21st century Gross World Product? Until we have a settlement plan it is premature to discuss costs, but I estimate that the U.S. space budget required would grow to about twice NASA's $20 billion annual expenditures in the Apollo lunar landing era. With a comparable sum spent by other technically-oriented nations, the Earth's investment in Mars would reach around 80 billion 1992 dollars per year (about the annual output of an automobile company). Assuming that the expenditure level will rise gradually over the century (although with peaks and valleys), this would represent a cumulative total Earth investment in a self sustaining Mars settlement of about $5 trillion. I believe that this price tag is not unreasonable. If the annual GWP in this period rises from today's $20 trillion to $100 trillion in 2085, this $5 trillion investment in Mars would represent less than a tenth of a percent of the cumulative GWP over the century. A prospering Earth can readily afford major investment in the extension of terrestrial life to Mars. Will it wish to do so?

DRIVING FORCES FOR MARS SETTLEMENT

Men need the mystery and romance of new horizons almost as badly as they need food and shelter. In the difficult years ahead, we should remember that the Snows of Olympus lie silently beneath the stars, waiting for our grandchildren.

— Arthur C. Clarke, *Author, Consultant to NASA, and former President of The British Interplanetary Society.*

Once Europeans mastered deep sea navigation, settlements were rapidly established on other continents for many reasons, attracting pioneers with equally diverse motivations. Men and women endured great hardships crossing stormy seas in sailing vessels, or opening the American West in Conestoga Wagons, impelled both by dissatisfaction with conditions behind them and visions of a better life ahead. Families undertook hazardous six to nine month trips as a matter of course. When space navigation and closed-ecology life support technologies have been mastered in the 21st century, Mars will be settled for a similar wide range of national and human motivations, including:

- *Economic Development* - Long term investment in a creative new growth economy, in which super-high-productivity industry utilizes unique extraterrestrial materials and energy to initiate the economic development of the inner solar system;
- *Limitless Growth Potential* - Access to virgin continents with vast untapped resources, opening an endless frontier that eliminates Malthusian limits to human aspirations;
- *National Pride and Leadership* - The desire of far-seeing leadership and citizens of affluent nations to participate in history's greatest high-tech human venture;
- *Religious, Ideological, and Humanistic Values* - The basic desire to preserve life and to expand and transmit deeply treasured human beliefs and cultural heritages to an expanding posterity;
- *Martian Descendants* - Opportunities for the transplantation and growth of genes, families, races and ethnic groups on new worlds;
- *A Fresh Start* - The attraction of an advanced, intellectually-based civilization working toward a limitless future for mankind free of the old world's diseases, ignorance, fears and outworn prejudices;
- *Technical Pilgrim's Haven* - The chance to build forward-looking, technically-oriented, frontier societies with advanced social compacts that place a high value on scientific vision, technical competence, individual responsibility and cooperative enterprise;
- *High-Productivity Systems Driver* - Initial labor costs on Mars will be a hundred times those on Earth, stimulating the development of reliable robotic production systems with a hundred times the productivity of terrestrial factories;

- *Research and Exploration* - A unique R&D base and technical society ideally located for astronomy, planetary research, spaceflight development and the exploration of the asteroids and outer solar system; and
- *Prototype Extraterrestrial Community* - A representative low-gravity, low-atmospheric-pressure, lifeless planet with rich resources (including water and carbon) for advanced engineering of self-replicating robotic mining, agriculture, industry and closed-ecology life support systems that future generations will employ throughout our solar system - and, eventually, on an expanding new frontier of remote planets circling other stars.

I am convinced that these driving forces will appeal to many individuals and national leaders, attracting more volunteers than can be initially accommodated. How then is the exploration and settlement of Mars likely to proceed over the next 100 years?

A TIMELINE FOR MARTIAN EXPLORATION AND SETTLEMENT

Only a miraculous insight could have enabled the scientists of the eighteenth century to foresee the birth of electrical engineering in the nineteenth. It would have required a revelation of equal inspiration for a scientist of the nineteenth century to foresee the nuclear power plants of the twentieth. No doubt, the twenty-first century will hold equal surprises, and more of them. But not everything will be a surprise. It seems certain that the twenty-first century will be the century of scientific and commercial activities in outer space, of manned interplanetary flight and the beginnings of the establishment of permanent human footholds outside the mother planet Earth.

— Wernher von Braun

Let's now project the achievements that we should expect to see in space exploration over the next ten decades if our three assumptions are correct:

- That major wars and nuclear Armageddon on Earth can be avoided,
- That steadily growing world capital, technological, and educated human resources will spur a major Mars settlement project, and
- That for a variety of reasons increasingly affluent nations will direct their space efforts toward Mars, attracting talented space pioneers willing to undergo hardships to participate in this inspiring adventure on the cutting edge of humanity.

The 1957-1992 Technical Foundation

We are considering Mars settlement today in the fourth decade of the Space Age. Let's begin by summarizing some significant achievements that provide the technical foundation on which the Mars venture will be built:

- Large rocket engines and heavy-lift launch vehicles (HLLVs) with capacities up to 120 tons to Low Earth Orbit (LEO), the U.S. Space Shuttle for round trips to LEO, European and Chinese commercial launchers, and Russian Energia HLLVs and automated LEO resupply freighters;

- High energy solid fuel and liquid hydrogen-liquid oxygen upper stages capable of propelling spacecraft from LEO into interstellar space, with advanced R&D on alternative high-specific-impulse spacecraft propulsion systems, from electric ion thrusters to thermal nuclear stages;

- Global tracking networks with large diameter antennas and tracking and data relay satellites capable of high-data-rate communication with spacecraft beyond Neptune and Pluto;

- Crewed orbital flights of many months duration in Salyut, Mir and Skylab space stations that demonstrate adaptation to weightlessness by cosmonauts and astronauts adequate for the long round trip to Mars;

- Life science advances, including closed-ecology life-support experiments by NASA, Biosphere II, and Russian laboratories;

- Robotic reconnaissance of the solar system from Mercury to Neptune, including fly-bys, orbiters, soft landers on three worlds, surface rovers, automated sample returns, and years of climate observations from two Viking landers on Mars;

- Earth observation satellite systems studying oceans, land masses, ice fields, the atmosphere, and their complex interactions;

- Reconnaissance satellite systems capable of monitoring arms control and peace keeping treaties;

- Orbiting astrophysical observatories scanning the universe across the spectrum from infrared through gamma rays; and

- Six manned lunar expeditions that demonstrated 23 years ago surface exploration with lunar landers, self-contained spacesuits, electric jeeps, automated lunar orbit rendezvous, and other systems and techniques usable on Mars.

In addition to this solid foundation of applicable space experience, we have many untapped sources of valuable information in the history of European exploration, in Magellan's and Drake's circumnavigations, in the colonization of "The New Worlds" of America and Australia, in the long voyages of New England Whalers, in stressful wartime submarine patrols lasting up to a year at sea, and in Antarctic research stations. Historical research to develop data on demonstrated human endurance, the settlement of new continents, and the theory of the American frontier can contribute greatly to planning the Mars program.

This is our heritage. What lies ahead as we build on our 35 years of spaceflight experience to pioneer the space frontier? Here is my proposed decade-by-decade future history of Mars settlement.

The 1990-2000 Decade

NASA's new management decides that in the fourth decade of the Space Age U.S. industry knows how to build space systems, and U.S. university teams know how to conduct space research. New contracting procedures are therefore instituted that give aerospace firms and academia new freedom and responsibility to carry out faster-paced programs at lower cost. Detailed specifications of every nut and bolt, followed by micromanagement of every operation, are eliminated in favor of performance contracts. This results in new launch vehicles and spacecraft being deployed on accelerated schedules, from families of expendable launch vehicles and direct broadcast satellites, to man-rated HLLVs and Earth observation systems. Fresh information on the planets is streaming in from Magellan at Venus, the U.S. Mars Observer and Russian Mars 4 spacecraft at Mars, asteroid fly-by missions, Galileo at Jupiter, Cassini at Saturn and other robotic platforms. NASA's Space Station *Freedom* is canceled in favor of a later HLLV-launched station designed to support the Space Exploration Initiative. The result is a new orbital maintenance and operations center, with life science laboratories, spacecraft maintenance, check-out, refueling and launch capabilities, experimental space construction test bed, tether experiments, and other 21st century-oriented capabilities.

The 2000-2010 Decade

The major spaceflight program of the decade is a return to the Moon 30 years after Apollo to establish permanent international lunar bases. Aerospace planes and man-rated HLLVs with crew-recovery capability from launch pad to orbit are in initial operation, greatly lowering the cost and risk of transporting men and women between Earth and LEO. The shuttle fleet is retired after another tragic accident. Additional international lunar base construction is supporting higher levels of activity in space, including high-resolution, multi-spectral, robotic surveys of the Moon and Mars. Proposed Martian habitats and workshops are being tested on the Moon. A crash program has been instituted to send prospector probes with penetrators to Phobos, Deimos and the Martian surface. Prototype systems under development and test include low-cost electric orbit transfer vehicles using large electric ion thrusters and aerobraking, closed-ecology life support habitations, robotic mining and primary metallurgy and ceramics industrial systems designed to process lunar and Martian resources, and 1-10 rpm artificial gravity spacecraft. A number of new university departments have received NASA contracts to step up work in such fields as planetary sciences, exo-agriculture, closed-ecology, robotics, artificial intelligence and exo-sociology. For example, NASA dedicates a new permafrost laboratory at the University of Alaska.

A man-rated space transportation system is extended to Geosynchronous Earth Orbit (GEO) and the Moon; man-tended orbiting bases constructed in LEO are deployed to GEO and lunar orbit using economical electric ion propulsion. Martian rovers with automated sample-return capability are helping to select the most promising sites for resource development. Meanwhile, back on Earth, the supporting R&D in academia is advancing rapidly under the stimulus of the accelerating Mars programs.

The 2010-2020 Decade

In this decade the first U.S.-led international Mars expedition shoves off from an Earth-orbital base to rendezvous with Phobos and Deimos and check out predeployed surface habitats. The first humans then land on Mars to activate bases and initiate exploration. Prototype extraterrestrial communities based on robotic resource development begin operation. A new *International Space Agency* is organized to manage extraterrestrial operations. All of these advances are eclipsed in public interest, however, by the dramatic announcement from Cal Tech, quickly confirmed by the Russian Academy of Sciences, that the first signals have been detected indicating the presence of an advanced extraterrestrial civilization in a remote arm of our galaxy.

The 2020-2030 Decade

The first automated five gigawatt nuclear electric generating station goes on-line to power Martian bases. Its energy is used to separate hydrogen and oxygen from permafrost to fuel surface transport and surface-to-orbit shuttles. Economical electric ion thrusters are providing propulsion for robotic freighters delivering heavy cargoes from Earth, and to emplace the first multipurpose space base in low Martian orbit. This facility serves as Mars Mission Control Center, orbital laboratory, transportation node, space maintenance base, and refueling station. One of its first missions is to experiment with advanced tether techniques to lower orbit transfer costs. Automated aerobraking freighters are using standardized nuclear generators to provide power for their electric thrusters; these one-way transports deliver structural materials and their automated electric power equipment for use on Mars. The first large cycling spaceships are emplaced into permanent orbits linking Earth and Mars. A number of planets are detected in orbit around nearby stars; one of them has an atmosphere rich in oxygen and water vapor, suggesting the presence of life. Simultaneous global holidays on Earth and Mars mark the birth of the first Martian baby.

The 2030-2040 Decade

The population of the Moon now numbers several thousand people, and the first extraterrestrial children are thriving. Closed-cycle lunar communities are now increasingly supported by robotic mining, metallurgy, ceramic processing, factory production and agriculture using indigenous resources, but hydrogen and carbon are still costly imports. Next-generation systems designed for future deployment on Mars are being tested at lunar test sites. Martian research bases are experimenting with high-productivity resource development and construction techniques adapted to Martian conditions. They are manned by several hundred men and women scientists, engineers and technicians who have volunteered for seven year tours of duty. First priority is given to launch pads, air strips and shelter. Additional cargoes to build Martian infrastructure continue to arrive from LEO during favorable Earth-Mars conjunctions.

The 2040-2050 Decade

Technologically this decade is advanced as far beyond 1992 as the 1940s are behind 1992. Closed-ecology lunar prototype communities and resource development systems now support 10,000 lunarians, but more advanced extraterrestrial life support systems utilizing indigenous resources are being emplaced on Mars. By 2045 there are a thousand people on Mars completing installation and initial operation of Martian infrastructure, with emphasis on lunar-demonstrated robotic agricultural and industrial facilities to reduce dependence on costly shipments from Earth.

The 2050-2060 Decade

Despite advancing technologies, teleportation has not been invented in the classic science fiction sense of *Beam me up, Scotty, there's no intelligent life here*. Because optical data communication between planets is so much cheaper than the transport of mass, however, robotic factories on Mars with direct software links to twin factories on Earth are achieving an effective "teleportation." Equipment like hydrogen-oxygen fueled jeeps with artificial intelligence guidance produced on Earth can be simultaneously duplicated on Mars, and vice versa. By 2055 this system of "teleportation via robotic factory software," in which bits are transmitted instead of objects, is creating a wealth of new opportunities for import-export and balance of trade among planets. The population of Mars has grown to 5,000 people through continuing immigration and strong encouragement of multiple births. Robotic Martian factories are working 24 hours and 37 minutes a day, 686 days a year, to turn out robotic Martian factories. The Arthur C. Clarke Observatory is dedicated atop towering Mons Olympus.

The 2060-2070 Decade

The population of Mars is passing 10,000 humans and 100,000 equivalent robots. Two-way Mars-Earth trade is beginning to move toward a balanced exchange and joint scientific research programs are underway. The space programs of Earth and Mars are cooperatively exploring the asteroids and the giant outer planets with their resource-rich moons; materials that are scarce on Mars and Earth's Moon are being sought on low-gravity worlds to reduce dependence on terrestrial resources. Mars-Earth astronomical interferometers have pinpointed the location of eleven other civilizations in the galaxy, but all are too distant to permit communication within a lifetime. The industrial policy of Mars is to reduce dependence on imported goods from Earth and attain self-sufficiency at the earliest possible date. On Earth fusion power is hailed as the space propulsion drive of the future.

The 2070-2080 Decade

As robotic mining, manufacture and construction extends Martian community facilities, immigration from Earth is stepped up and the human population of Mars nears 50,000 people. Trade with Earth is now nearing balance, with artistic products, engineering designs, scientific research results and other software products predominating. Specialized high value goods continue to be imported and exported, and tourism is becoming an important source of foreign exchange for the Martian economy. On Mars matter-antimatter drives are hailed as the space propulsion drive of the future.

The 2080-2090 Decade

The vigorous young Mars settlement contains 100,000 people and close to a million equivalent robots. Martian immigration policy is promoting a rich gene pool from every continent on Earth, while biological laboratories, zoos, gene libraries, museums, art galleries, and book and image libraries house extensive collections of terrestrial life. Although a lively trade is underway, the continuation of life on Mars no longer depends upon imports from Earth; the same is true of the Moon. Martians, lunarians and earthlings are now evolving independently on three resource-rich worlds.

Genetically engineered bacteria are under test to initiate the terraforming of Venus and Mars. Young Martians are pressing for additional settlements beyond the asteroid belt, and in response promising sites on Pluto, Charon, Triton, Titan Ganymede, Europa and other moons of Uranus, Saturn and Jupiter are being surveyed. Joint Mars-Earth-Moon research on matter-antimatter and photon propulsion to drive spacecraft at speeds approaching ten percent of the speed of light shows promise of opening the challenging "stellar frontier"; long range plans now call for launching robotic probes to temperate planets circling nearby stars before the end of the century.

ACKNOWLEDGMENTS

I am grateful to Carol Stoker for asking me to revisit this topic, and to Fred Paroutaud and Dr. Mark Fisher for invaluable assistance.

REFERENCES

Bernal, J. D., *The World, the Flesh and the Devil; an Inquiry into the Future of the Three enemies of the Rational Soul*, Paul, Trench, Tribner and Co., London, 1929.

Clarke, A. C., *2001: A Space Odyssey*, New American Library, New York, 1968.

Dyson, F., "The world, the flesh and the devil," Third J.D. Bernal Lecture, delivered at Birkbeck College, London, 16 May 1972; reprinted as Appendix D in *Communication with Extraterrestrial Intelligence*, Carl Sagan, ed., MIT Press, Cambridge, MA, 1973.

Huxley, A., *Brave New World*, Harpers, New York 1932.

Kahn, H., W. Brown, and L. Martel, (with the assistance of the Staff of the Hudson Institute): *The Next 200 Years: A Scenario for America and the World*, W.M. Morrow & Co., New York, 1976.

The National Commission on Space: *Pioneering the Space Frontier*, Bantam Books, New York, 1986.

Orwell, G., *1984*, Secker & Warburg, London, 1949, and Harcourt Brace Jovanovich, New York, 1949.

Paine, T. O., "A timeline for Martian pioneers," AAS 84-150, in C. P. McKay, ed., *The Case for Mars II*, American Astronautical Society, *Science and Technology Series*, 62, pp. 3-21, (Proceedings of the Second Case for Mars Conference held July 10-14, 1984), Univelt, Inc., San Diego, CA, 1985.

Snow, Lord C.P., *The two cultures and the Scientific Revolution*, Cambridge University Press, Cambridge, 1959.

THE THIRD ERA OF SPACE EXPLORATION*

Dwayne A. Day†

Only a miraculous insight could have enabled the scientists of the eighteenth century to foresee the birth of electrical engineering in the nineteenth. It would have required a revelation of equal inspiration for a scientist of the nineteenth century to foresee the nuclear power plants of the twentieth. No doubt, the twenty-first century will hold equal surprises, and more of them. But not everything will be a surprise. It seems certain that the twenty-first century will be the century of scientific and commercial activities in outer space, of manned interplanetary flight, and the establishment of permanent human footholds outside the planet Earth.

— Wernher von Braun

What does it take for Americans to do great things; to go to the moon, to win wars, to dig canals linking oceans, to build railroads across a continent? In independent thought about this question, Neil Armstrong and I concluded that it takes a coincidence of four conditions, or in Neil's view, the simultaneous peaking of four of the many cycles of American life. First, a base of technology must exist from which to do the thing to be done. Second, a period of national uneasiness about America's place in the scheme of human activities must exist. Third, some catalytic event must occur that focuses the national attention upon the direction to proceed. Finally, an articulate and wise leader must sense these first three conditions and put forth with words and action the great thing to be accomplished. The motivation of young Americans to do what needs to be done flows from such a coincidence of conditions......the Thomas Jeffersons, the Teddy Roosevelts, the John Kennedys appear. We must begin to create the tools of leadership which they, and their young frontiersmen, will require to lead us onward and upward.

— Harrison H. Schmitt

Although both Harrison Schmitt and Wernher von Braun were correct on the generalities, it is politics that has been responsible for the American space program in the past and politics which will determine its future. A historical analysis of the politics behind the American space program indicates that in the past, large projects have been undertaken for reasons that no longer exist today.[1] Unfortunately, there is no compelling rationale in existence today that will lead to a human mission to Mars within the fore-

---

* This paper builds on several ideas first advanced by other authors, most notably John Logsdon, Carl Sagan, and Ray Williamson. The author wishes to acknowledge their work and the assistance of a number of individuals whose comments were of considerable help in shaping this paper: Andre Bormanis, Michelle Heskett and Heather Young. Mistakes are, of course, entirely the fault of the author.

† Space Policy Institute, Washington, DC 20052.

seeable future—that is, the next thirty years. Until a convincing rationale is developed, dreams of the human exploration of Mars remain simply that—dreams.

THE FIRST AND SECOND ERAS OF SPACE EXPLORATION

Space occupies a rather unusual place in the sphere of governmental activities. It has strong ties to national security, both symbolic and substantial. For years the American space program has been justified by Cold War political concerns. The first era of space exploration (1957-1969) was driven by foreign policy considerations such as the desire for international prestige and national honor. This period, which began with Sputnik and reached its zenith with Apollo, ended once Richard Nixon entered office and rejected the recommendations of the Space Task Group report, which included a Mars mission, lunar base, space station and space shuttle. Numerous documents from the period during which Kennedy made his decision reflect the role of prestige and national honor in choosing to spend massive resources on space exploration.[2] The United States went to the moon not to explore or colonize or to establish a beachhead, the U.S. went to serve very real purposes of superpower competition here on Earth.[3]

The second era of space exploration (1972-1989) was driven more by domestic infrastructure and bureaucratic concerns, but still contained aspects of the superpower competition that dominated the earlier era. Foreign policy and nationalistic concerns still played a role in contributing to major decisions like the space shuttle and space station, but they soon were replaced by more domestic and bureaucratic concerns, like constituency politics, after a commitment had been made. They no longer provided the raison d'être of the programs or were a defining influence on their shape.

The space shuttle was, in large part, a result of a need for NASA to find a continuing mission. It was approved by President Nixon for a number of reasons, including national security and economic concerns, and a desire by the President not to be remembered as the person who ended the manned space flight program. Although it was in part an aspect of the Cold War competition between the superpowers, it also served several bureaucratic needs.[4] Similarly, the space station received a commitment from President Reagan because the Soviet Union possessed a space station and the United States did not.[5] Both projects were heavily influenced by domestic political concerns which, for instance, dictated that they be spread among key congressional districts and shaped other engineering design criteria.[6]

The defining characteristics of these two periods cannot be repeated. The Cold War is over and constituency ("pork barrel") politics is insufficient to maintain a large human space program by itself.[7] The old rules no longer apply.

THE THIRD ERA OF SPACE EXPLORATION

The Third Era of Space Exploration is only now being defined. The political environment in which space exploration must now be undertaken is fundamentally different from before in three major aspects: the Cold War is over, the budgetary situation is increasingly constrained, and the relationship between science and society has changed.

The Third Era's beginning was signaled by the total failure of the Space Exploration Initiative to garner political or public support. Clearly the justifications that worked in the past—leadership, bureaucratic direction, congressional pork—no longer work as significant factors in contributing to a decision to commit to major new endeavors in space. In addition, the cost of such an initiative was seen as prohibitive. Finally, Congress was unwilling to fund even small precursor missions on their scientific merits alone and the Bush administration was unable to demonstrate that significantly useful technological developments could be accrued from such an undertaking.

At the end of the Cold War there was an initial expression of enthusiasm among space advocates that defense funds could be redirected to the space program. But this hope soon faded and some analysts recognized that the end of the Cold War actually eliminated much of the justification for space exploration.[8] Space exploration owed much of its origins to the ballistic missile programs of the 1950's and the high level of tension of the Cold War. The political environment provided a clear, easy, and satisfying answer to the question of "why explore space?": To beat the Russians. Now that that answer is no longer relevant, others must be sought. If space was inexpensive, some of the answers so far proposed might be adequate. But spaceflight, particularly human spaceflight, is an extremely expensive undertaking by any measure. It is solely the realm of governments and constitutes among the most expensive single projects that they undertake.

Of course, the expense of spaceflight would matter quite differently if the United States was not running a budget deficit. In addition to operating in a new international political environment, the space program now also operates in a domestic political environment substantially different from that which existed in the past. Of primary importance is the budgetary crisis which will increasingly constrain spending on discretionary programs like space. NASA's budget is vulnerable for two reasons. First, unlike interest payments on the national debt or entitlement programs, it can be cut. The civilian space program receives its funding from the same committee that determines funding for Housing and Urban Development and Veterans funding. Thus, space funding does not compete directly with other aerospace or defense projects, but with popular social programs. Second, space is not viewed as necessary to the well-being of the country by the vast majority of the public. Although significant numbers of jobs depend on the space program, these numbers are not as large as those in the defense industry, which has already suffered significant budget cuts. As social welfare programs and other entitlement spending has increased over the past thirty years, NASA's domestic justification has eroded and its attractiveness to budget-cutters has increased.

The third major change in the political environment in which the space program operates is a shift in the relationship between science and society. This shift has two major aspects. The first is an increasing skepticism in Congress on the benefits of government funding of basic research. The second is the legitimization of a national industrial policy.

For evidence that the relationship between science and the government is changing, one need only look at the comments of George E. Brown (D-CA). Brown, the chairman of the Committee on Science, Space, and Technology, has said,

> For the past 50 years, U.S. government support for basic research has reflected a widespread but weakly held sentiment that the pursuit of knowledge is a cultural activity intrinsically worthy of public support.
>
> The U.S. space program was the principal metaphor in this tenuous linkage.
>
> ....we must test the hypotheses that link economic and societal benefits directly to advances in research. All research is not the same. What kinds of research offer the greatest probability of improving the quality of life of humankind throughout the world? Would not applied research, and policy research, be recognized as essential elements of the fabric of research? If scientists are not willing to rigorously and fearlessly confront—and answer—these types of questions, then they cannot claim, and surely will not achieve, a stronger grip on federal purse strings that other special interest groups.

Brown's comments demonstrate a shift toward government support of applied research and technology development for direct commercial concerns and away from basic research and engineering development with limited connections to commercial competitiveness. Space is vulnerable in this argument because so much of what is accomplished there has little relationship with events in American society in general. Astronomy, planetary geology and geochemistry, and solar physics may all advance the sum total of "knowledge," but they do little to save lives or enhance economic competitiveness. One need not agree with Congressman Brown to recognize that his words hold weight and signal a change in the attitude towards basic research in the United States.[9]

The second aspect of the change in the relationship between science and American society is the legitimization of industrial policy. Throughout the 1980's, the words "industrial policy" were forbidden within the Reagan and Bush administrations. The prevailing orthodoxy was that government sponsorship of high technology research with commercial payoffs was unwise because of the inability of the government to pick winners and the risk of wasting money on losers. The market was seen as a better determinant of the proper technological choices to be made.

However, even among staunch free traders, this argument did not extend to the defense industry. It was seen as vitally necessary for the government to fund specific technologies and to prop up faltering industries in order to preserve the defense industrial base. It was also accepted, albeit to a much lesser extent, that military technologies had commercial spinoffs—the civilian jet aircraft industry is largely a result of military funding of strategic bomber programs, time-sharing computers were a development of the Defense Advanced Research Projects Agency, etc. Funding of military technologies with civilian uses was seen as a legitimate practice of the government. The space program spinoff debate was made within this environment. Since direct funding of commercially useful technology was politically and philosophically unthinkable, indirect funding through the space program was considered legitimate.

But the Washington environment has changed. The success of foreign governments such as Japan and South Korea at directing technological development has had a profound influence on many within the U.S. government. The plight of American industry during the 1980's, particularly the consumer electronics and automobile industries, has also contributed to a change in attitude. The result is that American industry is no longer opposed, and is in many cases, in favor of an industrial policy. Furthermore, liberal advocates of industrial policy have now come to accept the inability of the government to pick technological winners and the need for industry involvement in planning and management of any industrial policy. The result has been organizations like Sematech, a joint industry-government cooperative research effort to develop advanced computer chip manufacturing technology, which was begun under the Reagan administration and is generally credited with the reemergence of the American microchip industry.[10] Sematech is a forerunner of future government-industry cooperative efforts now that the concept of an industrial policy is no longer illegitimate.

Space is increasingly being subsumed into the penumbra of a national industrial policy. The elimination of the National Space Council and delegation of space policy decision making to the Office of Science and Technology Policy is one indication of this shift. The words and actions of NASA administrator Dan Goldin, appointed by President Bush but retained by President Clinton, are another.

NASA has requested an 18 percent increase in its 1994 spending for aeronautics research, increasing it to a total of $1.02 billion.[11] This renewed emphasis on aeronautics research has been partly undertaken to correct decades of neglect and underfunding of this field, but also because of the importance of the civilian aerospace industry to the American economy and the emergence of international threats to that industry like Airbus. In addition, Goldin has moved the space agency towards technology development for commercial uses. A memo from Goldin to the agency's field centers indicates that industrial policy is now a goal of the American space program.

> You are, of course, aware of the Clinton Administration's program to stimulate the production of jobs in the United States by converting some defense-related activities to commercial uses. The program includes the Technology Reinvestment Project (TRP) wherein NASA has an excellent opportunity to give the Nation an increased return on our technology investment and make our technology the basis for new commercial products and services.
>
> I believe you have an excellent opportunity to strengthen your regional economic base by applying dual-use NASA technology, leveraging the TRP, and encouraging high quality industry participation.[12]

In addition to this new emphasis on technology development, Goldin has also moved the agency toward greater direct relevance to American society by placing higher priority on Earth sciences research by separating NASA's Office of Space Science into three offices, one of which is solely dedicated to Earth science research. Although the majority of the NASA budget is still spent on human spaceflight efforts, it is clear that the trend is beginning to shift toward aeronautics, Earth science, and above all, technological development.

The most likely direction for gaining support for space exploration now is serving the new international leadership paradigm which is defined by economic competitiveness fostered through a dynamic industrial policy. The American space program will, for the first time, be primarily centered upon providing clear economic spinoffs for commercial industry in such areas as aeronautics, electronics, computer software (particularly artificial intelligence), robotics, and materials research.[13]

Any program plan for future planetary exploration must provide commercially useful technologies to the private sector as a *primary aim*, not a spinoff. There are also signs that Congress doesn't want money simply thrown at "technology development" in a haphazard way, but wants assurances that the space agency has plans to carry it out.[14] According to Goldin the question is, "whether NASA can reestablish its 'can-do' reputation and relevancy with the American people, or whether we just continue to drift toward the fringe of national priorities and become a white-collar jobs program."[15] Both Congress and the President want demonstrable payoffs from space.

This requirement will apply better to some aspects of lunar and Mars exploration, such as robotic precursor probes and advanced sensors, than others, particularly human spaceflight. But since an overriding political imperative for returning to the moon or travelling to Mars is unrealistic, the value of such an endeavor must be exhibited in other areas. While this Third Era may still justify a fairly robust American space program, trying to cast space exploration as a subset of industrial policy will not get us to Mars.

THE CASE FOR MARS

There is a difference between those factors necessary for a decision by the political leadership to go to Mars and those necessary for carrying it out successfully. The Space Station *Freedom* case is a perfect example of this distinction. There were a different set of requirements for Presidential approval for the space station than for its successful implementation. Some of these, like international cooperation and Congressional support, were intertwined and changed over the years. Others, like the arguments placed before the President, were finite.

The Role of the President

A strong President committed to the goal is of vital importance in making any major decision on space exploration. It is important to look at the role of the President in past space decisions to gain a better understanding of why this is so.

One of the many reasons for the failure of the Space Exploration Initiative was the fact that the agenda was set by a group of White House officials and not the President himself. SEI's origins began in the newly created National Space Council, not with a Presidential Directive to propose a future direction for the space program. The decision was in large part a confluence of three factors: a perceived need to reinvigorate the civilian space program, the desire to capitalize on the 20th anniversary of the Apollo 11 lunar landing, and the desire of an ambitious Executive Secretary of the Space Council, Mark Albrecht, to make a name for himself in a field in which he had little experience.[16] George Bush committed to the program in an attempt to reap the political benefits of the symbolism of the endeavor without having to expend any political capital—much in the same way he labeled himself "the Environmental President" and "the Education President."

A President must be willing to do more than merely deliver speeches and sign executive orders; he must also be willing to sacrifice other aspects of his agenda for the program. He will only do this if it is a personal priority for him. It will only become so if he recognizes trends in society as a whole which indicate that this is both a desirable and achievable course of action.

In addition to a strong national leader committed to Mars exploration, the other requirements necessary for a decision to send humans to Mars include a receptive Congress, a national consensus that the goal is worthy, and an international consensus among potential partners that the project is important. All of these center on the argument for spending large amounts of money to send humans to Mars.

POLITICAL JUSTIFICATIONS FOR SPACE EXPLORATION

The cultural framework necessary for a decision to explore beyond Earth orbit is the most difficult factor to define and lies at the core of the Third Era of Exploration. In general, three different political justifications for the future exploration of space have been advanced. These can be summarized as the "realist," "idealist," and "logical" political justifications.

The "Realist" Justification

The decision to explore is essentially a political act. The Cold War made such a decision relatively easy given the context in which it was made. Since the Cold War provided such an excellent justification for the space race, great power competition is frequently advanced as a likely cause of a future increase in space exploration. A coup led by a hard line regime in Russia might lead to a new era of space competition. Similarly, the emergence of Japan as a new great power might lead that nation to spend more money on space and compete with the United States on a prestige basis. A United Europe or German superpower might also lead to the same situation.

This is the argument at the core of the "realpolitik" school of thought. The end of a unipolar world dominated by the United States is supposed to lead to a new era where the definition of a great power includes its ability to access space. Often this argument is made over a few too many beers, but when it is advanced in formal contexts, the dangers of European or Japanese "defection" to a cooperative partnership with the Russians is frequently invoked.

The "Idealist" Justification

Other arguments have been advanced for sending humans to Mars or conducting space exploration in general. This could be considered the "ideapolitik" school of thought. During the mid-80's The Planetary Society, led by Carl Sagan, advanced the argument that the United States and the Soviet Union should send a joint expedition to

Mars. The late Senator Spark Matsunaga also made a similar proposal in his book, *The Mars Project*.[17] There were a number of justifications for this, the two most commonly cited were the desire to build trust between the superpowers by engaging them in a common endeavor, and the possibly less altruistic purpose of siphoning money out of the Strategic Defense Initiative.[18]

It was in response to the U.S.-Soviet joint mission proposals of the 80's that John W. Garver proposed a joint U.S.-Japanese mission to Mars in 1988.[19] Garver argued that a mission involving the Soviets was unrealistic and that there would be a better chance of success and greater symbolism if the world's two most technologically advanced democracies united for a common goal. Mars was selected as a unifying goal, but the reason to go there was not based on science, but politics.[20] This argument still turns up even now that the Cold War is officially over. Its proponents are now more subdued and more practical and less enthusiastic about the chances of such a mission leading to world peace, but still holding the view that it is possible.[21]

The "Logician" Justification

A third argument most frequently advanced by engineers is that international cooperation in space is eminently logical. Advocates of this position argue that there are essentially four main reasons why international cooperation in space "makes sense": the "philosophical rationale," the "technological synergism rationale," the "elimination of redundancy rationale," and the "financial cost-cutting rationale."[22] The "philosophical rationale" is that the creation of a "global village" and the creation of new nations and democracies with "a stake in ensuring the continued evolution of the world into a heterogeneous, peaceful set of societies with a healthy international commerce"[23] leads to a confluence of interests. The "technological synergism rationale" holds that certain nations possess technological capabilities that others lack and therefore need. The "elimination of redundancy rationale" holds that current world space programs inefficiently cover many of the same aspects. The "financial cost-cutting rationale" holds that international cooperation offers reductions in costs that are attractive to financially strapped space agencies.

This argument has been advanced in a number of different contexts, including everything from Earth sensing to the international exploration of Mars. Because space cooperation is logical, these proponents assume it will therefore happen and, most importantly, will lead to better-directed spending on specific space goals.

Problems With the Arguments

These arguments essentially reflect the three camps in the debate over the political justifications for a human expedition to Mars and space exploration in general. The idealists see cooperation as offering tangible benefits for humanity and thereby justifying space exploration. The realists see space exploration as a result of the power struggle between nations. The logicians see cooperation as inherently logical and therefore virtually automatic. There is often overlap between camps, particularly the idealists and the logicians.

But there are problems with all of the arguments and these stem from a lack of understanding of the political system. Often these justifications have been advanced by scientists and engineers who have not studied international affairs and who also suffer from a belief that human behavior can be classified according to inherently logical and predictable principles.

The realist position fails to recognize that we cannot return to the Cold War days even if we wanted to. The conditions necessary for a great power space race were unique and will not be repeated. Even if tensions between the United States and Russia were to rise to the level they were at, at the height of the Cold War, this would not lead to a reemergence of a space race. It must be remembered that the space race essentially ended with Apollo, whereas the Cold War continued, through both détente and extreme tensions, for almost two decades. What led to the original space race between the countries was both the ideological and national security symbolism of the endeavor. It was significant that the two countries competed not only in highly visible technology—rockets—but technology which had clearly recognizable links to ICBMs and hence national security.[24] The fact that the space race ended at approximately the same time that the two countries achieved rough parity in ICBMs is not coincidental. This aspect of Cold War competition was singularly unique.

Today Russia is in dire economic straits and is currently witnessing the erosion of its scientific and technological base. Any future competition between the two countries will not take place in the technological area, where the United States has far more capable competitors in the Western sphere. Furthermore, a reemergence of the Cold War in the future will lack an ideological component necessary for international competition: Communism as a form of government has been thoroughly delegitimized. During the space race the need to prove the desirability of one form of government over the other drove both countries to spend large amounts of money on their space programs. Such an ideological competition is impossible today, especially considering that a resumption of Cold War tensions will only come about as a result of a totalitarian takeover in Russia. Totalitarianism in pure form has never been as appealing as a rather abstract communist ideal. A future Cold War between the United States and Russia, if it occurs, will probably involve vastly different regional competition than in the past. Rather than engaging in political or military conflicts over control of sections of the developing world, tensions will more likely manifest themselves through the sale of militarily sensitive technologies to independent states viewed as belligerents by the other side. If the two countries do compete in space technology, it will be for commercial profits rather than prestige or scientific accomplishment.

But what about the possibility of nationalism elsewhere fueling a return to the space race? Nationalism is clearly on the rise throughout the world, but it is assuming center stage primarily in newly emerging or liberated states, none of which have the resources necessary to undertake space exploration. There is little sign of renewed nationalism among the countries with the greatest space capabilities. It is unrealistic to expect either Japan or Europe (possibly in conjunction with the Russians) to assume a role as a great power adversary in space exploration, at least for the foreseeable future. Neither

Japan nor the Europeans have shown sufficient interest in space to spend as much on it as the United States as a percentage of their respective GDPs. In addition, because of structural differences—for instance, their lack of a robust military space program to underwrite some of the costs of civilian exploration—it is difficult for them to do so and therefore compete with the United States on a prestige basis, although they do compete on a commercial basis. In order to equal the United States in civilian space exploration, both Europe and Japan would have to spend considerably more than the United States does on the civilian space budget to make up for the benefits NASA accrues (such as a large stable of launch vehicles as well as tracking facilities) from the military space program. In addition, inefficiencies associated with international cooperation would mean that the Europeans would have to expend even more. In short, it would cost both Europe and Japan far more than it does the United States simply to develop and maintain the level of capability in civilian space that the United States now has. Such massive increases are unlikely.

It is also true that the Japanese and European programs, although equal to the American in some respects, lag far behind in others and generally have a different focus. The United States has always had a fascination with human spaceflight and exploration which was rooted in its culture. Europe and Japan lack this. Furthermore, while the American emphasis on human spaceflight and the lack of similar Japanese or European programs is apparent, there are other differences as well. Neither Japan nor ESA has any appreciable planetary exploration experience. In addition, the Japanese program is much more focused toward applications satellites such as communications and meteorology than its American counterpart, where communications satellites are the purview of the military and private industry. Both space programs would need to acquire considerable experience and refocus their efforts before they could challenge the United States in space exploration, if they wanted to.

Furthermore, emergence as a rival in the exploration of space would have to be accompanied by a profound shift towards a political, military, and ideological rivalry with the United States, which seems extremely unlikely. Those who believe that we will race the Japanese to Mars are deluding themselves. The Japanese and Europeans will continue to remain American allies—albeit possibly strained allies—for the foreseeable future.[25]

The idealist position fails to account for reality by misunderstanding why nations cooperate. Although political scientists do not completely agree on this, it is possible to generalize cooperation into two categories: security alliances and low-level cooperation. Security alliances are formed out of a mutual fear of a third power. NATO is an excellent example. Low-level cooperation occurs in areas which are not high on the national political agenda, such as fishing rights and Antarctica exploration.[26] It happens when neither the costs nor the consequences of the success or failure of cooperation are very high. In general, the idealist argument fails to recognize that the most successful cooperative efforts follow from a confluence of interests—they do not lead to them. Hoping that a successful cooperative effort in space will lead to peace and harmony on Earth is placing the cart before the horse.

The logicians, however, assume that because cooperation is sensible, it is therefore easy. But as former NASA Associate Administrator for International Affairs Kenneth Pedersen has noted, "cooperation can enhance the feasibility or affordability of space projects, but it cannot normally create or justify them."[27] Like the idealists, they see a confluence of interest in space that apparently transcends other interests like national security, autonomy, and intellectual property. They also, along with the idealists, ignore the fact that space still contains an important national security aspect. Cooperative proposals such as an international remote sensing agency fail to recognize that one of the reasons for redundancy of systems is the desire of individual countries to possess their own autonomous remote sensing systems, either for military or prestige reasons. Inefficiency results not out of oversight, but from independent political priorities.

It is generally accepted that, barring a return to the political situation of the early sixties, the United States cannot undertake any further major space exploration projects unilaterally. International cooperation is now taken as a given in most areas of space exploration. But simply recognizing this does not make cooperation any easier, or provide a justification for undertaking a large scale cooperative effort in the first place. In the past, other countries have cooperated with the United States because they felt they had much to gain from such cooperation. A human mission to Mars will not be undertaken simply because it is a good idea from an international standpoint. It is hard to find a satisfying political justification for the human exploration of the solar system in general and the human exploration of Mars in particular.

The theory popular today is that the increasing costs of space missions are a primary driver toward greater international cooperation But there is also another possibility: the increase in costs of a mission during the design stage can actually derail an international agreement. The result is similar to the law of diminishing returns in that, after a point, the situation is reversed: higher costs lead to cooperation, but if they continue to rise, cooperative agreements fall apart as partners withdraw.[28]

WHY GO TO MARS?

The decision to go to the moon was made out of a desire to demonstrate to the world that the United States was superior to the Soviet Union both in technology and in national organization. It was made in light of the public relations fallout of Yuri Gagarin's flight. There is unlikely to be a similar event in the future compelling the expenditure of billions of dollars to go to Mars other than the discovery of life or past life on the Martian surface. The failure to identify a reason to send humans to Mars that is satisfying to the public and to national decision-makers has been the weak link in the speculation on the human exploration of the red planet. No good justification for why Mars should be a national goal was provided in the Space Task Group report of 1969. The Mars goal was presented in response to a call for reevaluating the future of the American space program—it was justified mainly from the standpoint of continuing the exploration of space, not on its own merits. Twenty years later the Space Exploration Initiative made Mars a goal for many of the same reasons, invoking images of destiny and exploration. None of these reasons was extremely precise, nor convincing to a skeptical Congress.

But there have been other justifications besides politics advanced for space exploration. In general, six different reasons have been advanced for exploring space and sending humans to Mars.[29] These are: science, spinoffs, resources/trade, education, defense conversion, and bureaucratic purposes. It is now necessary to turn to an analysis of the other commonly cited reasons for going to Mars.

Science

Mars enthusiasts have long advocated the planet's exploration on the grounds of scientific return. They have cited the basic knowledge of Mars itself—its geology, chemistry, and meteorology—to be learned. In recognition that these arguments are not sufficiently convincing, some Mars advocates have extended them further by citing the relevance to understanding our own climate, in particular, atmospheric change and ozone depletion. Virtually all have cited the importance of determining the possibility that life may have once existed on Mars and its value to understanding what factors lead to the creation and destruction of life on our own planet.[30]

Of all these arguments, the one with the most validity is the search for past life on Mars. The discovery of past life on the red planet would revolutionize the study of biology and associated life sciences. It would certainly stand as one of the great discoveries of all time, profoundly redefining our understanding of life. Such a discovery would more than justify the costs of a mission. But this payoff (i.e., the value of such a discovery) will only be accepted by politicians after it has been made. It will be insufficient to justify the expense unless there is more evidence to support the existence of past life on Mars than currently exists today. Robotic missions may never be sufficiently sophisticated to totally disprove the possibility of past life on Mars, but their limitations will not necessarily justify the dramatically more expensive next step of sending humans to explore.[31,32]

There are other scientific projects that are more important, more relevant, and cost less than the Mars goal. Because Mars is interesting is a reason for going there, but it is not a reason for spending a large amount of money to go there. Mars science is, for the most part, basic science, and basic science is increasingly viewed in Washington as a luxury we can no longer afford. It will have to meet far more stringent requirements during the Third Era of Space Exploration than it did in the past when space was justified by politics and money was relatively abundant. The message is clear: you cannot go to Mars based on science alone.

Spinoffs

Another popular justification for the space program in general has been the value of technological spinoffs. Space has, according to this view, contributed to American technology, and hence, to American economic competitiveness. Everything from computers to medical scanners to Velcro, Teflon and Tang has been attributed to the space program. Space has served as the driving force in high technology in the past and can continue to do so in the future. Proponents of the spinoffs argument often point to the benefits to society provided by communications and meteorological satellites. But it is not more spending on these areas that they are advocating, but larger human exploration and science budgets, both of which have been far less productive in their generation of spinoff technologies.

Unfortunately, there has been a great deal of false information and misinformation associated with space spinoffs. Some of the developments attributed to the space program were actually already in existence before they were adapted for space use. Others were developed as part of either the ICBM or other aerospace programs and are therefore not directly attributable to the space program. Still others, like Tang, are of dubious value to society, or, like heart-monitoring equipment, may not justify their expense through the small number of people they help. It is also not clear that many of these developments, like error-free software (commonly attributed to the space program and now evident in automatic teller machines), would not have arisen somewhere else on their own if not funded through the space program. There are very few developments which are solely attributable to the space program which have more than justified the cost spent on them.

Furthermore, it must be recognized that space as a technology driver is a relative argument. New products, technologies, and concepts are constantly being developed around the world. How are we to judge the value of a technology developed for the space program versus one that was developed for commercial use? Apple Computer, it could be argued, has contributed far more to society in the past fifteen years through the development of the personal computer than NASA has in the past thirty-five, and Apple did not cost taxpayers any money unless they bought the company's products. Similarly, pharmaceutical research, one of the areas where the United States still maintains an undisputed lead, has helped far more people at less cost than, and largely unconnected to, the American space program.

Critics of the spinoff argument have noted that it is far more efficient to spend money directly on developing new technologies than to acquire them as the by-product of the space program. While true, they ignored the larger role of the environment in which the spinoff argument was made—an environment in which direct government support of technology development for commercial use was not legitimate.

Now that the government can fund such things as joint commercial R & D endeavors, space exploration is not needed as an excuse or substitute.[33] Spinoffs cannot get us to Mars, but the demand for technological benefits will redefine how we approach the goal. This will be discussed later.

Resources/Trade

The desire to exploit space resources is frequently used to justify lunar exploration. For instance, Helium 3, mined from the surface of the moon or the atmosphere of Jupiter, would provide the Earth with vast amounts of clean fuel for fusion reactors. Other resources, like nickel, iron and copper, could all be mined from near Earth asteroids, providing humanity with a virtually inexhaustible supply of metals. Other arguments have centered on the benefits of establishing trading centers throughout the solar system, drawing on the analogy of Columbus discovering the new world. As NASA Administrator Dan Goldin recently said, the Spaniards went to Latin America for "God, Gold and Glory" and came back with "corn and potatoes", having discovered a new world.[34] No one predicted what useful things they would find and no one envisioned at the time that only a few hundred years later, the economic power of the New World would be greater than that of the old.

The resources and trade justification fails on a number of accounts. First, there are no fusion reactors currently in existence requiring the use of Helium 3. Even if hydrogen fusion is perfected, and few experts working in the field doubt that this will happen in less than fifty years, it is still a long way from there to fusion reactions using Helium 3. Other resources suffer from rather basic economics: bringing enough iron, nickel or copper into Earth orbit to supply our needs for the next 500 years would instantly cause a collapse of the market, making the material relatively worthless and bankrupting the government or investor who had bankrolled the project. It is hard to see the world running out of any material for which it could not find an adequate substitute. Even oil will someday disappear with the dinosaurs and we will run our automobiles on something else. It is also true that Mars has nothing unique to offer in the area of resources that cannot be gained more cheaply and more easily elsewhere.

The trade analogy is more convoluted and fails to recognize that much of what the New World had to offer the Old for the majority of those first five hundred years was agricultural products. Now that the world has transitioned to an economy where value-added production is more important than raw materials, it is possible to produce greater wealth by sitting at home and thinking than by going somewhere else and digging. Although it is true that a vast space-based economy may someday exist, this does not justify destabilizing currently fragile economies already here on Earth. If something wonderful "turned up" on Mars, it might justify the cost of going there. But right now, nothing of such value is in view.

Education

The value of the space program to inspire the nation's youth has also been frequently cited as a justification for large expenditures on it. Advocates cite the fact that Ph.D.s in math, science and engineering surged during the Apollo era and that elementary school children are excited by the prospects of space travel. But here too the argument runs up against the question of efficiency. There are much more efficient ways to spend the billions of dollars directly on education, rather than to reap the benefits as a result of space exploration. There is also no way to quantify the benefits of the space program to education. Some observers, like former Executive Secretary of the National Space Council, Brian Daily, have noted that children are becoming less and less enamored of the space program as they view their chances of participating in it as increasingly remote.

Defense Conversion

The disarray in the American defense sector, particularly the aerospace industry, in the aftermath of the end of the Cold War, led to calls to embark on an ambitious civilian space program. There were two general thrusts to this argument. The first was to preserve jobs and therefore limit the effect on the economy that would result from the end of government subsidy of the aerospace industry. The second was to gradually convert these industries first to government civilian business, and ultimately to commercial enterprise.

Such arguments were more popular in 1990 and 1991 than they are today. It is now obvious that government leaders are more willing to suffer a slowdown in the aerospace industry and massive unemployment in key states, such as California and Washington, than to continue to spend money on unneeded projects. Although defense conversion is still a popular topic in Washington, there is a profound lack of consensus as to how to accomplish it. For instance, should the government actively intervene by propping up industries or concentrate instead on retraining workers? There is also the unfortunate problem that some skills, facilities, and technologies developed for the defense industry have little use elsewhere, even in the space program. Using the space program as an intermittent step to full commercial operations never enjoyed popular support among congressional lawmakers. Budget deficits were feared more than unemployment.

The Government Accounting Office has noted that allowing free market forces to guide the restructuring of the defense industrial base is "not a realistic strategy for ensuring that government decisions and industry adjustments will result in the industrial and technological capabilities needed to meet future national security requirements."[35] But despite this warning, there has been no indication that either the current administration or the Congress is about to take a more coordinated approach to the subject. Furthermore, it is not clear that if the goal is to maintain the defense industrial base, the space program is necessarily the best way to accomplish it.

It must be stressed that those hoping to see increases in the space budget as a result of defense reduction failed to recognize that the impetus for defense cuts came from a desire to save money and that the amounts involved were quite large. In other words, no increases in the space program could hope to match defense reductions. It would be impossible to spend the money fast enough in realistic terms and the political concern was in reducing budgets.

The failure of space to benefit from the need to convert the defense industry was also due in part to the fact that this argument was only a justification for spending large amounts of money immediately—not over the several years it would take to get a major program running. Defense conversion is not a justification for a decision to go to Mars a decade or even a few years from now. By that time, the disruption to the economy caused by the end of the Cold War will be almost completely resolved. Defense conversion cannot get us to Mars.

Bureaucratic Purpose

The aimlessness of the American civilian space program in the 1980's was in large part a contributor to the commitment to SEI. It was frequently proposed as a solution to NASA's ills. According to this argument, what was needed to get NASA back on track was a purpose and going to Mars was an excellent goal.

The advocates of this view, including many staffers on the now-defunct Space Council, ignored the fact that many of NASA's problems were not due to the lack of a goal, but to poor management and overall inefficiency. For instance, the lack of a central goal for the human spaceflight program could not explain the problems that plagued the GOES-Next meteorological satellite program or the Hubble Space Telescope.

But more importantly, this was an argument that held far more weight within the space community than outside it. The fact that a bureaucracy is bloated, inefficient, and aimless is not an argument that will convince politicians to give it massive amounts of money to fix it. From a political standpoint this is placing the cart before the horse: we create government organizations to fulfill certain tasks, we do not look for new jobs for drifting agencies that lack purpose. The mission should justify the organization, not the other way around.

Aggregate

Three of the non-political justifications for Mars exploration—education, spinoffs, and defense conversion—are second-level justifications. In other words, if your concern is primarily to find solutions to problems in these areas, you would be better to spend money directly on them than to use space exploration as a funnel. Ten billion dollars spent directly on education, technology development, or defense conversion, would be better spent than a $100 billion space program with an 8% impact on these problems. NASA's aimlessness and the resources/trade argument both arise out of inaccurate assumptions about bureaucratic politics and economics, respectively. Only science offers a coherent justification for space exploration.

It has been stated that, from a scientific standpoint, the most important robotic mission to Mars would be a sample return mission. Yet this is viewed as too expensive and is no longer being seriously considered by any of the major space powers despite the fact that it was clearly on the agenda of many of them only a few years ago. The reason is the high cost of such a mission—at least $1.5 to $2 billion in current dollars.

The scientific justification simply does not compare well with the extreme cost of space missions. A "cheap" space mission costs $100 million. Compared to other scientific endeavors, this is a tremendous amount of money. Half of that is often spent on transportation. A Delta rocket launch cost $50 million in 1991 dollars.[36] Additional costs are accrued designing instruments to weather the harsh environment of outer space and to avoid problems like interplanetary contamination. The need for high reliability further drives up the costs since, once launched, spacecraft are virtually impossible to repair. All of these costs must be paid before even one bit of data has been collected. Space is not cheap.

In 1993, NASA's budget was roughly $15 billion. By comparison, the National Science Foundation's budget was approximately $2.8 billion. It is hard to argue that the benefits to humanity from the space program were five times as great as those from the NSF. Science does not offer a sufficiently compelling rationale for the exploration of space, but it is the only one that we have.

The fact that there is no overwhelming justification for the human exploration of Mars at this time does not bode well for the prospects of such an undertaking in the near future. It is because of this that the other requirements necessary for a decision—Congressional support, national consensus, and international support—are not in existence, nor likely to emerge soon. A sufficiently powerful President who is strongly committed to the goal can make up for some of these shortcomings. It is not clear that, even in the exciting years of Apollo, there was ever very strong public support for human space exploration. The public, Congress, and international partners are at least partially manipulable by a President who understands the power of setting the agenda and leading by example. But this power, at least as it extends to large technological programs, is increasingly limited by the budget deficit in the United States and the recession worldwide. In other words, the time may already be past when a single individual could have set the nation on the path to Mars.

Finally, it must be understood that these factors necessary for such a decision are also highly linked to the perception of the success of its enactment. If Congress, the American public, and the international space community do not believe that the United States is capable of successfully undertaking such a mission, they will not agree to such a decision in the first place.

HOW DO WE GET TO MARS FROM HERE?

So far this paper has been very pessimistic about the political prospects for future large exploration efforts in general and a human mission to Mars in particular. Without a profound and unforeseeable change in the political climate, there will not be a decision to undertake such a bold and expensive mission in the near future. Unfortunately, much of the future of space exploration rests on a significant improvement in the budgetary situation. This may have to be done at the expense of space exploration. Until a more stable funding situation is achieved, pressure will continue to increase on the space program. Improvements in efficiency as a result of reform at NASA may offset some of this pressure, but not all of it.

The Third Era of Space Exploration is likely to see smaller projects producing less science. It may conceivably see the end of the human space program. It will certainly see space programs redirected away from scientific results and towards the development of commercially useful technologies. The Ballistic Missile Defense Organization's (formerly SDIO) Clementine lunar probe is a harbinger of things to come. Designed to test sensors intended for ballistic missile detection, Clementine will produce scientific data as a *secondary aim*. It is also relatively inexpensive.

Those who believe in space exploration will have to accept the new reality and learn to live with it. It is unlikely that a decision to send humans to Mars will be made in the foreseeable future. However, there are a number of things that can be done that may increase the chances for success of such a decision, if it does happen. These will not necessarily make it more likely, but will at least lower some of the obstacles to such a decision.

First, we must adapt to the new environment and accept it. This means a number of things. It means designing missions with the intention of producing technologies useful in areas other than space exploration and successfully demonstrating their usefulness. It also means improving ties with industry, not simply as a producer of equipment, but as a customer.

It means reducing the cost of space technology, particularly launch costs. While it is impossible to mount a human mission to Mars cheaply, it is important to identify the least expensive way of achieving it and developing both technologies and management techniques that will be able to somewhat lower the "pain threshold" for those required to fund it. Technologies developed for other programs, such as the Strategic Defense Initiative, may have tremendous usefulness for future space exploration.[37]

But by far the largest single cost is still that of transportation. This must be brought down substantially before large scale exploration becomes conceivable. Unfortunately, even if substantial reductions could be made in the cost of placing a pound of payload into orbit, its impact on a human mission to Mars is likely to be minimal.[38] Barring some unforeseen development, transportation costs will remain high. A Mars mission will require the development of a heavy lift launch vehicle, which will undoubtedly cost a great deal to design, build and operate.

The goal, then, is to economize on the margins with new and innovative technology. Some of this can be done with the robotic missions which will serve to expand our knowledge base of the Martian environment. The new philosophy of "faster, cheaper, better" which has its origins with SDI and has now been adopted both by the NASA administrator and the field centers, holds much promise for changing the culture of the way the United States approaches space exploration.[39]

Another important action that the space community can take is to justify robotic Mars exploration missions solely on their own merits and *not* as precursor missions for eventual human exploration. One of the lessons learned from the Space Exploration Initiative is that tying a robotic mission to a future human mission is often seen, correctly, as the "camel's nose under the tent" leading to future requests for large expenditures. Lawmakers do not want to have to face these decisions and will kill robotic missions even if they have their own legitimate scientific value.[40] The goal is to do what we can do now and learn what can be learned about the Martian environment. Linking these missions, even rhetorically, with future human exploration may cause them not to happen at all.

Second, we can work to change the environment for a decision. Grass roots educational efforts should have some effect at changing the cultural climate in which such a decision is made. The political leaders who decided to go to the moon and build the space shuttle and space station were not born and raised in the space age. Today's generation tends to take the space program for granted, as something that has always existed. This does not necessarily mean that they will automatically support it, however. But by exposing a large number of young people to the adventure, opportunities and challenge of space exploration, both through dedicated educational facilities and specialized academic curriculum, it is possible to make them more receptive to it.[41] A generation raised with the belief that Mars is a worthy pursuit will be more receptive to a commitment to the goal.

Third, we must devote more attention to the policy aspects of space exploration. Those factors which are necessary for the successful enactment of the decision are directly controllable although not necessarily easy. *But they are not yet well understood.* A great deal of time and effort has been expended on proposing engineering models of Mars exploration, but little on political models. Most of the discussion about human exploration of Mars has been initiated and perpetuated by engineers, yet it is not engineers, but politicians, who control government funds necessary for such an undertaking. Any argument for going to the red planet must be expressed in terms that they can understand and sympathize with.

We need more work by social scientists on potential international frameworks for cooperation. In the past, international agreements have often been created on the spot with little attention to previous successes and failures. The recent decision to establish an International Mars Exploration Working Group based on; the Inter Agency Consultative Group deserves close attention from policy experts. It offers a unique learning experience.[42] The space station shows that international cooperation is neither easy, automatic, nor always desirable.

In addition, greater emphasis should be placed on management reform with NASA and the lessons to be learned from such efforts. We must find out what works and be willing to accept failure along the way.

Finally, we must not only address the "why?" question, but deal with it more realistically. Too many in the space community feel that the failure to convince the public and the decision makers of the need for space exploration has been due not to a failure of the message but the medium. The problem, they argue, has been communication. But as already outlined above, many of the justifications advanced so far are logically flawed or unconvincing. The failure to arrive at a political justification as compelling as the Cold War is the primary reason we have not returned to the moon or sent humans to Mars. The only good reason is the presence of life or evidence of past life on Mars. But we must be willing to accept that the ultimate answer to the question of "why?" may not exist.

CONCLUSION

In the social sciences it is generally easier to make negative future predictions than it is to make positive ones. Political science has never been particularly effective at explaining change and so it is therefore hard to state with any degree of reliability that the political conditions for a Mars mission are not likely to emerge in the future. We do not know what the future holds, but it is clear that if we are to see a commitment to the human exploration to Mars, it will require a different political situation than exists today and one which is not presently predictable.[43]

Nevertheless, it is possible to reach some conclusions about the present environment. The Cold War is over, budgets are tight and space is increasingly being drawn into the penumbra of a still ill-defined industrial policy. The justification for the continuation of the human space program is slowly evaporating s it faces fiscal reality.

NASA has been extraordinarily lucky as far as funding is concerned. Despite some poorly managed projects, and in some instances, *because of them*, the space agency continued to gain increases in its budget throughout the 1980's. All of this may change with the result of the space station debate. It is entirely possible, that if the station is cancelled, there may be calls for eliminating human spaceflight completely.[44] The "buy-in" cost of human spaceflight for the United States, that is the cost of maintaining the capability of flying at least one shuttle a year, is at least five billion dollars. That is five billion dollars which has, so far, largely been unquestioned politically, except at the margins. The near term future, far from offering any optimistic chances of future exploration efforts, may even offer the possibility of the end of the American spaceflight program.

The only thing likely to change the situation and lead to an increase in human exploration efforts and a decision to send humans to Mars is the discovery by robotic probes of past or present life there. But such a discovery may never happen and, unless it does, the likelihood of a human mission to Mars in the near future is grim indeed.

The goal, then, is to do what can be done now to improve the environment and change it where possible. Management reforms, policy research (particularly on international cooperation), and the human factors research all deserve more attention than they have received. Scientific and technological approaches and philosophies must be changed. Technical breakthroughs, particularly in transportation costs, must be achieved.

None of this will be easy. Some of it may be impossible because the space program is being pulled in competing directions. On the one hand, there is pressure to make systems cheaper, faster, better, which implies the use of existing, off-the-shelf equipment. On the other hand, there is the drive for technology development, with all its attendant risks of abstract goals and challenges and cost overruns.

This paper began with a rather optimistic quote by Wernher von Braun on the impossibility of predicting the future. It ends with the pessimistic observation that the future may not be predictable, but that our inability to predict it is not necessarily cause for hope. The primary task in the meantime is to make those changes which will make the future more conducive to our vision. Late in his life von Braun also came to recognizc that it was not technical capabilities but political reality that prevented us from reaching our dreams when he said, "We can lick gravity, but sometimes the paperwork is overwhelming." Our job is to start fixing the paperwork.

REFERENCE NOTES

1. For a comparison of the historical underpinnings of previous major presidential decisions on space see John M. Logsdon, "Dreams and Realities: The Future in Space" W. W. Mendell, ed., *Lunar Bases and Space Activities of the 21st Century*, (Houston: Lunar and Planetary Institute, 1985) pp. 70l-709.
2. For instance, a memorandum from Richard E. Neustadt to President-elect Kennedy stated, "Since Sputnik we have been in a race to be "first" in physical achievements of a dramatic sort—the sort which has high visibility and thus makes an impression on mass opinion, especially abroad. This has got us into the business of pressing achievements for the sake of their psychological effect, regardless of concrete scientific or military utility." See Richard E. Neustadt, Memorandum for Senator Kennedy, "Memo on Space Problems for you to use with Lyndon Johnson, December 23, 1960, with attached: "Problems of Space Programs," p. 2. During deliberations on a response to Gagarin's flight only four months later, another memorandum from Secretary of Defense Robert McNamara to Vice President Johnson, chairman of the National Aeronautics and Space Council stated, "Dramatic achievements in space, therefore, symbolize the technological power and organizing capacity of a nation." See Robert S. McNamara, Secretary of Defense, Memorandum for the Vice President, "Brief Analysis of Department of Defense Space Program Efforts," April 21, 1961, p. 1.
3. For an explanation of the factors leading to John G. Kennedy's decision, see John M. Logsdon, *The Decision to Go to the Moon: Project Apollo and the National Interest* (Chicago: University of Colorado Press, 1970). Logsdon argues that Kennedy was primarily responding to Yuri Gagarin's flight and his feeling that the image of the United States had suffered greatly. By contrast, Michael Beschloss has argued that Kennedy was primarily responding to the humiliation at the Bay of Pigs. See Michael R. Beschloss, "Kennedy and the Decision to go to the Moon," paper presented at the "Presidential Leadership, Congress, and the U.S. Space Program," conference at The American University, Washington, D.C., March 25-26, 1993.
4. Although the history of the decision to build the space shuttle has not yet been written, perhaps the most important document from this process is a August 12, 1971 memorandum from then deputy assistant director, Office of Management and Budget Caspar Weinberger, to Director, OMB George Schultz which said, "We are being driven, by the uncontrollable items, to spend more and more on programs that offer no real hope for the future, Model Cities, OEO, Welfare, interest on the National Debt, unemployment compensation, Medicare, etc. Of course, some of these have to be continued, in one form or another, but essentially they are programs, not of our choice, designed to repair mistakes of the past, not of our making... America should be able to afford something besides increased welfare, programs to repair our cities, or Appalachian relief and the like." This memorandum contains the notation, "I agree with Cap" written by Nixon on the heading and is essentially the "smoking gun" for when and why Nixon committed to the space shuttle. It indicates that it was a decision, in part, by a conservative President to spend money on a civilian program that was not a social program.
5. See Howard McCurdy, *The Space Station Decision* (Washington, D.C.: Smithsonian Institution Press, 1990).
6. For instance, the decision to use segmented solid rocket boosters rather than solid cast motors on the shuttle and the addition and later deletion of the Flight Telerobotic Servicer on Space Station *Freedom* were both the result of Congressional decisions. Although the details of these decisions are too complicated to address here, the point is that politics

has had an effect at all levels of the space program in the past—from initial decision to final design—and will continue to do so in the future.

7. For evidence of the weakness of constituency politics at maintaining large, unneeded programs, one need look no farther than the defense budget, which has seen the recommended closing of several dozen unneeded military bases around the country and an industrial build-down which has had dramatic effects on the aerospace industry. Critics of this assessment point to the Superconducting Super Collider as evidence of the continued reign of pork, but the SSC, which has survived several attempts to kill it, benefits from a well-organized lobbying campaign. It is best understood as an anomaly. Pork barrel politics is most successful at creating and continuing relatively small projects.
8. Bruce Murray, "Can space Exploration survive the end of the Cold War?" *Space Policy*, February 1991, pp. 23-34.
9. See George E. Brown, Jr. "Rational Science, Irrational Reality: A Congressional Perspective on Basic Research and Society," *Science*, Vol. 258, October 9, 1992, pp. 200-201.
10. The American microchip industry recently surpassed Japan in its share of the world microchip market. See Brit Hume, "Lessons Learned by U.S. Computer Chip Manufacturers," *The Washington Post*, January 4, 1993, p. WB14.
11. Kathy Sawyer, "Reviving Aeronautics: Space Agency Focuses on Global Contest," *The Washington Post*, Thursday, May 27, 1993, p. A 23.
12. Daniel S. Goldin, Administrator, NASA, to Director, Jet Propulsion Laboratory, "Technology Reinvestment Project," May 5, 1993.
13. For a discussion of this, see Ray A. Williamson, "The US civilian space programme: Charting a new course," *Space Policy*, November 1992, pp.295-304 and U.S. Congress, Office of Technology Assessment, *Exploring the Moon and Mars: Choices for the Nation*, (Washington, D.C.: U.S. Government Printing Office, 1991).
14. Andrew Lawler, "House Panel Rejects Majority of NASA Technology Package," *Space News*, May 31-June 6, 1993, p. 16.
15. Peter Carlson, "Has NASA Lost Its Way?" *The Washington Post Magazine*, May 30, 1993, p. 15.
16. In recent months, Albrecht has stated that SEI was intended to serve as a form of defense conversion for the aerospace industry. See Albrecht's remarks during a debate at the 31st AAS Goddard Memorial Symposium, March 9-10, 1993, Hyatt Regency Hotel, Arlington, VA, titled, "Resolved: The Space Program Has Been and Continues To Be An Effective Stimulus To U.S. Technological Competitiveness."
17. Spark M. Matsunaga, *The Mars Project: Journeys Beyond the Cold War*, (New York: Hill and Wang, 1986).
18. Carl Sagan, "USA and USSR: Let's Go To Mars—TOGETHER" The Planetary Society brochure reprinted from *Parade*, February 2, 1986, pp.4-7. *The New York Times* also endorsed the proposal in an editorial.
19. John W. Garver, "Getting to Mars," *Orbis*, Summer 1988, pp. 403-409.
20. It is interesting to note that according to this line of reasoning, the one factor which should have made cooperation much more feasible, the end of the Cold War, also undercut the justification for cooperating in the first place. This highlights the political considerations for both a decision to go and those necessary for the successful performance of the mission, and the need to ensure that they are not contradictory.
21. See, for instance, John Pike, "United We Stand—and Fly," *The New York Times*, Thursday, April 8, 1993, p. 14.
22. See Subcommittee on the International Exploration of Mars of the IAA Committee on International Space Plans and Policies, International Academy of Astronautics, IAA Mars

Cosmic Study, International Exploration of Mars: A Mission Whose Time Has Come," Draft Version 5, December 15, 1992, pp. 16-18.

23. *Ibid.*, p. 17.

24. Indeed, Sputnik itself was a result of a concerted Soviet push to develop ICBMs which was a result of the Soviets' lack of a long-range bomber, due to Stalin's purges. The United States, possessing substantial superiority in long-range bombers, did not mount a crash effort to develop ICBMs and thus found itself behind in the ICBM race, but not in strategic nuclear forces.

25. Political scientist Christopher Layne predicts that the current unipolar world will give way to a multipolar one including Japan, Germany, and possibly China between 2000-2010. See Christopher Layne, "The Unipolar Illusion," *International Security* 17:4, Spring 1993. Of course, even if we do return to a multipolar world involving several superpowers, there is no reason to expect international competition to manifest itself in space exploration. For an excellent discussion on the prospects for a multipolar world as well as Japan's prospects for militarization, see the following articles in the same issue: Robert Jervis, "International Primacy"; Samuel P. Huntington, "Why International Primacy Matters"; Peter J. Katzenstein and Nobuo Okawara, "Japan's National Security"; Thomas U. Berger, "From Sword to Crysanthemum."

26. Anticipating a response, it must be stressed that Antarctica cooperation would not be very different from cooperation on the exploration of Mars because of the costs involved. Even relatively modest countries, like Argentina, are able to participate. Space, as has already been noted, is quite expensive and limited to only a few players. Furthermore, the costs of failure are high even for those with the funds to participate. The nature of cooperation is different, too, since those countries participating in cooperation in Antarctica are primarily agreeing not to fight each other, not to engage in extensive cooperative exploration efforts.

27. Kenneth S. Pedersen, "Thoughts on International Space Cooperation and Interests in the Post-Cold War World," *Space Policy*, August 1992, p. 219.

28. For an excellent discussion of the changing nature of international cooperation in space, see Space Policy Institute and Association of Space Explorers, "International Cooperation in Space: New Opportunities, New Approaches," April 1992.

29. Carl Sagan, in an address before the 8th Congress of the Association of Space Explorers entitled "Together to Mars: Does it Make Sense?" listed two types of reasons for going to Mars, the "tangible" and "intangible" ones. Among the tangible reasons for going to Mars and space exploration in general are: science, spinoffs, education, defense conversion, resources, population growth. The intangible reasons, according to Sagan, included the desire to join two formerly hostile superpowers in a common endeavor for the good of humanity. Sagan dismisses the education argument as an inefficient use of funds, stating that $100 billion would be better invested directly in education. He also notes that even substantial colonization of other planets would not relieve overcrowding on Earth. He concluded his talk by asking, "Can the sum of a set of individually inadequate justifications together justify the cost of a Mars mission?" He felt the answer was in the affirmative, but not a resounding yes.

30. All of these arguments have been advanced by Carl Sagan at one time or another, but he has become more skeptical of them in recent years. It is interesting to note that the end of the Cold War caused Sagan to have a change of heart, transforming his advocacy on the subject from "Let's Go To Mars—TOGETHER" to "Together to Mars—Does It Make Sense?" See Carl Sagan, "Why Send Humans to Mars?" *Issues in Science and Technology*, Spring 1991, pp. 80-85.

31,32. It is difficult to make a comparison between the efficiency of robotic exploration versus human exploration. Robots have been, and always will be, limited in what they can do by their inability to answer questions they have not been programmed to ask. Humans, on the other hand, are more flexible, but also require vast resources simply to keep them alive in hostile environments. A dollar spent on a robot may always buy less return in an absolute sense, but may be better spent in a relative sense. To this it is necessary to add Sagan's comment about the human versus robotic exploration controversy that it is unfair to compare smart humans and stupid robots: by the time a human mission can conceivably be mounted, vastly more sophisticated robotic probes will be available.

During the decision to commit to the Space Exploration Initiative, it is not clear that human exploration of Mars was even on the official agenda (i.e. mentioned in Bush's speech) until very late in the planning process. In the past there has been something of a "moon vs. Mars" controversy between space advocates. Those in favor of the moon pointed to the need to develop infrastructure in space and to avoid a situation similar to Apollo where, once the goal was achieved, the program ended and there was no long-term follow-up. They argued that the moon was a safer step on the path to Mars. Those in favor of bypassing the moon and heading straight for Mars, on the other hand, argued that the scientific return from Mars was far greater than from the moon, and that, politically, the moon would be a harder sell because of the fact that humans had already been there. One fact that was ignored even by the lunar advocates was that a Mars trip would cost several times more than a return to the moon, even if one were looking at "cheap" Mars missions and elaborate lunar base proposals. But the question of the moon or Mars was like the debate about how many angels could fit on the head of a pin. While all these scientific arguments have some validity (although the claims that studying the Martian climate can help us better understand global warming are highly circumspect), they all ignore one basic fact: the American space program has never been about science. Apollo was not and could not have been justified by science, and it cost less than a Mars mission is expected to cost in current dollars. For a discussion of the desirability of returning to the moon before a human mission to Mars see Michel B. Duke, Wendell W. Mendell, and Barney B. Roberts, "Lunar Base: A Stepping Stone to Mars," in Christopher P. McKay, ed., *The Case for Mars II*, Volume 62, *Science and Technology Series*, American Astronautical Society, (San Diego: Univelt, 1988), pp. 207-220. See also John Logsdon, "Resist the Pull of Mars," *Air & Space*, April/May 1988, pp. 44-45.

33. It must be stressed that this change is largely in means, not ends. Neither the Bush nor Reagan administrations was anti-technology. The Clinton administration is not necessarily more pro-technology than its predecessor. But it is much more utilitarian in its approach, viewing technology as serving tangible human needs, rather than abstract ones. This will lead to greater emphasis on applied research that basic research and more stress on demonstrable payoffs.

34. Comments of NASA Administrator at the Wernher von Braun Memorial Lecture at the National Air & Space Museum, April 29, 1993.

35. United States General Accounting Office, Office of Public Affairs, "Reports and Testimony: April 1993," GAO/OPA-93-7, p. 1. See United States General Accounting Office, National Security and International Affairs Division, "Defense Industrial Base: An Overview of an Emerging Issue," GAO/NSIAD-93-68, Mar. 29.

36. Steven J. Isakowitz, International Reference Guide to Space Launch Systems, (Washington, D.C.: American Institute of Aeronautics and Astronautics, 1991) p. 205.

37. For instance, SDIO has developed an infrared telescope with its own cooling unit and a star tracker, each weighing less than two pounds. Similar advanced systems are important because they will cost less to build and to place in orbit.

38. The reason is that such reductions are likely to be the result, not of fundamental technological breakthroughs, but of specific programs focused at certain heavily used payload ranges. For instance, if the SSTO Single Stage to Orbit vehicle proved to be all its supporters claim it will be, it will only provide low cost access to low Earth orbit in the 20,000 pound range. Whether it can be scaled up to the medium launch vehicle range (30,000 to

50,000 pounds to LEO) is questionable, and Mars would still require some heavy lift launch capability.

39. For an excellent discussion of the "culture" prevalent at NASA during the 1970s and 80s, see Gary D. Brewer, "Perfect Places: NASA as an Idealized Institution," contained in *Space Policy Reconsidered*, Radford Byerly, Jr., (Boulder, Co: Westview Press, 1989), pp. 157-173. For an example of an innovative new approach where cost, not science, is the driving factor in the development of the spacecraft, one need only look at the Mars Environmental Survey (MESUR) mission. See James R. Asker, "Cheaper, Faster Mars Missions in Sight," *Aviation Week & Space Technology*, April 12, 1993, pp.50-52.

40. Lunar Observer, which was not originally intended as the first step to a lunar base, and the lunar precursor missions, such as Artemis and Lunar Scout, which were intended to eventually lead to human missions, all fell to the budgetary axe because of their connections to human exploration. As a result, any science which could have been accomplished was lost.

41. A large number of space education facilities and programs emerged during the 1980s such as Space Camp, Challenger Center, and a project to bring remote sensing data directly into the classroom. These are part of a larger trend in American education (as well as museums) to give students hands-on experience. Perhaps the best example of these programs was the Jason Project, which allowed American school children to tele-operate undersea craft in the Mediterranean. A similar project using prototype Mars rovers in desert terrain would seem to be a logical progression for the space community.

42. See Roger Bourke to Tony Spear and John Beckman, Jet Propulsion Laboratory memorandum, "Wiesbaden Workshop on International Mars Exploration," May 18, 1993.

43. None of this is new. In a March 14, 1961 lecture at the University of California, Paul Dergarabedian, the Director of the Systems Research Laboratory of Space Technology Laboratory, noted "The prospect of traveling into outer space appears to be a near reality to many people. The fact that man may one day journey to the moon, Mars, or Venus is disputed by few people. The biggest questions are why, when, and how. The first question is very difficult to answer on a technical basis, but the answers to the last two questions can be attempted by considering the basic requirement placed on the design of space vehicles (ballistic missiles) intended for outer space exploration." Less than a month later, Yuri Gagarin orbited the Earth, providing a "why" argument that was satisfactory for President Kennedy. See Paul Dergarabedian, Director, Systems Research Laboratory, Space Technology Laboratories, Inc., "Engineering Extension and Physical Sciences Extension, University of California, Los Angeles, Space Logistics, Lecture No. 7, Space Vehicle Design," March 14, 1961.

44. Some observers have also suggested that this will follow any future accidents of the space shuttle.

AAS 93-806

MARKETING MARS: A PRELIMINARY SURVEY OF PUBLIC ATTITUDES

David T. Cadden*

Any United States commitment to a Mars landing program, either by itself or in conjunction with other nations, would require considerable and sustainable public support. Marshalling this support translates into a need to go beyond merely educating the public. It requires a deep understanding of the level of national willingness to pay the bill, support for international cooperation, and mission justification. Support would not be homogenous and it would be extremely useful for program managers to appreciate the segmentation of the public's attitudes.

This research examines the attitudes of (481) college students from two colleges in Connecticut. It focuses on three areas—attitudes toward the budget requirements for a Mars landing program, willingness to work with various international partners, and various mission justifications. The results indicate the manner in which the mission's costs are presented can have a significant impact on the level of support. There was also significant variation in student support for the proposed international partners and their perception of different justifications for a manned mission to Mars. This research is being extended to a national phone survey.

The author believes that these results indicate the need for a benefit-segmentation marketing approach to muster strong public support.

INTRODUCTION

For those who are interested, the word Mars has an almost magical ability to conjure profound images—Lowell's canals, John Carte traveling across the sands of Barsoom, Bradbury's golden-skin hidden race, Chester Bonestell's visions of Wernher von Braun's winged ships. For those who have shared these dreams and have been captured by these images there is no question that we must go to Mars; however, to make this a reality will require a long-term and sustainable commitment of national public opinion. This will require a conscientious effort on the part of NASA and space advocacy groups. This effort must recognize that the American public is not a monolithic entity, but rather a pluralistic entity with different groups having different agendas and interests. Recognition of this basic factor is more important than ever for the space program given the problems with the deficit, the pressing demands of health and social issues and technologically suspicious constituencies. Proponents of a mission to Mars must be able to present tailored arguments that are meaningful and cogent to particular groups.

* Professor, Quinnipiac College, Mt. Carmel Ave., Hamden, Connecticut 06518.

At the Case for Mars IV, the Mars working group issued a policy statement which provided the following justifications for a Mars mission:

1. **Social** - there are biological, sociological, and psychological drives to explore. Exploring Mars would marshall national pride in an epic undertaking.

2. **Scientific** - studying Mars would provide valuable insights into planetary evolution and global climate change. Mars could provide a second habitat for humanity.

3. **Educational** - a Mars program would provide inspiration for students to appreciate areas of science and technology and pursue these areas as careers.

4. **Economic** - a Mars program would preserve our national commitment to a large and successful aerospace industry in face of the changing geo-political world scene. It would be a stimulus to cutting edge technologies. Mars itself, could be a source of abundant resources.

5. **International** - the human exploration of Mars would allow for improved diplomatic relationships among space-faring nations and thus promote world peace.[1]

It is critically important for space advocates to understand how each of these justifications is viewed by different segments of the American public. This research investigates how college students view the support of technology, the international unity, and the educational justification arguments.

A major difficulty in organizing public support for a manned mission to Mars is the cost which varies according to the study. Lowell Wood, of Lawrence Livermore Labs developed a program with an estimated cost of only $10 billion. The Stanford US-USSR Mars Exploration Initiative arrived at an estimated cost of $60 billion. In 1987, Chandler Smith provided a cost estimate of $189 billion. Former Apollo astronaut Michael Collins assumed a cost of $200 billion. Other cost estimates rise to a value of $400 billion. Even though these values might be spread over a twenty to thirty year period, they are quite shocking to the average American. There may be more acceptable ways to express the cost to the public. This research uses a value of $300 billion dollars and a time period of twenty-five years. It also investigates how subjects respond when the project cost is expressed in terms of a percentage of government expenditures and as an annual per capita expenditure.

It has become increasingly popular to consider the necessity of incorporating international partners in a Mars mission. From a financial perspective such cooperation is attractive, indeed, it may be an absolute requirement. However, project managers would have to appreciate American public attitudes. This research investigates college students attitudes toward cooperating with just the Europeans, just the Russians, just the Japanese, and cooperating with all three.

SUBJECTS

The subjects (481) used in this preliminary survey came from two colleges in Connecticut - one private and one state university. Faculty members were approached and asked if they would disseminate the questionnaire in their classes. There was no attempt to produce a stratified sample - one in which the student representation corresponded to male/female enrollment or enrollment by major. A complete breakdown of student representation for the total sample is given in Table 1. Please note that some students failed to answer all questions; therefore, totals in some Tables do not equal the sample size of 481 students.

Subjects were asked to identify their undergraduate major. It was assumed that the selection and pursuit of a particular undergraduate major might have a significant impact on one's attitude toward the space program, a manned Mars mission and the programs expenditures. Undergraduate majors were grouped into four categories: Business, Technical, Health, and Liberal Arts. The Business group consisted of Accounting, Information systems, Marketing, Management, Finance, and Economics majors. The Technical group consisted of Engineering, Computer Science, Chemistry, Bio-chemistry, and Physics majors. The Health group consisted of Nursing, Occupational Therapy, Physical Therapy, and Radiology majors. The Liberal Arts group consisted of Psychology, English Literature, History, Art majors.

Across both schools the total number of subjects who were graduate students totaled sixty-two (62).

The subjects ages ranged from 18 to 66 years. The average age of the subjects was 22.69 years with a median value of 21 years.

Table 1
BREAKDOWN OF SUBJECTS BY MAJOR AND SEX

	Business	Technical	Health	Liberal Arts	Totals
Male	139	54	33	26	252
Female	86	32	72	25	215
Totals	225	86	105	51	467

QUESTIONNAIRE

The questionnaire used in this research was structured to be a precursor to a national telephone survey. It was limited to two pages so that it could be administered in class with minimal disruption.

Attitudes are measured by means of a five point Likert scale structured as follows: (1) *Strongly Support (strongly agree)*; (2) *Support (agree)*; (3) *Neutral*; (4) *Against (disagree)*; and (5) *Strongly Against (strongly disagree).*

Since the subjects were to be college students from Connecticut the number of demographic variables were limited to sex and undergraduate major. A copy of the entire questionnaire is provided at the end of this paper.

INTEREST IN THE SPACE PROGRAM

The first question posed to the subjects asked them "What is your level of interest in the American Space Program?" They could respond on a five-point Likert type scale. The range of responses went from *Strongly Support* which was coded as (1) to *Strongly Against* which was coded as (5).

All 481 subjects responded to this question. The average for the total sample was 2.158 with a median value of 2. A presentation of the number of responses for each of the five possible answers and the corresponding cumulative relative frequency is given in Figure 1. These results would appear to indicate relatively strong support for the space program. Nearly seventy percent of the respondents expressed support for the program while less than two percent responded negatively to the program.

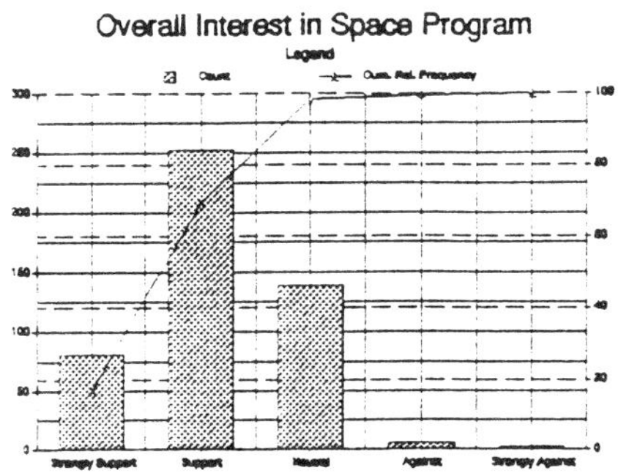

Figure 1

In comparing the two schools, by means of the Student t-test, there was no significant difference between the means.

Comparing the results for the four undergraduate majors - Business, Technical, Health Sciences, and Liberal Arts - the averages were found to be: 2.18; 1.97; 2.27; and 2.13, respectively. The median value was the same for all four groups: 2.

To test if there was a significant difference across the four groups a Chi-square test was performed. The Chi-square value was with 12 degrees of freedom.. The results were moderately significant (5% level); however, since eight cells had expected values less than 5 and seven cells had expected values less than 2, this result should be viewed with some caution. When one compares the difference in responses between men and women, again using the Student t-test, one finds a highly significant result. The mean score for the 256 men was 2.04 with a standard deviation of 0.73. The score for the 219 women was 2.30 with a standard deviation of 0.69. These numbers produced a t-value of 4.003 which was significant at better than the .001 level. Based on these findings, men have a significantly greater level of support for the space program than women.

BUDGET

A critical issue for those interested in promoting space activities is convincing the public that it is worth the expense. In order to justify the expense, it was seen as necessary to ascertain the public's estimate of that expense. The second question was - "What do you think was the budget for the American Space Program in 1992?" Only two-thirds (67%) of the respondents even attempted to answer this question. The answers varied widely and few came close to the actual value of $15 billion. The answers ranged from $100,000 to $200 trillion. For the 322 who responded the average was $1.45 trillion. That number is highly distorted and a more reflect answer is the median value of $4 billion.

The responses for the two schools were different. The private school had an average value of $196 billion and a median value of $5 billion. The state university had an average of $126 billion and a median value of $3 billion. A t-test performed on the averages demonstrated no statistically significant difference.

The 46 graduate students who provided an estimate had a mean of 110 billion dollars and a median value of $7.6 billion which was closest to the actual value of approximately $15 billion.

The average and median estimate of the budget for the four undergraduate majors is given in Table 2.

Table 2
MEAN AND MEDIAN VALUES FOR NASA BUDGET

	Business	Technical	Health	Liberal Arts
Mean	229.	4208.	3947.	98.
Median	4.5	5.	2.	2.

These results indicate that the public has a limited understanding of the expenditures of the space program. The median values, which in this case are more reflective of the subjects' estimates, are far below the actual 1992 NASA expenditures. Preparing the public for the additional expense of manned Mars mission will be a core activity of space advocacy groups.

INTERNATIONAL OPERATIONS

Economic, and perhaps technical, realities may dictate that future space activities have a large international dimension. The American public's attitude toward international partnership on projects such as a Mars mission is an unknown. This survey asked the subjects about their support of a manned mission with (1) just the Europeans; (2) just the Russians; (3) just the Japanese; and (4) a mission with three groups. The response to the four questions was the five-point Likert scale. The sample size, the mean response, and the median for the four groupings are given in Table 3.

Overall there appeared to be support for partnership with the Europeans. For the total population, approximately 63% were either supportive or strongly supportive (see Figure 2.) - A Chi-square test was conducted to see if there was a difference in the responses amongst the different majors. The test indicated no significant difference. However, when examining the differences in male-female responses the t-test value was 2.52 which is statistically significant at the 0.012 level.

Table 3
SAMPLE SIZE, MEAN AND MEDIAN VALUES FOR INTERNATIONAL PARTNERSHIP

	Europeans	Russians	Japanese	All Partners
Sample Size	473	470	469	478
Mean	2.37	2.61	2.74	2.07
Median	2	3	3	2

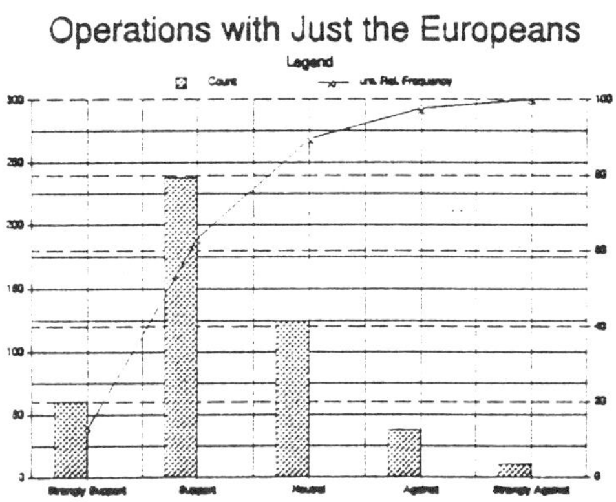

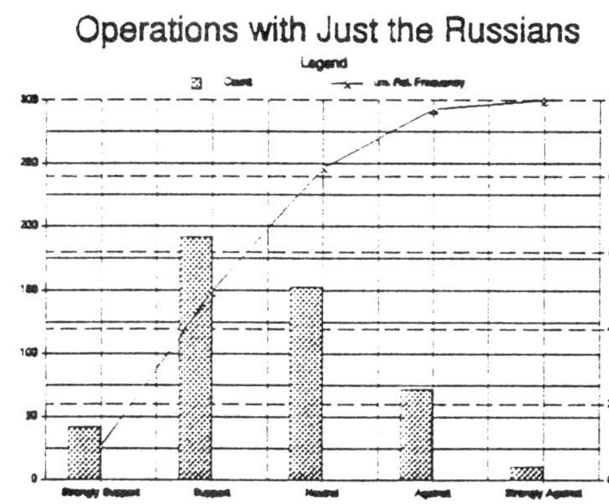

Figure 2 (left) and Figure 3 (right).

There is less support for going with just the Russians. For the total sample, slightly less than 50% were either listed strongly support or support (see Figure 3.). The Chi-square test indicated that there was no significant difference amongst the majors nor was there any statistically significant difference between men and women on this question.

The least support was found for going with just the Japanese. Less than 46% of the total population expressed strong support or support (see Figure 4.). The Chi-square indicated that there was no significant difference amongst the majors nor was there any statistically significant difference between men and women on this question.

Clearly the greatest support was for the inclusion of all three possible partners. Nearly 70% of the total population were either strongly supportive or supportive (see Figure 5.). The Chi-square test indicated that there was no significant difference amongst the majors nor was there any statistically significant difference between men and women on this question.

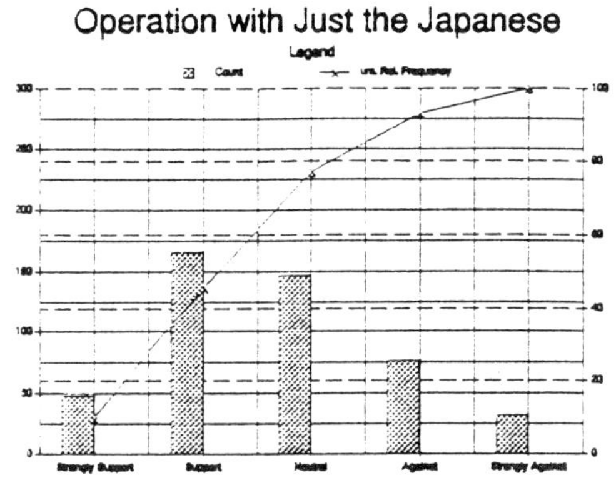

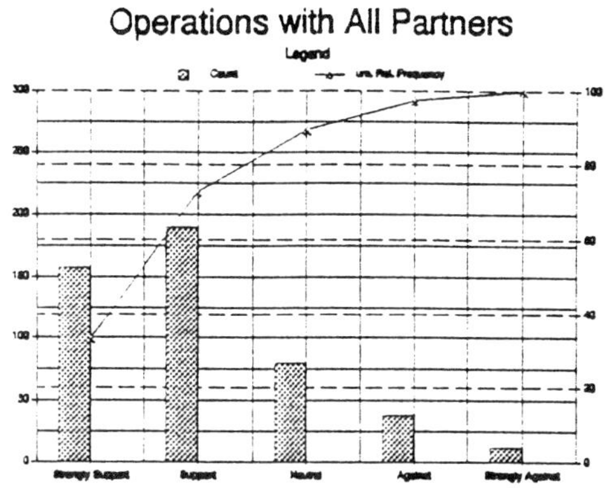

Figure 4 (left) and Figure 5 (right).

Using t-tests of the means (Table 4) of the total sample, one finds significant differences with respect to possible international partners.

Table 4
T-TESTS AND LEVEL OF SIGNIFICANCE FOR DIFFERENT INTERNATIONAL PARTNERS

	T-VALUE	Significance
Japan vs. Europe	5.85	0.0001
Japan vs. Russia	1.95	0.0500
Japan vs. All Partners	9.88	0.0001
Europe vs. Russia	4.16	0.0001
Europe vs. All Partners	4.74	0.0001
Russia vs. All Partners	8.54	0.0001

JUSTIFICATIONS FOR A MISSION TO MARS

Critical to winning public support for a mission to Mars is the ability to provide compelling justifications. As previously stated, the fourth Case for Mars identified several rationales. Each rationale, by itself, is compelling; however, the effectiveness of any particular rationale may depend upon the audience to whom it is presented.

This research focused on three subsets of the Case for Mars justifications: (1) International - namely, an international mission to Mars would unite the peoples of the world; (2) Economic - namely, a mission to Mars would be valuable in assuring American predominance in key areas of technology; and (3) Education - namely, a mission to Mars would be a way to interest the young in technology and thus increase enrollments in science and engineering.

The subjects were asked to respond to these three questions on the five-point Likert scale. On the question of a Mars mission uniting the peoples of the world, the mean response for the 479 subjects was 2.52 and the median was 2. Approximately 57% of the subjects expressed either strong support or support for this proposition (see

Figure 6.). The means and standard deviations for the four majors, males and females, and the undergraduate and graduate students are given in Table 5. A Chi-square test indicated that there was no significant difference amongst the majors nor was there a statistically significant difference between men and women on this issue. Interestingly, there was a statistically significant difference between the responses of undergraduate and graduate students. The t-test value was 3.39 - which was significant at the 0.0007 level. This last finding may be more reflective of a cynicism generated by an age difference rather than advanced schooling. The average age of the undergraduates was 21.6 years and a median value of 21 years. The average age of the graduate students was 30.5 years and a median value of 27 years.

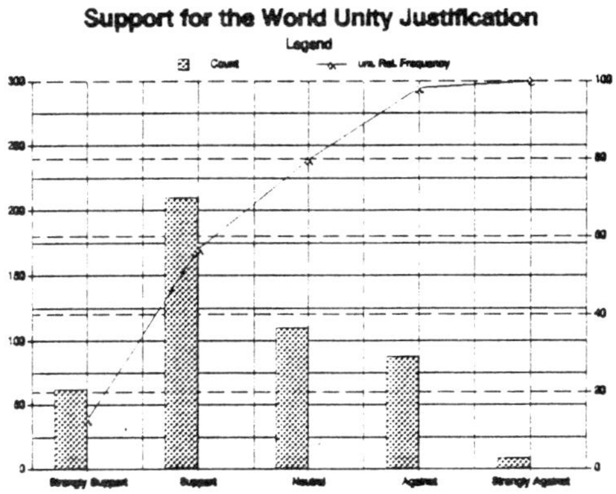

Figure 6.

Table 5

MEAN AND STANDARD DEVIATIONS FOR UNITY QUESTION BY GROUPS

	Mean	Standard Deviation
Business Major	2.58	1.02
Technical Major	2.57	1.02
Health Major	2.22	.86
Liberal Arts Major	2.54	1.12
Males	2.49	.87
Females	2.56	.87
Undergraduates	2.47	.98
Graduates	2.92	1.04

On the question of a Mars mission benefiting American predominance in key areas of technology, the mean response for the total sample was 2.46 and the median value was 2. Approximately 57% of the subjects expressed either strong support or support for this proposition (see Figure 7.). The average values for the four majors, males and females, and the undergraduate and graduate students are given in Table 6. A Chi-square test indicated that there was a statistically significant difference amongst the majors. The Chi-square value was 21.49 with 12 degrees of freedom. This value was moderately significant at the 0.04 level. It should be pointed out that 4 cells had expected values less than 5 and four cells had expected values less than 2; therefore, this result should be viewed with some caution. There was a statistically significant difference between the responses of men and women. The t-value was 2.66 - which was significant at the 0.008 level. There was no significant difference between undergraduate and graduate students.

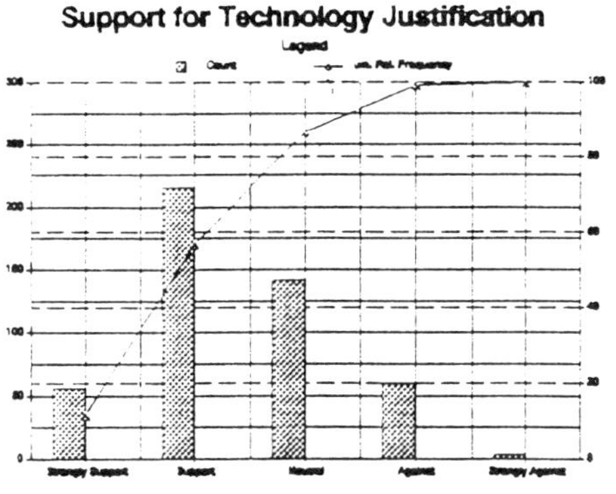

Figure 7.

Table 6

MEANS AND STANDARD DEVIATIONS FOR TECHNOLOGY QUESTION BY GROUP

	Mean	Standard Deviation
Business Major	2.38	.86
Technical Major	2.49	.88
Health Major	2.68	.86
Liberal Arts Major	2.32	.98
Males	2.36	.89
Females	2.58	.87
Undergraduates	2.46	.89
Graduates	2.54	.88

On the question of a Mars mission increasing young peoples' interest in technology and spurring enrollments in the sciences, the mean response for the 477 subjects was 2.16 and the median was 2. Approximately 72% of the subjects expressed either strong support or support for this proposition (see Figure 8.). The means and standard deviations for the four majors, males and females, and the undergraduate and graduate students are given in Table 7. A Chi-square test indicated that there was no significant difference amongst the majors nor was there a statistically significant difference between men and women on this issue. There was no statistically significant difference between the responses of undergraduate and graduate students.

If we make comparison of the three justifications across the entire sample we find, not surprisingly for these subjects, that the education justification is clearly viewed most favorably. Its responses were significantly different than the unity justification or the technology justification, while those two other justifications were not significantly different from each other. Table 8 presents the t-test results for the pair-wise comparison of the three justifications.

Support for Education Justification

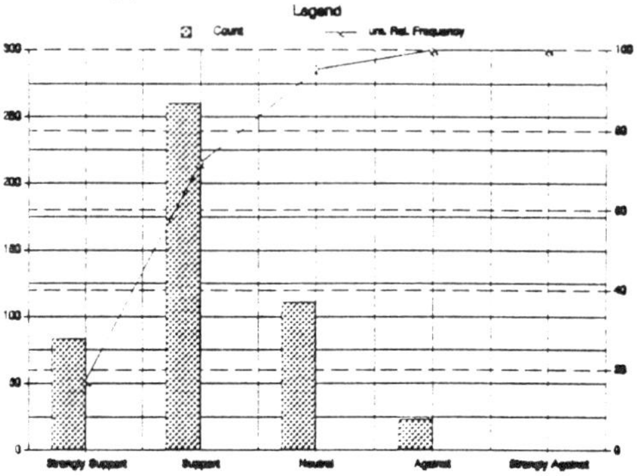

Figure 8.

Table 7

MEAN AND STANDARD DEVIATIONS FOR EDUCATION QUESTION BY GROUPS

	Mean	Standard Deviation
Business Major	2.13	.75
Technical Major	2.21	.86
Health Major	2.23	.71
Liberal Arts Major	1.96	.69
Males	2.13	.62
Females	2.18	.53
Undergraduates	2.14	.82
Graduates	2.23	.74

Table 8

PAIRWISE COMPARISON FOR THREE PROPOSED JUSTIFICATIONS

	T-values	Significance
Education vs. Unity	6.44	.0001
Education vs. Technology	5.65	.0001
Unity vs. Technology	1.10	.2710

COST OF A MARS MISSION

In an era of great concern over the Federal budget and the deficit, the cost of a mission to Mars will have a profound effect on the way the American public views of the mission's desirability. Most cost estimates have been in the hundreds of billions of dollars. Even though this expenditure might be spread over twenty or more years, such values contain considerable "sticker-shock" for the average citizen.

This study selected an upper level estimate of $300 billion which would be spent over a period of 25 years. The subjects were asked whether they would support a Mars mission at this cost. In order to determine if the level of support changed given a different presentation of that cost, the subjects were asked if they would support such a mission if its cost was approximately 1% of all government spending over a 25-year period. The subjects were also asked if they would support such a mission if it represented a per capita expenditure of $50 per year for 25 years. These last two values are approximately equal to the original value of $300 billion over the 25-year period. Subjects were asked to indicate their level of support for the three costs on the same five-point Likert scale.

The means for the total sample, the four majors, the undergraduate and graduate students, and males and females for the three costings are given in Table 9.

Table 9
MEANS FOR DIFFERENT COSTS LISTED BY GROUP

	$300 BILLION	1% OF BUDGET	$50 PER CAPITA
TOTAL SAMPLE	2.92	2.28	2.14
BUSINESS	2.97	2.27	2.11
TECHNICAL	2.62	2.25	2.08
HEALTH	3.15	2.41	2.34
LIBERAL ARTS	2.83	2.04	1.98
MALE	2.71	2.14	2.17
FEMALE	3.18	2.45	2.12
UNDERGRADUATE	2.93	2.27	2.15
GRADUATE	2.87	2.35	2.13

When expressed in terms of $300 billion, support for a Mars mission appears to be weak. Less than 31% of the subjects identified themselves as either strongly supporting or supporting the mission at this cost (Figure 9.). A Chi-square test was performed to see if there was a significant difference amongst the responses of the majors. The Chi-square value was 21.445 with 12 degrees of freedom. This is significant at the 0.044 level. Again, the reader is asked to consider this result with some caution because four cells had an expected value less than five. There was no statistically significant difference between the undergraduate and graduate students. There was a statistically significant difference between males and females on this question. Females were far less supportive of such a mission at this expression of cost. The t-statistic was -5.03, which was significant at better than the 0.0001 level.

When expressed in terms of 1% of government expenditures, support for a Mars mission rises significantly. More than 45% of the subjects identified themselves as either strongly supporting or supporting the mission at this cost (Figure 10.). A Chi-square test was performed to see if there was a significant difference amongst the responses of the majors. No significant difference amongst majors was discovered. There was no statistically significant difference between the undergraduate and graduate students. There was a statistically significant difference between males and females on this question. Again, females were far less supportive of such a mission at this expression of cost. The t-statistic was -3.67 which was significant at the 0.0002 level.

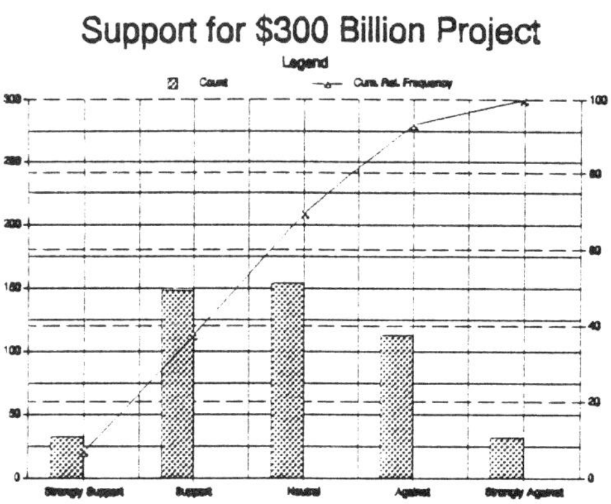

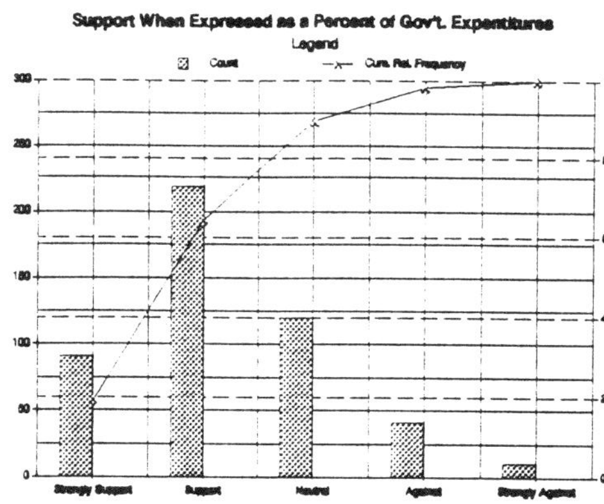

Figure 9 (left) and Figure 10 (right).

When expressed in terms of a $50 per capita annual expenditure, support for a Mars mission is also high. Nearly 43% of the subjects identified themselves as either strongly supporting or supporting the mission at this cost (Figure 11.). A Chi-square test was performed to see if there was a significant difference amongst the responses of the majors. No significant difference amongst majors was discovered. There was no statistically significant difference between the undergraduate and graduate students. There was no statistically significant difference between males and females on this question.

When examining the total sample, we find significant differences in the mean response for the three versions of a Mars mission's cost. When specified in terms of a flat dollar cost ($300 billion), the response was statistically significantly less supportive than either the percent of government expenditures or per capita specifications. When the percent of government expenditures is compared with the per capita specification, there was a small statistically significant difference. The results are presented in Table 10.

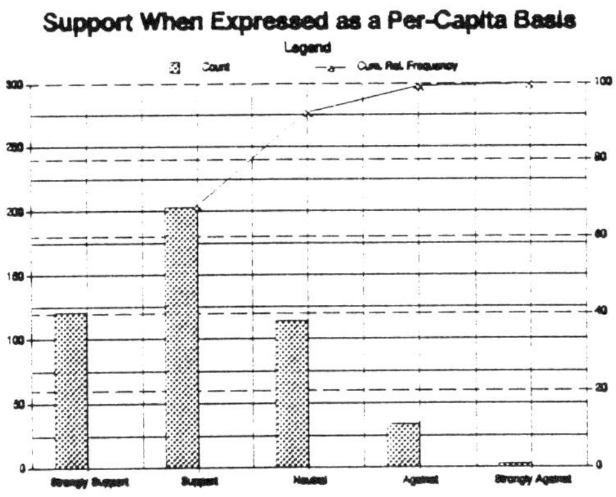

Figure 11.

Table 10
T-VALUES AND LEVELS OF SIGNIFICANCE FOR COMPARISONS OF EXPENDITURES

	T-VALUE	SIGNIFICANCE
$300 BILLION VS.1% EXPENDITURE	10.07	<.0001
$300 BILLION VS.$50 PER CAPITA	12.29	<.0001
15 EXPENDITURE VS. $50 PER CAPITA	2.28	.023

CONCLUSIONS

Marshalling and maintaining public support for a mission to Mars is critical to its success. This preliminary survey of Northeastern college students provides a number of interesting results. First, there appears to be significant interest in the space program and, to a lesser extent, support for the concept of a mission to Mars. This result should be rewarding to NASA and space advocates. Unfortunately, the same subjects' estimates of the budget of the space program varies widely and tends to grossly underestimate that value. This might pose a problem for maintaining support when individuals are informed of the actual value.

The results of this study indicate that there is strong support for a totally international approach to a mission to Mars and that attitudes vary significantly with respect to individual national partnerships.

Another interesting finding of this study is that public support of a mission to Mars is greatly impacted by the manner in which the cost is expressed. A flat dollar figure is far more likely to negatively affect an individual's support of a Mars mission than if it is expressed in terms of a percentage of government expenditures or on a per capita basis.

There are several possible justifications for a mission to Mars. Each is rational and can stand on its own merit. However, this study found that the subjects' attitudes toward three of these justifications varied and that students found the education justification most compelling. If we can extrapolate, one could easily argue that the different justifications might be more or less compelling to different groups of citizens. This is the logic of benefit segmentation marketing, that the same *product* can have different benefits to different market segments.

Even though a limited number of demographic variables were incorporated in this study, the results indicate that such variables have a *significant* impact on attitudes with regard to the space program. Clearly, the most important demographic variable with regard to differentiating attitude was sex. Women's attitudes were consistently, and in most cases, statistically, different than males. Women were less likely to support the space program or a mission to Mars. It appears that there is a dire need on the part of NASA and/or space advocacy groups to specifically focus on improving communications with the female population. The results also indicate that a college student's undergraduate major has an impact on his/her attitudes toward the space program.

The results of this study are limited and preliminary, but they point to a need to better understand the public's attitude toward the space program and the concept of a mission to Mars. By segmenting the public's opinion, the space program and its supporters could be significantly more effective in communicating its crucial role in human society. It would enable NASA to stop preaching to the converted and more effectively use its public relations resources. The author and the Quinnipiac College Polling Institute is currently planning a national survey of opinion on the space program and future missions to Mars.

APPENDIX A: QUESTIONNAIRE

QUESTIONNAIRE:

This survey has been designed to measure your attitudes toward the American space program and its possible future directions. Please take the time to read the following questions and select the response that most closely matches your belief.

1. What is your level of interest in the American Space Program?

1	2	3	4	5
Strongly Support	Support	Neutral	Against	Strongly Against

2. What do you think was the budget for the American Space Program in 1992? ______________________________

3. In 1989, President Bush proposed the Space Exploration Initiative. It called for Americans to return to the Moon and to send astronauts to Mars by the year 2019. To what extent would support this program?

1	2	3	4	5
Strongly Support	Support	Neutral	Against	Strongly Against

6. If you were informed that the cost of such a Mars landing was $300 billion dollars to be spent over 25 years - how supportive would you be of such a program?

1	2	3	4	5
Strongly Support	Support	Neutral	Against	Strongly Against

7. Many people have suggested that any trip to Mars should be an international undertaking with the cost being shared by many countries. To what extent would you support a joint program with the following partners:

a.) With just the Europeans:

1	2	3	4	5
Strongly Support	Support	Neutral	Against	Strongly Against

b.) With just the Russians:

1	2	3	4	5
Strongly Support	Support	Neutral	Against	Strongly Against

c.) With just the Japanese:

1	2	3	4	5
Strongly Support	Support	Neutral	Against	Strongly Against

d.) With the Europeans, the Russians, and Japanese:

1	2	3	4	5
Strongly Support	Support	Neutral	Against	Strongly Against

8. If you were informed that the cost of such a Mars landing was to be approximately 1 percent of all government spending over 25 years - how supportive would you be of such a program?

1	2	3	4	5
Strongly Support	Support	Neutral	Against	Strongly Against

9. Supporters of an international Mars program argue that it would help unite the peoples of the world. To what extent do you agree with this proposition?

1	2	3	4	5
Strongly Agree	Agree	Neutral	Disagree	Strongly Disagree

10. Many people argue that a commitment to a manned Mars program would be very valuable in assuring American predominance in key areas of technology. To what extent do you agree with this argument?

1	2	3	4	5
Strongly Agree	Agree	Neutral	Disagree	Strongly Disagree

11. Some supporters have argued that a manned Mars program would be a way to interest the young in technology and thus increase enrollments in science and engineering. To what extent do you agree with this argument?

1	2	3	4	5
Strongly Agree	Agree	Neutral	Disagree	Strongly Disagree

12. If you were informed that the cost of such a Mars landing was less than $50 per American per year spent over 25 year - how supportive would you be of such a program?

1	2	3	4	5
Strongly Support	Support	Neutral	Against	Strongly Against

13. What was (is) your Undergraduate Major? ______________________

14. Your sex: Male: _____ Female: _____

15. Your age: __________

AAS 93-808

IS THERE A SHORT TERM ECONOMIC AND SOCIAL JUSTIFICATION FOR MANNED SOLAR SYSTEM EXPLORATION?

Robert E. Becker*

The Space Exploration Initiative (SED) is now officially dead, totally unfunded, and widely viewed as unfounded. While one may disagree on the last point, we can all agree the case has not been made extremely well: Human wanderlust, or economic benefits thirty years down the road, pale in comparison with the world's quotidian economic and social problems.

Proponents of exploration sometimes have a tendency to declare "not my job" with respect to these problems. Yet, exploration is fundamentally a human endeavor, and should embrace solutions that may derive from it, even if this requires broadening the definition of that endeavor. Serious investigation of the short-term benefits of going full thrust with SEI provide a means for transforming a negative into a positive.

By this is meant something far more than a jobs program for the aerospace industry. Many of the benefits would derive from broadening the industrial base of space exploration during the development phase beyond the aerospace industry. This becomes possible because of the vast scope of SEI itself and the necessity to solve, for the first time on a large scale in space, many of the problems associated with life and society on earth: SEI is actually the space mission most closely related to the man in the street, not the least. In turn, SEI naturally encompasses those industries which developed over time to provide such solutions. And this opens the way for incorporation of "spin-on" technologies from the civilian economy, while naturally transferring aerospace workers much closer to the commercial sector.

This paper does not answer the question of the title. Rather, it provides a rationale for initiating coordinated efforts by the exploration community to convince the government and the public. A framework for this study, a generalization of the precepts of Concurrent Engineering, is proposed.

If the answer were affirmative, the resulting Mission from Planet Earth, what might be called "Mars Holistic" coupled with its mate, a similarly broadened Mission to Planet Earth, explicitly embracing terrestrial problems and solutions, would be announced with trumpets blaring, not with a bureaucratic whimper.

* 4-Sigma, Inc., 4111 Hunter's Ridge Drive, #4, Huntsville, Alabama 35802.

INTRODUCTION

Despite numerous ambitious strategies, and mission plans, and the enthusiasm of SEI proponents, even the latter now acknowledge that SEI might be better off dead than alive.[1] Given its checkered history as a much-maligned child whom nobody seemed to want to adopt, Louis Friedman[1] is undoubtedly correct in urging we not mourn this "stillborn."

Yet even someone as knowledgeable and influential as Friedman seems overly sanguine that new initiatives will readily be born. What killed SEI was not a paucity of viable mission concepts, applicable technologies, or a lack of long-term economic returns. Even new combinations of Precursor missions, innovative architectures like Zubrin's M^{2}ars Direct,[2] unveiled three years ago, and key enabling technologies, are unlikely to instigate this rebirth.

Rather, we must face the reality that what killed SEI was failure to affirmatively answer the question in the title of this paper. Sadly, the question was never really asked. Saddest of all, the answer might well be yes!

This paper does not definitively answer this question; nor does it try to. Instead, it provides a rationale to motivate the wider exploration community, perhaps starting with those assembled at the Case for Mars V, to gather data and form the arguments from which an answer can be derived, which this paper suggests, could well be positive.

To this end, we first look closer at our twinned questions, What killed SEI? and Is There a Short Term Justification for SEI? Exploration is a human endeavor, and next follows a brief historical excursion examining what ramifications of past explorations hold for the social aspects of our exploratory future.

Emphasizing the "Human" in human exploration through its relation to earthbound concerns offers the best chance of demonstrating the benefits of renewing SEI. This, along with the very size of an exploration program, plays an important role in potential solutions. In the next section, we examine these issues and others in the context of the developmental phases of a new form of Mission From Planet Earth (MFPE), here called Mars Holistic (MH). This is followed by a comparison of features of MH with those of other exploration strategies.

To answer our question will be a large undertaking. A general model for this effort is suggested in the next section, based on an extension of the concept of Concurrent Engineering.

We conclude with a Call to Arms to make MH a part of a perceived solution, noting that current initiatives emanating from the government might provide mechanisms for starting the effort.

SO WHAT DID KILL SEI?

As a practical matter, the demise of SEI could be anticipated[3] once it became clear the key question of this paper was not given much weight. Use of traditional tactics in selling SEI to Congress[3] only exacerbated the situation: slip the trunk (low-cost innovative Precursor missions) of the huge SEI elephant through Congress' budgetary door; and try to drag the rest of the bulk in afterward. This time, Congress was prepared for the argument, and thrust the trunk out the door.

If Daniel Goldin administered the actual *coup de grace*, Bowman Cutter, Deputy Assistant to the President for Economic Policy, delivered the unkindest cut of all[4]: "...The vista is important but the arguments are not worth a damn yet. If you can't concoct better arguments with a real bite, I am probably less willing to be an advocate. I would personally argue to have in NASA a continued level of support for exploration of the idea. It would take awhile for me to buy the argument that we ought to bet a large slug of resources - yet."

And why are the arguments "not worth a damn"? Most "outsiders" do perceive exploration as inviting bloated expenditures of little immediate value. That NASA's ability to carry out such a program under any circumstances is widely questioned, doesn't help. Diamandas recently[5] came close to, but didn't quite, hit the target, in recognizing that quotidian human problems loom large, and that SEI has not addressed them, thus dooming SEI for the time being.

The target is missed because the recognition of the problems SEI does not address is not transformed into a desire to see if SEI could address them. Instead, the usual solutions are rounded up: long-term economic benefits based on historic precedents; enabling, low-cost technologies like Single-Stage-To-Orbit, which though meritorious, offer terrestrial benefits without a driving need for exploration. and which "enables" access for a program which in totality may still hardly be "low-cost."

Overshadowing all are things we must spend our money on now. Potential economic or competitive benefits decades down the road cannot stand against these needs. As heretical as it might seem, SEI must first become "the" something upon which money must be spent as a major part of the solution to some of our problems.

That our country is notoriously short-sighted in comparison with, say, the Japanese, is nothing to be proud of. Nonetheless, it is true, and only when the question posed by this paper is answered in the affirmative, will the positive public mandate for exploration called for by Diamandas, whether signaled by voter referendum or the will of Congress, be forthcoming. Until Cutter's challenge is accepted, we might as well wait for an "Eureka" from the Search for Extraterrestrial Intelligence,[4] for that is surely the only other way a new wave of space exploration will be launched.

WHAT HISTORY CAN TEACH US

History has always been regarded as a friend of SEI. We all know of the wider economic benefits of our first Age of Exploration. Even more trenchant are arguments which point to the repercussions of ignoring exploration opportunities, for example, following the closure of China. Hawkins[6] notes the less permanent, but still serious effects, of the abandonment of exploration for over a century by the English.

If SEI is to address our short-term economic and social problems, it is logical to ask history to relate to us the experiences of the last Age of Exploration, for it is those endeavors which will most closely resemble our future Age.

Perhaps surprising to non-historians, the issue of who benefited from the explorations is still controversial.[7] Certainly, the maritime industries of the time gained as investment poured into them. Likewise, our aerospace industry would obviously benefit.

Less clear is the status of the peasants who were roughly analogous to our own poor and disadvantaged groups. A major difference between the two Ages is that alternate outlets for investment in the first were almost non-existent in the hinterlands, due to the primitiveness of the medieval economy.[6] We suffer under no such constraint. The mechanisms are now present or are being invented to take advantage of a similar influx were it to come. Furthermore, some forms of exploration, such as the exploitation of Alaska,[5] or Arctic voyages, could be carried out with relatively little investment beyond fortitude and persistence. Even the major voyages of Spain, though quite costly to the treasury, likely did not entail the enormous relative preparation time and expense of a "low cost" SEI. And the highly stratified societies of the time meant there was no need to secure a broad public mandate.

So exploration advocates today may face a harder task of persuasion than did their counterparts 500 years ago. On the other hand, in England, Spain, and France, in decreasing order,[7] prime beneficiaries of exploration were what today are called the disadvantaged classes. In the context of SEI, the challenge for history and its devotees is to translate these mechanisms, if possible, from the sixteenth century to the present so that exploration can once again succor the "lower" reaches of society.

Elimination of native cultures and slavery also characterized the Age of Exploration. Since the worlds we will explore are uninhabited some of the dangers of unrestricted exploration are obviated. Yet history exposes the dichotomy of those who did explore; the successful ones were magnificent navigators, but were often horrible administrators. Currently, astronauts are very closely scripted, perhaps too closely.[8] This cannot be so in the future. How autonomous can and should humans be in space exploration then becomes an important question.[9] There results another role for history and the social sciences[9-10]: to inform us of the dangers and lessons of selection and administration of voyagers and settlers far from direct control of the parent group, so we may avoid "Total Recall" scenarios.

WHEN BIGGER NEED NOT MEAN "BADDER"

Welcome trends to faster, cheaper, and better (FCB) in the space program have cast a long shadow over that which is "bigger". But it is the very size of SEI, when a strategy of extensive human settlement is selected, that would bode well for stimulating shorter-term economic and social changes.

Space missions are the natural purview of the aerospace industry. While contractor chains for large programs are long and many-tentacled, this industry has its own identity, and a not-so-imaginary boundary line is seen as separating it from "commercial" industries.

Always present has been the critical element of crew safety and well-being, and the systems which ensure them. These systems, whether they be food dispensers or waste disposal units, bear a distant relation to commercial equivalents used on earth. Earlier, more grandiose, versions of the Space Station pushed this relationship even closer as long duration habitability for larger crews became an issue. Mars transit phase crew complements typically proposed would be of the same order of magnitude.

Full-fledged communities to be planted in a robust human exploration strategy are a different matter. Antarctic research stations might be the best model for comparison; the Biosphere experiment gives an understanding of the physical magnitude and obstacles to be faced, though crew size is quite small. It is at this community stage that large-scale participation by commercial industry in the study development, and implementation of the systems needed to support settlements becomes logical and desirable, if not necessary. This may range from construction, transportation, agriculture, household electronics and appliances, edutainment, to economic activities such as mining. Requirements for lunar or Martian settlements are still unique, so traditional aerospace participation must continue.

In the optimum scenario, partnerships between commercial and aerospace industries are incubated in an effort to create "Mars Town". A whimsical example used to first broach this concept might help: Archetypical blue collar Rubbermaid worker Joe D. Sixpack may have no interest in SEI, indeed may be unalterably opposed to it due to perceived tax increases and his own employment pressures. But if he reads in his company newspaper that Rubbermaid received a contract to help study and develop "stick-em" gloves for low gravity operations, his attitude will likely start to change, as will his Congressman's.

Economies of scale, not in the classic Fordian sense of mass production (though this would come with time), but in the tremendous diversity of a "Mars Holistic," opened to our commercial sector, offers the promise of the widest direct economic and social benefit.

Many studies have been performed on settlement architectures, and like the formation of a nineteenth century Western town at the end of a wagon train route, it is well understood what kinds of facilities will be required, even if their exact form is still to be determined.

The difference is in the timing: most architectures now emphasize "gitting thar fustest with the leastest" so as to minimize development cost, thus holding out hope of retaining popular support. Intensive study and development of settlements is delayed until much later, after initial discoveries have been made, and budget opposition is likely to be muted.

Instead, the Mars Holistic concept kicks-off all three phases of space exploration, Precursor missions, initial Exploration, and Settlement, at about the same time.

Near-term implementation of Precursor missions, and early development and then implementation of early exploratory missions (e.g. Mars Direct) are clearly candidates for FCB technology and management techniques. This does keep costs down, and presents ample opportunities for technology spin-off, but FCB alone cannot create sufficient positive, short-term economic stimulus.

By simultaneously initiating near-term, coherent, large-scale studies of the broader settlement problem utilizing government-aerospace-commercial partnerships, Mars Holistic (MH) reaps a number of advantages:

- A mission and driving theme is provided, not only for the space program, but for the nation as a whole. Instead of a patchwork of independent programs attacking different problems, thrown together, an overarching framework, a thematic Mission From Planet Earth, is built under which solutions can be designed coherently.
- The base of the aerospace industry is immediately broadened, the general economy is stimulated, and rapid incorporation of "spin-on" technologies in the space program becomes possible, the slowness of which has long been a sore point in the community.
- Investment by commercial firms, independent of government capital infusion, is encouraged. This had started already, even in this country, as exemplified by Case for Mars IV; mining, heavy machinery, marine industry, and other traditionally non-aerospace firms participated in early studies, and seemed to feel that with proper backing, SEI could have represented the next phase in an economic expansion. Japanese firms did not waste any time in kicking off their own large-scale projects.
- An accelerated FCB development schedule for the early Precursor missions ensures the near-term importance of this work, without setting frequent milestones for purely political or public relations reasons. It also ensures rapid technology spin-off from traditional aerospace domains.
- MH must be inclusionary. There is both a desperate need and room for elite, FCB, Skunk Works organizations and programs. And de-emphasis of FCB for the Settlement phase is most definitely not an invitation to return to bureaucratic mismanagement of the space program. But by virtue of its scope and diverse objectives, MH as a whole, by definition, can't and shouldn't be "lean and mean" or elitist to the exclusion, alienation, or disenfranchisement of both the larger technical community and, more importantly, the public, which must support and pay for it. Everyone should have a useful, important role to play, while Skunk Works can still forge ahead in parallel, achieving breakthroughs.
- Mars Holistic is a natural means for transferring displaced aerospace and technical workers closer to the commercial world. Not directly into it, something which is proving extremely difficult for government to ordain in a free market economy, but much closer. These workers would straddle the government /commercial di-

vide and be in position to facilitate quick spin-offs from the early Precursor and exploratory phases.

- All social problems will not be overcome by MH, but bringing the space program directly into the venues where problems exist is a good start. If historical precedent is used and encouraged, one might, for example, set up mini-Mars Towns in inner cities, where training and education could be provided, and research data obtained, with an ultimate aim of signing-up future settlers.
- Because of the tremendous broadening of the contractual and indirect industrial base via MH, Congress will suddenly discover that MH is so vast that much of the traditional pork barrel politics, self-interest, and accompanying irrational decision making, become obviated when nearly everyone can legitimately receive a piece of the pie.
- Likewise, the divisive, wasteful and frequently destructive, competition and gamesmanship between and within government and industry can yield to more harmonious NASP-like cooperation. There will again likely be more than enough to go around.
- Conflicts between proponents of manned and unmanned missions, or between big and little missions, or big and little science, should be mitigated, since all will prove important, and will come forth naturally when conceived from a coherent perspective.
- Pay as you go is a prescription for disaster. ME must be defined, laid out and then budgeted, assuming a multi-administration, even multi-generational, time frame. It cannot be subject to major perturbations of funding as standard operating procedure. That is, preparation, decisiveness, and then commitment (what might be called the PDC principle) must take hold.

MARS HOLISTIC COMPARED TO OTHER STRATEGIES

At the highest conceptual level, the arguments of the previous section suggest reason for hope. Other positive attributes of ME are revealed by a conceptual analysis based on that of Pivirotto,[11] for what she called the Human Settlement Program strategy (HSP) . Table 1 is an extension of her Figure 2, adding the element of short-term economic benefit as a major potential purpose of exploration, and adding a column for ME.

ME and HSP, from which it is derived, obviously have similar goals as exploration strategies, but different time phasing and developmental scope and emphasis. ME is specifically proposed to enhance short-term economic benefits, while ESP as traditionally constituted, would bear economic fruit in its latter stages. Pivirotto assigns a “low” value to promotion of human spirit by HSP; since human spirit will be promoted by the collective mission, early economic revival, and cooperation between the social spheres, a “medium,” if not “high,” could be assigned to ME.

Both HSP and ME possess flexibility. Viable settlements must have some purpose, and this will usually be scientific research or economic development, as Pivirotto points out. once the paradigm of human settlement is begun, branch paths can readily be followed to exploit favorable conditions for a particular focus. Thus, lunar observatories might well be established even if the Moon were bypassed in initial settlement strategies. Likewise for asteroid mining operations.

Table 1
MARS HOLISTIC COMPARED TO OTHER STRATEGIES

PURPOSE \ STRATEGY	SPIRIT LIFTER	PLANET SCIENCE	ENERGY ENTERPRISE	TECH PUSH	HUMAN SETTLEMENT	MARS HOLISTIC
PROMOTES HUMAN SPIRIT	HIGH	LOW	MED	MED	LOW	MED
SCIENCE/KNOWLEDGE	LOW	HIGH	LOW	LOW	HIGH	HIGH
LONG-TERM ECONOMIC RETURN	LOW	LOW	HIGH	HIGH	HIGH	MED
SHORT-TERM ECONOMIC RETURN	LOW	LOW	LOW	MED	MED	HIGH
DESTINY/SETTLEMENT	N/A	LOW	LOW	N/A	HIGH	HIGH

From the perspective of science, ME also fares well. Budden and Spudis[12] analyzed the science content of different exploration strategies. They concluded that what would be called HSP or ME provided the most robust science, in the sense of offering a reasonable level of return to the most scientific disciplines, perhaps at some expense to geology. When one considers that there is no reason why "Precursor" missions and human exploratory elements can't continue as part of an overall ME strategy, geology should not be impeded much at all.

CONCURRENT CREATION

Space programs have not been judged on their potential economic and social benefits in their early planning stages, let alone designed to specifically enhance these attributes. Contractors must model and anticipate the economic returns of a given program to make informed business decisions, but these analyses are from an understandably parochial viewpoint.

By embracing societal problems and potential solutions, the proposed study of MH must bring to bear an unprecedented interdisciplinary outlook, comprising not only the engineering, science, and medical /psychological fields usually found in SEI studies, but economics, the social sciences, and the humanities. A new model for conducting studies of this kind is needed.

Generalizations of the methods of Concurrent Engineering (CE) may provide just that. CE[13] evolved from analysis of the factors causing complex aerospace and defense projects to exceed budget and fall behind schedule. It was found that disciplines involved in a program frequently did not communicate with each other, were not all involved in the early stages of a program, that mistakes made in early stages were only corrected at great expense in the latter stages, that organizations lacked flexibility or adaptiveness to the problems at hand, among other factors.

By organizing teams of engineers and support personnel from all pertinent disciplines at the very start of a program, built around a core designer or systems engineer who possesses an interdisciplinary outlook and responsibility, many of these difficulties are solved or at least ameliorated. There are many procedures and guidelines that go along with formation of a CE organization and ensure its success that need not be listed here.[13] These CE teams can be stacked hierarchically and can also be created and terminated on a temporary basis as needed to attack specific problems. One of the key features of CE is that the teams are maintained for the duration of the project, thus retaining collective continuity, knowledge, and experience of project personnel.

Figure 1 suggests how the CE concept can be generalized for MH. At the top-level, corresponding to the proposed study phase and the incipient developmental phases of MH, a Mission CE team encompassing the already unusually broad aspects of SEI, is just one element of the MH team. The MH team also consists of economists, social scientists, historians, technology and business development experts, surrounding a central core of generalists and overall program directors. The Mission CE team includes program management, administrative, engineering, science, medical/psychological, and architectural personnel. Since we have gone beyond engineering, or even what is usually termed "design,"[8] this approach is characterized herein as Concurrent Creation (CC) (unintentional metaphysical associations notwithstanding). In bringing together diverse disciplines, CC implements in the professional project development process, the space systems educational philosophy pioneered by the International Space University.

Though Mission Design is seemingly subsumed under the weight of "extraneous" disciplines, this does not in any way imply that the goals and objectives of MH are not human space exploration and settlement; it is precisely the responsibility of the top-level CC group to ensure that this objective is met. But since it cannot be achieved unless economic and societal objectives are also met, they must be balanced. One can envision a situation in which a major settlement structure is examined not only for fidelity to engineering requirements, but also for technology transfer issues and the ability to involve the commercial economy. In no case can fundamental engineering or life science requirements be violated, but within these constraints, choices can be made which enhance the overall system solution, where "system" is now broadened to include defined economic and societal objectives. The proposed study should provide a good proving ground and testbed for these concepts and this type of organizational model.

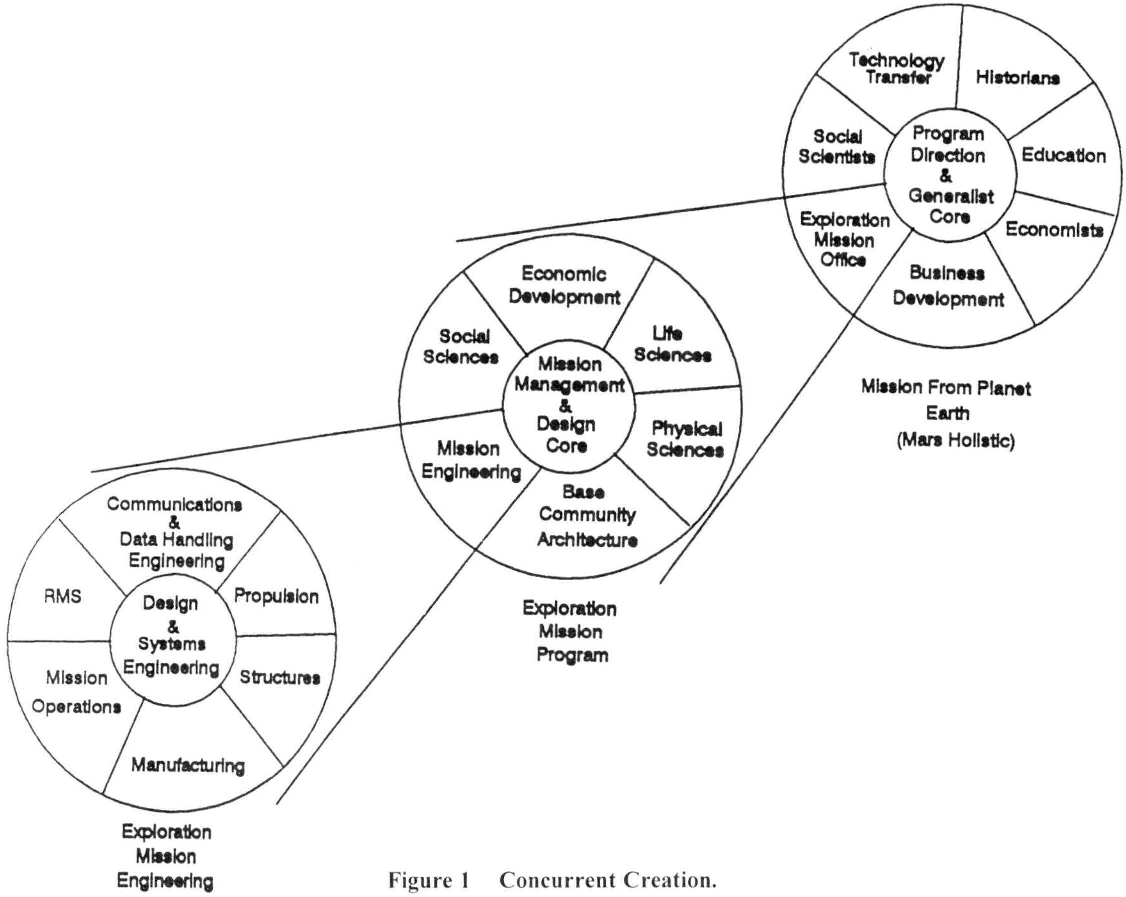

Figure 1 Concurrent Creation.

CALL TO ARMS

No longer is there sympathy for even "low-cost" exploration programs. Exploration supporters attribute much of this antipathy to confusion between newer, low-cost missions ($60-70B for Stanford,[15] Mars Direct has not been costed to a detailed level.[2]) and the original $400B "Battlestar" concepts.

Based on this, proponents believe "inexpensive" exploration architectures, combined with a public relations and lobbying campaign to eradicate any vestige of Battlestars from our collective memory, will succeed in reviving an exploration program. A key ingredient in this recipe is that the new program will actually fit within NASA's projected budget - if the shuttle and/or the space station are cancelled.[15] This last appears increasingly unlikely.

The fatal flaw in Battlestar concepts was that they yielded only that which could be obtained for $70B. That sum, spread over the more than two decades (Stanford) required to land humans on Mars, might yield only what any similar sized program might, and so must take its place alongside others in a competition for resources. It is not singular in cost, but may not be regarded (by the nation) as singular in results either.

Both Zubrin[2] and Lusignan[15] make the important point that their respective missions encompass sustained settlement, unlike the Battlestar concepts. Thus, not only are they too expensive, but Battlestars are simply not needed. True. But, this view also incorrectly assumes that larger scale missions necessarily entail wasteful Battlestar economics. MH does not revive Battlestars that only do the job of "Arctic dogsleds."[2] The exploration schedule might be accelerated and the scope broadened. And it is useful to recall for comparisons sake, that it has been many decades since the bulk of our air passengers moved by one and two-seaters alone!

Yet if MH, which might well cost the proverbial $400B of early conceptions of SEI, is to stand any chance of acceptance, demonstration of the countervailing shorter-term economic benefits had better be solid, if not ironclad. The crucial question is whether the initial investments in starting MH will diffuse through the economy rapidly enough so the benefit curve can overtake the cost curve in, say, three to five years - which is probably the limit of the patience of the American people. That is the purpose of the study.

There is a prominent movement afoot to bring government, industry and academia into partnerships to develop technologies (dual use technologies) and programs suitable for both government and commercial sectors, especially in manufacturing. Called the Technology Reinvestment Program (TRP), it is funded at about $500M, and represents a major departure from the usual ways of doing business with the government.

The philosophy behind MH is certainly fully compatible with this vision. Indeed, if it were not for the strong orientation of TRP towards Department of Defense interests, somewhat to the neglect of space and NASA, this study might be a prime candidate for the TRP program.

In any event, despite the budgetary crisis, cost cutting efforts, and a disinclination towards space exploration by the Clinton administration, this is the very administration likely to be the most receptive to the economic, technological, and especially, the "inclusionary" aspects of the MH concept, so long as it also offers solid short term benefits and solutions.

While rejecting traditional "not my job" attitudes in striving to become part of the solution, MH is not a panacea for all societies, problems; it will not, for instance, deliver a solution to the health care crisis. But if MFPE (MH) were simultaneously reborn with an equally broadened concept of Mission to Planet Earth (MTPE), including not only space-based earth observation and analysis designed to monitor and preserve our environment, but a coordinated assault on other earth-based problems, we are in a position to embark on space exploration in a manner which does not ultimately relegate our home planet and its inhabitants to the dust heap of history.

International partnerships for both MTFE and MTPE will naturally grow out of these steps. The possibility that humanity may take a great leap forward, leaving behind those nations which may be "anathema" to the human family (e.g. Serbia), may act as a powerful incentive for such states to adhere to humane norms of behavior.

Granted, these are only very top-level concepts, not implementation. But if there is an envelope of validity for these concepts, when buttressed with data and more cogent arguments, SEI could be transformed into a resplendent positive. Indeed, a move has begun" to establish a global exploration database; this could serve as the nexus for this kind of study. Case for Mars V can start this process. Thus, this "Call to Arms".

ACKNOWLEDGEMENTS

I am grateful to Nick Pocus of SCI Systems, Inc. for the artistic rendering of one of the figures in the original presentations, and to Rick Robinson of SCI for comments and production of the manuscript and the presentation accompanying this paper.

REFERENCES

1. Friedman, L., *Mars Underground News*, Winter 1993.
2. Zubrin, R., "Mars Direct: Humans to the Red Planet Within a Decade," presented at Case for Mars V, Boulder, CO, June 1993.
   See also Zubrin, R. M., Baker, D. A., and Gwynne, O., "Mars Direct: A Simple, Robust, and Cost Effective Architecture for the Space Exploration Initiative," AAS 90-168, in *The Case for Mars IV*, T. R. Meyer, ed., AAS *Science and Technology Series*, Vol. 89, 1997, pp. 275-314; and
   Baker, D. A., and Zubrin, R. M., "Mars Direct: Combining Near-Term Technologies to Achieve a Two-Launch Manned Mars Mission," AAS 97-366, in *From Imagination to Reality*, R. M. Zubrin, ed., AAS *Science and Technology Series*, Vol. 91, 1997, pp. 225-239.
3. Becker, R. E., Letter to Dr. Michael Griffin, June, 1992.
4. Cutter, B., Interview with *Space News*, April 5-11, 1993, p. 22.
5. Diamandas, P., Essay in *Space News*, April 19-25, 1993, P. 19.
6. Hawkins, W. R., Syndicated essay in *Huntsville Times*, April 28, 1993, p. B7.
7. Boucher, P., Dept. of History, Univ. of Alabama, Huntsville, April 1993, private communication.
8. Cohen, M. M., "Designing Space Habitats for Human Productivity," Summary Paper, 20th International Conf. on Environmental Systems, Williamsburg, VA, July 1990.
9. Becker, R. E., Special Session, "How Autonomous Can and Should Humans Be in Space Exploration," TABES '92 Symposium, Huntsville, AL, May, 1992.
10. Boucher, P., Presentation at TABES '92 Symposium, Huntsville, AL, May, 1992.
11. Pivirotto, D. S., "A Goal and Strategy for Human Settlement of the Moon and Mars: Part Two," Proc. Case for Mars IV, Boulder, CO, June 1990. Published as abstract only, AAS 90-178, in *The Case for Mars IV*, T. R. Meyer, ed., AAS *Science and Technology Series*, Vol. 89, 1997, p. 271.
12. Budden, N. A. and Spudis, P. P., "SEI Science: Measuring the Return," *Aerospace America*, March 1993, pp. 22-27.
13. Special issue on Concurrent Engineering, *Aerospace America*, April 1993, pp. 18-40.
14. T. Meyer, University of Colorado, private communication, April 1993.
15. B. Lusignan, Stanford University, private communication, May 1993.

AAS 93-850

IS BONE LOSS IN SPACE A PROBLEM?

Christopher E. Cann

The loss of skeletal mineral experienced by humans exposed to microgravity is considered to be one of the primary medical concerns which could limit mission duration or impact the design of interplanetary spacecraft. The current state of our knowledge is consistent with the hypothesis that there is an increase in the breakdown of bone with the onset of skeletal unloading, and that this leads to progressive bone loss. Results from experiments in growing animals also indicate a defect in the formation of new bone, but his has not been shown in the adult skeleton. The amount of bone loss, even on year-long missions, has been small compared to the bone loss required to place an adult at risk of development of osteoporotic fractures. This may be due, at least in part, to the use of exercise countermeasures on these missions.

From an operational perspective, bone loss during spaceflight has not been sufficient either to impact performance in microgravity or to cause osteoporotic fractures immediately on return to earth. Even if the loss of skeletal structure is irreversible, it should be considered as an occupational hazard of spaceflight, not as a driving force for spacecraft design or, at current levels, as a limiting factor in mission duration. However, the sustained elevation of blood calcium as a result of increased bone breakdown, and the apparent inhibition of normal bone formation processes which could impair fracture healing in space, could be cause for concern. Current countermeasures exist to treat some of these symptoms. However, continued research in bone and calcium metabolism in spaceflight should be directed toward determination of the fundamental mechanisms responsible for these alterations in bone cell function, in order that rational countermeasures aimed at these basic processes can be designed.

INTRODUCTION

The primary medical concerns of long duration space flight have often been cited to be radiation exposure and musculoskeletal deconditioning, especially bone loss. While many other physiological changes also occur in space, it is often thought that their long term effects are minimal (such as Space Adaptation Syndrome) or can be minimized using known countermeasures (such as lower body negative pressure or exercise for the cardiovascular system.) Some investigators dispute this traditional view, especially in light of the results of the Spacelab Life Sciences-1 mission in August 1991 where many traditional hypotheses were shown to be wrong, signaling that the effects of microgravity on the human body may need to be looked at in a more fundamental way than has been done in the past. For the purposes of this article, I make no use of the traditional view of bone loss in spaceflight, but instead approach the problem from an operational rather than a physiological perspective. I also reject out of hand the thesis that a minor risk of medical disability for an individual or a small group of persons is a reason for significantly complicating the design of interplanetary spacecraft. In fact, it is clear that other factors, such as the psychological well being of the crew, are probably more important driving forces in decisions such as whether or not to include artificial gravity in the engineering design, than is bone loss.

Bone loss and radiation can be compared from an operational perspective. The radiation issue poses an operational risk which cannot be eliminated, only minimized. The issue can be viewed in two ways, first as a risk to be dealt with during the time of the operation (the mission duration), and secondly as an occupational hazard with quantifiable lifetime risks of developing cancer or other problems. The Mars mission issue requires that the crew be protected sufficiently, that they can carry out their duties when they reach the Martian surface, and this requires providing protection from solar flares and minimizing galactic radiation exposure, both in transit and on the surface. The occupational hazards of radiation are better known for photons than for particles, but can be estimated for astronauts and are a known risk in addition to other risks such as riding atop several million kilograms of explosive material. I do not know any astronauts who refuse to fly because of a fear of radiation exposure, although because of the occupational career limits set for them by others they may try to avoid other sources of exposure so as not to jeopardize future flight opportunities.

The operational perspective of bone in space is that it must be able to perform its normal function of structural support and maintain its role as a calcium reservoir. This means that if bone loss occurs, it cannot be of sufficient magnitude to allow the bone to fracture under normal or somewhat increased loads during work in space or on planetary surfaces. In addition, the normal processes of bone repair must not be compromised, or an astronaut might break a bone in space just as on earth and it might not heal. The ability of the body to maintain the blood calcium level constant to provide normal muscle and nerve function is paramount, and any adaptations in reduced gravity must not affect this ability. In contrast, a small amount of bone loss which does not impair skeletal function acutely but may be irreversible and cause an increased risk of osteoporosis later in life can be treated as an occupational hazard in the same way as increased radiation exposure.

The loss of bone in humans due to exposure to microgravity is thought to occur by a process of accelerated bone breakdown, or resorption. This hypothesis is consistent with all spaceflight data, including the recent SLS-1 results. The basic physiological mechanism causing this increased breakdown is unknown; it is not sufficient to cite "unloading" as a cause without specifying which of the many postulated transducer mechanisms is responsible for signaling the active recruitment of the bone-resorbing cells. The traditional view that bone loss occurs late and is a slow, constant process is wrong; SLS-1 results showed that calcium metabolism is already dramatically affected by the second day in space, and that normal adaptations leading to decreased absorption of calcium from the diet occur within the first week. While a complete discussion of bone physiology is beyond the scope of this paper, some basic concepts are presented in the next section to show how bone could behave in space and how we might be able to alter its behavior for the duration of a Mars mission. With our current understanding of bone physiology we can treat bone loss in space similarly to radiation exposure, protecting against both acute and chronic effects. In both cases, however, we must consider the chronic effects as hazards of occupational exposure which, while not to be disregarded, should not unduly influence either engineering or policy decisions about inter-planetary exploration.

BACKGROUND

Calcium Metabolism, Bone Structure and Function

Bone is an organ system, and has two basic functions. It serves first as a reservoir for the calcium homeostatic system to maintain the blood calcium level constant within a very narrow range. This is necessary to maintain normal nerve and muscle function, including cardiac muscle, and this function overrides the second function, that of structural support. We know this because in diseases where the blood calcium is lowered, by impaired absorption from the intestine or by exaggerated excretion by the kidney, the body drains calcium from the bones to the extent that they fracture under normal loads. Calcium homeostasis is a complex process involving a large number of interrelated systems, including blood, bone, the intestine, the kidneys, the liver, and hormones produced by (at least) the parathyroids, thyroid, adrenals, ovaries, testes, and pituitary. When the blood calcium drops, adaptive responses occur to raise it back to normal. Because high blood calcium also causes problems when it rises due to an abnormal influx the homeostatic mechanisms also respond in order to lower it back to normal.

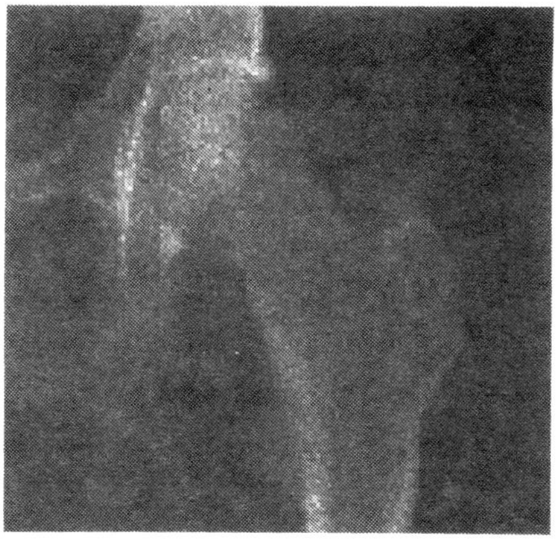

Figure 1 High resolution radiograph of the proximal femur. Note the complex internal architecture of the trabecular bone which undergoes specific tensile and compressive stresses during ambulation. The external geometry is also specific for ambulation, and is common among essentially all land mammals.

Bone homeostasis is related to calcium homeostasis, but is not controlled completely by it. Bone responds directly to a variety of influences, including mechanical forces, and adapts to its environment. The ability of bone to adapt, or "model," in response to forces placed on it was recognized as early as the 19th century.[1] Both the external and internal architecture of bones is designed to maximize strength for minimal mass, within the context of the function for each individual bone. Thus, a vertebral body is designed to support the weight of the torso, the posterior elements of the vertebra are attached to the muscles used to maintain erect posture on earth, and the proximal femur with its complex external and internal geometry is designed to support the stresses of ambulation (Figure 1). If the function of a bone is different, its architecture is also different; this is seen in the difference in vertebral geometry between an erect biped (human) and brachiating primate (rhesus monkey) (Figure 2). However, such differences have developed over millions of years, and a change in function in an individual results only in a change in internal architecture with a preservation of the macroscopic geometry, such as is seen in the bones of children who grow up with Marfan's syndrome, with lack of normal muscle function. In space flight, any changes in individual bones will occur as normal adaptations to a change in function of those bones. The only reason we have to prevent those adaptations is to maintain function for the original environment.

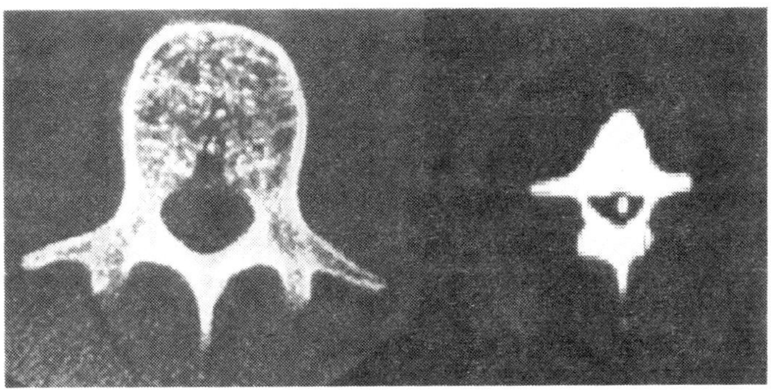

Figure 2 Cross-sectional images of human and monkey vertebral bodies. Note that the human vertebra has a cylindrical shape to take the normal vertical loading in the human upright posture, while the triangular shape of the monkey vertebra is better suited to the complex torsional loading of brachiation (swinging in trees).

Osteoporosis results when the amount of bone and its distribution are insufficient to support the loads encountered under normal conditions, and the bone breaks. Osteoporosis is thus relative; if the normal loads are very small, the amount of bone can be low without fracturing. Conversely, if normal loads are high, but the bone adapts accordingly, as in athletes, fractures should not occur. On earth, each individual's resistance to osteoporotic fracture is determined by the amount of bone developed during adolescence, the way the bone is distributed, the loading placed on the bones during normal activity, and how much bone is lost during adulthood due to a variety of factors. Bone is constantly being broken down and rebuilt, through a process known as remodeling. The two basic types of bone, cortical or compact bone such as in the arms and legs and trabecular or spongy bone in the spine, pelvis and ends of the long bones, can react differently to various stimuli. In general, the trabecular bone is more responsive because it is in close contact with the bone marrow from which bone cells are derived.

Certain factors, such as the removal of estrogen at the time of menopause in women, cause remodeling to increase. Remodeling is not a perfectly efficient process, and in some cases there is an imbalance between the amount of bone resorbed and the amount which is reformed, leading to a net loss of bone. The rate at which bone is lost varies among individuals, among different skeletal sites, and for different situations. The normal bone loss with aging can be only a few percent per decade, whereas paralysis can cause 2-4% per month loss from the spine. There does not appear to be a correlation between the amount of bone a person has and the rate of loss due to a stimulus, so that people who have more bone to start with but lose at the same rate do not develop osteoporosis as easily.

The Measurement of Bone Mineral Content

One way to estimate a person's risk for development of osteoporosis is to measure the amount of bone in the skeleton. This is done routinely using a variety of methods based on the differential absorption of x-rays by bone and soft tissues. Most techniques in common use measure the amount of bone in a selected site, such as the spine or hip, primary sites of osteoporotic fractures, using two-dimensional projection methods (Figure 3). Research methods such as 3-D quantitative computed tomography (3-D QCT) can obtain information about the distribution of bone as well as the amount, and when coupled with sophisticated structural modeling methods can be used to estimate directly the strength of the bone structure under defined loading conditions (Figure 4). This type of modeling approach can also be used to predict the change in the risk of fracture if a given change in bone density or distribution occurs, for example, if a significant amount of the central trabecular bone core is lost from the spine on a long term space mission.

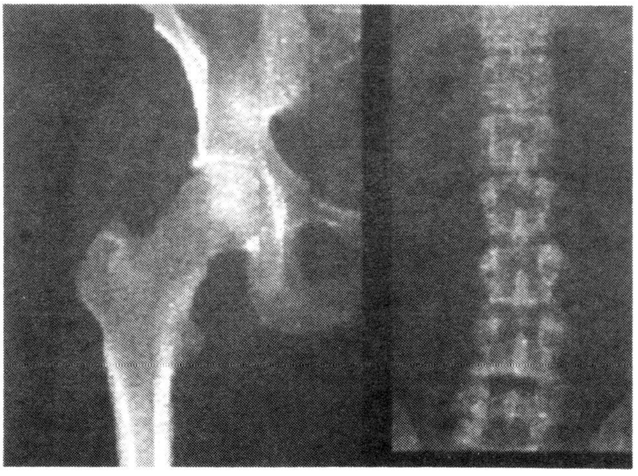

Figure 3 High resolution dual energy x-ray absorptiometry images of the proximal femur and the lumbar spine in a human. The total amount of mineral in regions of the bone can be determined by region-of-interest analysis of the images, and is commonly expressed in terms of projected mineral per unit area, or grams per square centimeter.

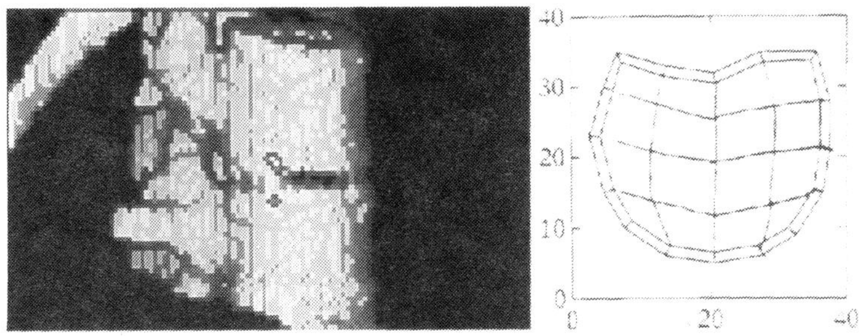

Figure 4 Three-dimensional quantitative computed tomography data set (left) can be used to determine the local volumetric density and the distribution of the bone mineral. These data are input into finite element models of the vertebra (right) and tested under specific loading conditions to determine structural properties such as strength and failure modes.

Traditional measures of bone density are interpreted as risk factors for osteoporosis, much as blood cholesterol level is used for heart disease. More complex bone structural assessments might be used in the medical evaluation of individuals in the same way as a 12-lead electrocardiogram is used. Most healthy persons under age 50 are not at risk for developing osteoporosis, and about 30-40% of the skeletal mineral must be lost on average before the risk of fracture increases substantially. However, individuals by definition are not represented by averages, so someone with a low bone density to begin with may not be the best candidate for a long term mission. On the other hand, we do not want to exclude astronauts based on flawed measures or interpretations, so we must be very careful in the use and assessment of such techniques.

Some researchers have tried to develop biochemical means of assessing a person's future risk of bone loss by static measures of parameters related to current bone remodeling. While this may work in some populations such as early postmenopausal women, it does not predict what type of response each individual's skeleton will have to a challenge such as microgravity. In addition, bone loss at least in paralyzed patients[2] and older women[3] is a self-limiting process related temporally to the bone remodeling cycle, proceeds at different rates in different people, and can change in a person over time. As such, at present there is no way to predict the final value of bone mass from a set of current measurements.

THE PERCEIVED SPACEFLIGHT PROBLEM

Bone loss as a biomedical problem for long duration space missions derived from early observations of bone loss in humans paralyzed by polio or spinal cord injury.[3-5] It was reasoned, rightly so, that unloading of the skeleton would cause the muscle and bone to atrophy from disuse. The first observation of alterations in calcium or skeletal metabolism due to spaceflight was a reported increase in urinary calcium in cosmonauts on board Vostok 2 and 3 in 1963.[6] US investigators used an x-ray densitometry technique to measure changes in the density of the heel bone in the astronauts from the Gemini IV, V, and VII flights and reported dramatic losses of 3-15% for these short 4-14 day spaceflights.[7] A puzzling problem arose in the analysis of these data because the greatest losses apparently occurred in the shorter flights. However, a complete reanalysis of the technique and results by Vose[8] showed that some substantial methodological errors had been made for Gemini IV and V. His analysis reduced the reported losses to the range of 2-4% for five of the six astronauts and 9% for the other.

The Gemini VII flight also was noted for the first attempt to do calcium metabolic balance studies in spaceflight. Considering the problems involved in doing metabolic studies in microgravity, it is not surprising that more studies were not done; it is surprising that any were done, necessitating relatively quantitative collection of urine, feces and diet records from astronauts or cosmonauts. The Gemini VII results showed urine calcium increased early in flight,[9] but no conclusions were reached regarding calcium absorption, and no inflight blood samples were obtained. Similar results were obtained for the 18 day flight of Soyuz 9, where both cosmonauts exhibited an increase of urinary calcium and an 8-10% decrease of heel bone density,[10] and for the Apollo 7 and 8 flights where slight bone density changes were noted.[11] Bone density was also measured in later Apollo flights up to 13 days duration using newer, more precise methods, but only on the Apollo 15 mission did any crew member lose a significant amount of bone from the heel.[12] A metabolic balance study conducted on Apollo 17 documented increased urinary and fecal calcium excretion, with an apparent decrease in intestinal calcium absorption as well.[12] More recently, the SLS-1 flight in August 1991 demonstrated a very significant increase in blood ionized calcium by the second day in flight, with a compensatory adaptive reduction in parathyroid hormone, followed by the expected decrease in the vitamin D hormone that promotes intestinal calcium absorption.[13,14] The combined result from all these studies done on short-term spaceflights is support for the hypothesis that there is an early primary increase in bone resorption, and that the body adapts to the subsequent rise in blood calcium in an expected way.

A significant amount of data has been generated about bone loss and calcium metabolism from longer-term flights, first from the US Skylab series of flights, and later from the Soviet Salyut-7 and Mir series. The only reported detailed metabolic studies have been for the Skylab flights,[15] and these three flights have provided the best overall data to date on changes in calcium metabolism due to spaceflight. These data were collected and reported in 6-7 day pools necessitating descriptive results, given our knowledge of calcium and bone dynamics. However, the Skylab results also showed that urinary calcium increased early (Figure 5). The published reports of Skylab metabolic data indicate that the urinary calcium output increases during the first 30 days to a level approximately 125-150 mg/day higher than preflight excretion, and then is maintained at this level for the remainder of the flight (for Skylab III and IV). The fecal calcium output has been published to increase substantially after the first month of flight and then to continue to increase with flight duration,[16] with a calculated total skeletal loss from excess urine and fecal excretion of 25 grams of calcium in 84 days. However, close inspection of the data from the individual flights reveals that when each crew member is used as his own control and the baseline data are not averaged, the total measured negative calcium balance is 12.8 grams, or about 1% of the 1250 grams of calcium in the average skeleton lost over the 3 month period. Reduced calcium losses seen during the

84-day Skylab IV flight have been attributed to an increased exercise program used by the astronauts for this flight.

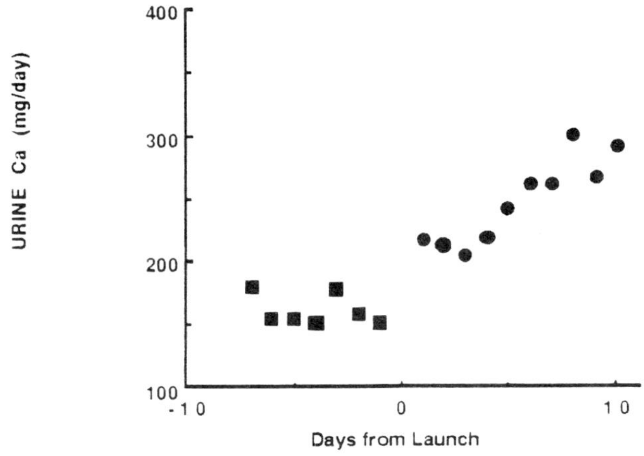

Figure 5 Average urine calcium output for the first 10 days of the Skylab missions. Note that there is an excessive calcium output for the first two days, possibly due to concommitant calcium and sodium excretion.

Direct measurement of changes in the bone mineral density of the calcaneus (heel) and the radius and ulna has been made in most long-term flights since 1973. Both in Soviet and US flights the bone loss from the calcaneus appears to average about 1% per month[16,17] with no observable loss from the upper extremities. Both US and Soviet mission planners have included a regimen of significant exercise for flight crew during spaceflight, including exercises such as running on a treadmill to load the skeleton. These exercises would be expected to confer their designed prophylactic effect on the calcaneus to minimize any changes, but if the exercises did not transmit appropriate loads to the rest of the weightbearing skeleton then bone could be lost from other sites. In 1979, a proposal was made to the Soviets for a collaborative US/USSR research project, using the then newly-developed method of quantitative computed tomography to measure mineral density in bones that were clinically important in the development of osteoporosis, in particular in the spine. The first postflight measurements were made in 1981,[18] followed by a series of preflight and postflight measurements for durations of up to 8 months. In contrast to the hypothesized loss of trabecular bone from the vertebral body due to unloading, the initial results showed that on average there was little or no bone loss at this site, apparently validating the effectiveness of the exercise regimen in preventing spinal bone loss as well as minimizing calcaneal losses.[19] However, further analysis of the data revealed some interesting results. The Soviets (now Russian) exercise countermeasure program was designed to simulate the vertical loads produced on the bones during normal ambulation or running, but it did not have a specific component to exercise the antigravity postural muscles in the back. These paraspinous muscles hold the body erect while standing or help return the torso to its normal position when twisting movements are performed. The loads exerted by these muscles on the spine are not transmitted to the vertebral bodies but to the transverse and spinous bony processes to which the muscles are attached (Figure 6), and consequently if the muscles are not exercised they will not load the bone. Analysis of the three-dimensional computed tomography data showed that even though the vertebral bodies did not lose bone, the posterior elements lost an average of 8%, and this loss correlated with a 4% decrease in the volume of the muscles attached to the bone[20] (Figure 7). Such changes in the bone are not of sufficient magnitude to raise a clinical concern about the development of osteoporosis, but clearly a decrease in the mass and strength of muscles in the back necessary to perform torsional tasks such as assembly during EVA is an operational concern and will need to be addressed in the development of new exercise regimens for flight.

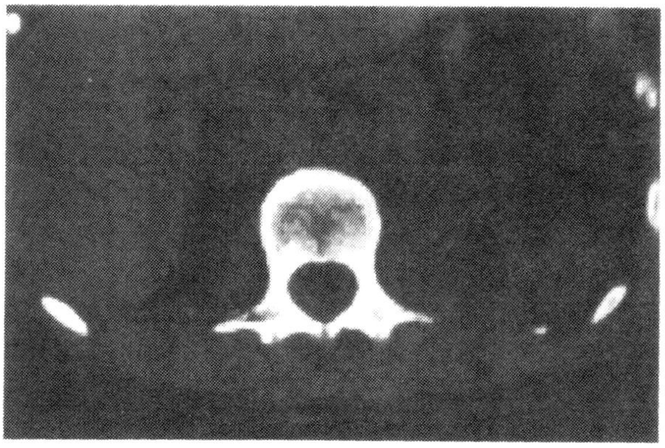

Figure 6 CT scan of the abdomen and first lumbar vertebra in a normal adult (anterior is to the top of the figure). The cylindrical vertebral body supports much of the weight of the body while standing, while the paraspinous muscles are attached to the lateral posterior elements and exert forces on the bone during torsion and while holding the body erect.

The analysis of bone mass by Russian investigators has continued on the cosmonauts who have flown on Mir. After the 366 day mission, one cosmonaut averaged 10% loss of trabecular bone from L1, L2, L3 (using computed tomography) while the other cosmonaut gained 5% mineral density when measured using dual photon absorptiometry (DPA).[21] However, the CT methods used were not the same ones used in prior flights and therefore the compartmental analysis previously described (Figure 7) could not be done; no mention was made whether the DPA data were corrected for possible soft tissue baseline inconsistencies pre and post flight due to the change in position of the internal organs or to the known change in tissue hydration and muscle composition. More recent studies using x-ray-based DPA have shown similar results, but with no evidence of whole body bone loss in flights up to 312 days.[22]

The bone loss seen to date in spaceflight is not an operational concern inflight or a clinical concern on return to earth. Exercise countermeasures minimize the losses, but it is not an operational benefit to have a cosmonaut or astronaut running two hours a day on a treadmill while someone else is trying to do a microgravity crystal growth experiment in the next laboratory. NASA has expressed a desire for exercises which can maintain cardiovascular and musculoskeletal fitness in 10-20 minutes/day. If we progress to living and working in space without consideration of return to earth, even this requirement may be eliminated (unless, of course, we wish to explore other planets).

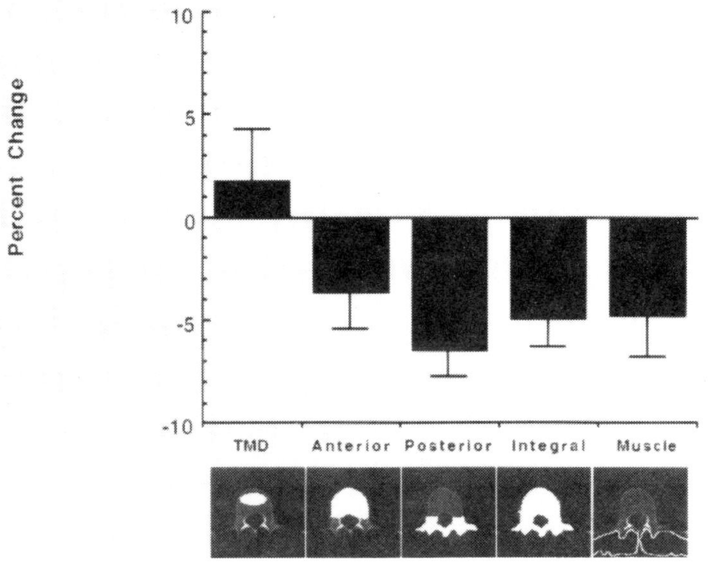

Figure 7 Regional spinal bone and muscle changes in Soviet cosmonauts (N=7) for missions from 3 to 8 months duration.

BONE CHANGES AND THE USE OF COUNTERMEASURES

All of the data obtained from the early studies on bone loss in spaceflight is observational rather than experimental. The small number of subjects available for study and the severe constraints placed on the gravitational physiologist by operational medicine concerns limits our ability to interpret the observations. However, we can use these observations to guide our design of experiments which test hypotheses about bone and calcium homeostasis both in short term flights and over extended durations.

Because bone is constantly being broken down and rebuilt in the adult, perhaps as a mechanism to repair local microfractures due to normal loading, an effect of microgravity that alters either bone breakdown (resorption) or bone formation can alter the balance between these processes and lead to a net change in the amount of bone. On a gross scale, our experiments are directed toward looking at these cellular processes of resorption and formation, or markers for them. On a more focused scale, the underlying mechanisms of action such as identifying the transducer causing cells to respond to mechanical stimulus or lack of it, or effects on transcription for production of specific proteins, are being investigated in animal models. It is necessary to understand both the macroscopic and microscopic processes which occur in order to begin to design rational countermeasures to prevent the bone changes, if in fact that is our objective.

A number of basic experiments have been done in rats flown in space, and the bottom line is that there is a defect in the way the cells which form bone work. It appears that the normal production of these cells from their precursors is impaired, and that the bone formed by cells already on the bone surface when introduced to microgravity may not produce normal collagen, the organic substrate to which the mineral of bone is bound. These effects reduce the rate of bone growth, and also impair the healing of fractured bones. This defect has been noted in the skeleton of rats, a species which is still growing, and it is not clear if the same effect will be seen in the adult human or primate skeleton. We do presume, however, that this may be the case because it has been seen on earth in paralyzed adult humans. The human data indicating a very rapid increase in blood calcium (24-48 hours) and increased urine calcium excretion at this time also are consistent with decreased movement of mineral into bone, although the rat experiments suggest that mineralization is not affected, only new bone formation.

Attempts to demonstrate an increase in bone resorption as a hypothesized mechanism underlying the observed bone loss have been less successful. In rats and juvenile monkeys, the skeleton is still growing rapidly, changing its architecture, and it is very difficult to increase bone breakdown in this situation. Results of experiments from SLS-1 in adult humans, however, are consistent with this hypothesis, showing a rapid and sustained increase in serum ionized calcium with a compensatory decrease in serum parathyroid hormone indicating that the blood calcium increase is primary. Experiments currently planned for flight on SLS-2 and SLS-3 will test this hypothesis directly.

At the present time, the data obtained directly from microgravity experiments in adult humans argues that bone breakdown increases dramatically early in flight. There are no data on bone formation, although we presume that it may be suppressed. Early investigators in spaceflight bone loss, notably Whedon,[15] suggested these mechanisms may apply, based on observations of immobilized subjects on earth. However, the data obtained from SLS-1 showed that the increase in blood calcium was so large, and occurred so early, that invoking the simple mechanism of skeletal unloading may not be sufficient to explain the results. It is possible that exposure to microgravity can alter bone homeostasis by more than one mechanism. For example, skeletal unloading will increase the recruitment of osteoclasts (bone resorbing cells) on the time scale of several days to weeks, similar to that seen in paralyzed patients, but at the same time may have a direct effect on cells comprising the bone-blood membrane, a layer of cells which separates the calcium-rich bone fluid compartment from the blood where calcium concentration is much lower, on a time scale of minutes. If in fact this does occur, our current concepts of bone and calcium homeostasis, as well as current approaches to the development of countermeasures to bone loss, may be inadequate.

Apparently rational countermeasures to bone loss in space include exercise and possibly drugs to suppress bone resorption. The rationale underlying use of these countermeasures is that the unloading of the skeleton produces an increase in the number of osteoclastic resorption sites, and that this will lead to bone loss. Exercise is also thought to promote bone formation, although all we know about the effects of mechanical usage on bone suggests that its effects on bone formation are adaptive, that is, it affects the "modeling" component of bone turnover, that responsible for the external and perhaps internal architecture of bone defined by loading patterns. This is not to say that mechanical forces do not prevent reduced bone formation in the presence of altered skeletal loading, but that the bone forming cells and their precursors may in addition be responsive to factors other than the direct loads placed on them. Besides the use of exercise, there are few other measures to stimulate bone formation; agents such as fluoride and anabolic steroids may do so, but are not highly recommended because of their other effects. Recombinant human growth hormone may also be useful in the future, but it is just starting investigative trials for bone.

Suppression of bone resorption is a viable avenue for attempts to prevent microgravity-induced bone loss in humans. Mechanical usage can do this, and there are numerous pharmacologic agents which also might be useful, in particular anti-resorptive agents such as the bisphosphonates, gallium nitrate, and the hormone calcitonin. Non-hormonal drugs to prevent resorption rely on poisoning the osteoclasts, by loading the bone mineral with an agent which prevents the osteoclasts from attacking that region of bone. These agents localize in the bone mineral and remain there for the lifetime of the bone, which can be months to decades. At the present time, these agents are being investigated for prevention of bone loss in established osteoporosis, where it is most important to prevent further loss of bone from an already fragile skeleton. Their use for prophylactic prevention of bone loss, for example, at the time of menopause, is not generally accepted practice because it is unclear what long term effects they might have on normal bone turnover and on fracture healing. Calcitonin is a hormone produced in the thyroid gland which reduces the blood calcium level over a short time period by suppressing resorption. This hormone has been investigated for postmenopausal bone loss, more recently in a nasal spray preparation rather than the older injectable form. It may also be useful for spaceflight, but requires daily administration.

ALTERED CALCIUM HOMEOSTASIS: THE OPERATIONAL PERSPECTIVE

It can be seen from the previous discussion that bone loss in space does occur, but that its magnitude with current countermeasures is small, and that there are pharmacologic measures available that may be able to reduce it even further if it is simply due to increased osteoclastic resorption as a response to skeletal unloading. From a purely operational perspective, then, it is not necessary to be concerned about the development of osteoporosis during a spaceflight, even of long duration, if the individual skeletal mineral is used as a flight selection criterion. Even if a crewmember had an exaggerated resorptive response to spaceflight, as documented by on-board urine analysis, drugs could be given to suppress this.

If osteoporosis is not an operational concern, need we be concerned about bone changes in spaceflight at all? One of our main concerns about bone turnover in spaceflight must center on the ability of the skeleton to repair itself, either from overt fractures or to prevent the accumulation of microfractures into a stress fracture. Repair processes in bone are complex, but rely on the same basic cellular processes involved in normal bone turnover, so our experiments looking at normal turnover will also help us to understand fracture healing. We must also understand exactly how the body handles an influx of calcium into the blood, to know if any of the many adaptations which occur will strongly influence other physiological systems. In particular, a high blood ionized calcium level will cause an increase in the amount of calcium that the kidney must filter and reabsorb, perhaps leading to kidney stones, and it can also influence the neuromuscular system. Our present knowledge is consistent with increased influence of decreased movement of calcium into bone as formation is suppressed. However, the magnitude of the increase in blood calcium and the rapidity with which it occurs suggests that there may be another mechanism adding to this release of calcium, perhaps a change in the permeability of the bone-blood membrane. If this is the case, then all of our countermeasures aimed at suppressing osteoclastic bone resorption may not be sufficient to solve the problem.

SUMMARY

Bone loss occurs in space, at a relatively slow rate compared to the time frame of current or planned mission operations. The magnitude of bone loss is not sufficient to believe that skeletal fragility will result during a spaceflight or even soon after return to earth, although a long-term effect of irreversible bone loss may be the development of osteoporosis in susceptible individuals earlier in life than had they not spent significant time in space. The hypothesis of a primary increase in osteoclastic bone resorption due to skeletal unloading is consistent with the observed data, and anti-resorptive agents exist which could be used to suppress this. The timing and magnitude of the rise in blood calcium suggests that other factors may also contribute. If this is the case, countermeasures directed at suppression of bone resorption may not be sufficient. The sustained elevation of blood calcium can lead to a variety of clinical symptoms associated with hypercalcemia, including kidney stones and neuromuscular irritability, and this may be one of the major operational reasons to try to prevent the increased bone resorption. In growing animals, bone formation is suppressed and fracture healing is impaired, and this may also result in operational problems if a similar effect is found in the adult skeleton. The bone loss observed in space to date should not be an operational concern limiting mission duration, but should be treated as an occupational hazard for those involved. The threat of massive bone loss should also not be used as a consideration in the design of a Mars spacecraft, except to the extent that other concerns such as prevention of hypercalcemia and fracture healing share the same design parameters. The issues and scientific questions surrounding bone changes in space are still unresolved, and the questions will need to be answered if and when truly long-term space missions (decades) are attempted. Current and planned experiments are designed to answer these questions, but those answers are not needed to go to Mars.

REFERENCES

1. J. Wolff, "Das Gasetz der transformation der knochen," Hirschwald, Berlin, 1892.
2. P. Minaire, P. Meunier, C. Edouard, J. Bernard, P. Courpron, J. Bourret, "Quantitative histological data on disuse osteoporosis," *Calcif. Tiss. Res.* 17, 1974, 57-73.
3. G. D. Whedon, and E. Shorr, "Metabolic studies in paralytic acute anterior poliomyelitis. II. Alterations in calcium and phosphorus metabolism," *J. Clin. Invest.* 36, 1957, 966-981.
4. J. E. Dietrick, G. D. Whedon, E. Shorr, "Effects of immobilization upon various metabolic and physiologic functions of normal men," *Amer. J. Med.* 6, 1948, 3-26.

5. R. P. Heaney, "Calcium metabolism is disuse osteoporosis in man," *Amer. J. Med.* 33, 1962, 188-200.

6. H. David, "Russians discuss space radiation in conference," *Missiles and Rockets*, October 21, 1963, p. 34.

7. P. B. Mack, P. A. LaChance, G. P. Vose, F. B. Vogt, "Bone demineralization of foot and hand of Gemini-Titan IV, V and VII astronauts during orbital flight," *Am. J. Roentgenol., Rad. Therapy, and Nuclear Med.* 100, 1967, 503-511.

8. G. P. Vose, "Review of roentgenographic bone demineralization studies of the Gemini spaceflights," *Am. J. Roentgenol., Rad. Therapy, and Nuclear Med.* 121, 1974, 1-4.

9. L. Lutwak, G. D. Whedon, P. A. LaChance, J. M. Reid, H. S. Lipscomb, "Mineral, electrolyte and nitrogen balance studies of the Gemini VII fourteen-day orbital space flight," *J. Clin. Endocrinol. Metab.* 29, 1969, 1140-1156.

10. N. Birykov, I. G. Krasnykh, "Changes in optical density of bone tissue and calcium metabolism in the cosmonauts," *Kosmicheskaya Biologiya I Meditsina* 4, 1970, 42-45.

11. P. B. Mack, F. B. Vogt, "Roentgenographic bone density changes in astronauts during representative Apollo space flight," *Am. J. Roentgenol. Rad. Therapy, and Nuclear Med.* 113, 1971, 621-633.

12. P. C. Rambaut, M. C. Smith, P. B. Mack, J. M. Vogel, "Skeletal response," In *Biomedical Results of Apollo*, NASA SP-368 (ed. Johnston, R. S., Dietlein, L. F., Berry, C. A.), NASA, Washington DC, 1975, p. 303-322.

13. C. Arnaud, C. Cann, B. Halloran, M. Hammond, D. Matsumoto, S. Sanchez, "Pathophysiology of mineral loss during spaceflight," *Bone and Mineral* 17, 1992, S-227.

14. C. E. Cann, C. D. Arnaud, B. P. Halloran, M. E. Hammond, D. Matsumoto, S. Sanchez, "Initial response of the calcium homeostatic system to spaceflight," *Aviat. Space Environ. Med.* 63, 1992, 412.

15. G. D. Whedon, L. Lutwak, P. C. Rambaut, M. W. Whittle, M. C. Smith, J. Reid, C. Leach, C. R. Stadler, D. D. Sanford, "Mineral and nitrogen metabolic studies, experiment M071," In *Biomedical Results from Skylab*, NASA SP-377, (ed. Johnston, R. S., Dietlein, L. F.), NASA, Washington, DC, 1977, pp. 164-174.

16. P. C. Rambaut and R. S.Johnston, "Prolonged weightlessness and calcium loss in man," *Acta Astronautica* 6, 1979, 1113-1122.

17. G. P. Stupakov, V. S. Kazeykin, A. P. Kozlovskiy, V. V. Korolev, "Evaluation of changes in human axial skeletal bone structures during long-term space flights," *Kosmicheskaya Biologiya I Aviakosmicheskaya Meditsina* 18; 1984, 33-37.

18. C. E. Cann, "Determination of spine mineral density using computerized tomography; a report," XII US/USSR Joint Working Group Meeting on Space Biology and Medicine, Washington DC, November 9-22, 1981.

19. V. S. Oganov, C. E. Cann, A. S. Rakhmanov, S. K. Ternovoy, "A computer tomographic investigation of the musculoskeletal system of the spine in humans after long-term space flight," *Kkosmicheskaya Biologiya I Aviakosmicheskaya Meditsina* 24, 1990, 20-21.

20. C. E. Cann, V. S. Oganov, "Direct measurement of spinal muscle atrophy in long term spaceflight," *Aviat. Space Environ. Med.* 58, 1987, 500.

21. A. I. Grigoriev, S. A. Bugrov, V. V. Bogomolov, A. D. Egorov, I. B. Kozlovskaya, I. D. Pestov, V. V. Polyakov, I. K. Tarasov, "Medical results of the Mir year-long mission," *Physiologist* 34 (suppl.), 1991, S44-S48.

22. V. Schneider, V. Oganov, A. LeBlanc, A. Rakhmanov, A. Bakulin, A. Grigoriev, L. Varonin, "Space flight bone loss and change in fat and lean body mass," *J. Bone Min. Res.* 7, 1992, S122.

AAS 93-822

RELEVANCE AND IMPACT OF PLANETARY ENVIRONMENT PROTECTION ON FUTURE MARS MISSIONS*

François Spiero†

Care and protection of the Earth environment have recently become major concerns. Indeed, human activities have often expanded at the expense of the preservation of natural balances, sometimes leading to irreversible damages.

Mars is, as of today, an almost virgin planet. Little—if any—pollution has yet taken place there. But an increasing number of automatic and later, manned missions to Mars are planned for the next decades. Eventually, human settlements will be built. Safe and comfortable settlements will be viable in the long term if they can develop in a clean environment. Such an environment may be achieved if, for all future Mars missions, adequate planetary protection measures are taken. Therefore, it would be most useful to soon define a long-term Mars environment policy.

The measures to protect the Mars environment may depend on the type of missions. For most automatic spacecraft, relaxed Viking-like planetary protection requirements could be applied. But later, for manned missions, a more stringent policy may be needed in order to avoid the contamination of Mars by human sub-organisms. Finally, a comprehensive set of procedures to protect both the Mars environment and 'settlers' might have to be enforced prior to the building of inhabited bases. In particular, such regulations would focus on how human activities on the planet should be performed in order to minimize the negative impact on the Mars environment.

PLANETARY ENVIRONMENT PROTECTION—DEFINITION AND RELEVANCE

Although Viking did not provide any evidence of the presence of organic materials on Mars, their existence cannot be ruled out. Therefore, in order to protect possible lifeforms on Mars, preventative actions during the development and assembly/verification/testing phases of a spacecraft to Mars must be performed. Such measures are specified in planetary protection documents.

As of today, the purpose of planetary protection is to avoid or at least limit the contamination of a planet (e.g. Mars) by terrestrial organisms. In the long-term, for human missions, environment protection might also deal with the issue of global human-planet biological interactions.

---

* This paper reflects the views of the author, and not necessarily the views of ESA.

† ESA/ESTEC, P.O. Box 299, NL-2200 AG, Noordwijk, The Netherlands.

Indeed, environment protection of Mars will fully develop with the start of the human exploration phase. With the development of human activities at the surface of Mars there is a risk of forward (planet contaminated), and possible backward (human beings contaminated). These two modes of contamination shall be evaluated so that appropriate measures can be applied.

Table 1

VARIOUS OPTIONS OF PLANETARY ENVIRONMENT PROTECTION MEASURES FOR AUTOMATIC SPACECRAFT

	Description	Advantages	Drawbacks
Dry heating / Baking	A few tens of cycles at high temperature (120-180°C); for Mars-94, experiments have been successfully performed at 110-130°C	Relatively easy to perform; experience gained from Viking	Batteries and solar arrays might suffer at a temperature superior to 120°C
Bioshield	Protective shield around the descending craft	Recontamination of the whole probe by Earth organisms avoided (if there is no leakage)	Might be too stringent, since no biological experiment is on-board Marsnet; for Mesur, a bioshield leads to a cost overhead of 30%
Gas flow	Radioactive gas (for ex:krypton)		Special equipment required
Radiation	Gamma-Rays		Very complex and expensive; memories destroyed (CNES experiment)
Chemical Sterilisation	Use of Ethylene Oxide		Too hazardous and corrosive

PLANETARY ENVIRONMENT PROTECTION—PAST AND PRESENT

Pioneer 10 and 11 (1972-1973) were the first planetary probes with a substantial planetary protection scheme. With Viking (from 1975 on), planetary protection requirements dramatically expanded. The major innovation was the use of a bioshield around the probe. There were also very stringent cleanliness requirements, baking of parts and heavy thermal cycling leading to a sterilization of the probe.

Planetary protection requirements have been relaxed since the Viking mission. The current approach is to have 5 major categories of planetary protection requirements, depending on the type of mission (ref: DeVincenzi and Stabekis). For automatic missions to Mars, the appropriate categories are III, IV and V:

- **Category III:** Orbiter mission: requirements; Documentation/traceability, clean-room, bioload reduction.
- **Category IV:** Lander mission; Requirements: those of Category III (but more stringent), plus partial sterilization, possible bioshield.
- **Category V:** Return-to-Earth mission; requirements: those of Category IV (but more stringent) with a continuous monitoring of the planetary protection parameters throughout the mission.

Table 2

IMPACTS OF PLANETARY ENVIRONMENT PROTECTION MEASURES ON THE DESIGN OF AUTOMATIC SPACECRAFT

	Thermal Design/ Mechanical Structure	Instrumentation/ Hardware Development	Materials, Parts and Processes
Dry Heating/ Baking	Thermal analysis to be performed for all units of the spacecraft prior to the development/ qualification phase	Limited number of options for power and other electrical subsystems; major constraints on technology selection, and in particular on solar arrays, batteries, electronic components and mechanisms	For Viking, electronic parts and materials had to sustain a temperature of about 135°C: this excluded the usage of polycarbonate films, paper capacitors, etc... (major impact on the development of instruments and subsystems)
Gas Flow		Could be applied to subsystems after baking individual parts (done on several subsystems of Viking)	Could be used for surface materials
Radiation			Could be used for batteries
Chemical Sterilisation		Could be used for mission-oriented vehicles	Might react with some metals
Bioshield	Thermal and mechanical interfaces with the probe/lander and associated ejection mechanism from the carrier need to be defined	Constraints on: . physical disposition of instruments and subsystems . launch vehicle fairing envelope . system mass . deployment kinematics . assembly, integration and verification implementation	The design of the bioshield shall focus on avoiding any leakage

PLANETARY ENVIRONMENT PROTECTION—FUTURE AUTOMATIC MISSIONS

The COSPAR meeting at the World Space Congress (August 1992) decided to present in 1994, a set of new planetary protection rules for future missions. They will likely be less stringent that those of Viking and also than those of DeVincenzi and Stabekis (Categories III, IV and V). The underlying assumption is that since no life has been detected so far on Mars, it is not worth spending unnecessary effort and money to "over-protect" the planet.

In the mid-term, several lander missions are forecast (MESUR, rovers...) The baseline so far for those missions is to perform relaxed Viking-like measures, such as assembly and testing in cleanrooms (usually class 100000) with some sort of sterilization (no full sterilization program for landers without life-detection experiments) and appropriate documentation. Later, sample-return missions will require more stringent measures to protect the returned sample from contamination by terrestrial organisms.

It is estimated that Viking or Category IV planetary protection requirements could add up to 30-40% of the cost of future lander missions. In the current period of economic uncertainty, that figure could incite project leaders to neglect planetary protection issues.

PLANETARY ENVIRONMENT PROTECTION—FUTURE HUMAN MISSIONS

For future human missions to Mars, both forward and backward contaminations are possible. Human beings will carry themselves organic materials which could affect possible Martian lifeforms. Even if no lifeform exists on Mars, the interaction between terrestrial organisms and water resources or sediments on Mars could trigger unknown biological reactions.

Conversely, the Martian environment might alter the well-being of human beings. It is impossible with current data on Mars to make any prediction in that respect.

In the long term, a global policy for the protection of the environment of Mars would have to be enforced to both protect the planet from human activities and allow a healthy growth of human settlements. Among other issues to consider are regulations for:

- Waste from human bases (recycled air, organic waste...)
- The use of nuclear energy at the surface (a nuclear accident on Mars could lead to a nuclear 'winter' on the whole planet since Martian winds could easily transport radiative elements in the atmosphere.
- Protection of water resources on Mars.

CONCLUSIONS

The tendency, in the short-term, is to relax past planetary protection requirements (e.g. Viking) for two major reasons: no evidence (yet?) of organic materials on Mars and limited financial resources. COSPAR is expected to provide a new policy for planetary environment protection in 1994 which will probably be a set of requirements less stringent than those used in Viking or proposed by DeVincenzi and Stabekis.

Later, a comprehensive policy to protect both the environment of Mars from terrestrial organisms and human beings and their settlements on the planet from environmental hazards could be defined. Such a policy would affect all interactions between the Martian environment and human settlers on Mars. In particular, protection of the resources of the planet (water...) and of the health of human beings from possible contamination shall take high priority.

REFERENCES

"Planetary Environment Protection—Implications for the Development of a Network of Surface Stations on Mars"; G. Scoon, A. Chicarro, G. Schwehm and F. Spiero; paper presented at the World Space Congress—Cospar, Washington DC, August 1992.

Revised Planetary Protection Policy for Solar System Exploration; D. DeVincenzi and P. Stabekis; *Adv. Space Res.* 4 (12);291-295; 1984.

AAS 93-866

ADAPTABLE SOFTWARE NEEDED FOR MARS MISSIONS

Ned Chapin*

Mars missions will require more software adaptation to be done than has been true of space missions generally. Some of this adaptation will arise from the greater role of software in the missions, and some from the need personnel will experience to change mission tasks, functions, and resources on short notice. Software adaptation often will be on the critical path for Mars missions. Hence, in these times of lean funding and high demand for performance, Mars missions will stand to benefit from adaptable software.

Drawing from accumulated experience, a fresh approach points to nine ways for Mars missions to improve software adaptation. These nine ways act synergistically to: 1--organize the software-hardware components as a network of nodes; 2--provide uniform human interfaces in limited variety; 3--transparently and routinely store data at two or more nodes; 4--allow transparent message response; 5--link command functions to personnel rather than to nodes; 6--provide common software for common tasks and functions; 7--use a uniform on-line help and documentation procedure; 8--actively manage the software; and 9--prefer adapting existing software to writing new software.

While those ways can be effective for Mars missions when taken in combination, six barriers rooted in tradition stand as hurdles: 1--no suitable software exists; 2--the software is too "......" to use; 3--the hardware is different; 4--the software does not fit the task, function, or environment; 5--it is "not good for the organization"; and 6--collaborative work demands better communication. Success in improving software adaptation will be critical in the success of Mars missions. Being able to implement changed mission tasks and functions with adaptable software at low cost, quickly and accurately, can enhance effectiveness and accomplishment for Mars missions.

SITUATION

Space missions have shown a consistent trend—the role of software in their systems has increased dramatically, while the success and breadth of performance of the missions increasingly depends on the software. This trend will continue and is characteristic for all Mars missions. Given the technology available, this trend has consequences for the planning and execution of Mars missions because of five facets of the nature of software:

- About three dollars will be spent to adapt (modify or change) the software after management has accepted it, for every dollar spent to bring the software to the point that management does accept it as meeting specifications.

* Information Systems Consultant, InfoSci Inc., Box 7117, Menlo Park California 94026-7117. E-mail: NedChapin@acm.org.

- Adapting aerospace software is a slow labor-intensive process requiring skilled personnel, specialized hardware, and committed infrastructure.
- Most software testing leaves some defects lying latent ready to cause performance impairments at unexpected times during use.
- Creating most software adaptations involves compromises among the needs of different people.
- Adapting the software alone rarely is enough, because people have also to adapt their behavior in carrying out mission tasks and functions in which adapted software has a role, as most NASA personnel who have trained for a task can testify. The rare exceptions will not be Mars missions.

Fortunately, since we are still getting underway on most Mars missions, we have the opportunity to build software adaptation provisions into the missions. All Mars missions must go with adapted software, or not go at all. The issue is not one of avoiding software adaptation. The issue is one of turning the necessity of doing software adaptation into a net benefit, or at least of minimizing the costs and unwanted side effects. Software adaptation can be managed in ways and with results that contribute positively to mission effectiveness. Neglected or mismanaged, software adaptation slows performance, raises costs, and impairs mission effectiveness. Doing software adaptation will be unavoidable for Mars missions.

Taking a fresh approach, this paper examines the major actions that managers can take or direct about software adaptation, that are most likely to yield positive results for Mars missions. Each action comes from accumulated experience with software and systems. For each, this paper briefly explains what it is, how it can be achieved, and what are its expected consequences. This paper also notes a few interactions among some actions examined, and considers some barriers. This paper is not exhaustive on such actions, barriers, or interactions, and deliberately concentrates on those of wider applicability and higher probability (for example, the paper passes over measurement matters [10]). To put a foundation in place, this paper now introduces some basic terminology and concepts.

UNDERLYING CONCEPTS

Software adaptation happens only to existing software. To have existing software, someone paid money to develop or build (write), or to acquire the software, and accepted that software as adequately meeting some specification. The reasons for software adaptation are usually to fit it to some new, modified, or deleted task or function, or to restore the software to the functionality it (was believed to have) had previously [1]. Two other reasons for software adaptation are to fit the software to changed hardware, and to fit it to work with other (new or modified) software.

In information systems technology, each of these four reasons has a technical process for achieving it [3]:

- "Enhancive maintenance" is the process of changing the existing functionality.

- "Corrective maintenance" is the process of restoring existing functionality, or "fix-it" process.
- "Adaptive maintenance" is the process of satisfying either changed hardware or changed software environments, or both (two reasons for one process).

A fourth process, "consultive maintenance" also exists, but personnel outside of information systems often do not recognize it separately. This fourth process involves interaction between mission personnel and information systems personnel to plan the role of software in meeting mission tasks and functions, to capture requirements, to justify changes, to evaluate results from testing, to train mission personnel, and to perform related activities. Collectively, the mix of all four of these technical processes goes by many names. Some common names are "software maintenance," "systems evolution," and "operational systems support." For convenience here, the term "software adaptation" refers to the mix of all four processes, and "adapted software" refers to the software resulting from the process of "software adaptation."

Software provides direction in the performance of activities, processes, or work by systems [4]. It typically specifies a sequence of steps that a computer, a person, or some special-purpose hardware follows to accomplish some task or function. Physically, software must be represented as data in some medium, such as magnetic spots on disk or ink on paper. Operationally, software serves as a tool in getting work done through systems. Systems are themselves deliberate combinations by people of the services of people, hardware, and materials to produce for a human purpose, responses to stimuli—i.e., outputs from inputs. Software's role in systems is to provide the method linkage between stimuli and responses, outputs and inputs, in various combinations and under various circumstances. For convenience here, "system" refers to any way of accomplishing the work of a Mars mission.

People working in space systems often talk in terms of the "-ilities" when discussing attributes of the role of software in systems [12]. Examples are useability, testability, etc. When doing software adaptation, the people working on the software have to take care to preserve the desired set of "-ilities." Even better, if possible they try to improve the presence of the desired "-ilities" in the adapted software and in the way it and the system it is a part of, perform.

Mars missions will be like other space missions—more resources and more time would be wonderful to have, but will not be available. Deadlines will crowd closely, funds will appear to sublimate, and the ideal mix of personnel skills will elude capture. We will wind up having to go with what we have. What we have for Mars missions will mostly consist of what management has accepted from what management has sought and been able to produce from what it has had access to. Hence, management must keep the certainty in consistently clear view, that someone will have to pay for software adaptation.

Table 1
GROUPING OF ACTIONS

Management Group	Non-management Group
5—Focus management	1—Organize as dual network
7—Have on-line aids	2—Use uniform interfaces
8—Manage the software	3—Store data redundantly
9—Prefer adaptation	4—Allow transparent communication
	6—Recognize commonalities

Fortunately for Mars missions, management can take or direct nine specific actions that can improve software adaptation and reduce the burden of software adaptation. Some of these actions have been rare in space systems, as can been seen from the attributes of the software [9]. The actions occupy two groups (see Table 1). One group emphasizes the actions that mission management itself can take at low cost. The other group emphasizes actions that mission management can direct project personnel to provide. This second group usually entails more resources (and hence cost and time) than does the first group. As summarized below, management can often simultaneously attend to both groups for the benefit of Mars missions.

WAYS TO IMPROVE ADAPTATION

1. Organize the Software-Hardware Components as a Network

It is an old and well-respected practice in aerospace to use communications networks [11]. However, from the point of view of software adaptation, the network should be viewed in a dual manner. As diagramed in Figure 1, the best network for Mars missions consists of two congruent topologies, one of the software and one of the hardware. Each network node consists of some software and some hardware. However, the complement of hardware and software at each node may be quite different, as explained below, but configured from a specific set of facilities. What is also common at every node is the capability to communicate quickly (send and receive messages electrically or via the electromagnetic spectrum) directly with at least two other nodes, and indirectly on a need basis with all other nodes. The network topology cannot take a fixed form, but the form must be stable, except when people are deliberately changing it. The network topology must accommodate a dynamically varying number of nodes. Physically, the nodes comprising the network may be mobile and may be anywhere, such as Denver, Los Angeles, Goddard SFC, a MESUR node, Mir 1, Candor Chasma station, Phobos base, a Mars rover, or a spacecraft inflight to Mars.

The network organization provides a basis for substituting one node for another, and altering node capabilities, while still potentially performing all mission tasks and functions. By itself, the network organization is insufficient to make this possible. The absence of node commonality and linkage greatly raises the burden of adaptation, given the long list of demands (tasks and functions) typically desired for a Mars mission.

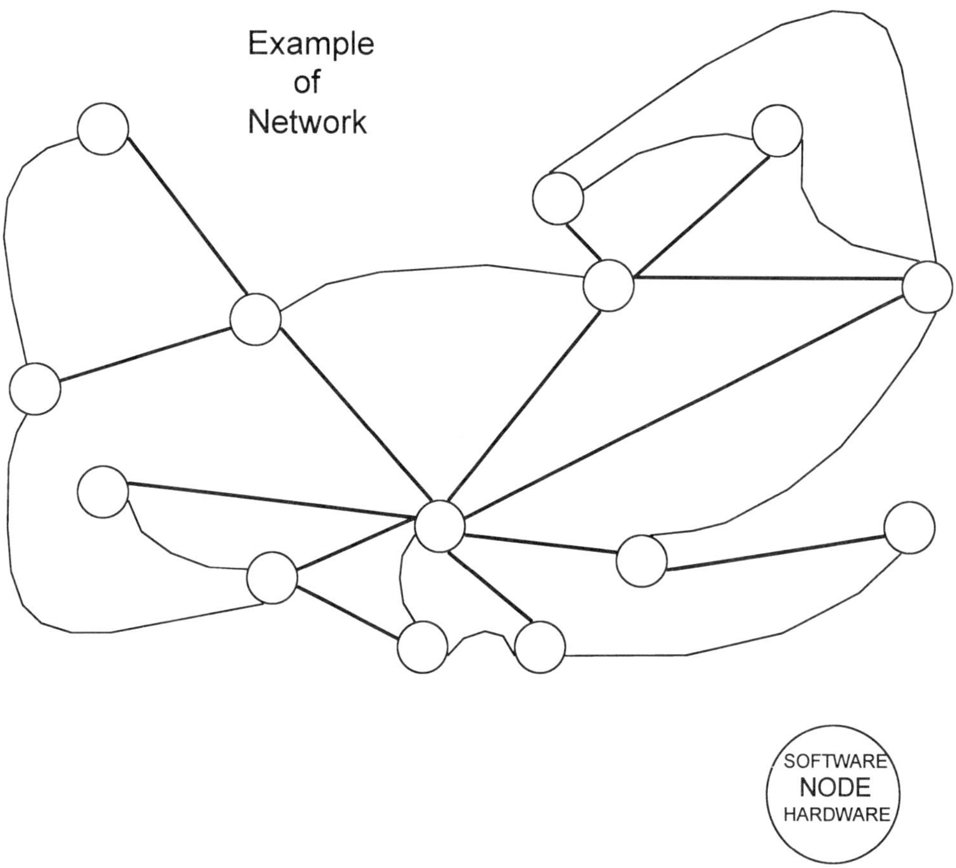

Figure 1 In this example of a network, nodes (shown by circles) consist of two parts, hardware and software. Major (usual) communication pathways (shown by straight heavy lines) link each node to at least one other node. Each node also links to at least one other node by alternate or secondary communication pathways (shown by curved light lines).

2. Provide Uniform Human Interfaces in Limited Variety

Each node has both a hardware and a software interface with people. Having multiple or idiosyncratic interfaces slows personnel performance and raises the risks of inappropriate performance. Task and function requirements can combine with human factors work and ergonomics to give a suite of useful interfaces. GUIs (graphic user interfaces) are especially appropriate where personnel use many different systems in different ways under varying circumstances. In some instances where hardware imposes especially limiting constraints for human interfaces, such as the Mars Balloon, the normal human interface may best be indirect via the capability to communicate with some other node.

The smaller the suite of interface forms and protocols, the easier is the software adaptation. With a limited suite, management can more quickly make substitutions of personnel, of hardware, and of software, to meet changed mission needs, and still be confident of adequate performance. A limited suite means that more people are familiar with the components, that fewer components in total have the potential to be adapted, and that people can test them more easily. Providing for human interfaces historically has been taking up ever more software. With regard to aerospace software, some people argue that hardware constraints should be permitted to limit the human interface in order to save hardware costs. Given the dominance of software adaptation costs, a more practical position with current technology is that more money should be spent on hardware if that added expenditure can pay for itself through an improved human interface and through a reduced number of different human interfaces involved in the mission.

3. Transparently and Routinely Store Data at Two or More Nodes

Action to provide this access and backup feature requires that the hardware capacity to store data and software be at least partially decentralized [2]. It also requires that the software in use at any node, can obtain a copy of data from or provide a copy of data to any node. The communication process is transparent to the people because the node initiates the actions and carries them to completion without human intervention or guidance, often as "background" work. This action also requires that the software at any node uses the storage capacity at other nodes as "backup" both for the data it generates and for the software it itself uses. When any node sends data or software, it sends only a copy of what it has. A node receiving software or data for use (not storage) deletes or erases those data and software after confirming that it no longer needs them. Not all nodes need to act as backup data stores. Often, the backup stores are best concentrated at a few of the larger nodes. This backup storage capability requires the use of the internode communication noted above, and is also a partial protection of mission accomplishment against communication outages and accidents to or at node sites.

One of the burdensome parts of software adaptation becomes much easier with this access and backup feature in place. This is the "fixit" recovery of data and software from hardware failures, or software failures, or both. This feature can result in reducing the personnel time needed and in improving the speed of achieving performance recovery at an affected node. The cost is added communication traffic, increased storage capacity demands, and added functionality in the "system software" used to support the mission. Mission personnel rarely sense this added performance burden.

4. Allow Transparent Message Response

Nodes are more effective in contributing to mission accomplishment when each node has both the software and the data to carry out the system's part of the local (using) personnel's tasks and functions. Delays get minimized and communication traffic gets reduced. However, the cost can be excessive redundancy in data storage, additional demands on a node's data storage capability, and sometimes a relatively slow speed of performance, when the node's hardware is not well matched to the demands upon the node. As noted above in action # 3 above, Mars missions should have basic redundancy in the storage of both data and software. Additional redundancy yields diminishing returns.

A common situation is for a person to make a request of a node for which the node lacks the data or the software or both to provide an appropriate response. In such situations, the network can transparently invoke the software on some other node, and have that node marshal the data needed. When the performing node requires guidance from the person, or has data for the person, the performing node communicates with the requesting node. Upon completion, the performing node stores any modified data and treats as garbage the used copies of the software and of the unchanged data. The results are that many nodes do work requested by mission personnel at other nodes, as diagramed in Figure 2. This allows distribution of the workload over the network. It is consistent with a "client-server" approach as well, with multiple servers [8]. Also, but with some hardware and security restraints, it permits all mission personnel to invoke from most nodes, the software to perform the tasks and functions of almost all nodes. The cost is a (usually minor) increase in network traffic, and an additional work load on some nodes.

The reduction in excessive data and software redundancy improves software adaptation. It also encourages task and function specialization for nodes, which tends to strengthen personnel skills and knowledge useful in software adaptation. It also often makes remote adaptation, where the information systems personnel doing the software adaptation work can avoid the time and cost of travel for accomplishing the adaptation work. That is, sometimes people can do the software adaptation work from a node different from the one that will use the data and software.

5. Link Command Functions to Personnel Rather Than to Nodes

As for space missions generally, the management of Mars missions will assign hardware configurations to match expected performance demands. For example, the hardware at fixed observer sites will not be the same as at mobile site, and both will differ from the hardware at an oxygen extraction plant on Mars. Clearly, this applies most firmly where specialized input hardware and output hardware are involved (for example, data acquisition in a hazardous environment). It also can apply to node assignments generally to attempt to minimize network traffic while providing the desired mission performance and backup. For such general node assignments, a common space mission practice has been to assign hardware capability to fit the rank or importance of the personnel likely to be at the node site. This in turn has often affected the software capability, especially with regard to command and management functions concerned with the direction of the mission and of major parts of the mission [6]. This introduces an inflexibility in the software that adds to the burden of adaptation.

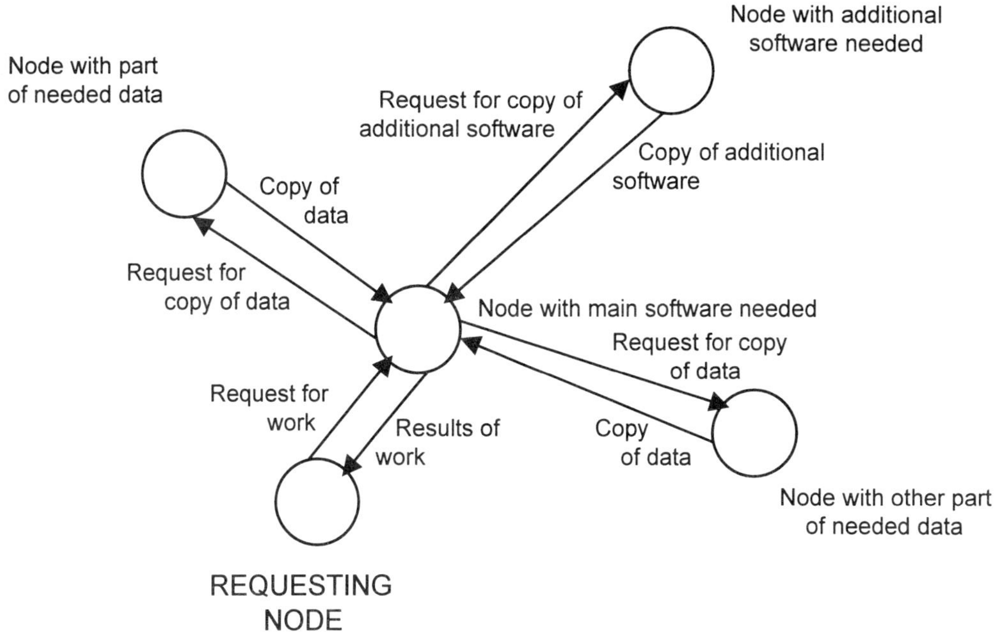

Figure 2 In doing work involving access to a system, a person at a node may make a request. The node receiving the request may call upon other nodes to do part or most of the work, and may call upon data stored at other nodes.

An alternative approach that eases adaptation, also adds to mission capability at only a minor cost of added functionality in the software used throughout the network. That approach is for the software to recognize the person who is using the node. The approach also permits that person to exercise his or her usual command and management functions from or at whatever node the person has access to or elects to use, provided the node hardware offers a suitable human interface. This means that persons may be dynamically granted or denied access to capability in the network, as well as that assigned capability travels with the person, largely independent of the nodes. This makes software adaptation easier by providing more consistency in software capability at nodes. It also makes software adaptation easier by encouraging possession and pride in performance among the personnel using the system. The better the people understand the system and its use, the easier is the software adaptation.

6. Provide Common Software for Common Tasks and Functions

This action has hardware as well as software aspects. For most nodes, the hardware has to have enough capacity to handle the software needed for the common tasks and functions. What that capacity is depends on the list of tasks and functions, and the system for each. For most nodes, the software has to provide send, store (queue), and receive messages, security-screen messages, security-screen personnel-requested actions, provide data storage and retrieval, preserve data integrity, keep a log of activity (usage), support the (GUI) interface, and operating system functions, such as managing node resources. Figure 3 diagrams some software aspects.

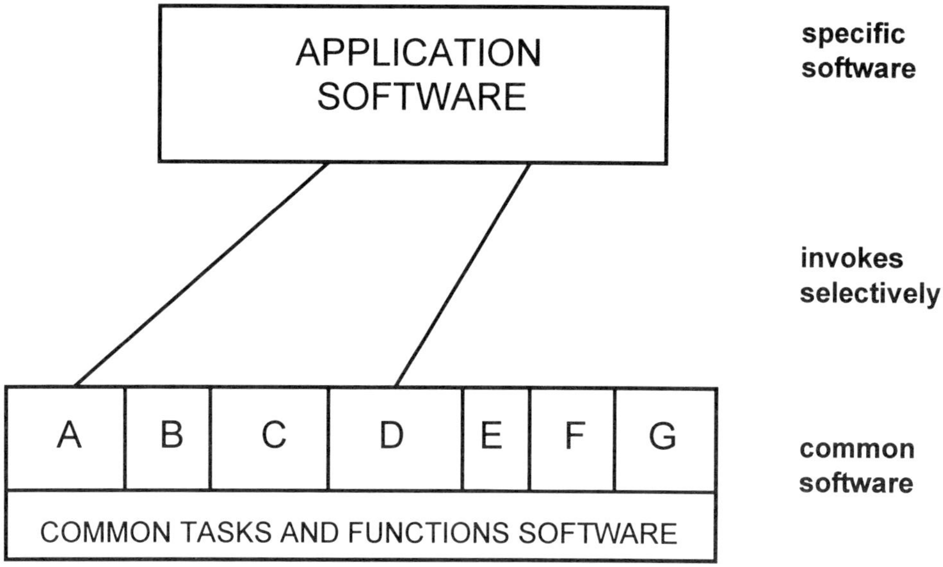

Figure 3 The software people use to get work done, the "application software," can invoke other software to perform common tasks and functions.

For adaptation ease, the software for the common functions should exist in as few versions as possible. Node hardware differences will force some variations, but the more commonality, the easier is the adaptation. One possible implementation route is to cluster the software for the common tasks and functions into a mission-specific special operating system. Another is to bud off specific varieties from a general Mars mission operating system. A third, and less attractive alternative is to have pieces of node-manager software operating under a generic operating system. Software related to specific (not common) tasks and functions but incorporating a selection of the common tasks or functions, should not embed the common ones in the specialized software. Instead, the software should invoke (call) the common capability from the common software. Consistency and reduced variation in handling common tasks and functions reduces the burden of software adaptation.

7. Use a Uniform On-Line Help and Documentation Procedure

Few mission personnel can become so familiar and expert with the software and hardware of a system that they rarely need to "look something up" about the software and its use. Most mission personnel deliberately cross-train and get assigned to use many systems involving software in order to build familiarity. Nearly all personnel use routinely only a small subset of a system's repertoire of facilities. They tend to forget how to use the infrequently used facilities, and often even what they are. Most personnel can benefit from having available on each node an on-line help and documentation capability, specific to each item of software.

Since the same data are needed system-wide for on-line help and documentation, only a minimum of two nodes (see action #1 above), have to carry the burden of storage for the majority of the help and documentation data. Sophisticated procedures are not essential [7]. Not all such data need be in the same two nodes, but at least one copy can be at a frequent use site. The adaptation advantage for the on-line help and documentation comes in the reduction in consultive work and in the simplicity of keeping current but one (master) copy and one backup copy. The hardware and software implications involve the variations in the storage capacity or in the user interface (see action # 2 above). For some nodes with limited interfaces or with especially limited storage, such as some mobile nodes or remote data acquisition units, the on-line help and documentation may be used indirectly via people at some other nodes.

8. Actively Manage the Software

The existing software serves as a major asset to a Mars mission. Its asset value justifies its management. The cost and effort involved in doing software adaptation is a major potential drain on a project's resources, and hence a potential liability to a Mars mission. Scant time availabilities compound the potential liability. Such liability potential justifies its management. As diagramed in Figure 4, management has to find a balance between the asset and liability aspects of software. If un-managed, software adaptation still occurs and still has to be paid for in time, personnel, money, and other resources. If unmanaged, the character and capability of the software becomes increasing unknown and invisible to mission management. Software's intangibility adds to the difficulty management faces in managing software, both for its role in systems and for software adaptation.

A workable solution for Mars mission software involves instituting deliberate by-product data capture about changes made to software during software adaptation, and about the character and functionality of the software during software development. The solution also involves deliberately keeping an inventory of data about software, its state, its status, and its history. The solution requires ready accessibility to the inventory records to see what data are there and to add to those data. It also requires limited accessibility to delete or modify those inventory data. The result is the potential for the better management of software adaptation because of the better visibility management can have of the software and of the changes to the software.

9. Prefer Adapting Existing Software to Writing New Software

Different systems are implemented with different software because the specific contribution of the software in the system is different for each system [5]. At the same time, the software in each system is composed of essentially the same components, but arranged in a different configuration. As a rough analogy, consider passenger automobiles. People have and use many different varieties, but all have four rubber tires, a driver's seat, a brake pedal, a windshield, a transmission, and headlights. Different brands of car show more variation among the components than do the components used in any one brand of car. Variation within a particular model within a brand is still less. Software shows similar characteristics. In aerospace systems, the programming language used and the hardware of the system limit the variation in the software's components. Within those limits, more than half of most software's components are the same, but arranged and connected differently.

Such similarity points to the potential of software reuse, both in software development and in software adaptation. When the people doing the software adaptation can recognize familiar components or the need for familiar components, they can then work more quickly and accurately in making adaptations. Management can take advantage of this by encouraging the use of familiar components. Management can also encourage meeting new and changed mission requirements by looking first to doing software adaptation on the existing software, rather than to developing new software. However, adding incompatible functionality to existing software typically impairs future software adaptation. Also typically, writing new software adds more to risk exposure than does doing software adaptation on existing software.

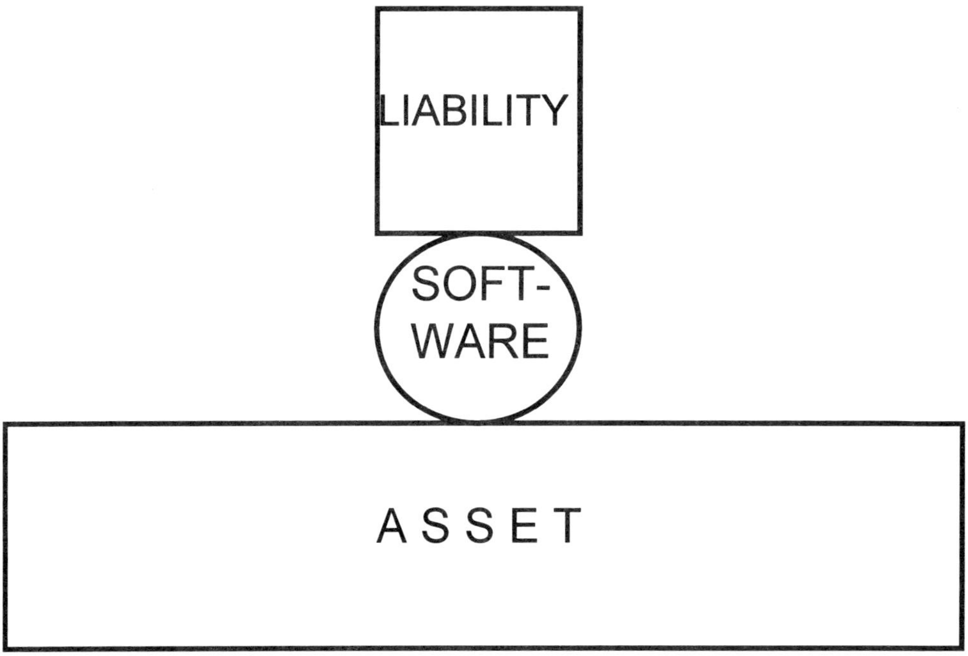

Figure 4 Mission management may have to maintain a precarious balance of the liability aspects of software in the mission, and the asset aspects. A common management goal is to get the software adaptation done so that the balance is less precarious.

BARRIERS TO IMPROVED ADAPTATION

Unlike some space missions, Mars missions are mostly newer and encumbered with less old intellectual baggage affecting software. One consequence of this is the greater role that software will play. A second consequencc is the different importance of barriers impeding the process and probability of adapting software effectively and quickly. The points below cover six of the major barriers.

1. No Suitable Software Exists

Mars missions will suffer as much as other space missions from NIH ("not invented here"). Poor communication about software availability will continue to result in redundant software development (writing) and in unnecessary software adaptation. Lack of personnel knowledge about reusable software and software components will obscure opportunities for effective adaptations. Because Mars missions will involve some new tasks and functions, suitable software may indeed not always exist. Mission management will often be tempted to accept that premise on faith.

2. Existing Software is Too ".........." to Use

Many terms can fill the blank, such as "old," "unreliable," "slow," "inelegant," "obsolete," "error-prone," or "costly," to cite a few. Some such claims are manifestations of NIH. Some are well founded on fact. The root underlying difficulty, however, often is a combination of three things:

- The personnel making the claim do not well know or understand the software.
- The documentation for the system and its software is both incomplete and not current.
- For what documentation that does exist, the personnel experience difficulty in understanding.

Software, it must be remembered, is the one human creation that is immune from deterioration from the ravages of time. The software may no longer direct the performance of a human-desired function or task in a currently desired way. If so, the action alternatives are to replace it or adapt it. In either case, at least portions of it may be salvaged effectively and reused.

3. The Hardware is Different

With some exceptions, when the hardware changes, the software must change (see "adaptive maintenance" earlier). Using new hardware, however, does not automatically require that all new software always be used. Sometimes people use the presence of new hardware as a basis for trying to justify NIH software claims. The major effects of hardware differences in a network of nodes are the different capabilities of the nodes in human interface, storage capacity, and speed. The usual justifications for using a minimum of hardware capability in space missions have been hardware cost, power demand, weight, and heat generation. With hardware technology already available, these justifications will be less significant for Mars missions than for prior space missions. The usual justification for using specialized hardware is the specific task and function to be done in an expected environment, as for example, a Mars rover. Variety in hardware, and hence in the associated software, is inevitable in Mars missions. Minimizing the variety improves software adaptation, however.

4. The Software Does Not Fit

When the fit discrepancy is in task or function, then either adaptation or replacement are the two options. When the fit discrepancy is in technical factors, then the options are more varied. For example, the lack of fit may be in the size of the software, usually too big for the hardware. Historically, space missions have used this situation as a justification for decreasing the adaptation potential of the software by a variety of technical expedients, such as using syntax drive. Even with current technology, bigger hardware is likely to be a more satisfactory choice to improve software adaptation. If the fit discrepancy is in language (such as C or Ada), operating system (such as Windows NT), internal organization (such as OOD), or services expected (such as from a database server), then a case-by-case weighing of the alternatives will be needed in Mars missions to pick a combination of adapt or replace the software, or change the software's hardware or software environment, or both.

5. It is Not Good for the Organization

The organization benefits most when it sells (uses) a lot of what is uniquely its own, the labor services of its own employees applying its own intellectual capital. Normally doing either software development (writing) or adaptation both qualify easily. Hence, most organizations doing work on Mars missions are going to propose to write (develop) software and to adapt software, or as a fall-back, to contract for one or both. In this age of lean funding for space missions, what is best for an organization may not always be best for a Mars mission. The same observation holds true even just for adaptation. Rarely is it the case that only one way is possible in doing software adaptation. Normally, many options are technically open. Examination will show some to be more labor intensive or otherwise beneficial to the organization or to the mission than others.

6. Collaborative Work Demands Better Communication

One organization might attempt to manage or do all work on a Mars mission in its own way of doing things, directly or via contracting. Alternatively, a Mars mission could be managed and done as a collaborative effort, perhaps even with a world-wide set of contributors. Doing work collaboratively places a heavier demand on communications, and deficiencies in communication are the common immediate cause of defects in performance in such arrangements. Collaborative work makes possible the more effective use of specialist personnel, methods, software, and hardware, but it makes software adaptation more difficult. A lot of the difficulty can be reduced by using vigorously the network of nodes described earlier. Some of the reduction has to come from people and organizations working not in ways that are best for them, but best for the Mars mission. The "consultive maintenance" noted earlier takes more effort in particular, an effort only partly offset by the richer base of hardware, software, and work practices to draw upon.

CONCLUSION

Software will play crucial roles in Mars missions. Nearly everything that people will do in and with Mars missions will involve software somewhere and somehow.

Each time people attempt to change (add, substitute, delete, or do differently) some mission task or function, they will also, with few exceptions, have to make some corresponding and implementing changes in the software. The cost of such software adaptation will exceed by several times over, the cost of acquiring the software initially and putting it into routine use. Since making software adaptations is very common, is labor intensive, and takes the time of skilled personnel, making software adaptations will often prove to be on the critical path for Mars missions.

This paper describes nine ways, based upon field experience, to improve software adaptation for Mars missions. These ways have individually been used in or with some aerospace software systems, but not in the rigorous combination which adds synergistic effects. Those nine ways are: 1--organize the software-hardware components as a network; 2--provide uniform human interfaces in limited variety; 3--transparently and routinely store data at two or more nodes; 4--allow transparent message response; 5--link command functions to personnel rather than to nodes; 6--provide common software for common tasks and functions; 7--use a uniform on-line help and documentation procedure; 8--actively manage the software; and 9--prefer adapting existing software to writing new software. The combined effect of these are to enable the Mars mission software to contribute better and more to achieving the mission, task, and function. At the same time, the combined effect is to:

- Constrain unbridled idiosyncratic and essentially redundant growth of software,
- Limit the number of different versions and variations of the software,
- Make the software more easily understood,
- Enable people to adapt software more quickly and accurately, and
- Reduce demands on mission personnel in using and learning to use the software.

All of these, individually and in combination, act to improve the effectiveness of software adaptation by making it possible for personnel to do more adaptation in less time with fewer defects. All of these (or even most of these) have rarely been characteristic of aerospace missions generally and of Mars missions in particular.

This paper also describes six barriers to improving software adaptation for Mars missions. Those six are: 1--no suitable software exists; 2--the software is too "......." to use; 3--the hardware is different; 4--the software does not fit the task, function, or environment; 5--it is not good for the organization; and 6--collaborative work demands better communication. These barriers are strongest when communications are weak, and where entrenched practices dominate how people have worked in preparing and carrying out aerospace missions. Technological advances have reduced the basis for some of the barriers to software adaptation for Mars missions, compared to earlier space missions. The basis for some of the barriers, however, continues to be traditional organizational practices and habitual human responses.

Being able to change mission tasks and functions accurately and on short notice can enhance mission accomplishment and effectiveness. Software adaptation is crucial to that process. Since it will usually not be wise to limit the demand for software adaptation for Mars missions, then an objective of management and other mission personnel is to make the software adaptation process faster, more accurate, and more cost-effective. The ways of doing that catalogued in this paper offer more benefit than added cost. They also can improve the performance potential for Mars missions.

REFERENCES

1. Arthur, Lowell J. *Software Evolution.* New York: John Wiley & Sons, Inc., 1988.
2. Bloor, Robin. "Information warehousing--mission critical view," *DBMS*, 5, 3 (March 1992), pp. 10-12.
3. Chapin, Ned. "Software maintenance: a different view." In *AFIPS Proceedings of the National Computer Conference*, Volume 54. Reston VA: AFIPS Press, 1985, pp. 507-513.
4. Chapin, Ned. *Information Systems.* San Carlos: Pea Press, Inc., 1992.
5. Chapin, Ned. "Software maintenance characteristics and effective management," *Journal of Software Maintenance*, 5, 2 (June 1993), pp. 91-100.
6. Cypser, R. J. *Communications for Cooperating Systems: OSI, SNA, and TCP/IP.* Reading MA: Addison-Wesley Publishing Co., 1991.
7. Fleeton, Nigel T., and Malcolm Munro. "Redocumenting software systems using hypertext technology." *In Proceedings Conference on Software Maintenance--1988.* Los Alamitos CA: IEEE Computer Society Press, 1988, pp. 54-59.
8. Khoshafian, Setrag; Arvola Chan; Anna Wong; and Harry K. T. Wong. *Client/Server SQL Applications.* San Mateo CA: Morgan Kaufmann Publishers, 1992.
9. Mueller, Karl Ludwig. "Payload operations for precursor mission E1" (in *Proceedings of Columbus VIII Symposium on Space Station Utilization*). *Space Technology*, 13, 2 (March 1993), pp. 155-160.
10. Munson, John C., and Taghi M. Khoshgoftaar. "Software measurement for the space shuttle HAL/S maintenance environment." In *Proceedings Conference on Software Maintenance.* Los Alamitos CA: IEEE Computer Society Press, 1992, p. 387.
11. Newman, T. E. "Dispersed flight control computer systems." In *Proceedings of the National Aerospace and Electronics Conference.* Los Alamitos CA: IEEE Computer Society Press, 1980, pp. 204-210.
12. Peercy, David E.; Eric Tomlin; and Gary Horlbeck. "Assessing software supportability risk." In *Proceedings Conference on Software Maintenance--1987.* Los Alamitos CA: IEEE Computer Society Press, 1987, pp. 72-80.

AAS 93-807

FOOTSTEPS TO MARS: AN INCREMENTAL APPROACH TO MARS EXPLORATION*

Geoffrey A. Landis†

A program for the exploration of Mars will fail unless it can pass three critical feasibility criteria: (1) technical feasibility, (2) political feasibility and (3) economic feasibility.

Most discussions of programs for Mars exploration have focussed on only the first of the three necessary criteria. Political and economic feasibility, although equally important, are usually ignored.

In terms of political and economic reality, a Mars exploration plan must learn from the lessons of the Apollo program and of the Shuttle program. The lesson of Apollo: If you accomplish your goal, your budget will get cut. The lesson of Shuttle: Infrastructure is boring. If you do the same thing over and over, the public will focus on your failures and ignore your successes.

A middle course is suggested, where a series of well-defined intermediate goals can be accomplished, resulting in the incremental development of the technology required for full-scale Mars exploration.

---

* This article appears in full as paper number AAS 97-372, in *From Imagination to Reality: Mars Exploration Studies of the Journal of the British Interplanetary Society* (Part I: Precursors and Early Piloted Exploration Missions), R. M. Zubrin, Ed., AAS *Science and Technology Series*, Vol. 91, pp. 59-73, 1997. The article was first published in the *Journal of the British Interplanetary Society*, Vol. 48, No. 9, pp. 367-372, September 1995.

† Ohio Aerospace Institute, NASA Lewis Research Center 302-1, 21000 Brookpark Road, Cleveland, Ohio 44135.

CHAPTER 2:
SURFACE EXPLORATION

AAS 93-817

MESUR PATHFINDER MICROROVER FLIGHT EXPERIMENT: A STATUS REPORT*

Donna Shirley†

MICROROVER BACKGROUND

Mars Rover missions have been studied since about 1985, primarily in conjunction with Mars Sample Return (References 1,2) or as precursor explorers for human missions (References 3,4,5). Until 1989 the rovers envisioned for these missions were large (800-1000 kg), highly autonomous and were intended to rove as much as hundreds of kilometers, returning a geologically diverse set of samples to a large earth return vehicle. Mars Rover Sample Return (MRSR) was conceived as having an imaging orbiter to provide high resolution pictures to ensure safe landing sites for the costly rover and sample return landers. The total cost of MRSR was estimated to be about $10B.

Large rovers were also thought of as site characterization precursors to human missions, autonomous vehicles which could scout out safe human landing sites, locate resources, and even prepare sites for human exploration. Rovers were also envisioned as field geology assistants for human outposts on Mars, providing "telepresence" at remote sites for geologists located at a base camp (Reference 6).

In 1989-1992, despite President Bush's announcement of a Space Exploration Initiative, and despite efforts such as those of the Synthesis Group (Reference 7), funding for human missions and their robotic precursors was not only not increased to the scale required for the large MRSR missions, but was steadily cut by Congress. This funding shortfall forced innovative concepts to be developed for Mars Rovers, and the most promising of these concepts focused around small ("mini" to "micro") rovers in the 5 to 50 kg class.

Four factors contributed to the feasibility of small rovers:

1. The worldwide miniaturization of electronics.
2. The availability of small, extremely mobile chassis developed as test models for the large rovers at JPL (Ref. 8).

* The work described in this paper was performed at the Jet Propulsion Laboratory, California Institute of Technology, under a contract with the National Aeronautics and Space Administration.

† Mesur Pathfinder Rover Team Leader, Jet Propulsion Laboratory, 4800 Oak Grove Drive, Pasadena, California 91109-8099.

3. The development of insect-like autonomous control algorithms, initiated by Rodney Brooks of MIT with his "subsumption architecture" and extended by David Miller and others at JPL (Ref. 9).

4. The development of the "Computer Aided Remote Driving" technique of remote operator control at JPL (Ref. 10).

These factors were combined at JPL in a series of experiments where increasing smaller computers and increasingly sophisticated sensors were mounted on increasingly capable "Rocker-Bogie" chassis. Called "Rocky" for short, this series culminated in a demonstration in June 1992 of "Rocky 4" in the Arroyo Seco next to JPL. The Rocky 4 demonstration, led by Dr. Lonne Lane, proved the concept of an integrated microrover system (7 kg including instruments) which could conduct useful science experiments in rugged terrain, with a combination of earth operator designation of targets and tasks, and autonomous, on-board control for task execution, path following and hazard avoidance.

The success of this demonstration led to a submission by JPL and OACT (Office of Advanced Concepts and Technology) to the FY94 NASA budget of a MESUR Pathfinder Microrover Flight Experiment (MFEX). MFEX is to take advantage of the opportunity to be carried by the Mars Environmental Survey (MESUR) Pathfinder mission to the surface of Mars, where it will perform technology, science and MESUR mission engineering experiments.

MESUR MISSION BACKGROUND

The MESUR concept was an extension by the Ames Research Center of a mission set envisioned to place a network of meteorology and seismology stations on the surface of Mars. These missions were to extend the results of Mars Observer to continue to global characterization of Mars. The uniqueness of the MESUR concept was that the landers would go directly from the earth to Mars, enter the atmosphere and land without an orbiter.

The MESUR Pre-Project was transferred to JPL in 1992, where it was divided into two activities, Pathfinder and Network. Because the concept of a direct Earth-Mars mission is new, and because the capital investment in a multi-lander MESUR Network is large, MESUR Pathfinder was conceived as a technology demonstration. MESUR Pathfinder is planned to be started by NASA OSS (Office of Space Science) in FY94 and is cost-capped at $150M in FY92$ for Phase C/D as the first in the Discovery series of low-cost planetary missions.

MESUR Pathfinder will launch in November or December 1996, with a single spacecraft which will fly directly to Mars, enter the atmosphere with a Viking-derived heat shield, and land with the aid of parachutes and airbags. Landing will occur in November or July 1997, depending on the choice of trajectory.

The MESUR Pathfinder experiments are primarily engineering (atmospheric measurements during entry, assessment of parachute and airbag performance, etc.) However, for science Pathfinder will include a stereo camera system (to be provided by the University of Arizona) which will provide a panorama of the Martian surface, thus increasing the small scale knowledge of Mars by 50% over the two Viking sites. Pathfinder will also make temperature and pressure measurements at the site.

Other important science measurements desired for Pathfinder are to identify the elemental composition of rocks and soil, and to measure the seismic background of Mars for designing seismometers for the Network mission. The Max Planck Institute of Germany, aided by the University of Chicago, is providing an Alpha Proton X-Ray spectrometer for the composition measurements. A small rover is a desirable means of instrument deployment since it provides a high likelihood of being able to reach rocks in the vicinity of the lander. The rover can also deploy other instruments, provided they are small enough.

However, a question exists, given the uncertainties in the Martian soil and terrain characteristics, of a small rover's ability to move about competently enough to perform these tasks. NASA OACT has funded rover research for many years. Therefore OACT agreed to provide a rover to MESUR Pathfinder as a proof of concept of small rover capabilities, and as a mechanism for technology transfer to MESUR Network and following missions. MESSUR MFEX is cost-capped at $25M in real year dollars for Phase A through operations, and is not included in the $150M MESUR Pathfinder costs.

The primary requirement of the MFEX is to accomplish its objectives within the $25M cost cap. The second most important requirement is that the rover must not cause MESUR Pathfinder to exceed its cost cap. The first requirement led to the selection of the basic Rocky 4 design as the starting point for MFEX rover design. This features 6 powered wheels attached by a set of "bogie levers" to a single body. The front and rear wheels are steered via Ackerman steering (like an automobile). Control is shared between the earth (operator target designation based on stereo images sent from Mars) and the rover (autonomous "behavior" control using on-board sensing for path following and hazard avoidance).

While other designs (e.g. legs, on-board target selection, etc.) may eventually prove superior, the years of experience with, and approximately $3M of investment in the Rocky series, combined with the time and cost constraints of MESUR Pathfinder, prohibit the consideration of other, less developed concepts. The information gained from MESUR Pathfinder will allow selection of the optimal concept for future Mars rovers for Network, Sample Return, etc.

The "Rocky" based design is described below in more detail, including decisions made to minimize the impact of the rover on the MESUR Pathfinder mission cost.

MESUR PATHFINDER MISSION DESCRIPTION

MESUR Pathfinder will be launched on a Delta. The rover will be mounted on one of three panels in the tetrahedral lander and tied down with a separation connector. During cruise (launch through landing) the rover will be provided structural and thermal support and limited data collection/transmission by the lander/cruise stage. The entry and landing are planned to minimize g-levels on the payload to < 50 g's with the combination of parachutes and airbags, so the rover will be qualified to 100 g's. After impact the lander will open (see Figure 1), positioning the panels so that solar energy can be collected and the rover is positioned for deployment.

The lander will immediately transmit the engineering data collected during the descent. Then its camera will take a panoramic image of the surroundings and begin transmitting it directly to the earth at a few hundred bits per second. (MESUR Pathfinder has no orbiter, and Mars Observer cannot be relied on a communication relay). The initial portion of the panorama to be transmitted will be in front of and behind the rover panel to allow the choice of rover deployment direction and initial operations to be made.

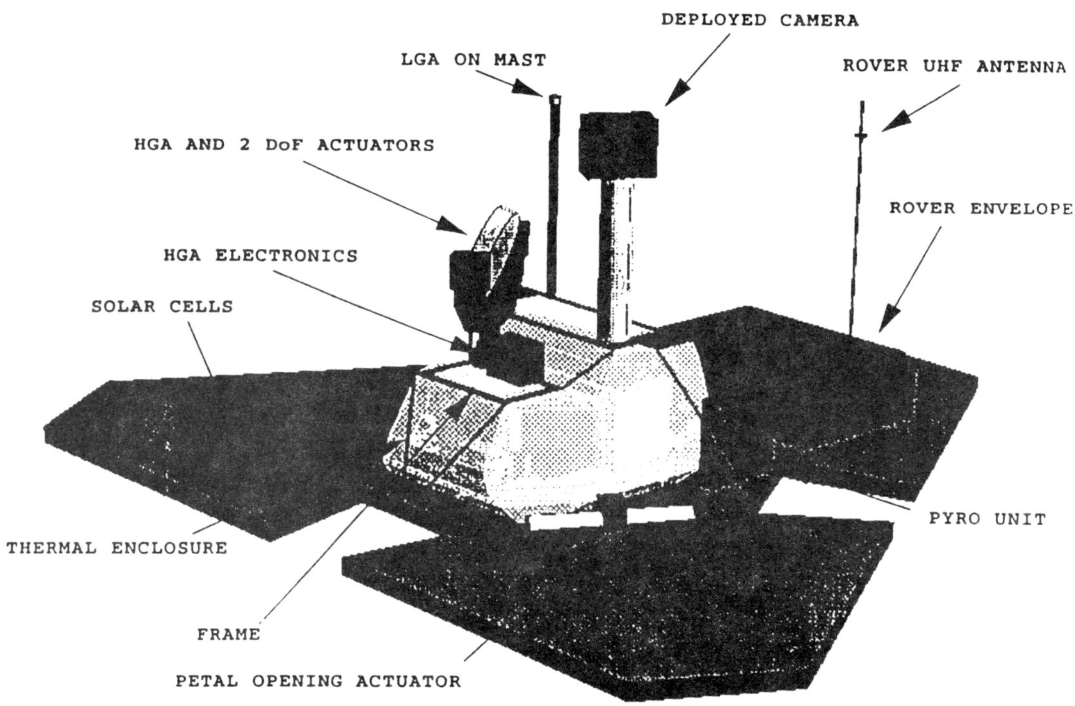

Figure 1 MESUR Pathfinder Deployed Configuration.

The rover will be deployed from its attach point on the lander. Since the lander is volume constrained, the rover will be carried in a stowed configuration with the body lowered, and will erect itself after release and before deployment. The rover will move down deployment ramps to the surface, and will thereafter be independent except for using the lander data and communications functions for command and telemetry. After rover deployment and transmission of the panorama, much of the lander mission is focused on supporting the rover with imaging, telecommunications, and data storage.

ROVER MISSION OBJECTIVES

The rover has three primary mission objectives:

1. Technology Experiments
2. Science Experiments
3. Mission Experiments.

The rover is a NASA OACT flight experiment for autonomous mobile vehicle technologies, whose prime function is to determine microrover performance in the poorly understood Martian terrain. The primary means of collecting this technology information is by instrumenting the rover mechanisms to determine wheel-soil interactions, detect hazards, determine navigational errors, etc.

Because the rover is part of the MESUR Pathfinder payload it will also gather science data by deploying an alpha-proton-x-ray spectrometer against one or more rocks, and possibly soil.

Finally, because MESUR Pathfinder is an engineering test of a transport and landing system for Network, the rover will image the lander to allow its condition to be assessed.

Technology Experiments

The MFEX is primarily a technology experiment to determine microrover performance in the poorly understood Martian terrain such that future rovers can be designed to be effective in navigating and moving over the surface of Mars.

The technology experiments to be performed by the rover were selected by two workshops in the fall of 1992 and the spring of 1993. The workshop participants were from NASA centers, universities, industry and other countries (France and Russia). At the first workshop experiment objectives were defined and experiment concepts were developed. Experiment concepts were fleshed out and presented at the second workshop and the final set was selected.

The selected technology experiments are listed in Table 1. They were chosen to meet the most important objectives within the constraints of the MFEX mission. The microrover will conduct the experiments and telemeter experiment data to the lander for return to Earth, in order to enable the mobility, controllability, and robustness of the microrover design in the Mars surface environment to be assessed and to provide data for the design of future Mars rovers.

Table 1

PRIORITIZED TECHNOLOGY EXPERIMENTS

1. Terrain Geometry Reconstruction from Lander/Rover Imagery
2. Basic Soil Mechanics
3. Dead Reckoning Sensor Performance and Path Reconstruction/Recovery
4. Sinkage in Each Soil Type
5. Logging/Trending of Vehicle Performance Data
6. Rover Thermal Characterization
7. Rover Vision Sensor Performance
8. UHF Link Effectiveness
9. Material Abrasion
10. Material Adherence
11. Rock Hardness (to be performed if feasible within cost/capability constraints)

---

1. *Mars Terrain Geometry Reconstruction from Imagery*—Lander and rover imagery will be analyzed on the ground to determine terrain feature classes (soil, rocks, hills, etc.) as well as statistical size and location distributions.

2. *Mars Basic Soil Mechanics*—Ground analysis of rover telemetry will be used to determine basic Martian soil mechanics parameters needed for future Mars rover design, such as cohesion, internal friction angle, slippage, and drive resistance. The rover will be instrumented to provide relevant telemetry such as motor currents, wheel and steering motor encoders, bogie angle encoders, etc.

3. *Mars Dead Reckoning Sensor Performance and Path Reconstruction/Recovery*—Position errors on Mars arising from using dead reckoning (internal state) sensors to control mobility will be measured, including 3-axis accelerometers and a heading gyro. Uncertainties will be characterized as a function of distance over various Martian terrain types determined from the lander and rover imagery. It will also be determined if the dead-reckoning system can recognize terrain types by observing vehicle behavior, for example, by traversing a closed or open path, noting vehicle behavior through engineering telemetry, and noting differences between the lander and/or rover imaged position and that position output from the dead reckoning system. Also, visual sensing of terrain types will be correlated with the behavior of the vehicle measured from telemetry and an array of several proximity sensors mounted on the rover.

4. *Sinkage in Each Martian Soil Type*—Wheel tracks will be viewed with the rover camera or proximity sensors to estimate sinkage. In particular, the rover will lock 5 wheels and rotate the sixth, measuring the motor current as the wheel breaks free, and then imaging the hole left by the rotated wheel. For experiments done close to the lander the lander camera may be used to image the wheel tracks. A goal is for the lander camera to image the total trail left by the rover during its mission.

5. *Logging/Trending of Vehicle Performance Data on Mars*—All measurable engineering parameters (drive torques/current, rpm, voltage, etc.) will be logged and

time tagged. Terrestrial analysis of the logged data determine performance, degradation, etc. This data will also be used to determine the values in Basic Soil Mechanics, above.

6. *Rover Thermal Characterization*—Rover thermal behavior as a function of time and operating situation will be monitored by thermal sensors in the wheels and some of the sensors, and inside the Warm Electronics Box (see below for a description of the WEB). The data will be analyzed on the ground to characterize rover thermal behavior.

7. *Rover Imaging Sensor Performance*—Terrain, motion and track measurements derived from the rover imaging sensors will be correlated with direct measurements from proximity sensors and engineering instrumentation.

8. *UHF Link Effectiveness*—This experiment will determine whether the UHF Link will function effectively on Mars by observing signal strength and noise a function of distance between the rover and lander, and/or when the rover is occluded by terrain features. This measurement will be made by tracking the number of retransmissions required between the rover and lander.

9. *Material Abrasion*—The abrasive qualities of Martian soil and dust will be measured. The proposed measurement involves painting one wheel with a fluorescent paint, covering it with an opaque paint, and observing the wheel with a photocell to see when the outer paint wears away, revealing the fluorescent paint. Lewis Research Center is providing this experiment.

10. Material Adherence—The tendency of Martian dust to adhere to rover surfaces, especially solar arrays and detectors, will be observed. Lewis Research Center has designed a simple experiment to periodically uncover a reference solar cell to determine the difference in power between the clean reference cell and a (possibly) dusty cell. The experiment also uses a quartz microbalance to determine the mass of the dust.

11. *Rock Hardness (if feasible)*—Rock hardness is of interest for correlating the materials abrasion experiment, and for shedding light on the design of material sampling technology for future missions. However, the constraints of the rover prohibit it from carrying a rock chipping or coring tool. A concept has been proffered for the rover to "claw" at a rock with one of its wheels, using the wheel grousers or an abrasive strip to erode the rock face, however, implementation of this concept awaits testing of the rover capabilities.

The mission design allows at least one instance of each of the technology experiments to be accomplished within the first three days of surface operations. Soil characterization experiments will be conducted in at least three soil types, provided three soil types are within range of the rover at the landing site.

Science Experiments

Although the information gained from the rover technology experiments (soil mechanics, materials adherence, etc.) are of scientific interest, the rover's main science objectives are to deploy the APXS and take close-up images of Martian features.

1. *APXS*—The primary science objective is to place the alpha-proton-x-ray spectrometer (APXS) against a rock, collect a spectrum, and transmit the spectrum to the lander for return to Earth. A full APXS measurement of a rock requires 10 hours. The 10 hours of measurement need not be contiguous but the rover must keep the APXS fixed on the rock until 10 hours of data is obtained.
   As a goal, if surface geometry and time permit, the rover will place the APXS on a rock particularly designated by the science team. However, if terrain conditions prevent the best rock from being contacted early in the surface operations, the rover will measure any rock it encounters.
   An APXS deployment mechanism is currently being designed. It appears most likely to be located on the back of the rover where real estate is available without interfering with the hazard detection sensors. If feasible, this mechanism will allow the rover to collect soil measurements. However, a rock measurement has priority over a soil measurement. As a goal, if time permits and the APXS has the appropriate deployment mechanism, APXS measurements of three different soil types will be made.

2. *Imaging*—The microrover will take a close up image of any rock of which an APXS spectrum is obtained and transmit the image to the lander for return to Earth. The image will be black and white and monoscopic but will be intended to show structure at higher resolution than obtainable by the lander camera. The rover computer is capable of compressing and storing a single image on-board. Imaging of rocks may be done by turning around and using the front cameras, or a rear-mounted camera may be included as part of the APXS deployment mechanism.

3. *Other Science Objectives*—The rover may be asked to carry other science instruments, i.e. a neuron spectrometer, and may be asked to assist in the deployment of a seismometer. The ability of the rover to carry out these tasks depends on the mass, power and configuration of the instruments. Design trades are currently being made to determine whether the experiments are affordable by MESUR Pathfinder, and whether their design can meet the rover capabilities.

Mission Experiments

Because MESUR Pathfinder is primarily an engineering demonstration, assessment of the lander's condition is very important. The microrover will image the MESUR Pathfinder lander at least once to assess its condition after landing, and will transmit the image(s) to the lander for return to Earth. The microrover's image(s) of the MESUR Pathfinder lander will be black and white, monoscopic, and will encompass the entire lander cross section.

As a goal, the rover will acquire three images of the lander from three rover vantage points spaced at 120° apart around the lander. If necessary, the nominal mission could be revised to allow the rover to take close-ups (stereo, if necessary) of damaged areas on the lander.

ROVER MISSION DESIGN

Because of cost constraints, both the lander and the rover have limited redundancy. In addition, the thermal environment of the Martian surface is very harsh (cycling between about 0 to –100°C daily). Therefore the surface mission design is focused on achieving the important mission objectives in a relatively short time.

The mission design is severely constrained by the lander-to-earth direct communications link (limited to a few hundred bits per second, even using a lander high gain antenna). The landing site only faces Earth for 12 hours per day or less. The lander is solar powered (RTG's being prohibitively expensive) and therefore transmits mostly during daylight hours, although it does have rechargeable batteries. However, transmission power is so high that the lander's batteries are excessively drained by more than a few hours of transmission per day.

The rover navigation scheme involves the earth operators selecting a target and "way-points" on the way to the target, and uplinking the way-point designations and commands for tasks on the way to the target. Because of telecommunications limitations the rover can only be commanded once per day. Since terrain conditions can only be inferred from the lander images operators will be reluctant to initially send the rover on traverses of more than ten or so meters per day. All of this means that the surface operations must be designed to be very efficient in accomplishing objectives.

The tool for surface mission design is scenario development, with timelines built around data transmission capabilities. A cartoon of an example scenario is shown in Figure 2. Here, the rover is deployed on the first day (or "sol," a Martian day which is nearly the length of an Earth day). The cartoon shows that the rover's first activity is to deploy a seismometer. However, this now appears likely to be too heavy for the rover to carry. Next, the rover is directed toward a target rock, pausing to perform soil mechanics experiments and image the lander.

This scenario assumes that the rover gets near the target rock on the first sol. The next sol is consumed directing the rover to contact the rock and place the APXS on it. Cost constraints make it unlikely that the rover will be able to autonomously reach the rock and place the APXS on it with only one Earth command. This, plus the probability of navigation errors (e.g. wheel slip preventing accurate dead reckoning) have led to a scheme where the rover position will be determined relative to the target at the end of each sol by imaging it from the lander.

The APXS spectrum in this scenario is taken overnight, to conserve the daylight hours for rover movement. The overnight spectrum will be powered by the rover's batteries. An APXS soil measurement may be taken first to increase the probability that some key science data will be returned, even if the rover or lander "die" early.

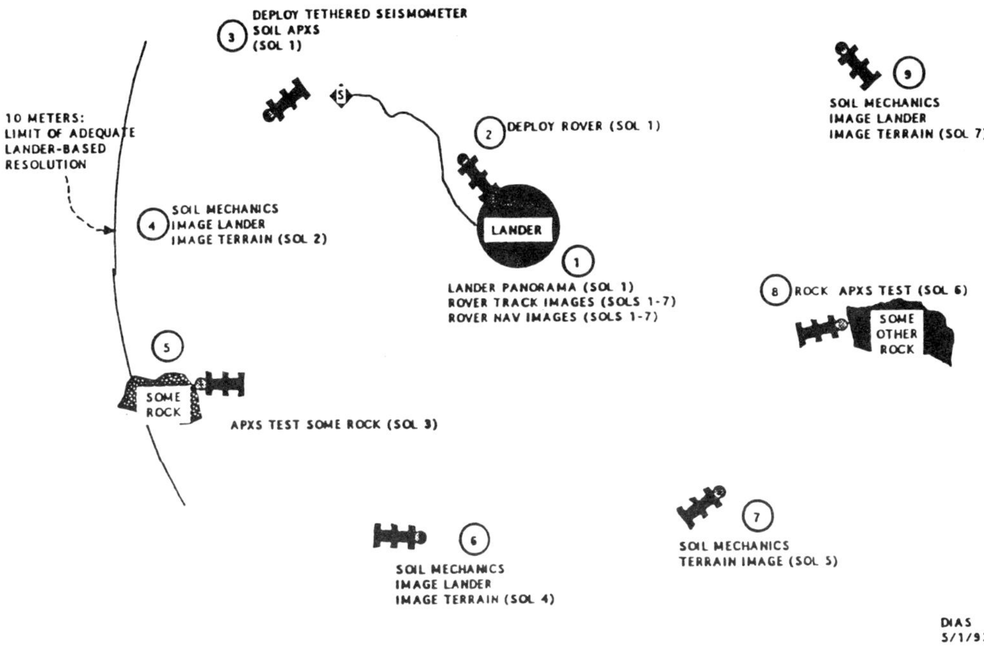

Figure 2 MESUR Pathfinder Rover, Nominal 525 BPS 7-Sol Scenario, 15°S.

Having accomplished its most important objectives in the first three sols or so, the rover is directed around the lander, conducting technology experiments in different soil types, imaging the lander from other vantage points, and collecting APXS data from another rock. By the end of the seventh or eighth day of operations the full lander panorama has been transmitted, the primary rover experiments accomplished, and the extended mission phase begins. The prime mission is envisioned to be accomplished within 10 meters of the lander to take advantage of the highest resolution lander images. For the extended mission the rover can be risked on longer traverses. It can even head "over the horizon" (out of range of the lander's camera) by using its own images for navigation.

The scenario described above is an example. A number of alternative scenarios will be constructed before the landing to account for differences in circumstances: non-nominal landings, problems with panel deployment, unexpectedly rough (or rockless) terrain, etc.

ROVER OPERATIONS DESIGN

The rover is controlled by an Earth operator who views a work station with a stereo display of the lander's image of the terrain through 3-D glasses. Figure 3, from Reference 11, shows this arrangement. The work station's software allows an icon of the rover to be placed in the scene and the coordinates of this placement determined. These coordinates form the basis of the rover traverse commands. Commands which enable rover tasks, e.g. "perform a soil mechanics experiment," "place the APXS on a rock" are interspersed with traverse commands and sent to the rover through the MESUR

Pathfinder Mission Operations System (MOS). The commands are sent shortly after sun and earth rise on Mars, and are received by the lander. The lander then stores them until notified by the rover that it is awake and ready to receive its commands. The lander forwards the commands to the rover over the lander-rover UHF link, and the rover stores them for execution.

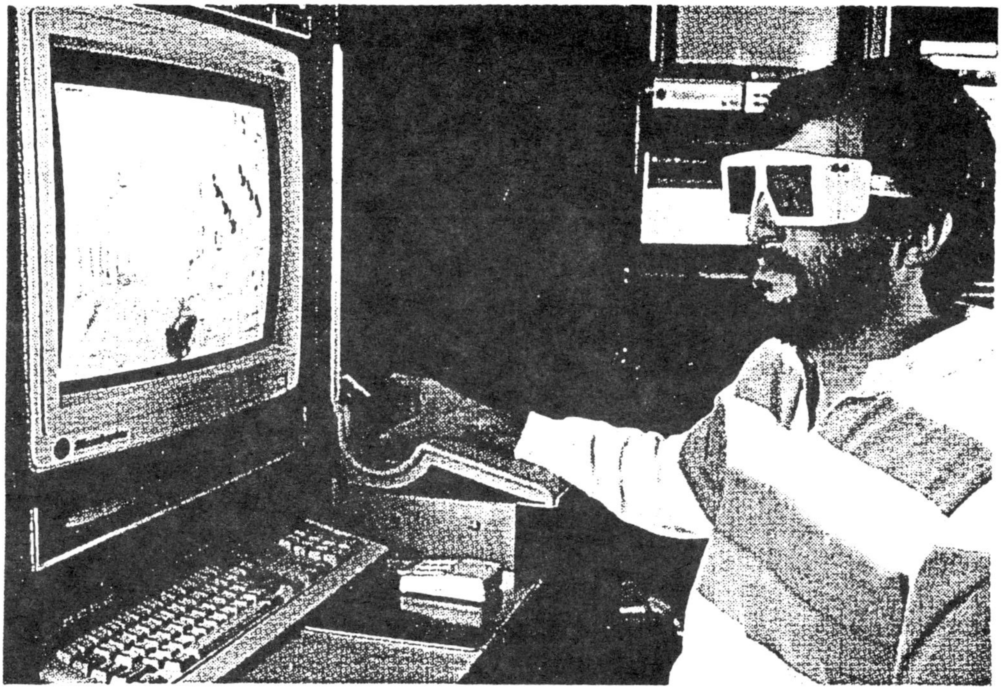

Figure 3 Operator Work Station for Rover Control.

Because the rover reacts to Martian terrain conditions the times of traverse and experiment execution are non-deterministic. Therefore, the rover activities are commanded asynchronously in contrast with normal spacecraft sequences. "Time-out" limits will be provided for each function so that if, for example, a destination is not reached within a pre-specified time, the rover will cease trying and continue with the next planned activity or "call home" for further direction.

Rover telemetry, including images, is stored, compressed and packetized on the rover. The rover notifies the lander when it is ready to transmit, and the lander o.k.'s transmission when it is ready to receive. Telemetry packets are downlinked over the rover-to-lander UHF link and the lander stores the packets for transmission to Earth. The lander is not required to process either rover command or telemetry packets, other than to read the packet headers and transmit the packets. This greatly simplifies the rover/lander interface and is a cost savings on both sides.

On the Earth, the rover images and engineering telemetry can be displayed on the operator workstation and/or sent to auxiliary workstations for analysis.

The total cycle for rover (and lander) commanding is designed to be done overnight. That is, from the time the lander images of the rover and the rover telemetry are received at the end of the sol's Mars operations, to the time the next sol's commands are transmitted, is about twelve hours. In the absence of anomalies this timeline is feasible, provided that review and approval is minimized. Experience shows that designating the rover path and selecting the tasks to enable only requires a few minutes. Since the rover and lander commanding is relatively independent and asynchronous, interweaving of commands and contention for resources is minimized.

The highest risk area, in terms of timing, is the first sol after landing. The landing will occur in the morning and the ground operators must react to the first lander image within two hours to allow rover deployment on the first sol. That is, the representatives of technology experiments, science and lander health assessment must be prepared to select targets and activities with little contention. The development of operations team structures to minimize the number of decision makers and approvers is a challenge.

As part of the low-cost process for MESUR Pathfinder the End-to-End Information system (EISS) is being developed using a concurrent engineering process. The rover operations development is an intimate part of this process and is the focus of the earliest design and test activities.

ROVER DESIGN

Design Philosophy

MFEX is an example of a "Better, Faster, Cheaper" NASA activity. Consequently its design and new processes are being developed concurrently, as illustrated in Figure 4. The focus of the cost constrained activity is a deliberate constraint on computational power, coupled with new autonomous control software technology and sensing/control architecture. Performance and risk are variables to achieve the costs constraints (for example, low-speed computing is enabled by slow rover speeds).

The selection of affordable parts and a cost-constrained design are built around the low performance core. Around that are wrapped cost effective processes: a small, low overhead team; rapid prototyping/concurrent engineering; selected industrial participation for cost-effective technology transfer; new processes for parts qualification and procurement; etc.

Although only in Phase A, MFEX is currently on schedule and within budget. A top level schedule, with the projected funding profile, is shown in Figure 5.

The NASA Office of Safety and Mission Quality (Code Q) has selected MFEX as an example of "Class D" project and MFEX is documenting its risk mitigation processes and their results to be used in the formulation of future standards for such projects. New processes for parts qualification, procurement, and program control are being developed by JPL with an MFEX focus.

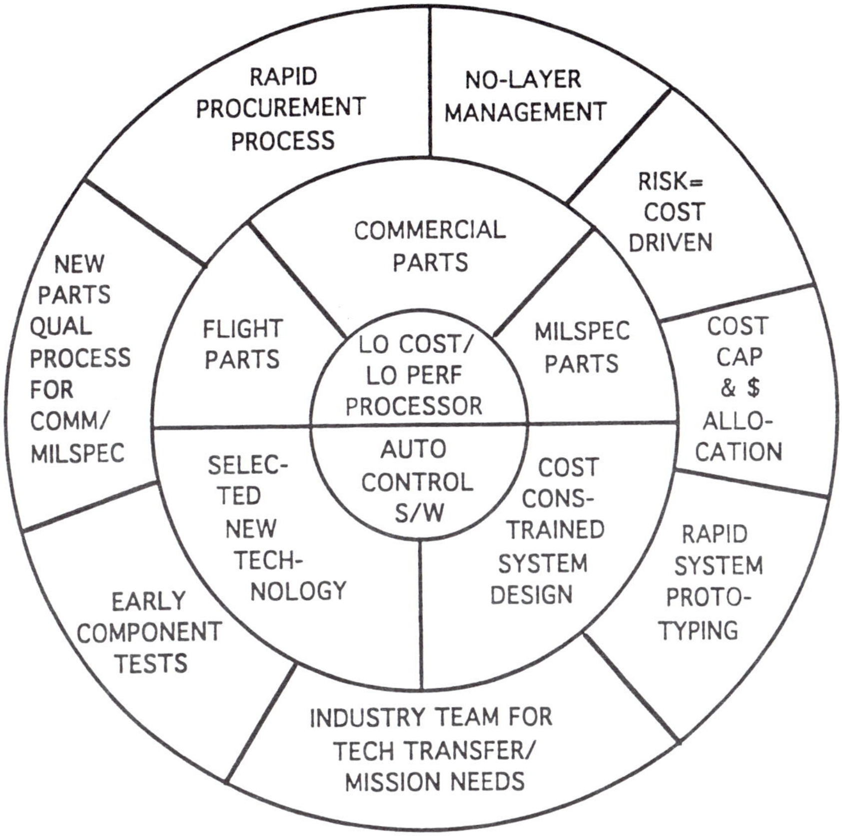

Figure 4 Rover Design Philosophy.

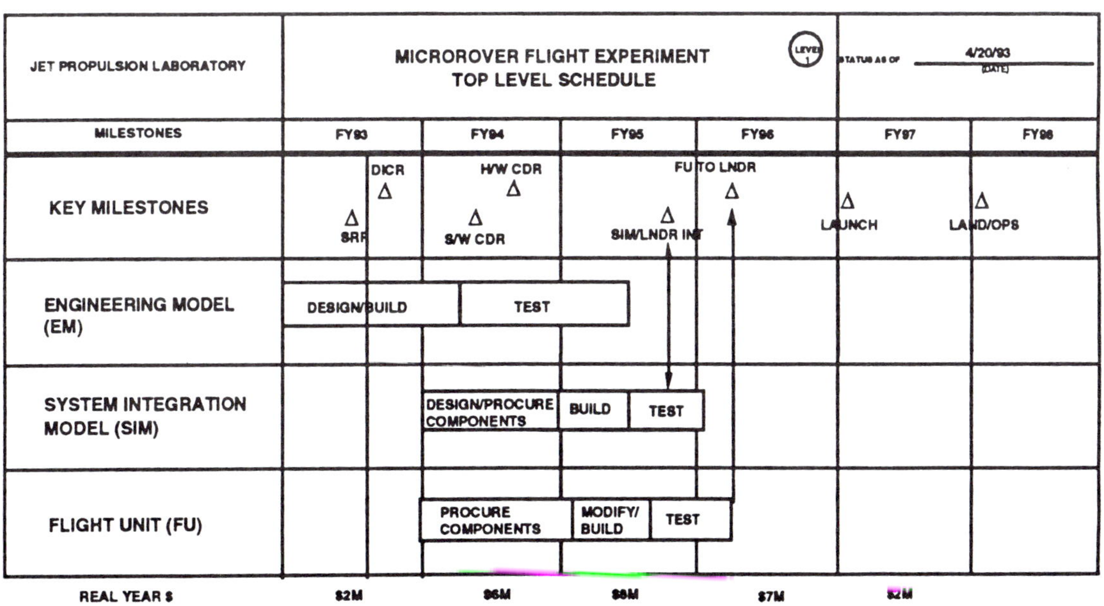

Figure 5.

The rover design is governed by the $25M cost constraint. Costs were allocated to system and subsystems, including operations, before the design was selected. The initial design exceeded the cost constraints, so it was descoped. For example, extra test hardware was removed and mission performance was lowered.

The rover design is also limited by the need to minimally impact the lander. Consequently, the rover mass and volume are limited, and the rover is "free-ranging," utilizing a UHF link for communication even during cruise, to minimize the interfaces with the lander.

In order to meet the cost constraints the rover will take selected design risks. One of these is to maximize the use of off-the-shelf commercial and Mil-standard parts. This is enabled by the fact that the mission duration is relatively short, and that the radiation environment is relatively benign. New classification and qualification procedures are being developed at JPL to minimize the cost of using such parts. Class S flight qualified parts are used where ever cost effective, e.g. the CPU is Class S. Figure 6 is a matrix of parts type vs. flight readiness.

TECHNOLOGY LEVEL	ACTIONS USE AS-IS	MODIFY	QUALIFY	DEVELOP
COMMERCIAL	• 80C85 S/W DEVELOPMENT ENVIRONMENT • SOME MOBILITY COMPONENTS	• THERMAL-VACUUM ENCLOSURE • SOME MOBILITY COMPONENTS • PROXIMITY SENSING H/W	• SOME MOBILITY COMPONENTS	
MIL-SPEC	• UHF MODEMS • POWER CONVERTERS			
FLIGHT QUALIFIED	• 80C85 CPU • SOLAR CELLS • BATTERIES			
NEW TECHNOLOGY			• MOBILITY CONCEPT • SYSTEM CONCEPT	• CONTROL S/W • I/O BOARD • PROXIMITY SENSING S/W • SOME MOBILITY COMPONENTS

Figure 6 Microrover Technology Mix.

Another risk is the limitation on redundancy. The rover design is as failure resistant as possible within the cost and mass constraints. Block redundancy is practically nonexistent, but considerable functional redundancy is used, e.g. solar panels and primary batteries.

Test Philosophy

The design makes maximal use of the "Rocky" series heritage, but includes several features that were not necessary for the previous ground demonstrations. Table 2 illustrates some of these differences. The MFEX activities are built around a rapid prototyping approach. "Rocky 4" was stripped to the bare chassis for mobility testing (Rocky 4.1). Motors and mechanisms closer to flight were added, and the chassis was given to the control subsystem. Control breadboarded a complete suite of sensors and a processor and I/O board, for use in software development (Rocky 4.2 is now operating in the sandbox.). The next step is to replace the breadboard components with more flight-like "brassboard" components to allow the software to be tuned for flight-like operations, including integration in the MESUR EEIS development (Rocky 4.3).

Table 2
ROCKY 4 DEMONSTRATION VS FLIGHT ROVER CHARACTERISTICS

	ROCKY 4 DEMO	FLIGHT ROVER
SYSTEM		
SIZE	60 X 46 X28 CM	SAME
MASS	7 KG	SAME
PERFORMANCE	TAILORED DEMO	ROBUST
QUALIFICATION	NONE	FLIGHT QUAL
ENVIRONMENT	BENIGN	HARSH
COMMAND CYCLES	MANY	1 PER DAY
BUDGET/SCHEDULE	TIGHT	TIGHT
SUBSYSTEM		
CHASSIS	FIXED	STOWABLE
MECHANISMS	RADIO SHACK	MIL-SPEC
COMPUTER	WIRE WRAP	CLASS S
S/W	HACK	FLIGHT QUAL
THERMAL	NONE	WARM BOX, HEATERS
POWER	COMMERCIAL BATTERIES	SOLAR ARRAYS, FLIGHT BATTERIES

In parallel with this system development, component selection and testing are proceeding. For example, commercial motors and gears are being evaluated in a small thermal-vacuum chamber for their resistance to frigid Martian conditions.

These rapid prototyping activities have been facilitated by a new rapid procurement process developed by JPL's procurement organization to support the "Better, Faster, Cheaper" way of doing business.

The next step in the rapid prototyping process is to procure two sets of flight hardware, storing one set in clean conditions and using the other set to construct a "System

Integration Model" or SIM. The SIM will be subjected to field and environmental testing, and will be used to do interface checks with the MESUR Pathfinder spacecraft during its Assembly, Test, and Launch Operations (ATLO). Based on the SIM results the Flight Unit will be built up, perhaps with some parts modifications, in clean room conditions. This is more arduous than normal spacecraft cleanliness because of planetary protection restrictions.

The Flight Unit will be integrated with the flight spacecraft and will be subjected to flight environmental testing in conjunction with the spacecraft.

Rover Design Description

The MFEX rover configuration is shown in Figures 7, 8 and 9. The rover, including its mounting and deployment equipment, is constrained to mass less that 14 kg. The rover has an anticipated mobile mass of 9.0 kg, without science instruments. Another 5.0 kg is allocated for lander-mounted rover telecommunications equipment, structural support of the rover and its deployment mechanisms. (The MESUR lander mass is limited by the relations between ballistic coefficient, g-loading of airbag impacts and parachute size. Significant launch mass margin exists.) Volume is constrained by the folded lander's tetrahedral shape. The rover competes for space with the lander thermal enclosure, parachutes, and other systems. While the rover has a nominal height of 280 mm (with 130 mm ground clearance) the available lander volume allows only 200 mm, forcing the rover to stow at a static height of 180 mm (see Figure 10). The rover is 630 mm long by 480 mm wide.

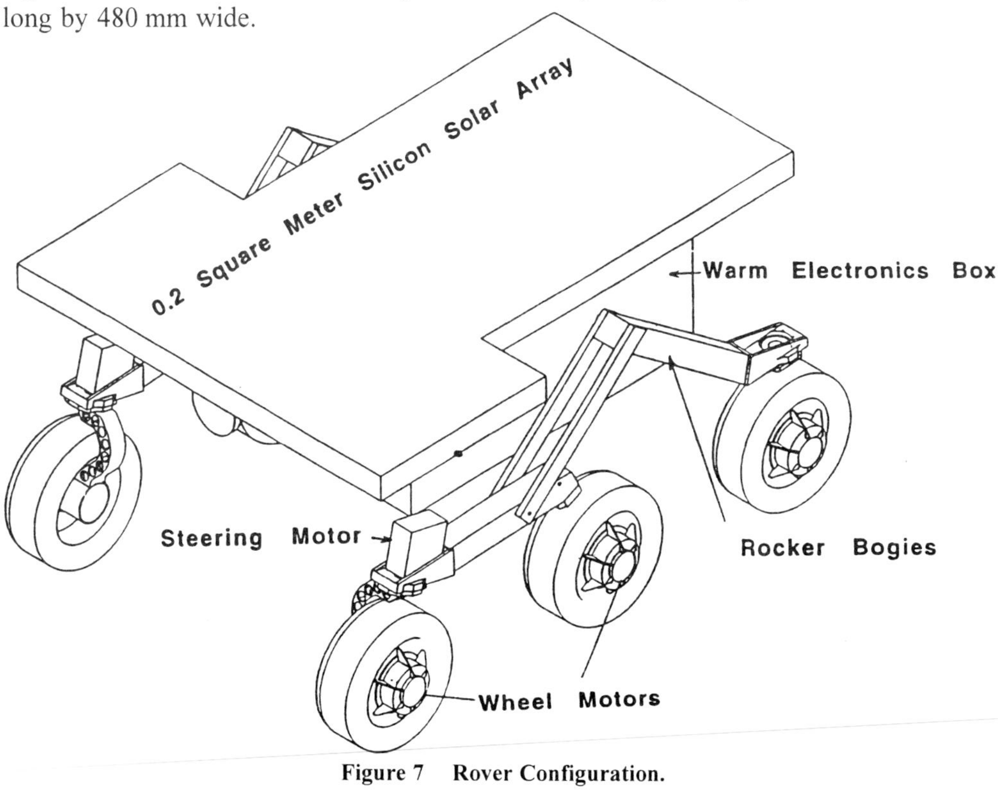

Figure 7 Rover Configuration.

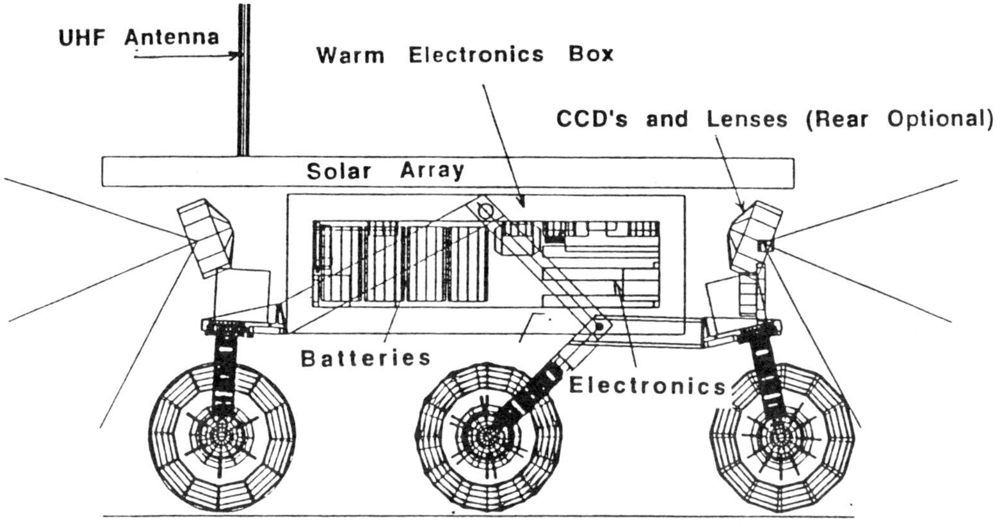

Figure 8 Rover Configuration Side View.

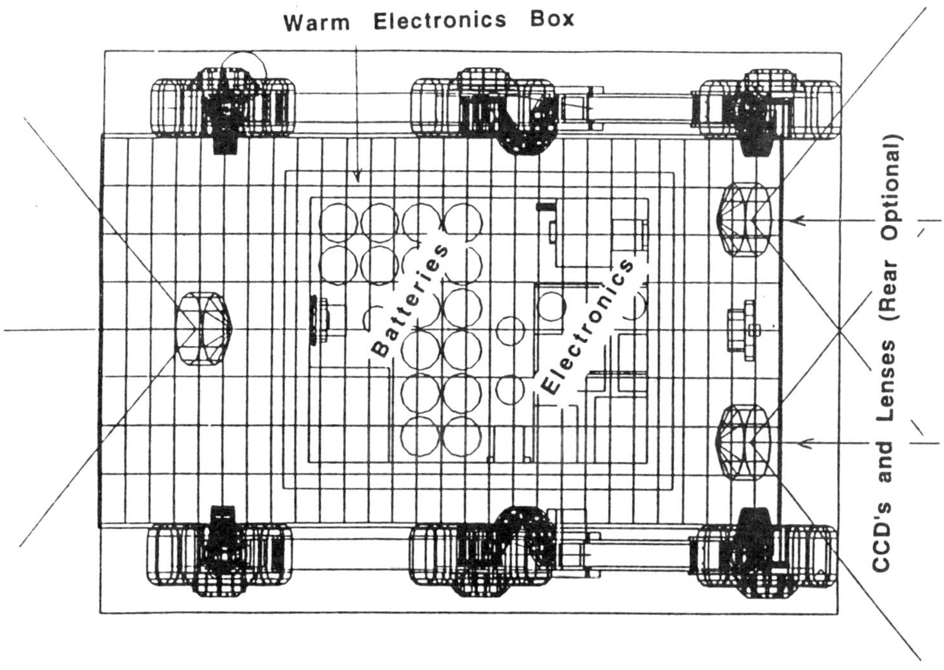

Figure 9 Rover Configuration Top View.

This small size poses considerable challenges for mobility, computation, thermal control and power.

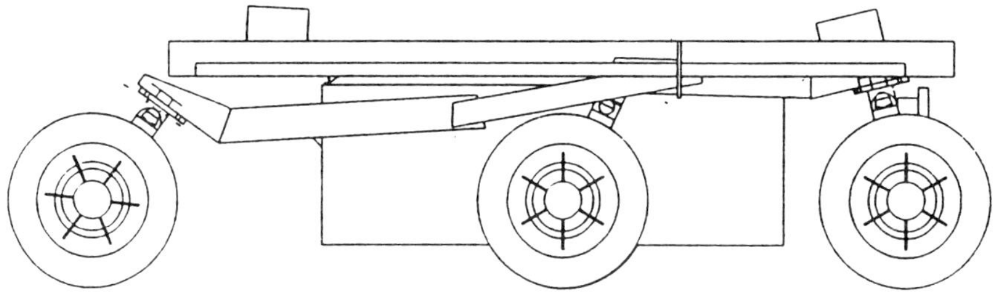

Figure 10 Stowed Configuration.

Mobility/Structure/Mechanisms

The rocker-bogie mobility system was developed over several years using custom computer models developed at JPL, as well as analysis tools such as NASTRAN (FEM) and ADAMS (dynamics code), verified through the development of scale physical models. When overcoming obstacles, the lever arms of the suspension reduce the incline angle of the master rocker. The single body is mounted to the rocker through a differential, cutting the incline angle in half again. This results in a very stable platform (when compared to more conventional three body designs) for science, imaging, and proximity sensing. The vehicle is steered with its four outer wheels, allowing it to turn in place, or to Ackerman steer (like an automobile) about a point outside the vehicle.

In the past few months, mobility was tested in a tilting sandbox filled with dry sand and then with lunar simulant. The 130 mm diameter wheels were widened to 60 mm contact to reduce ground pressure and therefore limit sinkage in loose soils such as are expected on Mars. Sandbox testing was done with a stripped chassis to simulate performance at the 3/8 g Martian gravity. The results were that the system could climb slopes of 32° in dry sand (within 2° of the angle of repose, which is the maximum slope at which loosely deposited material will be stable), and 17° in lunar simulant. Additional testing may be performed in dry clay material, if resources permit, which is believed to be a better simulant of Martian soil. The results of the tests demonstrated some difficulty with repeatability, which is not unusual in soil mechanics experiments. But they do indicate the Rocky configuration is likely to function adequately on Mars to acquire the necessary terrain data, provided sufficient soil mechanics experiments can be done.

The Rocky 4 structure was severely lightened to achieve the desired 7 kg. This resulted in some problems with insufficient stiffness which could affect the navigation sensing, and the MFEX rover structure is likely to be somewhat heavier to handle these stiffness concerns as well as landing loads.

Component selection and qualification is proceeding in parallel with system testing. Maxon rare earth motors have been selected for propulsion and steering. Mobility sensors have been designed and parts selected for testing. Gears with plastic bearings are being tested with different types of lubricants at low temperatures and pressures. Wear does not seem to be a problem, but viscosity is undesirably high at the present time. The use of commercial parts is enabled in part by the slow design speeds of rover mobility (< 2 cm per second). Simplifications in the mobility approach have been taken to minimize power and control subsystem costs. For example, the motors are powered through a "bang-bang" control system rather than through proportional speed control.

Control

As described earlier, the rover control system features operator designation of targets and autonomous control to reach the targets and perform the tasks. The on-board control system (shown conceptually in Figure 11) is built around an Intel 80C85 processor, selected for its low cost and Class S radiation/Single Event Upset resistance. This is only an 8-bit processor which runs at about 100 kips, in contrast to new technology flight processors. However, development to date indicates that it is ample for the rover's needs, provided that rover speeds remain slow. (Note that it is also vastly superior to all the computers used by planetary spacecraft before Galileo.) A commercial version of the 80C85 was acquired inexpensively for system and software development, and flight parts are also available at JPL in the form of Cassini spares available for purchase by the rover.

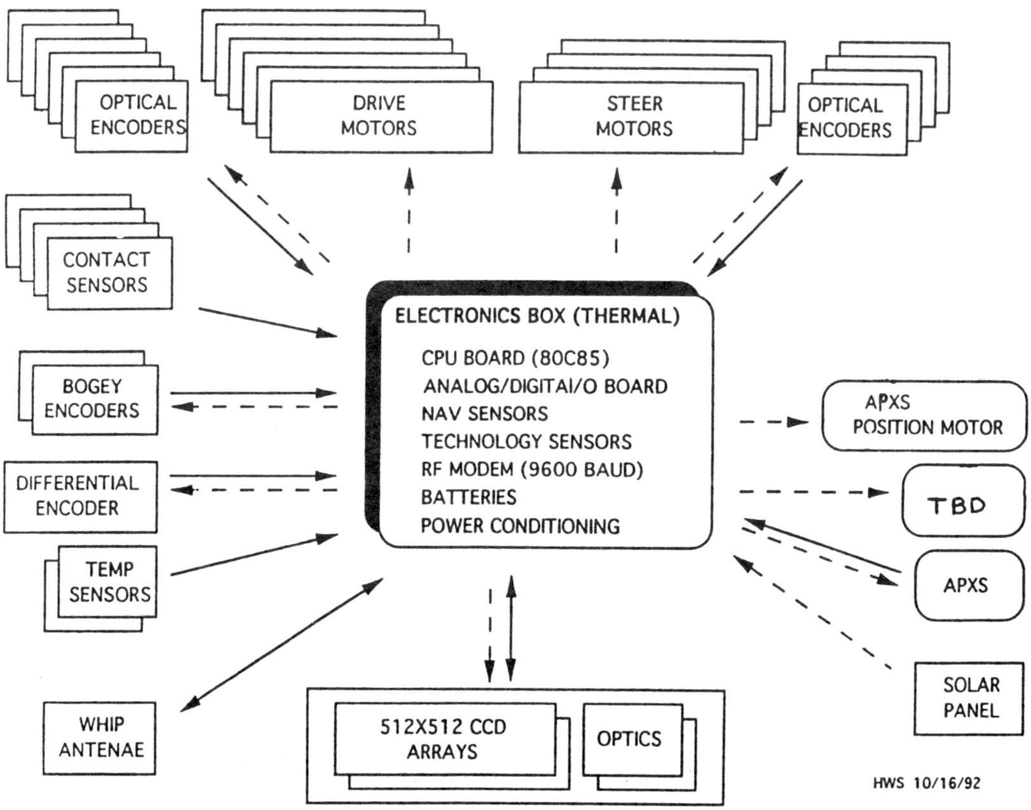

Figure 11 Control Schematic.

An I/O board is being constructed which can be cheaply manufactured in Class S technology, and in a small, low power package. The processor and I/O board package form a computer which will perform all the computational functions of the rover, from command decoding, to sensor data collection, to control response calculations, to telemetry collection and packetizing. Almost all elements of the computational system are Single Event Latchup (SEL) resistant, with the possible exception of an "alarm clock" which will wake the rover up in the morning and maintain a clock reference. The rover design currently incorporates triple redundancy in the alarm clock. Another design feature allows the rover to wake up when the solar panels are activated by the morning sun. Clock updates are also maintained by communicating with the lander.

All electronics are thermally protected by being enclosed in the Warm Electronics Box described in the Thermal Control section, below.

The rover's sensors are primarily commercial/Mil-spec parts. For example, the rover imagers are two commercial CCD's with wide angle lenses mounted on the front of the rover as shown in Figures 8 and 9. The sensors which must be exposed to ambient Mars conditions are being environmentally tested at the component level.

A new proximity sensing technology which is being developed by an OACT technology program is a candidate for use by MFEX. This utilizes the central processor to read out CCD images by direct control of the CCD's clock line. Commercial laser light stripers project vertical light planes ahead of the rover as it moves. When laser light strikes an object, bright points show up in the CCD image. By proper choice of CCD look angle and correlation of the laser points with triangulated range to points in the scene, the rover can identify obstacles, such as large rocks, overhangs and holes. Then autonomous hazard avoidance "behaviors" will drive the rover around the obstacle. Use of this technique also appears promising for ranging to obstacles and for "true ground speed" measurements.

The rover's other sensors are used for hazard avoidance, e.g. if the bogie angle encoders indicate that a front wheel has dropped beyond acceptable limits, the rover can back up and avoid the detected hole. Other sensors for hazard detection include contact sensors on the solar panel, three single-axis accelerometers for vehicle orientation detection, and temperature sensors in the Warm Electronics Box, wheel motors, solar panels, etc. The CCD's, of course, can be used for stereo or monocular imaging as well as for hazard avoidance, providing an alternative to the lander imagery for navigation, and providing images for the various experiments. The rover computer design allows switching between the two cameras for efficient power usage.

The navigation and hazard avoidance software, as well as the data compression, packetization, etc., software, forms the application layer around a simple operating system kernel especially designed for the rover. Navigation utilizes a tuning fork rate gyro and sensors such as the steering potentiometers and wheel motor encoders for dead reckoning. The on-board "behavior control" automatically follows the operator designated heading until an anomaly (such as an obstacle) is encountered. Then higher level "behaviors" modify the "bee-line" traverse heading algorithm to react to the situation.

For example, if the proximity sensors indicate a hazard to the rover's left, it will turn right to avoid it. After avoiding the obstacle the navigation system again heads directly for the target.

Data from the sensors is sampled at intervals compatible with rover control and hazard avoidance, technology experiment requirements, and computational capability. The computer will compress the imaging data, probably using a block truncation coding algorithm, and packetize it with other data for transmission to the lander.

Telecommunications

Rover telecommunications is planned to be by means of Mil-Spec, half duplex UHF modems operating at about 460 MHz and broadcasting over 1 meter high whip antennas. one modem/antenna is located on the rover, the other on the lander. Both modems and their associated whip antennas are provided by MFEX. The lander data system interface with the modem is through an RS 232 port for simplicity. The modem currently being investigated is from the Motorola R-Net series. It was selected primarily for its low power (it has 100 mW and 300 mW operating modes), and for its Mil-Standard qualification. Because the modem is half duplex the rover/lander telecommunications link has been designed to be controlled by the rover. That is, the rover notifies the lander when it wishes to communicate.

Analysis of the modem has shown that communication is possible over several hundred meters in the worst case, even if the rover goes "over the horizon" from the landers view. Communication distances of up to 2 km are feasible in the higher power mode. In order to ensure the maintenance of a communication link the rover will conduct a "heartbeat" check with the lander every time it traverses one-half its body length. If the "heartbeat" signal is not detected the rover will automatically back up one-half body length and try again. The rover will try this several times before invoking other fault recovery actions, which may include continuing the mission assuming that the rover to downlink may still work and/or the lander will continue to image the rover.

The Motorola modem is currently the focus of the new parts JPL "Class D" qualification process. The modem is qualified for operation at –40°C and for g-levels far in excess of the 50 g's required for landing. Analysis has shown that almost all the modem parts can be expected to be Single Event Latchup (SEL) resistant, which would allow the modem's use in a space flight application. The exceptions are two microprocessor chips. These parts are currently being tested for SEL hardness. If acceptable, the modems can be flow without modification. If the chips fail the SEL test several alternatives are being investigated, including selection of other modems or replacement of the unacceptable parts. Another solution is to operate the modem sparingly, merely risking the relatively remote chance of SEL.

During cruise, rover state information is communicated through the UHF modems and antennas to the spacecraft data system, which retransmits this information to earth. This design eliminates the need for a costly and risky umbilical interface between the lander and rover. The rover status is communicated after launch and before entry into the Mars atmosphere so that the state of the system is known after and before the most traumatic events. Since the rover cannot actually be operated in cruise the state information will be limited to a computer memory readout, thermal measurements, and a check of the UHF data link itself. In order to allow the UHF interface to be used the antennas will be tied down and shielded to prevent excessive EMI from reaching the rest of the spacecraft/rover system.

Power

Rover power is provided by 0.2 square meters of solar array. This is sufficient to power the rover for several hours per sol, even in the worst conditions of atmospheric dust. The array cells are standard, space qualified silicon, and will be purchased in a common buy with the lander solar arrays for cost savings. As a backup and augmentation, 150 MW of primary, Lithium Sodium Dioxide D-Cell batteries are enclosed in the Warm Electronics Box. These batteries are flying in the Galileo probe and will be flown in the Cassini probe so they are flight ready and very inexpensive. Primary rather than rechargeable batteries were chosen because the mass of rechargeables is prohibitive.

The batteries are not required for the nominal mission, which can be conducted entirely on the solar array. However, the batteries are used to power rover communications during cruise so that a power link between the rover and the spacecraft is not required. The batteries are also used to power night operations of the APXS, and for "emergency boost" if the rover is navigating very rough terrain. The batteries are sized such that the most important rover functions can be accomplished in a few days even if the solar array fails.

Power control is relatively complex. The use of a variety of commercial, flight and Mil-Standard parts means that regulated power at a number of voltage levels is required. DC voltages include ± 12 v at 2.5 w, ± 5 v at 2.5 w, + 24 v at 10 w, + 5 v at 10 w, and + 9 v at 2.5 w. The independent alarm clock battery is a small $LiSOCl_2$ cell which can run the alarm clock for several years.

Redundancy in power switching is provided by cross strapping between the regulators and batteries, however the complexity thus introduced is being investigated for potential failure modes.

Power is managed to maximize power margin and simplify system design. For example, motor startups are done sequentially to minimize power surges and the CCD proximity sensing system is momentarily powered off during motor startup. The UHF modem is only powered when the rover is standing still. All power switching is controlled by the central computer.

Thermal Control

Control of temperatures in the harsh Martian environment is a severe challenge, especially when combined with the need for using low-cost components. Consequently, temperature sensitive elements (electronics and batteries) will be enclosed in a thermally insulated compartment called the Warm Electronics Box, or WEB. Elements outside the WEB will be qualified to withstand ambient temperatures.

Cabling out of the WEB is minimized and all cables between WEB equipment and external equipment are pigtailed at their source. An insulated "igloo" tunnel is provided for cables exiting the WEB. A connector panel is provided outside the igloo tunnel and WEB equipment connectors mate to the external equipment connectors at that panel.

The WEB is planned to be a thermal vacuum enclosure to minimize mass. A proprietary Owens-Corning material is being evaluated for the WEB. The WEB will be heated by the operation of the rover electronics. The idea is to operate the computer and modem from the time the solar panels have enough light to generate power until the sun goes down, raising the temperature inside the WEB. Ideally this temperature peak, plus the insulating capability of the WEB would prevent the WEB temperatures from dropping below –40°C, the operational qualification temperature of the batteries and electronics. However, preliminary analysis shows that the WEB temperature swings down to between –55° and –60°C at night unless additional heat is provided. The electronics and batteries are unlikely to be able to operate at those temperatures, so the rover would be too cold to wake up in the morning.

Resistive heaters may be used to consume excess solar array power and overheat during the daytime to bias the nighttime temperatures upward. The higher thermal losses at the higher temperatures would result in greater temperature swings which are hazardous to electronic packages. A preferred option is the use of two 1-watt each Radioisotope Heater Units which supply a constant power at all times and limit the day-to-night swings.

Two RHU's are a small enough radiation issue that the launch approval process will only require an Environmental Assessment, which is already required because of the APXS's radioactive alpha particle source. This is in lieu of a very costly Environmental Impact Statement which would be required for RTGs. Additional costs for the RHU's and the Environmental Assessment are being bookkept as a lien in the rover budget pending the final selection of RHU's as a thermal stabilization technique.

Cruise thermal control will be maintained by the MESUR spacecraft as part of its normal thermal control design. This may include a small heater in the rover WEB powered by the spacecraft through a simple umbilical. If RHUs are included in the WEB the heater will probably not be necessary, further simplifying the rover/lander interfaces.

MFEX IMPLEMENTATION/INDUSTRY INVOLVEMENT

MFEX is being conducted as a JPL in-house project because limited funding in the early years plus the short project cycle and the relatively unproven technologies involved where JPL's expertise prevent an intelligent system contracting approach. Members of the mobile robotics technology community are being involved as potential experimenters, and can analyze the data returned from Mars to improve future technology directions.

Much of the fabrication of the SIM and Flight Unit will involve contracted items. An issue remains of how to effectively transfer the technology to industry, which will be building future small rovers for MESUR Network and other Mars surface missions. The MESUR Network RFP was released this spring and proposals are to be submitted by the end of May, 1993. Two associate contractors will be selected and will provide personnel to the MESUR Pathfinder Project for purposes of spacecraft technology transfer. MFEX is investigating the feasibility of funding additional personnel to participate in the development of the rover, thereby transferring rover technology at the same time.

MFEX is also a candidate for technology transfer through industrial partnerships for other space mission and commercial applications, and industry interest is solicited.

REFERENCES

1. AIAA-89-0419, "Mars Rover 1988 Concepts", D. Pivirotto, T. Penn and W. Dias, 9-12 January 1989/Reno, Nevada.
2. JPL Publication 90-6, "United States Planetary Rover Status-1989," C. S. Pivirotto, W. C. Dias, 15 May 1990.
3. Pivirotto, D. S., "A Goal and Strategy for Human Exploration of the Moon and Mars, " *Journal of Space Policy*, Vol. 6, number 3, August 1990.
4. Pivirotto, D. S., "A Goal and Strategy for Human Exploration of the Moon and Mars, Part Two," Proc. Case for Mars IV, Boulder, CO, 4-8 June 1990. Published as abstract only, AAS 90-178, in *The Case for Mars IV*, T. R. Meyer, ed., AAS *Science and Technology Series*, Vol. 89, 1997, p. 271.
5. AIAA-90-3785, D. S. Pivirotto, "Site Characterization Rover Missions", 25-28 September 1990/Huntsville, Al.
6. Taylor, J. and Spudis, P. "Teleoperated Robotic Field Geologist for Use on the Moon or Mars," Space '90: Engineering, Operations and Construction in Space, September 1989.
7. "America at the Threshold, America's Space Exploration Initiative" The Synthesis Group, May 1991, (U.S. Government Printing Office).
8. Bickler, D. B., "A New Family of Planetary Vehicles," International Symposium on Missions, Technologies and Design of Planetary Mobile Vehicles, 28-30 September 1992, Toulouse, France.
9. Gat, E., *et al.*, "Behavior Control for Planetary Exploration: Interim Report" *Proceedings of the IEEE Conference on Robotics and Automation*, 1993 (to appear).
10. AUVS-86, Wilcox, B., Cooper, B., Salo, R., "Computer Aided Remote Driving," 21-23 July, l986, Boston, Ma.
11. Reynolds, K., "JPL Rocky IV Road Test," Road and Track Magazine, Vol. 44, number 8, p. 92, April 1993.

CONTRIBUTORS: The work summarized in this paper was performed by the MESUR Pathfinder Rover Team (most working part time): Leon Alkalaj, Ron Banes, Chuck Barnes, Don Bickler, Gary Bolotin, Dick Caputo, Jim Clawson, Brian Cooper, Jim Coss, Rick Crisp, Bill Dias, Howard Eisen, Carl Gilchriest, Shawn Goodman, Murray Hirschbein, Denise Hollert, Katie Kilgore, Dave Lavery, Bill Layman, Randy Lindemann, Jake Matijevic, Don Meyer, Andy Mishkin, Bob Miyake, Jack Morrison, Tam Nyguyen, Angel Olivera, Mike O'Neal, Betty Preece, Tom Rivellini, Henry Stone, Gayle Shinn, Scott Stride, Lin Sukamoto, Lee Sword, Wallace Tai, Curtis Tucker, T. V. Venkataraman, John West, Brian Wilcox, Jesse Wright, Y. C. Wu.

AAS 93-818

THE MARS AERIAL PLATFORM MISSION: A GLOBAL RECONNAISSANCE OF THE RED PLANET USING SUPER-PRESSURE BALLOONS

Robert Zubrin, Steve Price, Terry Gamber and Ben Clark[*]
Jim Cantrell[†]

The Mars Aerial Platform (MAP) mission is a conceptual design for a low-cost, Discovery class mission whose purpose would be to generate tens of thousands of very high resolution (20 cm/pixel) pictures of the Martian surface, map the global circulation of the Martian atmosphere, and examine the surface and sub-surface with ground penetrating radar, infrared spectroscopy, neutron spectroscopy, and other remote sensing techniques. The data would be acquired by instruments which are carried by balloons flying at a nominal altitude of about 7 km over the Martian surface. Because new balloon and micro-spacecraft technology is now available, the balloon probes could be quite long-lived, lasting hundreds or even thousands of days, producing an immense science harvest in the process. Together with the Mars Environmental Survey (MESUR) surface network science mission, MAP would revolutionize our knowledge of the Red Planet.

MISSION DESCRIPTION

The MAP mission will be carried out as follows: A low-cost launch vehicle such as a Delta 7925 (1000 kg TMI) is used to propel a small spacecraft carrying 8 entry capsules onto trans-Mars injection. Approaching Mars, a spinner is used 10 days prior to arrival to release the capsules outward so that they enter Mars' atmosphere at widely dispersed locations. Each capsule then enters the atmosphere, and deploys a parachute which slows it down to the point where a balloon can be inflated. The inflation is accomplished during descent, so that no landing system is required (the practicality of accomplishing this has been demonstrated on Earth at altitudes of up to 150,000 feet, where atmospheric density is similar to Mars). After the balloon is inflated, the parachute, capsule, and inflation equipment is jettisoned, and used to land a meteorology package consisting of pressure and temperature sensors, battery and transmitter on the surface. Each of the balloon probes will then commence their float around Mars at an altitude of about 7 km (23,000 ft) above the mean surface level.

The probes, which will never land in the course of their long duration aerial cruise, will carry a gondola payload of 8 kg, which includes 2 cameras, 1 kg of atmospheric science and additional instruments, data recording and transmitting equipment, a rechargeable battery and solar array.

* Martin Marietta Astronautics, P.O. Box 179, Denver, Colorado 80201.
† Space Dynamics Lab, Utah State University Logan, Utah 84322-4140.

When it is daylight two pictures are taken simultaneously every 15 minutes. One is taken with a high resolution CCD black and white camera with a nominal resolution of 20 cm per pixel; with a 1024 x 1024 pixel field, this gives a field of view 204 m on a side. The other is taken with a color camera at a moderate resolution of 10 m per pixel; this gives a field of view 10.2 km on a side. The two cameras are aligned so that the high resolution image is located at the center of the medium resolution picture, whose features in turn can be used to locate the region studied on a map of the planet. The data is stored and then is periodically uplinked to an orbiting satellite, which could either be Mars Observer or the MESUR comsat, and is then transmitted back to Earth.

It is anticipated that the science return of this mission will be large, with 32,000 high resolution pictures returned to Earth for each 100 days of operation of an eight balloon fleet, plus an equal number of pictures at resolutions superior to the best Viking images. The high resolution pictures will have a factor of 7 greater resolution than the best pictures returned from Mars Observer. Additional science return will result from use of other instruments, such as a ground penetrating electromagnetic sounder. Tracking the balloons will yield knowledge of Mars' global atmospheric circulation. The cost of the mission is anticipated to be low, falling below the $150 million (without launch) ceiling of the Discovery-class guidelines.

A mass breakdown for the mission is given in the tables below. As can be seen in Figure 1, each probe can be packaged in a 1.1 meter diameter aeroshell and 8 such probes together with their carrier spacecraft can be fitted into the launch fairing of a single Delta.

MISSION MASS BREAKDOWN

Table 1
DELTA 7925 LAUNCH

(8) Probes	728 kg
Spacecraft	150 kg
Margin	122 kg

Table 2
PROBE MASS BREAKDOWN

Aero-Shield	12.9 kg
Base cover and Drogue	9.5 kg
Main Parachute System	15.9 kg
Surface Meteorology Package	3.0 kg
Despin Device	1.0 kg
Tanks, lines valves	18.3 kg
Balloon Container	3.7 kg
Balloon and Associated Systems	16.5 kg
Gas Mass	2.1 kg
Gondola Payload	8.0 kg
Total	90.9 kg

Table 3

GONDOLA PAYLOAD MASS BREAKDOWN

Battery (Nickel-metal Hydride, 18 W-hrs)	0.5 kg
Solar Array (40 W-hrs, 0.22 m^2)	0.5 kg
Radio Transmitter	1.1 kg
Computer	1.0 kg
Thermal Protection and misc structure	1.5 kg
Science	
Cameras	0.6 kg
Atmospheric monitors	0.1 kg
Other Instruments (EMS, Neutron Spec, IR Spec, etc.)	1.3 kg
Cabling	0.3 kg
Contingency	1.1 kg
Total	8.0 kg

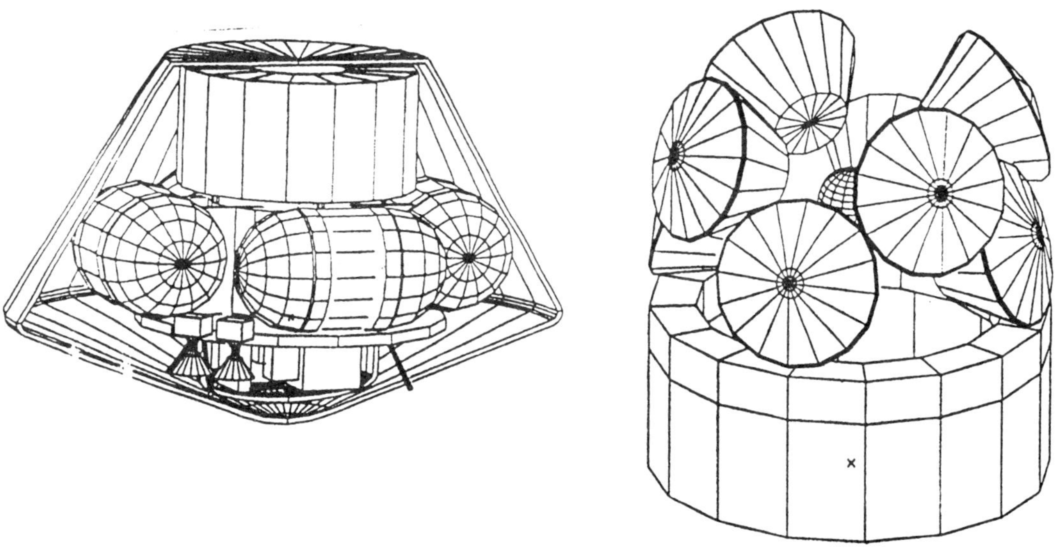

Figure 1 The MAP entry capsule has a diameter of 1.1 meters. Eight such capsules together with their spinning carrier vehicle can be fitted into a Delta fairing and launched to Mars. Shown within capsule, from top, is the parachute container, the balloon container surrounded by gas cans, and the gondola.

SCIENCE RETURN

The MAP mission has high potential for a large science return. Some of the areas that will be benefited include:

1 Geologic Science

The greatest benefit from high resolution images is the information that could be gained on surface processes. In contrast to Mars Observer, whose highest resolution images will show objects as small as a mid-sized car, the cameras on the MAP gondola will be able to reveal surface features the size of a cat. Such images will give us submeter surface morphology, block size distributions, and surface textures, all of which provide insight into volcanic processes, sedimentary deposits, fluvial history, and aspects of impact cratering. In addition, aeolian features such as ripple fields, individual dunes, small channels, etc., will be visible, providing direct evidence for the link between the atmosphere and the surface. The scale of some of these is below Mars Observer resolution. Assessment of these and other features that are submeter in size will provide a whole new suite of information on surface/atmosphere interactions.

2. Atmospheric Science

The MAP mission provides a unique in-situ platform for atmospheric science. Each balloon will float at a fixed density level (0.008 kg/m^3, or about 7 km above the mean surface) for at least 100 days. If the balloons were to average 10 m/s, a conservative value, then thousands of kilometers will be traversed. Depending on the circulation, it is possible that some of the balloons will circumnavigate the planet several limes. Thus global sampling of the atmosphere is possible.

Measurements that can be taken aboard the MAP balloons include temperature, pressure, horizontal wind speed and direction (via tracking), dust size and composition, and water vapor concentration. Atmospheric science goals that could then be addressed include the critically important surface boundary layer and its diurnal variability, atmospheric dynamics, circulation systematics, aerosol composition and properties. and the seasonal cycles of dust and water. Tracking of the balloons would give information on winds and circulation systems which could be used to test the computational global circulation models currently being developed by NASA. Slope wind systems may be particularly important on Mars and the MAP balloons should be able to help characterize their structure and dynamics. Aerosols, dust in particular, are known to be ubiquitous in the Martian atmosphere, yet their composition, size, and shape are uncertain. MAP balloons could measure these properties directly and thereby greatly increase our understanding of these climatically significant particles. Water vapor concentrations could be measured, furthering our understanding of the nature and distribution of near surface water reservoirs.

3. Exploratory Science

Infrared detectors on the balloons would be able to detect local geothermally heated hotspots on the Martian surface, should any exist and fall within several kilometers of the probe's ground track. Neutronic devices would be able to detect local concentrations of ice, hydrated minerals, or deposits of carbonates. Ice layers 10 meters or more below the surface may also be revealed by electromagnetic sounding using the radar altimeter, which can function as a simple ground penetrating radar. The discovery of such features would be of very high value for exobiological investigations.

4. Development of Mars Engineering Models

Numerous U.S. and international groups have a high interest in developing engineering models of the Martian surface. A recurring theme is the need for imaging of the surface at the scale that could be provided by MAP. In particular, knowledge of sizeblock distributions, slopes, surface discontinuities, etc., at a scale important to landers, roving vehicles and other landed mission operations is of high interest. The advantage of MAP is that a wide variety of terrains could be sampled to provide input to engineering models. The sub-meter scale imaging provided by MAP is also important to assess landing hazards to certify safe sites for future landers. The Mars Observer camera has a resolution of about 1.5 m/pixel, allowing it to resolve features about 4 m across. Such resolution is too gross for use in assessment or avoidance of landing hazards facing Mars surface science missions. Submeter scale imaging is also needed to provide context for the activities of a landed rover mission. This context would permit significantly richer surface operations activities, and a concurrent improvement in the caliber of science conducted.

The results from the MAP aerostats will be obtained concurrently with data returned by the meteorology packages delivered to the surface by the entry probe parachutes. This will enhance the potential science return by the synergy that will result from concurrent measurements from the surface network with the drift measurements on the balloons. This will provide more information on the atmospheric structure than could be obtained by either a balloon network or a surface network acting alone.

The large array of potential science instruments that could be carded by MAP are summarized in the table below. There are far too many to be all carried in the same gondola, however if the probes sent to Mars are divided into 4 "science sets" (which we label "A,B,C, D,") with 2 probes per set for a Delta launch, then the full range of potential investigations could be carried out. To minimize instrument and computer power, all instruments would be cycled on and off once every 15 minutes in coordination with the camera operation.

THE CAMERA

The primary instrument carried by the probe will be a CCD camera such as that currently under NASA-sponsored development by Malin Space Science Systems (Figure 2). This camera has a mass of 1.0 kg, including two sets of optics (additional sets can be added for 0.3 kg each), all electronics, data compression, and 40 MBytes of data storage. (In the candidate instrument table above, part of the camera mass is accounted for by the gondola computer.) The compressor reduces the data so that 0.6 bits are required for each pixel of black and white photography, and 0.7 bits for each pixel of color. Thus a pair of pictures, each 1024 x 1024, one color and one black and white will require a total of 1.3 Mbits and each balloon probe can store a maximum of 248 pairs of pictures at a time.

Table 4
CANDIDATE MAP INSTRUMENTS

Instrument	Mass (kg)	Power (W-h/sol)	Data (Mb/sol)	Science Set (example)
Camera BW or 3-color	0.6	10	26.0	A,B,C,D
Temp and Pressure	0.1	0.01	0.02	A,B,C,D
Sun Sensor	0.1	0.01	0.02	A,B,C,D
Accelerometer	0.1	0.1	0.2	A, C
Neutron Spectrometer	0.2	0.03	0.06	A,B,C
Gamma Ray Spectrometer	1.0	2.0	5.0	B
Electromagnetic Sounder	0.4	8.0	1.0	A, C
Microwave Radiometer	0.3	0.1	0.02	A
Magnetometer	0.2	2.0	0.02	A D
Infrared Radiometer	0.1	0.01	0.02	A, ,C,D
Infrared Spectrometer	0.5	2.0	5.0	C
Nephelometer	0.2	0.5	1.0	D
Hygrometer	0.2	0.02	0.02	D
Dust Composition Analyzer	0.4	0.1	0.02	D

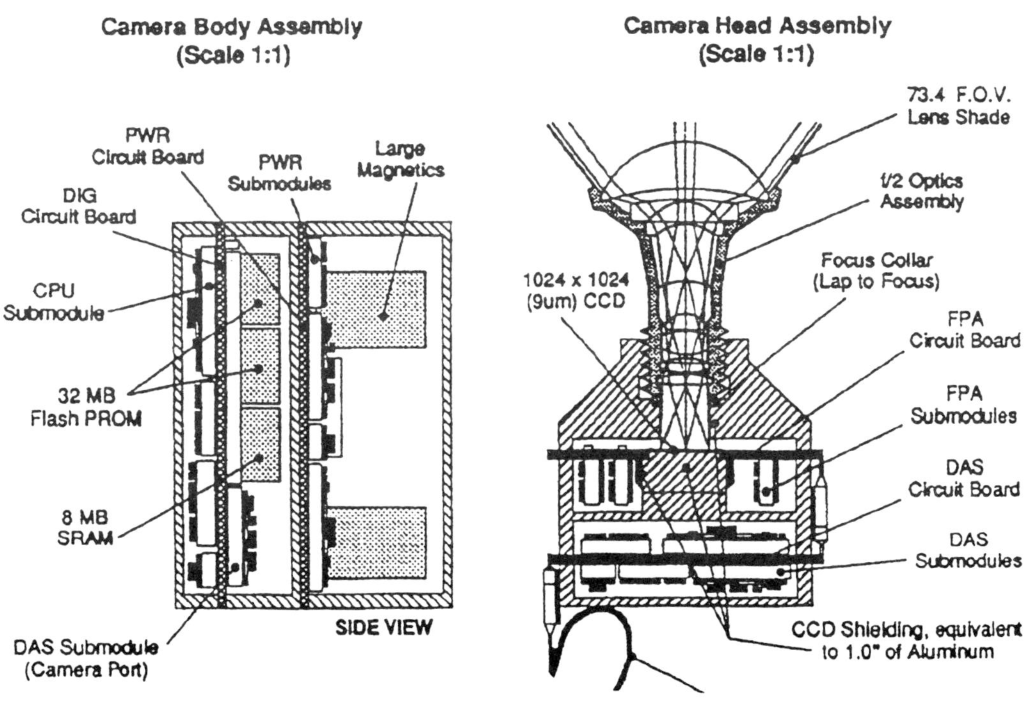

Figure 2 A miniature CCD camera such as that designed by Malin Space Science Systems would be the primary instrument carried by the MAP probe. This camera has a mass of 1.0 kg, and a volume of 400 cc, including 2 sets of optics and a computer for data compression and storage.

To enable a significant advance in the understanding of Mars surface geology, a resolution about an order of magnitude better than Mars Observer (20 cm/pixel) is desired. As it is not possible to find an area only a few 100 meters across in orbiter images, a "context" or low resolution frame is also needed. This context image must be around 10 km on a side. based upon attempts to locate the highest resolution Viking Orbiter frames (about 10 m/pixel, 1200 pixels across) in the Mars Global Digital Image Mosaic (250 m/pixel resolution).

The baseline MAP camera system consists of two camera heads bore-sighted to take simultaneous high and medium resolution images. The camera heads are placed on a pan with 1 degree of freedom to allow downward looking pictures to be obtained, horizontal pictures (allowing the observation of sedimentary layers), or any angle in between. The medium resolution camera can also be used to acquire images of the Sun to determine atmospheric opacity and dust particle size information.

DATA RETRIEVAL AND POWER BUDGET

The aerostat transmitter system is generally kept silent while its radio receiver is cycled on and off listening for a beacon on the orbiting communications satellite. Once such a beacon signal is acquired, the gondola computer is awakened and a return signal is sent up to the orbiter, allowing a communications link to be established.

Using 18 watts of power, 5 watts of UHF (401 MHz) can be produced, which will result in a transmission rate of 64 kb/s to the Mars Observer uplink, or to a similar relay carried by a MESUR comsat. Processing of data requires about 0.4 W-hr per picture pair, while transmission to the comsat requires about 0.1 W-hr. Thus a total of 0.5 W-hr is required for the processing and transmission of each pair, with 80% of this expended during the daytime when the pictures are taken.

The balloon probe carries a power system with a storage of 18 W-hrs. and a solar array that can produce 43 W-hr/sol (a sol is a Martian day of 24.6 hours) under moderately below average Martian conditions (100 W/m^2 diurnally averaged solar incidence.) If half of this power is available for image processing and transmission, then each probe can transmit 40 picture pairs per sol. The total link time to Mars Observer required for this transmission is about 813 seconds per sol, which is only about half of what will be typically available to each aerostat on a daily basis.

Of the aerostat's remaining power, 8 W-hrs/sol is can be used in the ground penetrating radar, 3 W-hrs/sol for the computer, and 10 W-hrs/sol is extra and available for contingency utilization. The thermal protection power needed by the MAP gondola is provided by 3 Radioisotope Heating Units (RHUs). These require no electric power. Thus the gondola is survivable even if solar incidence should drop far below anticipated levels for an extended period. Should solar levels drop, it simply means that imaging activities will be curtailed in accord with a reduced power budget; should solar levels rise, imaging can be expanded beyond the 40 picture-pairs/sol baseline.

BALLOON

Recent advances in balloon materials and microelectronics give us the hitherto impossible prospect of building lightweight, high-altitude super-pressure balloons that would last at least two Martian years . Balloons of this nature are being currently flown on Earth and most of the technologies needed to fly these balloons on Mars (remote inflation equipment and descent modules) have already been demonstrated.

The balloons required by the MAP probes would be 9.1 m in radius and made of a new biaxial nylon 6 film, 12 microns thick, that is much stronger than the 6 micron mylar used in the French Mars 96 balloon and guaranteed by the nature of its manufacturing process to be free of pores. Superpressure balloons made of this material are currently being manufactured by Winzen Balloons of San Antonio, Texas. Because the balloon material is so much stronger, venting will be unnecessary (i.e. the fabric can react to the pressure differential caused by day-night temperature changes), and according to Winzen, the estimated mean life of the balloon (before leakage caused sufficient buoyancy loss to terminate flight) would be many years. Since the balloons in the MAP mission avoid the surface, unlike the French Mars 96 balloon which will land its snake every night, the probability of crash or snag is much lower. Mission termination for each probe is thus most likely to occur as a result of failure of the instrument package. Such packages can be designed for lifetimes of thousands of days; we assume a conservative average mean time to failure of 100 days. This is in contrast to the payload on the French Mars 96 Balloon, which is powered by a primary battery which limits mission life to 10 days, even in the absence of a balloon mishap.

To date, Winzen has constructed and successfully flown super-pressure balloons made from biaxial nylon 6 on two separate flights. More long-duration testing is needed, however. Plans for such a series of long duration tests are currently being developed by NASA's Jet Propulsion Lab.

The MAP baseline balloon design assumes an atmospheric density, of 0.008 kg/m^3. This corresponds to a seasonal minimum geometric altitude of 7 km which will clear 93% of the terrain on the planet. The baseline design also assumes a floating mass of 26.5 kg and a 12 micron thick biaxial nylon balloon material. This design was analyzed using Global Circulation Model data over all latitudes and solar longitudes (seasons). Balloon super-temperature. relative balloon skin stress, and lifetimes were all calculated. Based on this data, the balloon can be expected to operate satisfactorily anywhere on the planet. The northern hemisphere yields the most temperate conditions (showing a large operational margin) and the southern atmosphere yields less margin. Lifetimes are also maximized in the northern atmosphere, around 3000 sols.

BALLOON DEPLOYMENT AND INFLATION

Remote inflation of balloons is a challenge, but it has been successfully demonstrated at Venus and on Mars-like conditions on Earth by the French Mars 96 project. The procedure used (Figure 3), in both cases, is to use the combination of an aeroshield and parachute to decelerate the vehicles to about 35 m/s, after which the balloon is deployed out of the bottom of the descent module by the gondola and entry-shield.

The balloon is then inflated from the top, first slowly (20 g/s), to prevent erosion damage to the adjacent balloon fabric and then 20 seconds later at a faster rate (170 g/s) to quickly inflate the balloon.

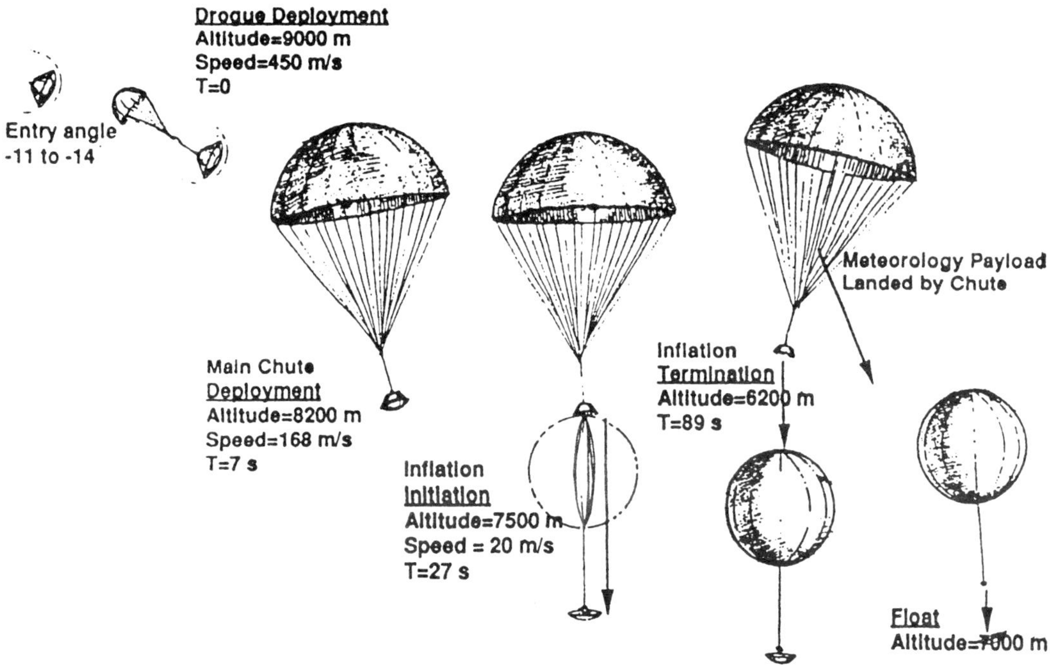

Figure 3 The entry and balloon inflation sequence that will be used on the MAP mission. Such dynamic inflation techniques have been demonstrated by the French Mars 96 team on Earth at altitudes of 150,000 feet using a larger balloon made of weaker material than will be employed on MAP.

Once the balloon is inflated, the balloon and inflation equipment is separated. Due to the relative size of the parachute (20.8 m diameter), the inflation equipment slows compared to the balloon which still has the aero-shield attached. After a waiting period of about a minute, the balloon drops the aero-shield and the balloon attains its stable float altitude. The inflation equipment, having been at a different altitude, will have experienced a different wind shear pattern and will not directly fall onto the balloon when their altitudes cross again.

BALLOON FLIGHT PROFILE

At the time of entry, the balloons may be targeted to widely dispersed locations, typically spread over 130 degrees of latitude. The balloons will then commence to float at the velocity of the Martian winds at their attitude. Current models suggest that the wind will carry the balloons primarily in the West-East direction with typical velocities on the order of 30 to 60 km per hour. At such speeds each aerostat could be expected to circumnavigate the planet every 10 to 30 days. There will also be some North-South migration which will probably cause some balloons to wander into and out of Mars' polar regions during portions of the mission. This may afford an opportunity to examine regions of the planet that cannot be targeted for direct probe entry during a given launch opportunity. If an aerostat should wander into a polar region during the Arctic night, gondola operations will be curtailed until it leaves. Characteristic aerial velocities of the balloons suggest that such a shutdown should last no more than a few days at most.

The balloons have a nominal float altitude during the Northern summer of 7 km, with a diurnal attitude variation of a few hundred meters. However because of the evaporation of CO_2 from Mars' south polar cap, the average atmosphere during the course of the year is thicker than that of the northern summer, and so the balloons on the average will actually float at a typical altitude of about 8.5 km, clearing over 97% of all Martian terrain. Having such a large fraction of the planet safe for travel gives each balloon fair odds for a long duration flight. Although an eventual crash is certainly possible, chances for balloon loss through collision with terrain obstacles is diminished by the tendency of balloons to follow the streamlines of air which must, of necessity, go around all obstacles the airflow encounters.

PROBE NAVIGATION

For science purposes, it is important that the sites photographed by the MAP probes be identifiable.

Latitude can be determined by the time of comsat overpass. In addition, the perpendicular distance of the probe from the comsat groundtrack can be determined by the Doppler shifting of the signal the comsat receives from the Probe. Thus a navigational fix for each probe can be determined 2 to 5 times each sol. In addition, longitude can be determined twice a day by noting the time of sunrise and sunset. If longitude is known, then a sun-sensor can be used to determine latitude and gondola orientation. As a third navigation technique, the aerostat location can be determined by comparing its 10 km x 10 km medium resolution images with Mars Observer and Viking surface imagery. Since the position of the probe is periodically known precisely, either by comsat tracking or identification of surface features, the position of the probe at all times can then be known precisely if overlap can be maintained between the areas captured by the medium resolution images. Since the probe can know its own location using such techniques, probes can be preprogrammed to take horizontal photographs when they find themselves in a certain region and pointing (the balloons are expected to slowly rotate) in a given direction. thus it will be possible to obtain horizontal or oblique panoramic images of many of the major sights on Mars, such as Olympus Mons or the Valles Marineris.

CONCLUSION

The MAP mission will produce a glorious science return for Mars geology and meteorology, and provide engineers with the knowledge of the surface of Mars needed for the design of rovers and safe landers. It will provide future explorers with information about where precious water may be found, and may identify locations that offer improved prospects for a search for indigenous life. By charting the planet's winds, MAP will provide thc routes for future mobile science missions - for the skies of Mars are its highways. But perhaps the greatest return from the MAP mission will be the least tangible - its impact on the intellectual life of humanity at large. Today, nearly 500 years after Copernicus, most people still think of Earth as being the only world in the universe, and the other planets, despite academic knowledge to the contrary, as mere points of light in the sky. The MAP aerostats will take humanity's eyes to another planet in a way that has never been done before. Through the gondolas' cameras, we will see Mars in its spectacular vastness - its enormous canyons, its towering mountains, its craters, its plateaus, its dried up lakes and rivers, its rocky plains and frozen fields of water and dry ice. We shall see that Mars is truly a world, and for the first time, our knowledge that the Earth is not unique shall really be made tangible. What finer return can a mission have?

ACKNOWLEDGMENT

Portions of the above report were contributed by Mike Malin, Ron Greeley, and Robert Haberle. Important analytical contributions were provided by Larry Adams, Nick Smith, John Ford, Charles Rosbach, Jeff Summers, and Khalid Sharmit. Useful suggestions or review was provided by Roger Bourke, Ralph Eberhardt, Al Schallenmuller, Gary Olhoeft, Larry Epley, Geoff Briggs, and Chris McKay. Cost estimates were provided by Craig Mogensen. Supporting artwork was provided by Robert Murray. The CAD work was done by John Hupp.

AAS 93-844

MARS EXOBIOLOGY SITES FOR FUTURE EXPLORATION

R. Landheim,[*] R. Greeley,[†] J. D. Farmer,[‡] D. J. Des Marais[‡] and H. P. Klein[**]

INTRODUCTION

The major goals of Exobiology in future Mars exploration can be summarized to include the following: to determine whether a past/present biosphere existed/exists; to define the nature of Martian environments, especially those which are ancient and/or favorable for sustaining life; to understand the biogeochemistry of the biogenic elements (C, N, O, S, P) and organic matter. In pursuing the search for sites that may have been hospitable to life, the objectives in exploration are to search for organic compounds, water and/or repositories of biologically important minerals, a fossil record, and extant life. All general exploration objectives are based on the assumption that liquid water is required for living systems to exist and that a hydrologic cycle persisted long enough for life to develop and evolve. The strategy that is used involves identification of land forms created by water-driven surficial processes and/or their sedimentary deposits. The objectives of this study were to detect geological environments where water may have been active, to assess the duration of aqueous environments, and to prioritize sites for high resolution imaging and for future surface exploration. The geological environments of primary interest to Exobiology are fluvial, lacustrine, hydrothermal, and periglacial. The data that were used in the study include the Mars Charts (MC) photomosaics (scale 1:2M), the Mars Transverse Mercator (MTM) maps (scale 1:500K), high resolution Viking Orbiter images (~40-250 m/pixel), and Mosaicked Digital Image Models (MDIM; 231 m/pixel). All sites were screened for morphologic features and deposits indicative of water related sedimentary facies (lacustrine, fluvial, thermal spring, and periglacial). Geological features were assigned scores based on visibility on image (0-3), specificity of feature/process relationship (1-3), and importance to Exobiology (1-3). Within each facies, site priorities were determined based on total scores. Sites were selected for high resolution imaging and/or sampling on future missions. A detailed description was developed for the area around each site, emphasizing geological evidence for relative duration of hydrologic processes. The value of each site was stated for meeting defined goals of Exobiology (e.g. search for extant life, a fossil record, organic matter, biologically relevant mineral deposits.)

* Departments of Botany and Geology, Arizona State University, Tempe, Arizona 85287.

† Department of Geology, Arizona State University, Tempe, Arizona 85287.

‡ NASA Ames Research Center, Moffett Field, California 94035.

** Department of Biology, Santa Clara University, Santa Clara, California 95053.

CONCLUSION

Of the 83 sites evaluated from the current site catalog, 13 were assigned a high priority. An additional 5 sites were proposed for the next edition of the site catalog. Inability to identify geological features, and to specify surface compositions presently limit and confirm the need for high resolution visible and infrared remote sensing data for high priority Exobiology sites during the Mars Observer and Mars '94/'96 missions. Viking Lander sites received zero scores by our criteria.

In summary, this study identifies a preliminary site list for Mars Exobiologic exploration and outlines a conceptual framework for the objective evaluation and prioritization of sites to meet Exobiology science objectives. Exobiology shares important goals in Mars site selection with other planetary science disciplines, illustrating the advantage of a multidisciplinary approach in developing site selection strategies for future Mars missions.

CAPTIONS FOR SITES

Figure 1. Mangala Valles. The site is located at the base of a relatively steep delta foreslope that formed where a small channel enters the crater from the south. This location may provide access to potential lacustrine and prodelta deposits.

Figure 2. Gusev. The site is located on the floor of Gusev crater (~160 km). The crater contains fluvio-lacustrine sediments derived from Ma'adim Valles, an 800 km channel that enters the crater from the south. Ancient shorelines may also be present, which would indicate the former presence of a paleolake.

Figure 3. Parana Valles. The site is in ancient cratered terrain that has been heavily dissected by numerous dendritic valley networks. Channel networks surround a central basin that may have been a depocenter for fluvio-lacustrine sedimentation. The site was also proposed by S. W. Squyres (in 1) as a potential site for sampling volcanics and channel deposits.

Figure 4. White Rock. The site is located in an ~80 km crater in ancient cratered terrain and contains a high albedo feature that has been interpreted to represent a playa deposit (2) or a salar deposit (3).

Figure 5. Dao Vallis/Hadriaca Patera. The site is located on the floor of Dao Vallis. The simple channel morphology and enlarged source area indicate that it may have formed by spring sapping. The close proximity to a potential heat source (i.e., magma chambers underlying Hadriaca Patera) suggest the possibility of hydrothermal activity. The site was also proposed by F. R. West (in 1) as a potential site for sampling volcanics and channel deposits.

Figure 6. Ismenius Lacus. The target is located amidst debris aprons of cliff faces in the fretted terrain. The debris has been described to mix with ice and form "rock-glaciers" (4). The possible presence of ground-ice may enhance the possibility of detecting a fossil record that could have been preserved in the ice and protected from the highly oxidizing Martian atmosphere (5).

Figure 1

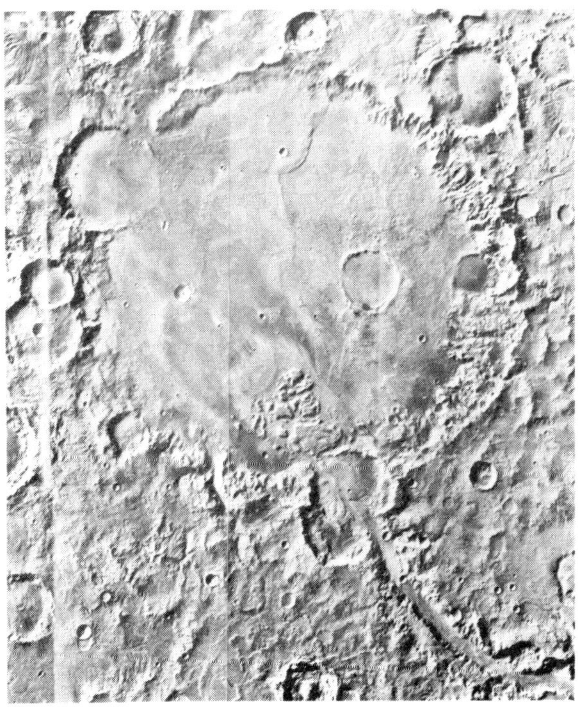

Figure 2

Figure 3

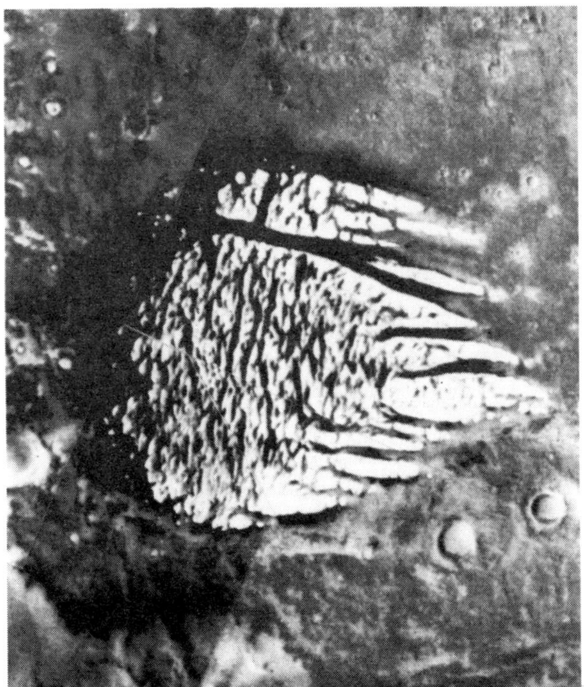

Figure 4

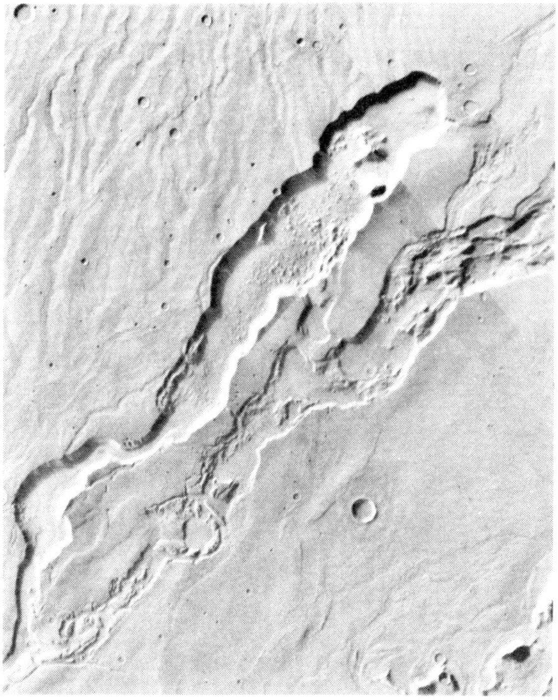

Figure 5

Figure 6

Table 1
MARS SITES FOR EXOBIOLOGY

Data Source ______________________________ Site __________

______________________________ Ref. __________

Diagnostic Criterion	**Visibility* *[0,1,2,3]*	**Value to Exo.* *[1,2,3]*	**Process Uniqueness* *[1,2,3]*	*Total*
Lacustrine deposits				
a) basin fed by simple channels (fluvial), or	____	x 3	x 2 =	____
b) basin fed by extensive channels (fluvial)	____	x 3	x 3 =	____
c) layering (possibly aeolian, volcanic, etc.)/terraces	____	x 3	x 2 =	____
d) circumstantial (inferred water eroded impact crater rims with no other potential process)	____	x 3	x 1 =	____
e) impact crater ejecta (for subsurface sampling)	____	x 3	x 1 =	____
			Subscore	____
Spring deposits (point source)				
a) point source feature	____	x 3	x 1 =	____
b) with heat source (volcanism)	____	x 3	x 1 =	____
			Subscore	____
Fluvial channel (deposits)				
a) tributaries/dendritic patterns	____	x 2	x 3 =	____
b) floodplains	____	x 2	x 3 =	____
c) terraces	____	x 2	x 2 =	____
d) deltas	____	x 2	x 2 =	____
e) widening downslope	____	x 2	x 2 =	____
f) meandering	____	x 2	x 1 =	____
			Subscore	____
Periglacial				
a) alases	____	x 2	x 3 =	____
b) laminated terrain	____	x 2	x 3 =	____
c) pingos	____	x 2	x 2 =	____
d) polygonal terrain	____	x 2	x 1 =	____
			Subscore	____
Debris flow (mass wasting)				
a) possible presence of water, but not definitive	____	x 1	x 1	____
Comments:			Score	____

* 3=highest, 1=lowest

Table 2

POTENTIAL MARS EXOBIOLOGY SITES FOR FUTURE EXPLORATION

*MLSC site no.	Location of interest area (lat., long.)	Target (lat., long.)	Name of site region	Potential deposits	Relative level of priority	Suggested reference
142	8°-9°S, 158.5°-160°W	8.5°S, 159.5°W	Mangala Valles	Waterlain sed.	High	MTM -10157
143	33°-35°N, 340°-344°W	33.5°N, 342.5°W	Ismenius Lacus SW	Waterlain sed.	High	MC-5SW
1	35°-38°S, 227°-231°W	37°S, 230°W	Eridania NW	Waterlain sed.	High	MC-29NW
2	21°-23°S, 9°-14°W	22°S, 11°W	Margaritifer Sinus SE	Waterlain sed.	High	MC-19SE
8	22°-23.5°S, 229°-231°W	22.8°S, 230.6°W	Mare Tyrrhenum SE	Waterlain sed.	High	MC-22SE
10	24.5°-26.5°S, 264°-266.5°W	24.8°S, 265.8°W	Mare Tyrrhenum SW	Waterlain sed.	High	MC-22SW
137	5.6°-6.4°S, 149.1°-149.9°W	6.3°S, 149.5°W	Mangala Valles	Waterlain sed.	High	MTM -05147
79	13.5°-15.5°S, 187°-190°W	15.5°S, 188.5°W	Aeolis SE	Waterlain sed.	High	MC-23NE, SE
138	13.5°-15.5°S, 183.5°-185.5°W	15.5°S, 184.5°W	Aeolis NE (Gusev)	Waterlain sed.	High	MC-23NE, SE
140	6.5°-9°S, 302°-305.5°W	7.3°S, 305°W	Iapygia NW	Waterlain sed.	High	MC-21NW
144	35°-40°N, 130°-143°W	39.5°N, 135.5°W	Diacria SE	Waterlain sed.	Moderate/High	MC-2SE, SC
145	11°-15°S, 173.5°-175.5°W	14.5°S, 175°W	Memnonia NW	Waterlain sed.	Moderate	MC-16NW
146	30°-40°S, 136°-149°W	37.5°S, 146°W	Phaethontis NC	Waterlain sed.	Moderate	MC-24NC
5	10°-11.5°S, 277°-280°W	11°S, 279.5°W	Iapygia NE	Waterlain sed.	Moderate	MC-21NE
7	41.5°-43.5°S, 208°-212°W	43.2°S, 208.1°W	Eridania NC	Waterlain sed.	Moderate	MC-29NC
22	5.9°-6.2°S, 73.7°-74°W	6.05°S, 73.75°W	Candor Mensa	Waterlain sed.	Moderate	MTM -05072
141	0°-2°S, 73.5°-78°W	1.5°S, 76.5°W	Hebes Chasma	Waterlain sed.	Moderate	MC-18NW
147	5°-10°S, 333°-337°W	8°S, 335°W	Sinus S. (White Rock)	Evaporite dep.?	High	MC-20NE
148	20°-25°N, 5°-10°W	21.3°N, 8°W	Oxia Palus NE	Evaporite dep.?	Moderate	MC-11NE
139	22°-23.5°N, 34.5°-36.5°W	22.1°N, 36.7°W	Oxia Palus NW	Evaporite dep.?	Moderate	MC-11NW
32	33°-35°S, 264°-268°W	33.2°S, 266.4°W	Dao Vallis/Hadriaca P.	Hydroth. dep.?	High	MC-28NE
77	2°-4.5°N, 15.2°-17°W	2°N, 16°W	Ares Vallis	Hydroth. dep.?	Moderate	MC-11SE

R. Landheim 4/15/93

REFERENCES

1. Greeley, R. (Ed.), 1990), NASA Reference Publication 1238.
2. Ward, A. W., 1979, *J. Geophys. Res.* 84, 8147-8166.
3. Forsythe, R. D., 1990, *Lunar Planet. Sci. Conf.* 21, 379-380.
4. Squyres, S. W., 1978, *Icarus* 34, 600-613.
5. Squyres, S. W., 1989, *Icarus* 79, 279-288.

AAS 93-837

GENETIC REINFORCEMENT LEARNING IN A ROBOTIC SYSTEM FOR SPACE EXPLORATION

Robert E. Smith[*] and Lia Noemi Tusanotte Cisnero[†]

The objective of this paper is to suggest that a general reinforcement learning controller achieved through the application of genetic algorithms could be an interesting control tool for robots for space exploration.

Space is a dynamic and unstructured environment with perpetually novel data stream. Space robotic activities could require the continuous execution of control strategies in real time. These control actions must be directed at multiple goals, possibly with conflicting priorities. Given these complexities, it is difficult to specify a sufficient fixed-structure controller.

Classical control methods require accurate plant models. In effect, these models are used to predict plant behavior so that appropriate control decisions can be made. However, as indicated above, application domains for robots in space exploration make *a priori* plant model specification difficult. Space robots are constrained to interact with an unstructured environment through noisy sensors and effectors. A learning method could allow adaptive specification of control strategies, without an explicit plant model specified beforehand. Such methods could allow for more effective, robust robotic applications in space systems.

Reinforcement learning techniques allow an automatic system to refine control strategies towards a long term goal based on an immediate reward or cost signal. These techniques do not require that desired *actions* be provided to the controller, only that the controller can evaluate the effectiveness of actions that it provides. A reinforcement learning controller acquires a control strategy by exploring, possibly refining an internal plant model, and adaptively updating its control strategy. The generic nature of reinforcement learning could allow a robot to learn in unknown environments, to adapt its learning to the particulars of a task, and choose actions based on the currently adapted evaluation of its perceived state.

Most of the reinforcement learning techniques found in the literature are based on a lookup table mapping of sensory inputs and possible control actions to evaluations that allow for the determination of a control strategy. In practical systems, such a table-based approach is likely to be computationally infeasible. The approach proposed in this paper is a learning method using genetic algorithms to vary the structure of a computationally realizable reinforcement learning controller.

* Assistant Professor, The University of Alabama, Department of Science and Mechanics. E-mail: rob@comec4.mh.ua.edu.

† Assistant Research Engineer, The University of Alabama, Department of Science and Mechanics. E-mail: lia@comec5.mh.ua.edu.

Genetic Algorithms (GA's) (Goldberg, 1989; Holland, 1975) are robust, globally oriented search procedures whose mechanics are based on those of natural genetic evolution. Their most distinctive feature is that they do not require any explicit information about the relationship between input parameters and their effect on performance. Instead, they implicitly evaluate these relationships through a population-based approach involving simple selection, recombination, and mutation operators. GAs are not restricted by requirements of continuity, modality, etc. that are common to other search techniques. The robustness of GAs has been demonstrated in a large number of practical applications (Goldberg, 1989).

The broadly applicable character of GAs and their ability to operate on performance-only feedback, make them uniquely suited to restructuring reinforcement learning controllers. This presentation will focus on how GAs can be integrated with reinforcement learning techniques, and how this combination can be used to supplement robotics for space exploration. The resulting systems hold promise for adaptive robots that can play a more substantial role in ongoing tasks in the perpetually novel environment of space.

INTRODUCTION

Classical control methods for dynamic systems require mathematical models to (in effect) predict their behavior before appropriate control decisions can be made. However, in many systems complexity or insufficient knowledge make such modeling possible. Robots for space exploration are constrained to interact with a dynamic and unstructured environment with a perpetually novel data stream, through noisy sensors and effectors (Astrom & McAvoy,1990). The objective of this paper is to suggest that a reinforcement learning controller achieved through the application of genetic algorithms can be an interesting framework for intelligent control.

The proposed implementation deals with any arbitrary control problem where learning would be advantageous, for example, consider an intelligent robotic manipulator for in-orbit activities. There are many goals for the in-orbit activities of intelligent robots. These include the assembly and disassembly of objects in space, space structure assembly or the removal of orbital replacement units, assistance of astronauts during EVA activity, inspection of components and maintenance tasks, etc. Common to these goals is the fundamental requirement for manipulation (Murphy & Wen, 1991). Manipulation is a fundamental task that must be solved for the effective development of an intelligent robot. In autonomous robot, manipulator control algorithms must operate without human intervention, on arbitrary objects taken through arbitrary trajectories. We have selected the autonomous system for manipulation of such arbitrary objects as a workbench for our reinforcement learning research.

REINFORCEMENT LEARNING CONTROL PROBLEM

Reinforcement learning control problems are useful as models of the elements of control problems that often require human intervention. Consider a black box environment (Figure 1) with two output channels and one input channel. The black box environment represents the entire, initially unknown world outside the controller. The first output channel provides partial information about the internal state of the environment.

The input channel allows an external control system to affect the environment's state. The second output channel provides a reward signal that indicates the utility of input signals applied to the environment. The effect of input signals on the environment's state is not known beforehand. A control system applied to this environment can only gain information through its interactions with the environment. The environment provides no direct information about input signals that the system should have applied. The goal of a control system applied to this environment is to optimize the reward signal (Barto, 1990)

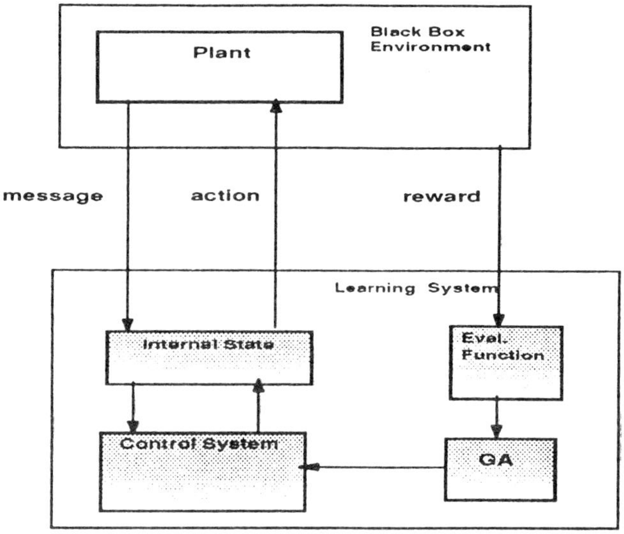

Figure 1 A General Reinforcement Learning Task.

The limitations of this black box environment embody a reinforcement learning control problem. This problem is a worst-case scenario for control: no initial knowledge, and the need to improve performance based on sparse, noisy information that can only be acquired on-line.

The reinforcement learning controller's components are an internal state, an update function that modifies the internal state, and an evaluation function for choosing actions. Such a controller attempts to choose control actions based on its current and past sensor values such that it maximizes over a reward function over time (Barto *et al.*, 1991). A Reinforcement learning algorithm can be seen as an instantiation of the general scheme:

1. Initialize the learner's internal state to S_0.
2. Do forever:
   - Observe the current world state.
   - Choose an action (function of the internal state and the world state) using the evaluation function.
   - Carry out the action in the world.
   - Let the immediate signal for executing the action in world, be reward.
   - Update the learner's internal state using the update function.

Updating the control policy in a reinforcement learning controller is clearly a type of optimization problem. Genetic algorithms are a class of optimization procedures that can be applied to this problem. In fact, GAs have a robust, global, population-oriented character for mathematically ill-conditioned search problems that make them well suited to refining reinforcement learning policies.

GENETIC ALGORITHMS

GA's are computerized, globally-oriented search procedures whose mechanics are based on those of natural genetic evolution. They search by starting with a sample, or *population*, of structures that represent encodings (usually binary) of points in the search space. GA's most distinctive feature is that they do not require any explicit information about the relationship between parameters of points in the search space and the effects of those parameters on performance. Instead, they implicitly evaluate these relationships through a population-based approach involving simple selection, recombination, and mutation operators.

In its simplest form, a GA proceeds as follows:

Evaluation: A positive measure of utility called fitness, is determined for all members of the GA population.

Selection: Populations members are assigned a number of copies in a mating pool that is used to construct a new population. Structures in the current population are selected for high fitness. The higher a population member's fitness, the more copies it receives. These copies are processed by other operators to form the new population. Selection is the emphasis phase of the GA.

Recombination: Copies of high-fitness structures from selection are recombined to form new structures. A common recombination method is single-point crossover. Recombination is the exploration phase of the GA. Selection and recombination account for most of the GA's search power.

Mutation: Point mutation as an insurance policy to keep the population diverse during the search.

Analysis shows that in explicitly applying these 3 simple operators, the GA implicitly emphasizes and recombines certain design features that correlate to high utility. The central assumption of GA's is that recombination of features that correlate to high utility will lead to overall solutions with high utility. This is a broadly applicable assumption, that has been demonstrated to be correct in a wide range of search problems (Davis, 1991; Goldberg, 1989).

Because GA's are based on assumptions that are applicable to a broad class of unknown search spaces, they are very robust:

- GA's are not limited by restrictions of linearity, quadratic character, continuity, modality, etc.
- Because of their population bases, GA's are global (they resist local optima).

- GA's are efficient methods for effective search in a broad class of problems.

GA AND CONNECTIONIST ARCHITECTURES

The simplest form of GA described above has proven itself as a powerful technique in a broad class of problems. However, natural extensions of the GA yield interesting new abilities. One of this extension is to use the GA to design connectionist architectures. Most of the reinforcement learning techniques found in the literature are based on a lookup table mapping of sensory inputs and possible control actions to evaluations that allow for the determination of a control strategy. In practical systems, such a table-based approach is likely to be computationally infeasible. The approach proposed here is a learning method using genetic algorithms to vary the structure of a computationally realizable, network-based, reinforcement learning controller.

In this approach GA uses a population of hidden nodes in a single network. Hidden nodes that demonstrate high utility are recombined to form other units with similar connectivity. There is an important distinction between this use of GAs and GA optimization applications. In optimization, the GA must find a single individual that specifies a solution. When finding a set of nodes in an interconnected network, the GA must simultaneously form a set of individuals that jointly perform useful computation. These nodes from a computational ecology (Smith *et al.*, 1992). Typical studies where GA's are used to form connectionist networks use more optimization-based approaches (Schaffer *et al.*, 1992).

EXPERIMENTAL FRAMEWORK: A ROBOTIC HAND

We have chosen a multifingered robotic hand (Murray & Sastry, 1990) as a system for experimentation with evolutionary connectionist architectures. The proposed system has a gripper with three links fingers. These fingers are considered to be three separated robots joined to a base (the palm) (Mason & Salisbury, 1985). The fingers must cooperatively manipulate an object in a working space. The proposed hand has visual and tactile sensors and is capable of recognizing and securely grasping a variety of rigid objects. The goal is to design the system such that it is stable, manipulable, and prehensile.

We propose to provide the system with the following sensors:

- A tactile array of 4 per 8 sensing elements in each finger.
- A pair of stereo cameras.
- Sensors indicating the angular positions of gripper drive motors.

The controller must learn the behaviors that the articulated hand performs according to their kinematic structure, and the constraints imposed by contacts between bodies including the effect of friction.

Our experimental task is the coordination of motor activity via sensor information. The fundamental structure for our cognitive architecture is an adaptive, noise tolerant dynamical neural network, that must be trained through reinforcement learning techniques. This network must gather input data from the sensors and determine how to set the hand for the next time step.

Reinforcement learning takes place through the success or failure of incrementally refined solution attempts which are based on a solution hypothesis adapted by the learning system. As shown in Figure 2, the aim of teaching here is to optimize the solution of the hypothesis with respect to a given evaluation (or reinforcement) function.

The learning cycle can be divided into three phases:

- Generation of a solution attempt.
- Evaluation of the solution attempt.
- Alteration of the solution hypothesis.

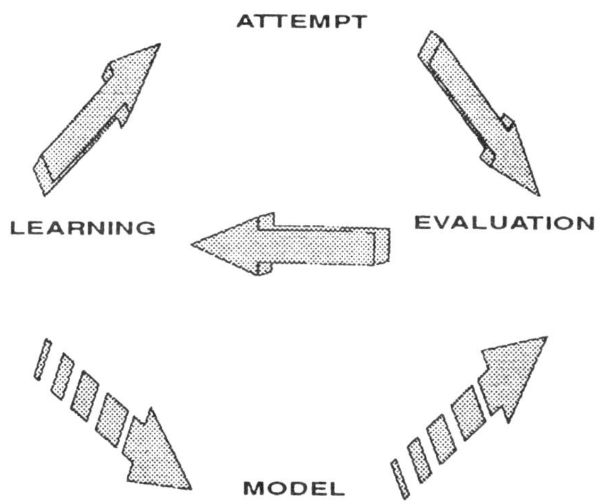

Figure 2 Teaching a Behavior for Recognizing Different Objects.

INCREMENTAL ARTIFICIAL EVOLUTION USING GENETIC ALGORITHMS

The artificial evolution approach (Figure 3) will maintain a population of network nodes in our network-based cognitive architectures. Selective pressure in the GA will be controlled by a task oriented evaluation function.

Rather than attempting to hand design a system to perform a particular task or range of tasks well, the evolutionary approach will allow their gradual emergence. There is no need for any assumption about means to achieve a particular kind of behavior as long as this behavior is directly or implicitly reinforced by the evaluation function (Cliff *et al.*, 1992). Furthermore, it is not necessary to be dogmatically committed to an exclusively behavioral decomposition. In the suggested approach, the GA will evolve a network that implicitly decomposes the task according to reinforcement signals.

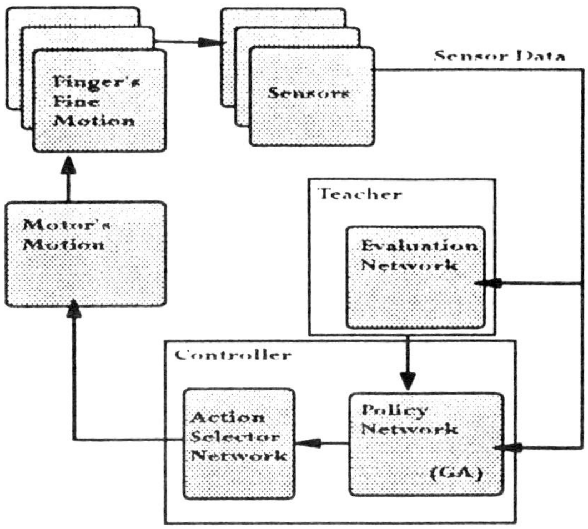

Figure 3 Artificial Evolution Approach.

SUMMARY

We have proposed that effective learning systems for robotics in space exploration may be realized through connectionist networks, that are trained by reinforcement learning, and structured by GA's.

Reinforcement learning has the potential to allow for online refinement of robotic control in situations that could not be foreseen by the robot's designers. These include interaction with arbitrary objects and environmental conditions, as well as coping with unforeseen system damage.

GA's are a robust and effective search procedure. The GA's offer substantial computational leverage, efficiency, and robust search in noisy, mathematically ill-conditioned search problems like those found in structuring network-based reinforcement learning controllers.

The authors believe that incremental artificial evolution of a connectionist control network, coupled with reinforcement learning techniques, may be a viable approach to practical robotic learning in space applications. We have selected a robotic hand that must manipulate arbitrary objects as a workbench for experimentation. We expect that refining techniques in this experimental environment will have implications for more general control situations where learning would be advantageous.

ACKNOWLEDGEMENTS

The authors gratefully acknowledge support by the National Space Foundation under contract number ESS-9212066.

REFERENCES

Astrom, K., & McAvoy, T. (1990). Intelligent control, an overview and evaluation. In D. A. White & D. Sofge (Eds.) *Handbook of intelligent control* (3-24), New York: Van Nostrand Reinhold.

Barto, A. G. (1990). *Some learning tasks from a control perspective* (COINS Technical Report No. 90-122). Amherst, MA: University of Massachusetts.

Barto, A. G., Bradtke, S. J., & Singh, S. P. (1991). *Real-time learning and control using asynchronous dynamic programming* (COINS Technical Report 90-57). Amherst, MA; University of Massachusetts.

Cliff, D., Harvey, I., & Husbands, P. (1992). *Incremental evolution of neural network architectures for adaptive behaviour* (Cognitive Science Research Paper). England, UK: University of Sussex.

Davis, L. (1991). *Handbook of genetic algorithms*, New York: Van Nostrand Reinhold.

Goldberg, D. E. (1989). *Genetic algorithms in search, optimization, and machine learning*. Reading, MA: Addison-Wesley.

Holland, J. H. (1975). *Adaptation in natural and artificial systems*. Ann Arbor: University of Michigan Press.

Mason, M., & Salisbury, J. (1985). *Robot hands and the mechanics of manipulation*. Cambridge, Massachusetts, The MIT Press.

Murphy, S. & Wen, J. (1991). A perspective on the development of intelligent robots for manipulation in space. In *Intelligent robotic systems for space exploration* - third annual conference in Troy, New York (128-130). Los Alamitos, CA: IEEE Computer Society.

Murray, R., & Sastry, S. (1990). Grasping and manipulation using multifingered robot hands, *Proceedings of Symposia in Applied Mathematics*, 41, 91-127.

Schaffer, J. D., Whitley, D., & Eshelman, L. J. (1992). Combinations of genetic algorithms and neural networks: A survey of the state of the art. In L. D. Whitley & J. D. Schaffer (Eds.), *International Workshop on Combinations of Genetic Algorithms and Neural Networks*, 1-37, Los Alamitos, CA. IEEE Computer Society Press.

Smith, R. E., Forrest, S., & Perelson, A. S. (1992). *Searching for diverse, cooperative populations with genetic algorithms* (TCGA Report No. 92002). Tuscaloosa: Univ. of Alabama.

A "JUMPER" FOR MARS SURFACE EXPLORATION

G. Savu*

An autonomous vehicle having the possibility to make jumps over the natural obstacles of the rugged Martian soil is presented. The low solar radiation existing on the Martian surface is converted by photovoltaics into electrical power which drives a compressor by an electric motor. The rarefied Martian atmosphere is compressed in a storage reservoir which ensures sufficient volume and pressure for the isentropic expansion through a Laval nozzle. The thrust obtained by this expansion can move the robot by spatial jumps to the desired places, combining the advantages of the rovers with those of the flying machines.

INTRODUCTION

Due to the rugged soil of Mars, a vehicle with vertical take-off/landing capabilities is proper for Mars surface exploration. Because of the severe Martian environment, aerodynamic devices like pilotless airplanes or rotorcrafts must have low mass and large aerodynamic surfaces in order to fly in the Martian atmosphere [Refs. 1,2].

A jet type propulsion seems to be more efficient, especially if the "in situ" disposable energy is used: solar radiation and atmosphere. In this paper a solution using only the mechanical energy obtained by the expansion of CO_2 through a Laval nozzle from a storage reservoir is proposed, in contrast with other solutions such as using chemical reactions to get the energy requirements for propulsion [Ref. 3]. Some characteristics of the Martian environment, normalized to terrestrial values, are listed below:

-day duration:	Md=1
-atmospheric composition:	95% CO_2, 2.7% N_2, 1.6% A
-gravity acceleration:	g=0.39535
-pressure:	p=0.007
-density:	ρ=0.01469
-solar constant:	k_s=0.4316
-temperature:	T=0.868
-gas constant (state eq.):	R_g=0.658
-molecular mass:	M_m=1.56051
-adiabatic exponent:	κ=0.92028
-sound speed:	a=0.665
-Mach number:	M=1.523
-dynamic viscosity:	μ=0.72124
-kinematic viscosity:	ν=49.1
-Reynolds number:	Re=0.0204
-laminar skin friction coefficient:	C_{fl}=7.0
-turbulent skin friction coefficient:	C_{ft}=2.18

* Senior Scientist, Institute of Fluid Mechanics and Flight Dynamics, Bd. Pacii 220, 77538 Bucharest, Romania.

The components of the Martian atmosphere are pumped into a tank by a motor-compressor unit by electric power, collected from the solar arrays (Figure 1). The nozzle flow produces sufficient lifting force and command thrust for the vehicle movement. The reservoir charging-nozzle expansion cycle continues as much as the local solar radiation ensures the energy requirements for the pump.

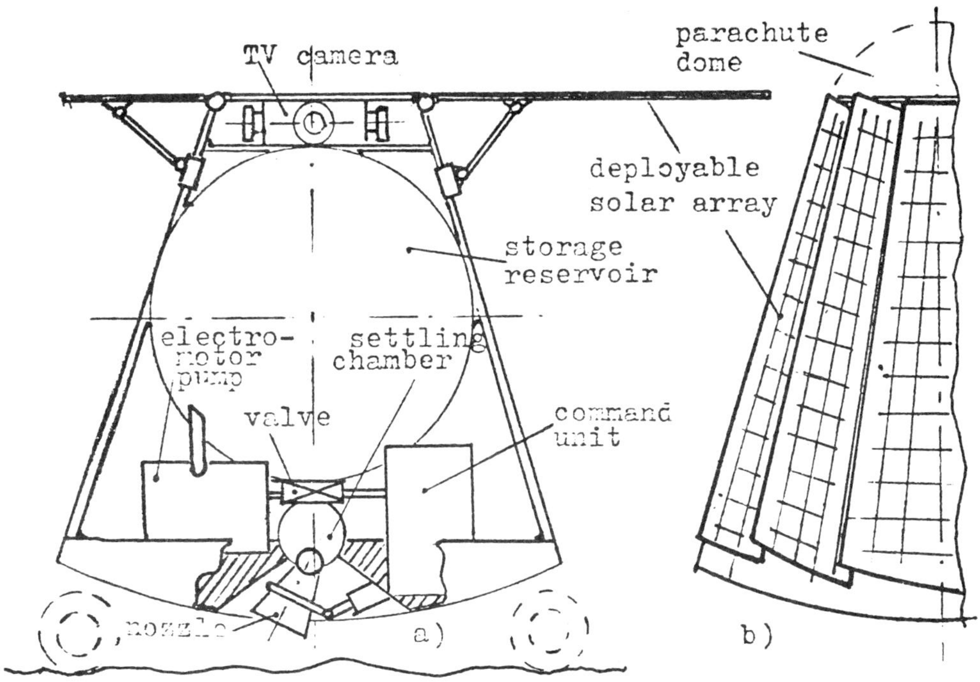

Figure 1 Solar Powered Martian Jumper:
a) Deployed Configuration; b) High Altitude Entry Configuration.

The thrust obtained due to the expansion through the nozzle can be easily controlled if a settling chamber with constant total pressure, similar to the blown down wind tunnels, is mounted between the main tank and the nozzle. The constant total pressure p_0 is obtained using a pre-programmed or a loop-controlled regulatory valve.

The lateral force for longitudinal displacements can be ensured by a directional main nozzle or by separate auxiliary nozzles. The duration of jump is equal to the discharging time of the reservoir until the value of the pressure permits a slow landing.

THE JUMP DYNAMICS

In order to avoid structure damages, the jumper must take-off vertically, overpass the natural obstacles and land also vertically with minimum touch-down velocity (Figure 2). In the present paper only planar trajectories (zOx plane) are analyzed.

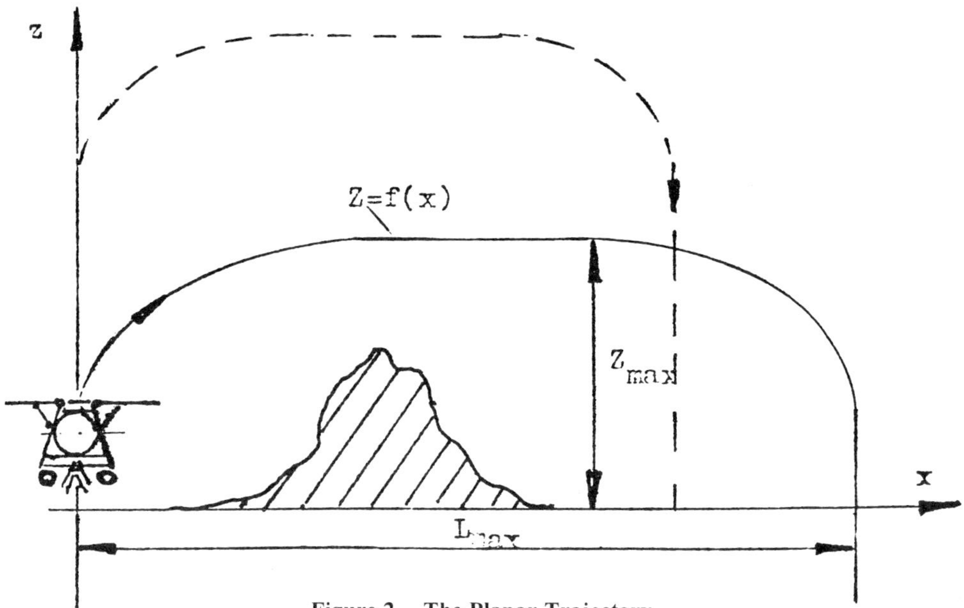

Figure 2 The Planar Trajectory.

There are several ways to study the dynamics of a vehicle with jet propulsion, as that proposed before:

1. Imposing the trajectory and continuously modifying the vector of the propulsion force T (Figure 3a).
2. Imposing the constant jet force and integrating the equations of motion in order to obtain the trajectory (Figure 3b).
3. Imposing the trajectory and using a properly chosen pulsed jet force with different time steps (Figure 3c).

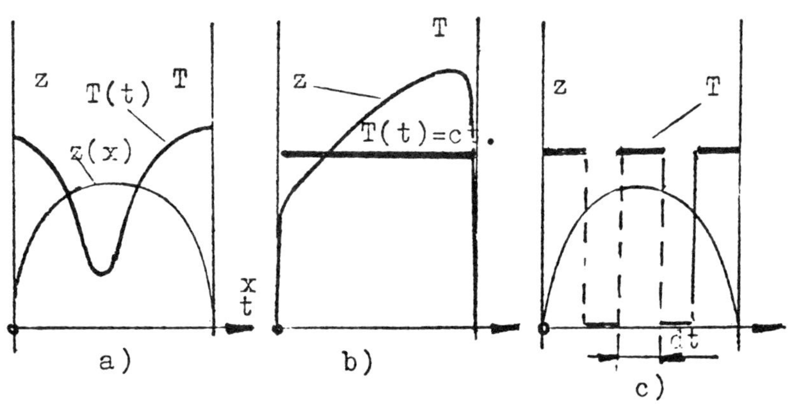

Figure 3 Different Jump Profiles.

The first alternative (case 1) implies some technological difficulties but could ensure much more autonomy and control possibilities for the vehicle. The variation of the jet force magnitude is obtained by the pressure control valve (different p_0) and the vector angle is controlled by the command unit.

The equations of motion:

$$\begin{cases} m\ddot{x} = F_x \\ m\ddot{z} = F_z \end{cases} \tag{1}$$

are solved by imposing the profile of the trajectory:

$$z = z\,(x) \tag{2}$$

and the variation of the horizontal acceleration:

$$\ddot{x} = \ddot{x}(t) \tag{3}$$

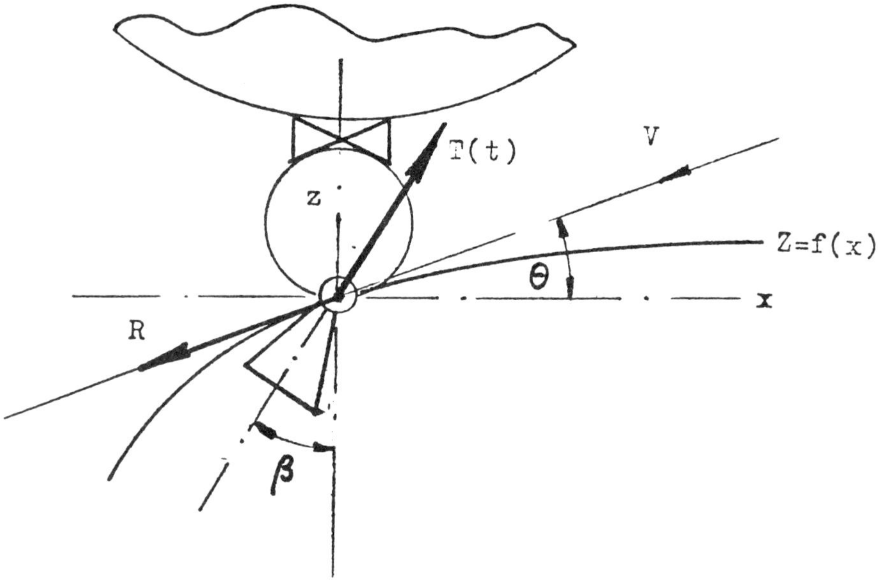

Figure 4 The Dynamics of the Trajectory.

and from Figure 4:

$$\begin{cases} F_x = T\sin\beta - \dfrac{\rho}{2} S V^2 C_D \cos\theta \\ F_z = T\cos\beta - G - \dfrac{\rho}{2} S V^2 C_D \sin\theta \end{cases} \tag{4}$$

where we denoted:

$m = G/g$ - the variable mass of the vehicle

$\ddot{x}, \ddot{z}$ - the horizontal and vertical second derivatives respect to time (acceleration)

β - the command angle of the nozzle

S - the reference surface

$V = \dot{x}^2 + \dot{y}^2$ - total velocity

$\theta = a\tan\left(\frac{\dot{z}}{\dot{x}}\right)$ - the trajectory angle

T - the nozzle thrust

The nozzle thrust can be expressed as follows:

$$T = \frac{\dot{G}V_e}{g} \tag{5}$$

where:

$\dot{G}$ - gas weight rate expanded through the nozzle in unit time

V_e - the velocity of gaseous jet through the exit area of the nozzle

G - the acceleration gravity

The total weight G_0 of the jumper can be expressed as follows:

$$G_0 = G_{inert} + G_{fuel} \tag{6}$$

$$G_{inert} = G_{structure} + G_{payload} \tag{7}$$

$$G_{fuel} = G_{0fuel} - \int_0^t \dot{G}dt \tag{8}$$

G_{0fuel} = the weight of the gas compressed in the storage reservoir.

It is more convenient to normalize the parameters and to calculate in dimensionless form:

time - $\bar{t} = t / \tau$

lengths - $\bar{x} = x / (g\tau^2)$

velocities - $\dot{\bar{x}} = x / (g\tau)$

accelerations - $\ddot{\bar{x}} = \ddot{x} / g$

forces - $\overline{F} = F / G_0$

masses - $\overline{m} = m / m_0$ $\quad$ *where* $m_0 = G_0 / g$

where we denoted τ the discharging time of the tank.

The Eqs (4) becomes:

$$\begin{cases} \overline{m}\ddot{\bar{x}} = \overline{T}\sin\beta - \overline{R}\cos\theta \\ \overline{m}\ddot{\bar{z}} = \overline{T}\cos\beta - 1 - \overline{R}\sin\theta \end{cases} \tag{9}$$

where:

$$\overline{R} = \frac{\rho}{2}\frac{S\left(\dot{\overline{x}}^2+\dot{\overline{z}}^2\right)U_{\max}^2 C_D}{G_0} \tag{10}$$

$$\overline{m} = \left(1-\int_0^1 \dot{\overline{G}}d\overline{t}\right) \tag{11}$$

If the trajectory of the jump and the horizontal velocity are imposed $\overline{z}=\overline{z}(\overline{x})$, $\overline{x}=\overline{x}(\overline{t})$ the values of the thrust and the angle of the nozzle can be obtained:

$$\overline{T} = \left[\left(\overline{m}\ddot{\overline{z}}+1+\overline{R}\sin\theta\right)^2+\left(\overline{m}\ddot{\overline{x}}+1+\overline{R}\cos\theta\right)^2\right]^{\frac{1}{2}} \tag{12}$$

$$\beta = \arcsin\frac{\overline{m}\ddot{\overline{x}}+\overline{R}\cos\theta}{\overline{T}} \tag{13}$$

with:

$$\dot{\overline{z}} = \frac{d\overline{z}}{d\overline{x}}\dot{\overline{x}} \tag{14}$$

$$\ddot{\overline{x}} = \frac{d\dot{\overline{x}}}{d\overline{t}} \tag{15}$$

$$\ddot{\overline{z}} = \frac{d^2\overline{z}}{d\overline{x}^2}\dot{\overline{x}}^2+\frac{d\dot{\overline{z}}}{d\overline{x}}\ddot{\overline{x}} \tag{16}$$

The variable thrust vector given by Eq. (12) must be ensured by the regulatory valve mounted between the tank and the settling chamber.

THE THRUST OBTAINED BY THE EXPANSION OF THE COMPRESSED GAS THROUGH THE NOZZLE

During the supersonic expansion through the nozzle the mass flow rate is maximum and constant until at the nozzle throat the sonic conditions are satisfied:

$$\dot{m} = A^*\left(\frac{2}{\kappa+1}\right)^{\frac{\kappa+1}{2(\kappa-1)}} P_0\sqrt{\frac{\kappa}{R_g T_0}} \tag{17}$$

A^* being the critical area of the nozzle.
The subscript "o" denotes the stagnation conditions in the settling chamber.
The exit velocity V_e is also a function of the stagnation conditions and the area ratio:

$$\left(\frac{A}{A^*}\right)^2 = \left(\frac{2}{\kappa+1}\right)^{\frac{\kappa+1}{\kappa-1}}\left[\frac{V_e}{a_0}\left(1-\frac{\kappa-1}{2}\left(\frac{V_e}{a_0}\right)^2\right)^{\frac{1}{\kappa+1}}\right]^{-2} \tag{18}$$

We denoted:

A - the exit area of the nozzle

a_0 - the speed of sound in the settling chamber

The variable thrust $T = \dot{m}V_e$ can be performed by the variation of the values of p_0 in the settling chamber or by the variation of the critical area A^* (Figure 5).

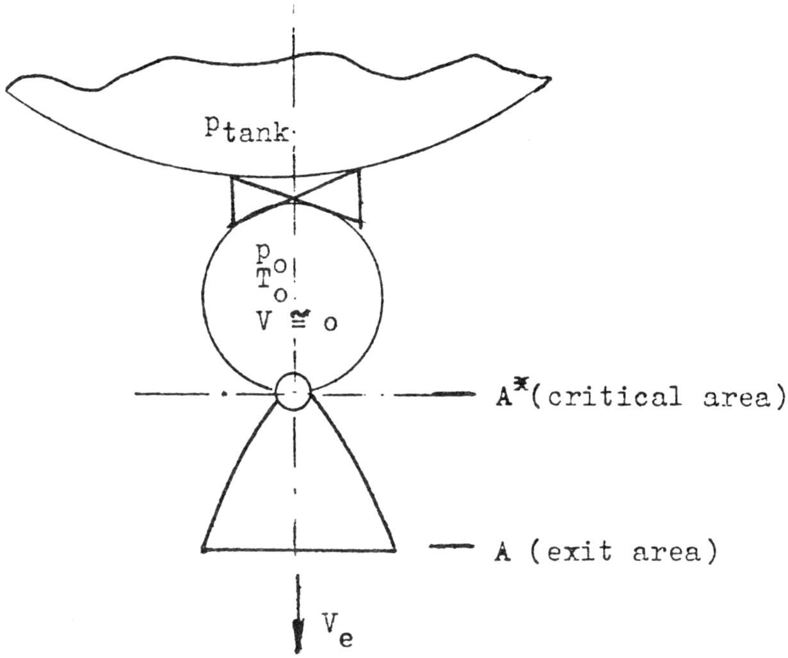

Figure 5 Variable Thrust Performing.

In case of a constant rate of mass flow the discharging time t of the reservoir is [Ref. 4]:

$$\tau = \frac{\rho_i^V}{\dot{m}}\left(1 - \frac{\rho_f}{\rho_i}\right) \tag{19}$$

$$\rho_i = \frac{p_i}{R_g T_i} \tag{20}$$

$$p_f \cong p_0$$

$$\tau = \frac{p_i^V}{RT_i\dot{m}}\left[1 - \left(\frac{p_f}{p_i}\right)^{\frac{1}{n}}\right] \tag{21}$$

subscripts "i" and "*f*" denoting the initial and final values and "n" is the polytropic exponent ($1<n<\kappa$) and V is the volume of the storage tank.

If the thrust must be variable, the discharging duration τ is numerically calculated from the relation:

$$\int_0^{\tau} \dot{m}\, dt = \sum_{j=1}^{N} \dot{m}_j \Delta t = \frac{\tau}{N} \sum_{j=1}^{N} \dot{m}_j \tag{22}$$

$$\tau = \frac{N_{p_i} V}{R_g T_i \sum_{j=1}^{N} \frac{T_j}{V_e}} \left[1 - \left(\frac{p_f}{p_i} \right)^{\frac{1}{n}} \right] \tag{23}$$

N being the number of integration time steps.

SUPPLYING THE RESERVOIR

If the reservoir is filled by a compressor driven by an electric motor, the necessary time to reach the final pressure p_f, starting from an initial pressure p_i is given by (in SI system of units):

$$t_s = 0.0047828 \frac{V}{Q} \left(p_f - p_i \right) \tag{24}$$

Q being the capacity of the compressor [Nm^3/min] for normal conditions.

The theoretical work "L" for the compression in a multistage compressor (z stages) is:

$$L = -z \frac{n}{n-1} QRT \left[\left(\frac{p_f}{p_i} \right)^{\frac{n-1}{zn}} - 1 \right] \quad [daN \cdot m] \tag{25}$$

and the actual effective power of the compressor drive is:

$$P = \frac{L}{60 \eta_e} \qquad [daN \cdot m / \sec] \tag{26}$$

with:

$$\eta_e = \eta_i \, \eta_m \tag{27}$$

η_i - the indicated efficiency

η_m - the mechanical efficiency

The required power to drive the compressor must be ensured by photovoltaics:

$$P = \eta_c S_c k_s \tag{28}$$

where:

η_c - the photovoltaics efficiency

S_c - the solar array surface

k_s - the solar constant

Combining Eq. (26) with Eq. (27) the necessary surface of the solar array for supplying the reservoir is obtained.

A NUMERICAL EXAMPLE

A real mission of a Martian jumper was simulated, taking into account the numerical values for the mission parameters.

• the trajectory profile:

$$\bar{z} = \bar{z}_{max} \sqrt{\bar{x}\left(\bar{x}_{max} - \bar{x}\right)}; \qquad \bar{z}_{max} = 03 \tag{29}$$

• the horizontal acceleration:

$$\ddot{\bar{x}} = \bar{t}(1 - \bar{t}) \tag{30}$$

• the payload weight (scientific equipment, TV camera) was estimated as 10 kg on Earth.

• the structure, electric motor, compressor, valve, and the command module were estimated to have the payload weight (10 kg).

• the reservoir weight was deduced from the Eq. (31):

$$\frac{G_{res}}{G_{gas}} = 3\rho_m R_g T_f \sqrt{\frac{1}{0366 p_f E}} \tag{31}$$

where:

ρ_m - the material density (4530 kg/m^3 for Titanium)

E - the elasticity modulus (10520 kg/mm^2 for Titanium)

R_g = 19.25 m/°K

T_f = 250°K

• the thickness of the reservoir wall (δ) is given by:

$$\delta = r\sqrt{\frac{p_f}{0366E}} \tag{32}$$

r being the radius of the spherical tank (0.5 m).

For p_i = 2 MPa it results:

δ = 0.0036 m

G_{res}/G_{gas} = 2.35

G_{gas} = 21.73 kg (Earth)

G_{res} = 51 kg (Earth

• total weight of the jumper is G_o = 92.73 kg (Earth) or 36.67 kg (Mars).

• adiabatic exponent κ = 1.293

• polytropic exponent n = 1.20

• nozzle exit Mach number M = 3.0 (V_e = 678 m/sec)

• mean mass rate through the nozzle $\dot{G} = 0.211$ kg/sec

• pressure ratio (for M=3.0) $p_o/p=40.9$ ($p_o=0.0286$ MPa=p_f; $p_{tank}=2$MPa)

• tank volume $V=0.523$ m^3

• discharging time of the reservoir until $p_f=p_o$ (Eq.(23)) τ = 100.6 secretary.

• power required by compressor for reservoir supplying (Eq.(26)) P=0.1128 kW ($\eta_e=0.5$; z=4 stages)

• time for reservoir supplying (Eq.(24)) $t_s=5.1$ min. (Q=0.01 Nm^3/min)

• the necessary surface for solar array $S_c=2.61$ m^2 (for a solar cell efficiency of $\eta_c=0.10$)

• diameter of circular solar array d=1.82 m

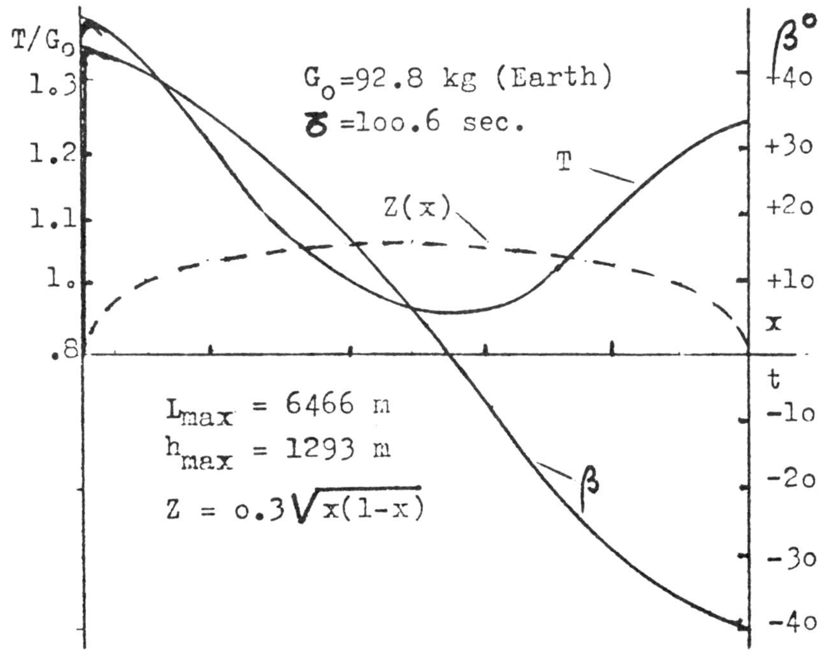

Figure 6 The Thrust (T) and the Nozzle Angle (β) Variation for a Given Trajectory z=z(x)

A cycle supplying-discharging of the reservoir has 6.67 minutes, that means, for a given trajectory (Eq. 29) a covered distance of 58.1 km from the initial point.

A MISSION PROFILE FOR THE MARTIAN JUMPER

Landing on the Martian surface and the first jumps can be performed in the following stages (Figure 7):

1. Detachment from Earth-Mars carrier spacecraft on a low Martian orbit.
2. Entry as a blunt body in the upper Martian atmosphere.

3. Thermal shields detachment.
4. Aerodynamic deceleration by parachute.
5. Parachute detachment.
6. Slow landing due to jet force (reservoir was charged on Earth).
7. Deploying of solar arrays and supplying of reservoir.
8. First jump with the Martian compressed gas.

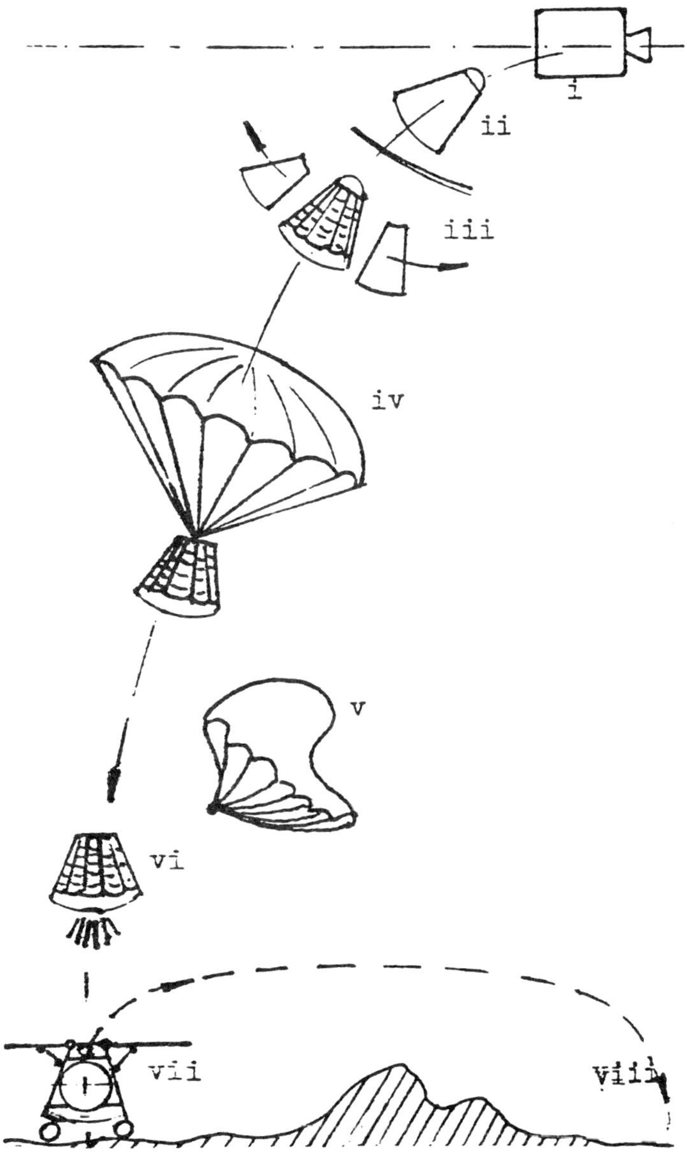

Figure 7 A Possible Mission Profile for a Martian Jumper.

During the Martian nights or during storms, the solar arrays could be folded as in the entry configuration (Figure 1). The power required by the wheels for the displacement as a rover is not analyzed here.

CONCLUSIONS

The numerical example has demonstrated that a Martian autonomous robot, using the disposable local sources of energy, is feasible with actual technology.

The geometrical dimensions and the weight are reasonable (~1 m^3 volume and ~100 kg total weight on the Earth for 10 kg payload; a multicriterial optimization could produce a substantial diminution of these parameters [Ref. 1]). Also the trajectory can be optimized, in order to cover a maximum distance during the jump.

Similitude criteria can be deduced for testing the Martian jumper in the terrestrial environment.

Due to its very long life cycle (the covering of the solar cells with dust during the Martian storms could be a possible limiting factor), such a jumper could go around Mars.

The main advantage of this robot is the possibility of vertical take-off landing at the obstacles overpassing and its energetic autonomy.

REFERENCES

1. Savu G., Trifu O., Oprisiu C., "Solar Pilotless Plane Optimized to Fly in the Atmosphere of Mars," IAF paper, IAF-93-U.4.568, 44th Congress of IAF, Graz, Austria, 1993.
2. Savu G., Oprisiu C., Trifu O., "An Autonomous Flying Robot for Mars Exploration," IAF paper, IAF-93-R.3.445, 44th Congress of IAF, Graz, Austria, 1993.
3. Bowles J. W., Tauber M. E., Anagnost A. J., Whittaker T., "Propellent Production from the Martian Atmosphere," *Journal of Propulsion and Power*, Vol. 8, No. 4, July-August 1992, pp. 907-909.
4. Pope A., Goin K. L., *High-Speed Wind Tunnel Testing*, John Wiley & Sons Inc., New York, London, Sydney, 1965.

AAS 93-891

OVER-THE-HORIZON COMMUNICATIONS ON MARS VIA HF RADIO PROPAGATION

C. D. Fry* and R. Yowell†

INTRODUCTION

Various Mars mission scenarios envision establishing widely distributed networks of fixed observing stations, free rovers and balloons on Mars. One such mission, the Mars Environmental Survey (MESUR) will begin to put a chain of small stations on Mars at the end of this decade to collect meteorological, seismological and other surface and atmospheric data for transmission back to Earth (Hubbard, 1992; McNamee, 1992). This project will consist of 16 stations and several micro-rovers globally distributed on the Martian surface. Another concept developed by the European Space Agency is a mission called MARSNET to put three surface stations on Mars during this same time frame, as part of an international global Mars network (Scoon, 1992).

Both the MESUR and MARSNET missions will likely rely upon relay satellites to bring the data collected at multiple sites back to Earth for processing. Direct transmission of the data from each site to Earth is also a possibility. Relay satellites contribute a sizable portion of the overall cost of planetary missions, and direct transmission of data from a large number of Mars sites may place unacceptable scheduling demands upon the Deep Space Network. On the other hand, if the collected data and telemetry from each station and rover could be forwarded to one or two central nodes on the Martian surface for retransmission back to Earth, then a much smaller number of stations would need to be equipped with costly high gain transmission systems. This might reduce the overall expense imposed by the network communications requirements. We propose the use of the ionospheric refractive propagation mode for over-the-horizon network communications on Mars. We used the MESUR scenario, and observed Martian ionospheric conditions to investigate the feasibility of connecting a chain of remote landers via shortwave radio for the relay of collected data.

BACKGROUND

Early in this century, communicators demonstrated that the earth's ionosphere could be used to transmit shortwave radio signals far beyond the visible horizon. See, for example, Rishbeth and Garriott (1969), Davies (1990), and McNamara (1991). The path of an upward directed radio wave, typically in the medium frequency to high frequency (MF to HF) range of the radio spectrum, will be refracted back toward the earth

* P. O. Box 5128, Herndon, Virginia 22070.

† 14614 Laurel Shadows Court, Houston, Texas 77062.

as it travels through the ionospheric medium. The wave may eventually travel downward to a distant point on the ground, several hundreds or thousands of kilometers from the transmitter. If the radio wave is modulated by a continuous wave (CW), voice or digital signal, this mechanism provides a fairly reliable means of communicating. In addition, the radio wave may be reflected off the ground at the distant point and back into the ionosphere to repeat the refractive propagation process again. In this way around-the-world communication occurs via "multi-hop" propagation.

Even with the advent of satellites, shortwave radio continues to play a major role in long distance communications on earth because of its low cost and fairly consistent performance. Although high power levels and directional antennas are commonly used in commercial HF applications, recreational users find that low power and simple low gain antennas normally allow communications during at least part of each day, with the right choice of operating frequency. The effective radiated power at the transmitter needed for good communications often ranges down to the one watt power level (Martin and Newkirk, 1993).

The same successful HF radio propagation phenomena we on earth depend upon might also occur in any planetary ionosphere, as long as the electron densities are sufficient to refract the radio wave back to the surface, and absorption of the radio wave does not cause excessive signal attenuation at the operating frequencies. Early planetary missions discovered Mars also possesses an ionosphere, with electron densities similar to those of the earth's ionosphere. See the review by Kliore (1992). Therefore, over-the-horizon HF radio propagation should be possible on Mars (Yowell, 1991).

MARTIAN IONOSPHERIC PROPAGATION

Mariner, Soviet Mars and Viking probes measured peak electron densities of about 1-2 x 10^5 cm^{-3} in the daytime Martian ionosphere at a height of about 135 km (Hanson *et al.*, 1977; Kosolov, 1979). Zhang *et al.*, (1990a,b) re-analyzed all radio occultation measurements from Mariner 4, 6, 7, and 9, and the Viking orbiters. Their analysis showed the bottomside ionosphere was well established and fairly consistent from orbit to orbit. This condition is a prerequisite for reliable HF communications. The daytime Martian ionosphere is similar to a Chapman layer model on the sunlit hemisphere out to large zenith angles. The bulk of the Mariner and Viking electron density profiles were obtained during the declining phase of the solar cycle (Mariner 9) or at solar minimum (Viking). This implies that electron densities at solar maximum, when MESUR flies, will be somewhat higher.

The Martian ionospheric densities are comparable to terrestrial conditions at mid latitudes during summertime at solar minimum, although the terrestrial height of peak electron densities range from 350-400 km, on the average (for example, Davies, 1990). Figure l, from Hanson, *et al.*, (1977) shows the Martian daytime ion density profiles obtained during the Viking 1 lander descent. The Martian ionosphere does not entirely disappear during the nighttime. Peak electron densities often remain above 10^3 cm^{-3}, due to convection of ionospheric plasma driven by the solar wind, or energetic particle precipitation (Verigin *et al.*, 1991, Zhang *et al.*, 1990b).

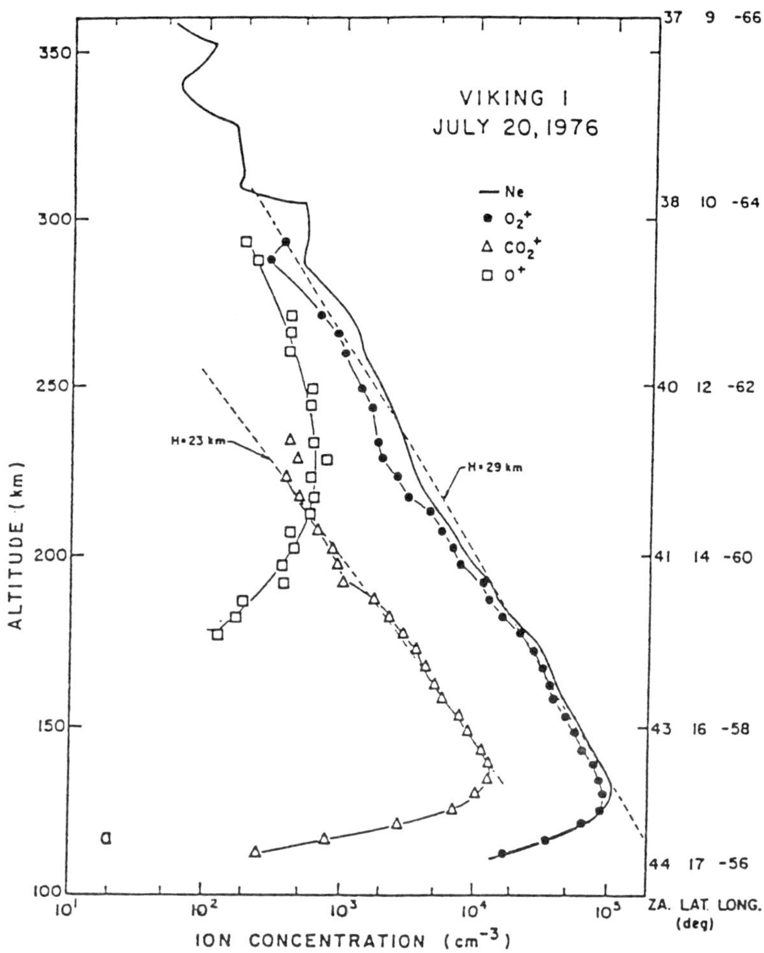

Figure 1 Observations of ion concentrations in the Martian ionosphere during the Viking 1 lander descent. The solar zenith angle, labeled ZA, ranged from 37° to 44° in this data set. This figure is from Hanson *et al.*, (1977).

HG radio propagation conditions on Mars might actually be better than on earth. The interaction of the solar wind with the earth's magnetosphere deposits large amounts of energy in the thermosphere. Geomagnetic activity often causes large variations in ionospheric parameters, leading to "ionospheric storms" which degrade HF radio communications. The earth's strong magnetic field also causes radio waves to split into ordinary and extraordinary components, again complicating the propagation situation. The earth's ionosphere is a noise-rich environment, with contributions from auroral, atmospheric and artificial sources. The HF spectrum is also crowded with a number of users, may interfering with each other.

Mars lacks a strong magnetic field, and although its upper ionosphere is directly exposed to the solar wind, the bottomside profiles are relatively stable from observation to observation, as shown in Zhang *et al.*, (1990a). This implies a fairly consistent radio propagation environment on Mars. The magnetic field at lower altitudes on Mars, whether intrinsic or induced by the magnetized solar wind flow, is very weak, so birefringent refraction effects will be much smaller than on earth. Potential noise sources expected on Mars include extraneous emissions from electronic equipment at the observation sites (which can be eliminated by proper design), and possible electrostatic discharge noise due to atmospheric wind-driven dust or ice. Noise sources external to the planet, such as solar and galactic radio emissions cannot penetrate through the ionosphere to the surface at frequencies below the local ionospheric plasma critical frequency. However, it is possible galactic radio noise may pass through the nightside ionosphere, propagate over the horizon and contribute to the received noise at a sunlit site.

HF NETWORK STUDY

We conducted a tentative investigation of expected HF propagation on Mars using the MESUR landing site scenario. The 16 site locations are listed in Table 1, and the great circle distances, in kilometers, between sites are in Table 2. For example, by referring to the chart for the distances between sites 1, 2, and 5 in Valles Marineris, the three locations can be seen to form a triangle with sides of 178, 243, and 335 km. The radius of Mars (3393 km) is about half that of Earth, so the ratios of ionospheric heights to planetary radii are nearly the same. Ionospheric refraction geometry allows one-hop ray paths to a distance of about one half of a planetary radius along the surface of both planets.

Table 1

Candidate MESUR observation sites from McNamee (1992). Latitudes are in degrees North, longitudes are in degrees East, and elevations are in kilometers.

#	Site	Latitude	Longitude	Elevation (km)
1	Valles Marineris	-6.00	58.00	3.50
2	Valles Marineris	-5.00	54.00	3.50
3	Chryse Plan. VL-1	23.00	48.00	-2.20
4	Olympus Mons	13.00	130.00	2.00
5	Valles Marineris	-2.00	54.00	3.50
6	Hadriaca Patera	-32.00	268.00	2.00
7	NW Hellas	-40.00	310.00	-2.00
8	Argyre Planitia	-37.00	44.00	3.50
9	South Pole	-86.00	315.00	4.00
10	Sirenum Terra	-45.00	185.00	4.20
11	Northern Plains	60.00	50.00	-2.70
12	Gusev Crater	-15.00	185.00	0.50
13	Syrtis Major	5.00	295.00	3.50
14	North Arabia	38.00	309.00	2.20
15	North Pole	82.00	55.00	0.00
16	Chasma Borealis	-66.00	66.00	2.50

Table 2

Great circle distances from each of the 16 candidate MESUR sites listed in Table 1, to each of the other sites, in kilometers.

1 Valles Marineris	0															
2 Valles Marineris	243	0														
3 Chryse Plan. VL-1	1813	1694	0													
4 Olympus Mons	4382	4594	4603	0												
5 Valles Marineris	335	178	1521	4551	0											
6 Hadriaca Patera	7845	7750	8506	8124	7887	0										
7 NW Hellas	5904	5768	6541	9062	5885	2048	0									
8 Argyre Planitia	1986	1971	3561	5606	2144	5904	4138	0								
9 South Pole	5029	5072	6702	6336	5249	3277	2725	3141	0							
10 Sirenum Terra	6541	6729	8220	4522	6873	3755	4841	5376	2822	0						
11 Northern Plains	3927	3854	2193	4370	3677	8438	7614	5753	8890	8860	0					
12 Gusev Crater	7311	7551	8216	3625	7618	4519	6216	6891	4596	1777	7371	0				
13 Syrtis Major	7301	7063	6451	9277	7052	2668	2788	6405	5404	6374	5790	6550	0			
14 North Arabia	6434	6212	4898	7640	6102	4723	4620	6821	7345	8193	3716	7451	2095	0		
15 North Pole	5213	5153	3498	4445	4975	7616	7795	7056	10170	8278	1305	6518	5273	3233	0	
16 Chasma Borealis	3569	3647	5333	5438	3824	4770	3744	1877	1522	3528	7496	5174	6527	8000	8772	0
	1	2	3	4	5	6	7	8	9	10	11	12	13	14	15	16

Croft and Hoogasian (1968) presented exact equations for ray path parameters, including ground range, for a quasi-parabolic model ionosphere. The ray path geometry is shown here in Figure 2. We hand-fit quasi-parabolic curves to observed Martian profiles and used Croft and Hoogasian's methods to evaluate HF single-hop propagation distances along the Martian surface. Figure 3 shows the daytime and nighttime quasi-parabolic electron density profiles. Here we use a peak electron density of 1.7 x 10^5 cm^{-3}, a peak height of 135 km and a semi-thickness parameter of 35 km during the daytime. The nighttime parameters used were 1 x 10^3 cm^{-3}, 150 km, and 50 km, respectively.

The achievable ground ranges for various takeoff, or elevation angles and operating frequencies are displayed in Figure 4 for the daytime ionospheric conditions and in Figure 5 for nighttime conditions. For a given frequency, the same ground distances can be achieved at several takeoff angles. This indicates the potential for multipathing and resulting degradation of signal performance, but by analogy with the terrestrial situation, this should not be an overriding communications limitation. Results indicate that with the appropriate choice of operating frequency, 2000 km paths are possible during both daytime and nighttime conditions. From Table 2, it can be seen that a number of sites are within this one-hop range. For example, in this scenario data could be relayed from Site 9 at the South Pole to Site 3 at 28'N via Sites 16, 8 and Valles Marineris (Sites 1 or 2).

If a peak density height of 150 km is assumed, typical for larger solar zenith angles, the one-hop distance increases to 2200 km. Now both the North and South Pole (Sites 15 and 9) can communicate with the Valles Marineris sites, and therefore also with each other. If two-hop propagation is supported, the ground range increases to 4000 km. Now data can be exchanged between any given site and any other site by HF radio relay, and connectivity becomes global.

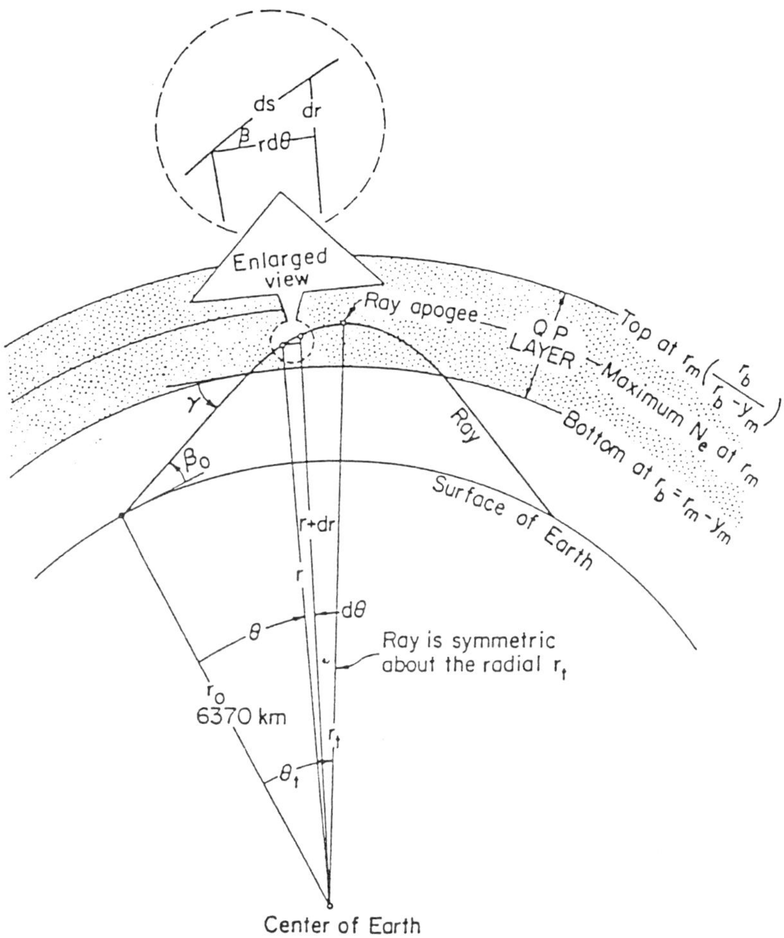

Figure 2 Ray path geometry for the quasi-parabolic ionospheric representation, from Croft and Hoogasian (1968).

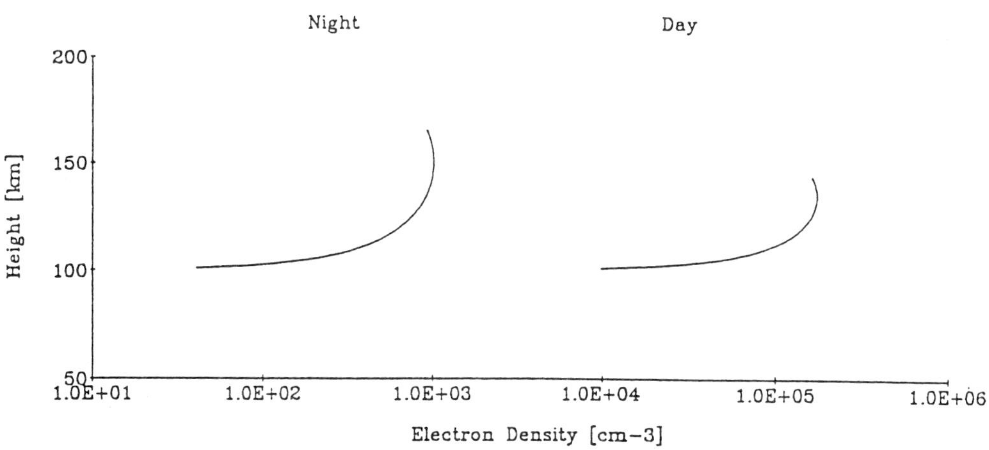

Figure 3 Electron density profile for the quasi-parabolic ionosphere representation, with daytime parameters of Nm = 1.7 x 10^5 cm^{-3}, ym = 35 km, and Hm = 135 km, and nighttime parameters of Nm = 1 x 10^3 cm^{-3}, ym = 50 km, and Hm = 150 km.

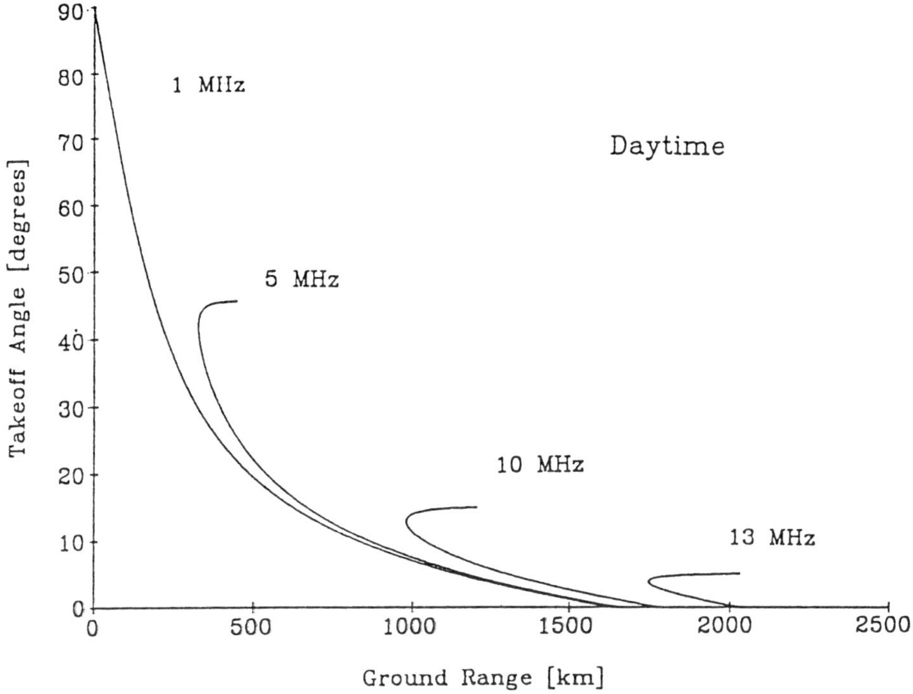

Figure 4 Ground ranges traversed by a radio ray as a function of takeoff angle (elevation angle) for 1, 5, 10, and 13 Mhz waves, using the quasi-parabolic daytime electron density profile shown in Figure 3. Peak electron density is 1.7 x 10^5 cm^{-3} at 135 km.

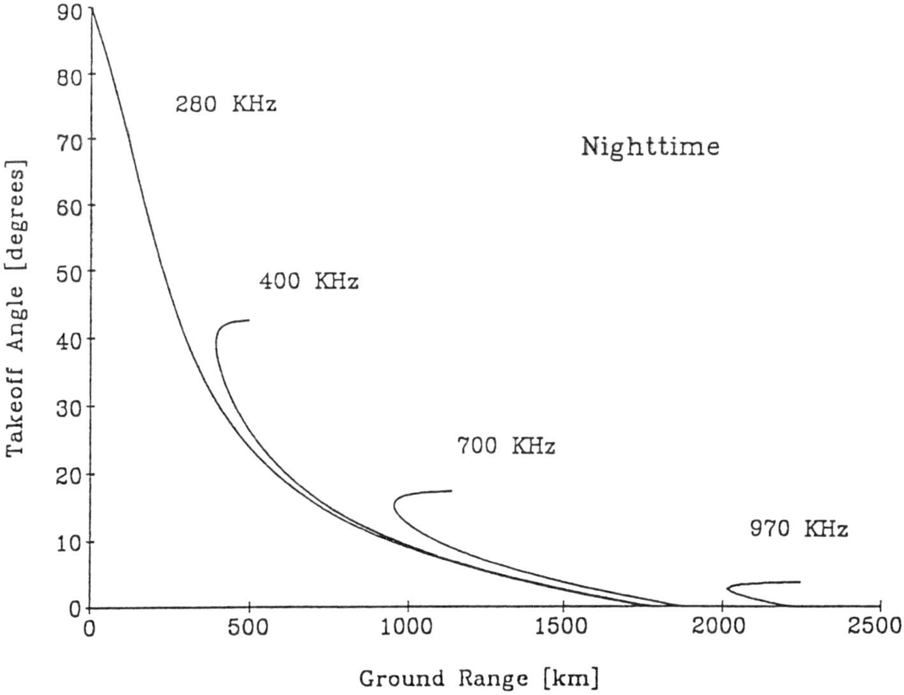

Figure 5 Ground range versus takeoff angle for 280, 400, 700, and 970 Khz waves, using the quasi-parabolic nighttime electron density profile shown in Figure 3. Peak electron density is 1 x 10^3 cm^{-3} at 150 km.

In this tentative study, we ignored small scale ionospheric structure, the existence of D and E layers, and we assumed a spherically symmetric ionosphere with no horizontal electron density gradients. We ignored the presence of a weak Martian magnetic field. We also did not consider the effect of ionospheric absorption of the radio signal, but a more thorough analysis must do so. In the terrestrial ionosphere, absorption can range from 0-20 dB per hop (McNamara, 1991). Several researchers have investigated the absorption of HF and VHF radio waves traveling through the Martian ionosphere (Agarwal, 1972; McCormick, 1973). For example, Agarwal (1972) looked at the effect of electron collision frequencies and concluded that trans-ionospheric propagation was not possible below a frequency of about 100 kHZ. He assumed that O^+ was the dominant ion in the Martian ionosphere. This analysis should be revisited since we now know that O_2^+ is the major ionic constituent in the bottomside Martian ionosphere and we have a better understanding of expected ion densities.

COMMUNICATIONS REQUIREMENTS

In light of our present knowledge of the Martian ionosphere, a thorough investigation of the feasibility of the HF communications mode on Mars should be conducted. Several key issues must be considered when investigating the potential of communications on Mars via ionospheric refraction.

1. The required hardware must not increase the complexity, mass or power demands on the mission.
2. The peak Martian ionospheric electron densities must be sufficient to refract obliquely incident radio waves back to the surface at some distance from their source.
3. Ionospheric plasma collisional processes must not cause excessive absorption of the radio wave.
4. The natural and machine-generated radio noise environment must be fairly quiet at the sites.
5. The one-hop or multi-hop coverage zones should be sufficient to allow mutual access between adjacent sites. That is, every lander must be within communications distance with at least one or two of its neighbors to allow the relay of data from the most remote sites to the network nodes.
6. The lower and upper usable frequencies must be far enough separated to allow the entire bandwidth of the transmitted signal to fit in the ionospheric band window.
7. Band openings must be of sufficient duration to allow the transfer of data and telemetry at rates necessary to meet mission requirements.
8. Ionospheric irregularities and traveling disturbances must not cause excessive scintillation, dispersion or multipathing which can result in distortion or loss of the radio signal.

9. Effective radiated and received power levels, as determined by transmitted power, antenna efficiencies and receiver system sensitivities, must provide adequate received signal strengths above the noise floor for reliable communications.

10. For multi-hop propagation, ground reflection losses must not attenuate the signal below link margin requirements. Otherwise the network will have to rely upon one-hop propagation between landers for communications. Terrestrial surface losses range from .5 to 3 dB per reflection (McNamara, 1991).

FUTURE INVESTIGATIONS

Several investigations might be conducted as precursors to actual radio experiments in the Martian environment. These investigations include Martian ionospheric radio propagation simulations and analogous terrestrial radio experiments which replicate conditions expected in the Martian environment. This will provide valuable insight into whether the ionospheric circuit can support reliable long distance communications on the Martian surface for upcoming missions.

HF Propagation Simulations

Recent research has focused on the interaction of the solar wind with the Martian ionosphere, whether Mars has an intrinsic magnetic field or one induced by solar wind interaction, and the structure of the topside Martian ionosphere at and above the electron density peaks. However, the bottomside Martian ionosphere is important for HF propagation, because radio waves will be refracted back to the surface only at altitudes below the height of electron density peak.

An ionospheric model which characterizes the current theoretical and empirical understanding of the Martian ionosphere should be coupled with an accurate radio wave ray tracing algorithm. Anderson (1987) discussed the development of a terrestrial semi-empirical low latitude ionospheric model, and Watkins (1978) introduces a terrestrial high latitude physical model. Several investigators have developed Martian ionospheric models. See Tan and Wu (1977), Shinigawa and Cravens (1989) and the review by Cravens (1992). Zhang *et al.*, (1990a) presented a modified Chapman model for the Martian ionosphere in their reevaluation of the Mariner and Viking radio occultation data. A thorough propagation study would include a full three-dimensional numerical or piecewise analytical ray tracing algorithm. Candidate ray tracing routines are described in Jones and Stephenson (1975) and Reilly and Strobel (1988).

A realistic simulation must account for signal losses due to absorption, although theoretical models of ionospheric absorption based upon collisional processes are not known for accurately predicting observed absorption conditions in the terrestrial ionosphere (McNamara, 1991). Future Mars missions might collect absorption data using radio occultation, ionosonde, oblique sounder, riometer or other techniques. Much is yet to be learned about the morphology of the Martian ionosphere, and the roles solar wind interactions, ionspheric convection, neutral winds, atmospheric gravity waves and airborne dust play in the relevant processes.

Equivalent Terrestrial Radio Experiments

HF radio experiments conducted at terrestrial sites and under environmental conditions which replicate those observed on Mars can be used to predict HF communications system performance. This would allow evaluation of HF propagation for its ability to support the transfer of data and telemetry between remote stations and rovers on Mars. Sites should be selected which support oblique single-hop HF propagation with geometries similar to those anticipated on Mars. Geometries with 1000-4000 km propagation paths on earth correspond to equivalent Martian separations of 500 to 2000 km. Candidate sites include regions where ground conductivities and dielectric constants are similar to those expected on Mars, and where ionospheric conditions are similar to those observed by previous Martian missions. Analogous regions on earth might include the high latitudes when the ground is well frozen, with experiments conducted at low levels of geomagnetic activity. Evaluations of the performance of simple antenna configurations, low power levels and new HF modulation schemes for the transfer of data between sites should be conducted. This would determine the minimum power requirements, optimum antenna configurations and whether sustainable data rates are sufficient to meet likely Mars mission scenarios. Micro-rovers might be tested for their ability to deploy antenna systems.

Recent advances in adaptive digital modulation techniques for HF radio allow true error correction, and increased data rates (Henry and Petit, 1992; Horzepa, 1993). Error-free data transfer rates of up to 750 bps are achievable using the CLOVER modulation scheme (Henry and Petit, 1992). Available power will be a limiting factor in future planetary missions, so the average transmitted power must be kept as low as possible. Automatic link establishment systems are commercially available at six watts output (BR Communications, 1991). High power communications systems are a potential possibility, as long as the transmission duty cycles can keep power consumption within acceptable margins.

SUMMARY

The existence of a Martian ionosphere similar to the earth's implies that over-the-horizon communications via HF radio should be possible on Mars. Our study using a quasi-parabolic ionosphere approximation with exact analytical raytracing shows that one-hop propagation out to 2000-2200 km is possible. Approximately one half of the landers could be connected via HF radio during at least part of the day in the one-hop case, and all landers would have connectivity if two hop propagation were supported. Further investigations and link margin analyses require three-dimensional ray tracing through more realistic ionospheric models. Analogous terrestrial propagation experiments would be beneficial. The results should be analyzed in terms of the implications for Martian radio science experiments, and in terms of HF radio as a viable communications mode on Mars. These studies will help determine the expected ionospheric circuit characteristics for likely scenarios of lander locations. The tentative results presented in this paper, using the MESUR scenario of 16 landers, are encouraging. For the ionospheric HF propagation mode to be practical for network communications on the Martian surface, it must not require more complexity, cost or power than other communications options, including the option to directly transmit observational data back to Earth from each site, or via relay satellite. In-depth investigations of these issues will shed light on unanswered physical and engineering questions, and may significantly benefit future planetary missions..

REFERENCES

Agarwal, D. C., "Electromagnetic wave propagation in the Martian ionosphere," *Institute of Telecommunications Journal*, Vol. 18, Jan 1972.

Anderson, D. N., "A semi-empirical low-latitude ionospheric model," *Radio Sci.*, Vol. 22, pp. 292-306, 1987.

Tactical Chirpcomm (R) Model 4280 Transceiver product brochure, BR Communications, 1249 Innsbruck Drive, Sunnyvale, CA 94089, 1991.

Cravens, T. E., "Ionospheric models for Venus and Mars," in *Venus and Mars: Atmospheres, Ionospheres, and Solar Wind Interactions*, Geophysical Monograph 66, American Geophysical Union, pp.277-288, 1992.

Croft, T. A., and H. Hoogasian, "Exact ray calculations in a quasi-parabolic ionosphere with no magnetic field," *Radio Sci.*, Vol. 3, pp. 69-84, 1968.

Davies, K., *Ionospheric Radio*, Peter Peregrinus Ltd., London, 1990.

Hanson, W. B., Sanatani, S., and Zuccaro, D. R., "The Martian ionosphere as observed by the Viking regarding potential analyzers," *J. Geophys. Res.*, Vol. 82, pp. 4351-4363, 1977.

Henry, G. W., and Petit, R. C., "HF radio data communications, CW to CLOVER," *Communications Quarterly*, pp. 11-24, Spring 1992.

Horzepa, S., "CLOVER: The future of HF data communications," *QST*, p.107, Jan. 1993.

Hubbard, G. S., "Mars Environmental Survey Concept," in *Mars: Past, Present and Future*, edited by E. B. Pritchard, pp. 141-152, American Institute of Aeronautics and Astronautics, Washington DC, 1992.

Jones, R. M., and Stephenson, J. J., *A versatile three-dimensional ray tracing computer program for radio waves in the ionosphere*, Office of Telecommunications Report 75-76, Department of Commerce, October 1975.

Kliore, A. J., "Radio occultation observations of the ionospheres of Mars and Venus," in *Venus and Mars: Atmospheres, Ionospheres, and Solar Wind Interactions*, Geophysical Monograph 66, American Geophysical union, pp. 265-276, 1992.

Kosolov, M. A., "Investigation of radiowave propagation in the solar system," *IEEE Transactions on Antennas and Propagation*, Vol. AP-27, #1, Jan 1979.

Martin, D., and Newkirk, D., "Transmitter Power: What it is, what it does, and how to use it," *QST*, pp. 28-31, March 1993.

McCormick, P., "On the effects of electron-neutral particle collisions upon the refraction of high frequency radio waves by the lower atmosphere and ionosphere of Mars," *Planet. Space Sci.*, Vol. 21, May 1973.

McNamara, L. F., *The Ionosphere: Communications, Surveillance, and Direction Finding*, Krieger Publishing Company, Malabar, FL., 1991.

McNamee, J., *The Mars Environmental Survey (MESUR) Mission*, Poster presentation at the American Geophysical Union Meeting, San Francisco, CA, Dec. 1992.

Reilly, M. H., and Strobel, E. L., "Efficient ray tracing through a realistic ionosphere," *Radio Sci.*, Vol. 23, pp. 247-256, 1988.

Rishbeth, H., and Garriott, O. K., *Introduction to Ionospheric Physics*, Academic Press, NY 1969.

Scoon, G. E. N., "MARSNET—A Mars network of surface stations," in *Mars: Past, Present and Future*, edited by E. B. Pritchard, pp. 153-168, American Institute of Aeronautics and Astronautics, Washington, DC, 1992.

Shinagawa, H., and Cravens, T. E., "A one-dimensional multispecies magnetohydrodynamic model of the dayside ionosphere of Mars," *J. Geophys. Res.*, Vol. 94, pp. 6506-6516, 1989.

Tan, A., and Wu, S. T., "Post-Viking Martian ionospheric model," *Acta Astronomica*, Vol. 28, pp. 187-194, 1978.

Verigin, M. I., Gringauz, K. I., Shutte, N. M., Haider, S. A., Szego, K., Kiraly, P., Nagy, A. F., and Gambosi, T. I., "On the possible source of the ionization in the nighttime Martian ionosphere 1. Phobos 2. HARP electron spectrometer measurements," *J. Geophys. Res.*, Vol. 96, pp. 19, 307-319, 313, 1991.

Watkins, B. J., "A numerical computer investigation of the polar F-region ionosphere," *Planet. Space Sci.*, Vol. 26, pp. 559-569, 1978.

Yowell, R., *Investigation of Radio Propagation in the Martian Ionosphere*, Draft Thesis Proposal, U. of Houston, Clear Lake, TX 1991.

Zhang, M. H. G., Luhmann, J. G., Kliore, A. J., and Kim, J., "A post-pioneer reassessment of the Martian dayside ionosphere as observed by radio occultation methods," *J. Geophys. Res.*, Vol. 95, pp. 14,829-14,839, 1990.

Zhang, M. H. G., Luhmann, J. G., and Kliore, A. J., "An observational study of the nightside ionospheres of Mars and Venus with radio occultation methods," *J. Geophys. Res.*, Vol. 95, pp. 17,095-17,102, 1990.

AAS 93-861

SYSTEMS FOR MARS EMU DESIGN

George William Herbert*

The surface of Mars presents new challenges to space suit designers, including high surface gravity and an atmosphere. Previous EMU solutions will not function on Mars, and new concepts and technologies must be explored to produce viable systems. These include passively cooled, low and variable insulation suits, materials including dense membranes which may allow sweat-through-suit cooling, and PLSS system selection to minimize mass.

INTRODUCTION

There are five primary justifications for manned extravehicular surface operations on Mars: Geological Exploration, Exobiology, other Mars Science, Base Construction, and Maintenance. Accomplishing these first three goals is the primary intent of any manned Mars mission.

The EVA Systems, the Space Suit and Portable Life Support System (PLSS) together known as the EMU, or Extravehicular Mobility Unit, are required for accomplishing those surface operations goals. To the extent possible, those goals should drive the concept and design of Mars EVA systems, to maximize the successful returns from any manned Mars mission.

This paper discusses a portion of the results of a 1989-1991 study performed on issues involved in designing Mars EVA systems. It concentrates on technology and systems issues involved in Mars EMU design.

EMU REQUIREMENTS

An EMU must provide a safe, comfortable environment for its occupant, maintaining atmospheric pressure, temperature, acceptable oxygen and CO_2 levels, protect from contamination (physical and biological), handle other metabolic wastes, and provide physical protection for tears and punctures.[1] To maximize useful work per EVA, 8-hour endurance is required. A 30 minute backup life support system should be installed to maintain current safety levels. The EMU should provide constant communications with other suits, rovers, or bases. The occupant should be able to walk at 6.5 km/hr and drive a rover while wearing the EMU.

The EMU must provide all of these in a package which is light and mobile enough to allow the primary EVA goals to be accomplished as easily as possible.

* President, Retro Aerospace, 2240 Blake Street, #101, Berkeley, California 94704. E-mail: gwh@retro.com.

EMU Weight

The Mars environment presents a unique challenge to EVA operations: significant surface gravity. Its surface gravity is 0.36 Earth's. The Moon, the only other surface EVA location to date, has less than half this gravity.

EMU systems have traditionally been quite massive. The Apollo suits massed 38 kg, with a 62 kg PLSS for a total of 100 kg. A total EMU mass of 100 kg would be overwhelming on the surface of Mars; obviously system mass must be reduced. The baseline rule determined by the study group was that the Mars EMU perceived weight (mass x gravity) on Mars should not exceed the Apollo EMU perceived weight on the Moon. The target system mass was therefore determined to be 45 kg.

This massively lower allowable EMU mass required a fundamental rethinking of suit design concepts.

PASSIVE EMU COOLING

Around 14 kg of the Apollo EMU's mass is in its thermal control systems. This is a particularly attractive area for reduction of mass because the Martian atmosphere provides a cool sink which can be used to passively cool an EMU.

The general concept behind passive cooling is to design a suit such that the thermal balance of heat generated metabolically and absorbed from sunlight is balanced by convective and radiative losses to the Martian atmosphere, maintaining a stable and comfortable temperature for the user. Current suits seek to insulate the user from the environment as much as possible, removing metabolic heat via a liquid-cooled garment and heat rejection system. These systems, and the required insulation in the suit itself, are quite massive.

To determine the initial feasibility of passive suits, a simple suit thermal model was integrated with Mars surface conditions data and various metabolic loads.[2] We determined that a suit approximately 8 times as thermally conductive as the Apollo suits is well balanced for passive thermal control. At this insulation level and a 250 watt metabolic rate, the user will remain comfortable over nearly the whole yearly cycle of day EVA operations on Mars. A suit with variable layers of insulation from 60% more to 75% less of this value (an overcoat and no insulation except tear/dust and pressure restraint layers, respectively) can maintain comfort over the whole spectrum of operations. In some hotter cases, the optimal solution is for the user to sweat to remove some metabolic heat, but this was accomplished with Earth-normal amounts of sweating. Follow-up work with other models has confirmed these general results.[3]

To verify that the assumed metabolic heat loads were reasonable, detailed breakdowns of EVA task type were made across a spectrum of EVA types (Exobiology, Geology, and Construction/Maintenance) based on discussions with planetary scientists and field geologists. Our group assumed that the Mars metabolic rates (0.36 G) for these tasks could be estimated by averaging Earth values (1 G) and measured lunar (0.16 G) values. The average EVA metabolic loads ranged from 258 W to 268 W, with maximum peaks of 800 W. This was within the working area of the thermal modeling.

The assumed average of Earth and lunar metabolic data for Mars metabolic rates was generally confirmed by partial-gravity metabolic profiles generated in various testing at NASA Ames.[4]

Passive Cooling Technologies

Basic thermal modeling established that it is possible to manufacture a passively cooled EMU. In addition to variable insulation suits (removable suit layers to tailor insulation level to climate and time of day, similar to Earth outdoors clothing today), several technologies to enable actually building passive suits were investigated: dense membranes, to enable "sweating suits": insulating properties of materials in Mars conditions; chemical and isotope heaters for extreme cold conditions; and phase change materials to stabilize temperature shifts. The following sections address these areas.

Dense Membrane Technology

The accumulation of water produced by sweating within a passive suit is undesirable. In addition to incompletely removing heat from the EMU system, it will become uncomfortable and potentially dangerous to have large amounts of collected water within the suit.

One potential solution to this problem is the Dense Membrane polymer.[5] These are a family of polymers which are solid membranes, able to resist pressure and mechanical force, which actively transport selected molecules via diffusion.

These dense membranes are solid sheets, giving high biological isolation in both directions and able to function as a pressure bladder (with suitable restraint layer for mechanical support). Microporous membranes (for example Gore-Tex) use porous materials treated to prevent liquid leakage by using surface tension to keep the liquid out of the pores. They are not pressure tight, and contamination or pressure can force liquids through the pores anyway. Dense membranes, with a solid structure, hold gas pressure and will not leak unless torn. The active diffusion will continue under basically any circumstances.

Dense membrane polymers able to transport sweat water at metabolic production rates exist, such as PTS High-HVTR Polyurethane. This material enables the construction of sweat-through passively cooled suits.

Insulation Characteristics

Our group was unable to locate good references on thermal properties of insulation materials under Mars surface conditions. Some data from the Viking program was available. Other studies indicate that traditional Multi-Layer Insulation (MLI), used on current suits, degrades significantly at pressures found on the surface of Mars. Alternative insulation materials must be tested and identified to provide a good database for suit materials selection.

Chemical and Isotope Heaters

Passively cooled suits exhibit a potential problem in extremely cold conditions and low metabolic workloads: the extremities of the suit occupant will cool quickly. Extra insulation in the suit gloves severely impedes user dexterity. To solve these problems, and other cold-climate situations, the ability to add heat to specific areas of the body is highly advantageous.

Three methods were investigated: electrical heating, chemical heaters, and isotope heaters. All three offer localized heating in areas that may require it, can be activated only when needed, and do not impede dexterity by adding additional insulation around the fingers.

Further investigation of these localized heating methods and of the development of heat pipes within gloves to transfer heat along glove fingers is considered useful, but was beyond the scope of our study. Some research in this area was performed at Stanford University as part of a concurrent Mars mission design. It suggested that a thermally conductive silver mesh be used to transfer heat within the glove, and that a radioisotope heater element, about 23 grams of alpha-emitting fuel with additional shielding, be used to generate 2 watts of thermal energy per glove.[6]

Phase Change Materials

It is possible to incorporate into a suit materials which change phase (freeze) at temperatures which will be encountered within the suits operating range. As materials undergoing a phase change absorb heat during that freezing (at a constant temperature), the incorporation of phase change materials can stabilize suit temperature at that point for an extended period depending on the materials' enthalpy. Phase change materials such as Glycerol Triacetate may be of great use in Mars EMU design. They are already being used in ski clothing manufactured in Japan, to good effect.[7]

ELECTRICAL SYSTEMS

The electrical power system is another area which may produce significant mass reductions. The Apollo EMU had an average electrical power usage of 47 watts over its eight-hour lifetime. Most of this was used to drive ventilation and the liquid cooled garment cooling system. The Apollo suits (and shuttle suits today) had 5 kg batteries to provide this electrical power.

Power Requirements

As has been stated earlier, passive suits remove the cooling system requirements. This then directly reduces the battery mass significantly. Improvements in communications equipment power efficiency will provide further mass savings, Reduction of ventilation airflow rate may be possible.

Battery Options

The Apollo suits used silver-zinc batteries with one cycle lifetimes and an energy density of 83 W-hr/kg. The current Shuttle spacesuits use silver-zinc batteries with an energy density of 102 W-hr/kg and 8 cycle lifetimes. Current silver-zinc technology extends life to as much as 50 cycles at energy densities of 90 W-hr/kg.

A Mars mission may well stay for on the order of years, with EVAs on a nearly daily basis. Several hundred EVAs can be expected. As can be seen, several sets of EVA batteries would have to be brought along with silver-zinc EMU batteries. There are alternative battery technologies which can potentially reduce these replacement requirements, and potentially increase battery efficiency.

Nickel-cadmium batteries have an energy density of only 36 W-hr/kg, but lifetimes in the range of 35,000 cycles. NiCad is a very heavy solution with life significantly higher than required.

Nickel-hydrogen batteries have energy density of 60 W-hr/kg, and about 15,000 cycle lifetimes. These are improved solutions, but still heavier than current batteries in the suit.

Sodium-solid electrolyte batteries are available which have energy densities up to 150 W-hr/kg, an improvement over current technology, and lives up to 1000 cycles. This is approximately the desired lifetime, and the improved energy density is helpful. However, these batteries operate at temperatures ranging from 440 to 700 K, quite hot for use in an EMU system.

Lithium-Titanium Sulfide batteries are the most recent development. They operate at normal temperatures, have energy densities around 100 W-hr/kg, and have lifetimes of 300 cycles. While not a mature technology at this time, Li-TiS2 batteries appear to be a good balance of lifetime and energy density.

Power System Design

Ideally, the Mars EMU will be a passively cooled system with power usage of at most 25 watts, around half what the Apollo suits used. Over an eight-hour EVA, this amounts to 200 W-hr of power. A number of battery options providing 100 W-hr/kg of storage capacity exist, so it is safe to say that the battery on a Mars EMU can be as light as 2 kg.[8]

CO_2 REMOVAL SYSTEM

Current PLSS designs utilize lithium hydroxide to scrub carbon dioxide metabolic waste from the EMU air supply. About 2 kg of LiOH are consumed per 8-hour EVA. While this is the most EMU mass-efficient option, it is also a large mass impact on the overall mission; a mission staying 600 days, with a 2-man EVA every day, would thus consume about 2400 kg of lithium hydroxide (plus packaging factor) during its ground stay. Lithium Hydroxide can be regenerated with a complex process, but it is energy intensive and complex. Lithium Hydroxide regeneration was ruled as impractical for initial missions.

Alternatives to Lithium Hydroxide are available. Magnesium Hydroxide $Mg(OH)_2$, will work nearly as well as LiOH and requires only about 50% more mass, or about an additional kilo, in the PLSS. It can be regenerated by heating to 500 K in a vacuum or near-vacuum for 20 or more use cycles.

Another option is to use KO_2 as both the oxygen gas supply and carbon dioxide scrub chemical. It appears to be viable, based on initial information, though further work is needed.

Other alternatives including Ag_2O, molecular sieves, and membrane diffusion all appear to be overly massive for Mars surface applications, requiring at least five times as much mass.

PLSS MASS ESTIMATES

A preliminary system tradeoff analysis was made to help select general EMU configuration.[9] It assumed 8-hour EVAs, 0.6 kg of oxygen consumption and 0.6 kg CO_2 production, factors of safety of 1.4 in gas supplies and 2.5 in chemical supplies, moderate suit leakage to the environment, and a packaging factor of 2.2 for all subsystem masses (the Shuttle EMU packaging factor). Eight suit configurations were examined. Three involving overboard dump were immediately ruled out as the projected PLSS mass exceeded the total allowable EMU mass. The surviving five options included one demand-breathing configuration, using tanked oxygen and lithium hydroxide for carbon dioxide removal, and four helmet circulation options. Three of the helmet circulation PLSS configurations used LiOH, $Mg(OH)_2$, and Ag_2O respectively for carbon dioxide scrub. The fourth used KO_2 as both oxygen supply and carbon dioxide scrub. The estimated PLSS masses for these five options are listed in Table 1, along with the estimated resupply mass per 8-hour EVA in kilograms of material brought from Earth.

Table 1
PLSS CONFIGURATIONS

Configuration	System Masses (kg):				
	O2	scrub	Elect.	Total	Resupply
Demand/LiOH	3.2	3.0	0	13.6	1.9
Hlm./LiOH	3.2	3.0	2.6	19.4	1.9
Hlm./Mg(OH)2	3.2	7.3	2.6	28.8	0.10
Hlm./AgO2	3.2	3.8	2.6	21.1	0.08
Hlm./KO2	4.7	0	2.6	16.1	2.9

Notes: Demand is demand breathing through mask, with seperate pressurization. Hlm. is full circulation in helmet of breathing gas. Total mass includes packaging factor.

Following the initial tradeoff study, a more detailed design was done on a sample helmet circulation/gaseous $O_2/Mg(OH)_2$ PLSS. The results of that design study are shown in Table 2, compared to initial project target masses and a 2-hour endurance demand breathing open system PLSS which was conceived as a special purpose or emergency PLSS.

Table 2
BEST CONFIGURATIONS

System	Target mass (kg)	Mg(OH)2 PLSS (kg)	2 hr (kg)
Thermal	5.0	0	0
Vent/Press	2.0	3.2	0.6
Main O2	3.0	3.2	6.9
Emerg O2	6.0	2.8	2.8
CO2 Scrub	5.0	3.8	0
Elect./Com.	6.0	5.0	5.0
Misc.	2.0	2.0	2.0
TOTAL:	28.0	20.0	17.3
Allow. packaging factors:		1.4	1.6

PACKAGING FACTOR

One critical improvement needed to design successful Mars EMUs is that the PLSS system packaging factor must be controlled. Current Shuttle EMU systems have a packaging factor of 2.2: every kilogram of actual life support or other system in the PLSS has 2.2 kg of structure and other packaging. A careful re-examination of the design philosophy of PLSS units is required. Current engineering will produce unbearably heavy PLSS systems for Mars.

DIRECTIONS FOR FURTHER STUDY

A number of items were identified as areas for further work in defining Mars EMU design:

- More detailed task analysis, possibly including field work
- Mars gravity metabolic rates for various tasks
- Improved thermal modeling
- Materials research
- Materials test chamber development
- Thermal control optimization
- Glove development
- Metabolic waste removal systems
- Helmet configuration
- Carbon dioxide scrub systems
- Emergency systems configuration
- PLSS packaging factor
- PLSS mechanical design

At least two of these areas have been worked on in the two years since the project at UC Berkeley was completed: more advanced thermal modeling (Ref. 3), and Mars gravity metabolic rates for various tasks (Ref. 4).

CONCLUSION

The Mars environment presents a very different challenge for EMU design than those environments man has already explored: the higher gravity demands lighter EMU designs, and the atmosphere provides a relatively stable thermal environment capable of being used in EMU design. Much lighter, passively cooled EMU designs appear possible, and other advanced technologies can contribute to optimized EVA systems for use in Mars exploration and development.

ACKNOWLEDGEMENTS

The work presented in this paper was accomplished by a team at UC Berkeley consisting of Dr. Lawrence Kuznetz, Professors Boris Rubinski and Linda Hays, coordinated by Lissa Sabia, with Mike Itamura, Tracy Madden, George Herbert, Frank Crary, Lee Thompson, and Alan Leibovitch. It was supported by Dr. Bruce Webbon at NASA Ames Research Center. Additional contributions were made by Robert Ward of The Polymer Technology Group and Professor Bruce Lusignian of Stanford University, among many others.

REFERENCES

1. *Space Suits and Life Support Systems for the Exploration of Mars*, project report, L. Kuznetz *et al.*, 1991, Overview section.
2. *Ibid.*, Computer Analysis Results section.
3. F. Crary, "Passive Cooling of Mars EMU," presented at The Case for Mars V, May 28, 1993, Boulder, CO. Abstract published as AAS 93-863, Appendix A, Section 15, in this volume.
4. D. J. Newman, personal communications, summer 1992.
5. R. S. Ward, "Use of Moisture-Permeable Dense Membranes In Space Suit Design," presented at Mars EMU Final Presentation, June 10, 1991, NASA Ames.
6. B. Lusignian, personal communications, spring 1991.
7. L. Kuznetz, personal communications, 1990-1991 & May 29, 1993.
8. Space Suits...., Power Requirements section.
9. *Ibid.*, Suit Comparisons and Design Recommendations section.

AAS 93-867

PROJECT EMA: A PLAN FOR THE EXPLORATION OF MARS

Philip A. Turek*

Our EMA (Earth Mars Associated) concept offers many advantages over traditional proposals for the exploration of Mars. Possessing enhanced societal involvement, the high profile mission expands public support thereby improving mission success. By addressing both technical and political barriers, the concept shows tremendous promise for its use in future exploration programs. The system architecture is designed for optimal science yield while maintaining sufficient flexibility for accommodating all cultural objectives. The Orange County chapter of the National Space Society has played a significant role in the advancement of this EMA design. This paper provides background information for understanding an EMA mission.

We have targeted key elements of society for inclusion in our Mars exploration mission. The three primary groupings whose involvement is solicited are: The US Congress, environmentalists, and students from kindergarten through college. An indispensable feature of the EMA Mars exploration mission is the inclusion of the formal exploration of Earth, both before and after the manned time spent on the surface of Mars. This mechanism allows the public to make a direct connection between their everyday lives and the otherwise remote subject of space exploration.

INTRODUCTION

Our objective is to identify potential advancements in space exploration. We have drawn upon our involvement with the general public and the space community to propose the evaluation of an exploration mission that is fully responsive to the need for reduced bureaucratic inertia and cost in the next generation of space missions.

The Orange County chapter of the National Space Society (OCSS) is pleased to submit this proposal to provide the Office of Science, Technology and Space and the National Aeronautics and Space Administration (NASA) with assistance in the development of advanced concepts for Mars exploration missions. OCSS proposes a unique mission to Mars which includes the formal exploration of the planet Earth. This plan is based on the so-called ABA scientific methodology, wherein a calibration phase is carried out both before and after the exploration of the unknown. In this scenario, astronaut candidate teams are dispatched to explore Earth as part of their Mars exploration preparatory training program. Importantly, the team that goes to Mars is sent to explore Earth a second time upon their return to Earth. This three phase mission (Earth, Mars, Earth, or ABA) ensures that the best qualified team is sent to Mars, and it provides a mecha-

---

* Project leader and Secretary of the Orange County Space Society, c/o Discovery Museum, 3101 W. Harvard, Santa Ana, California 92704. OCSS is a chapter of the National Space Society.

nism for reintroducing the returning Mars astronauts to Earth. Exploring Earth also allows a plethora of disparate elements of society to become directly involved with a manned Mars mission.

In support of future exploration programs, OCSS brings to this effort a combination of a clear vision for our nation's space program and a strong commitment to innovation and rigorous scientific methodology in the design process. We offer to the Office of Science, Technology and Space and NASA a small but effective group having an in-depth background in exploration technologies including spacecraft design, system architecture design, human factors, and multidisciplinary optimization. The OCSS will utilize this team to further refine the EMA exploration concept.

PROJECT BACKGROUND

Advantage of Including Earth

Exploration missions are driven by scientific curiosity, political agendas and public interest. These parameters are intertwined, however, and historical trends indicate that an intensification of each will ultimately contribute to an enhancement in total mission performance. The majority of past designs met the requirements of one or two parameters but rarely satisfied all three. The Apollo Moon landings are often cited as our space program's greatest achievement, but even this project was strong in only two of the three parameters cited. (Apollo project science objectives seemingly were little more than an afterthought.) A mission which emphasizes all three requirements would be of tremendous value because of its greater performance capabilities and sustainability. EMA's ABA methodology means models of planetary weather systems would improve, allowing better weather forecasting on Earth. Examination of the sufficiency and necessity of conditions for life on Mars would lead to heightened awareness to the conditions required to maintain life on Earth. Congressional representatives would view the process of selecting the Earth exploration routes symbolic of democracy as an opportunity to draw recognition to their districts and themselves. In the area of foreign relations, consideration could be given to the idea of forming international exploration teams that study at selected world universities. A data collection system could be established in which local students continue the measurements initiated by an exploration team that passes through their area, thereby building a baseline of Earth data that would facilitate the comparison of Earth data to Mars data. Local communities could tackle local environmental problems, improving their preparation to receive the returning Mars astronauts during the second Earth exploration phase of the mission. An EMA mission scenario emphasizes scientific objectives in combination with a strong effort to stimulate the public's interest in the mission. Because of this direction of stress, an EMA mission, in comparison to a conventional Mars mission, is more likely to succeed.

The parameter which is fundamental to an EMA mission is stimulating the public's interest. As a result, both scientific and political agendas are framed in the context of the public interest. A strong emphasis on building and sustaining the public interest around an EMA mission will translate into increased political support, a critical element required for a multi-year project such as this. As a practical matter this emphasis is done by expanding the mission parameters to include the formal exploration of Earth, both before and after the manned time spent on the surface of Mars. This mechanism allows direct, widespread, personal involvement of large numbers of an Earth-bound population in the exploration of Mars. Figure 1 is a thematic image of project EMA.

Figure 1 Project EMA.

Understanding EMA

The term 'EMA' stands for 'Earth Mars Associated,' and it serves as a reminder that planet Earth is to be included in our total exploration scenario involving Mars. To place the EMA plan into perspective, it is instructive to look back over an Apollo mission profile. In a conventional Apollo mission, the general public first tuned in on their television sets just in time to watch a Saturn V rocket blast off from Earth. Less than a week later the general public ended its involvement with the mission as it watched the Apollo capsule splash down in the Pacific ocean. This is illustrated graphically in Figure 2. The vertical line represents public interest/involvement as a function of time. Note that public involvement with an Apollo mission ramped up suddenly during blast off, and ended just as suddenly at splashdown. This was a major flaw with the Apollo program, although it was not recognized as such at the time.

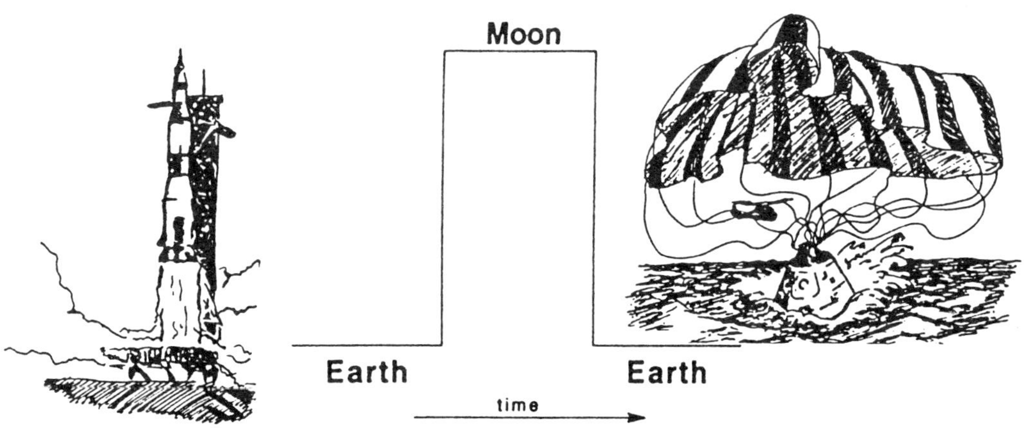

Figure 2 Public involvement/interest in an Apollo mission ramped up and down abruptly.

The general public had difficulty becoming instantly connected to an Apollo mission in progress. It had developed an air of being just another television show, and as a television show it suffered from a boringly repetitive plot. The public turned the channel. When that happened, federal support for the Apollo program dried up so fast the last three Saturn Vs built weren't even launched. The program was cancelled.

An EMA mission strives to build and sustain public involvement with the mission over time. This is uniquely done by expanding the mission to include Earth as well as Mars as an object of exploration. Other mechanisms are employed as well. Figure 3 illustrates public interest/involvement with an EMA mission over time. Note that the gentler EMA ramp profile reflects a mission design strategy of allowing the public to become closely connected with the mission over time. This connection is built up through conventional media methods such as television broadcasts, nightly news reports, and newspaper articles, but EMA goes beyond that. Connections are built up with the public through the use of interactive technologies, similar in nature to the type of public participation currently advocated by Dr. Robert Ballard in the field of undersea exploration. Dr. Ballard employs satellite links with remote terminals located at universities and high schools to enable the public to control the actions of probes located miles below sea level. An EMA mission is much more fully integrated with the public's day to day activities. As a result, it becomes much harder for the public to completely lose interest in seeing the project through to completion. This is especially important in consideration of the long trip times associated with a Mars mission.

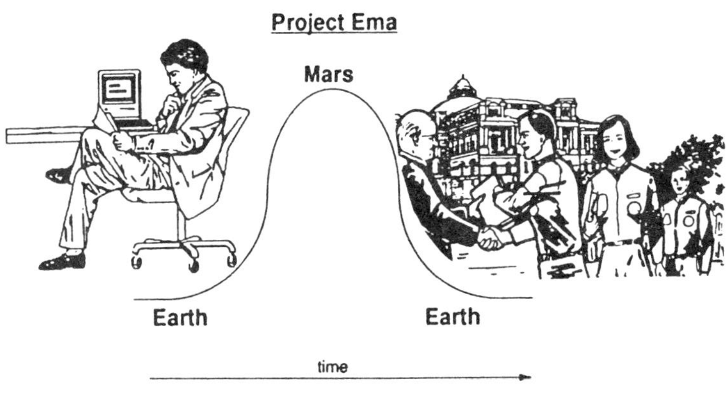

Figure 3 EMA timeline of public involvement/interest.

EMA uses a number of tactics to involve the public with the mission. One technique is the so-called '80% rule.' This signifies that 80% (as opposed to the conventional 100%) of the mission time is to be spent conducting NASA-directed scientific research. The remaining 20% is to be devoted to activities defined by other interested groups, such as educational and public interest groups. For example, the public may wish to direct a rover to visit the Viking 1 landing site. Under this rule, mission controllers would take this mission objective into consideration. This is somewhat analogous to the Get Away Special (GAS) canister program available for student-sponsored research aboard the US space shuttle. Figure 4 illustrates a rover "paying respects" to the Viking 1 lander.

Figure 4 Rover visits Viking 1.

Another EMA innovation is in the arena of education. In a conventional space mission, professional scientists design and fly complex instruments whose functioning is beyond the level of comprehension of most ordinary laymen. The scientists interpret their data and the results filter down through the educational system. The information reaching young students is never better than the worst filter in the flow of information. An EMA mission strategy is to reverse the flow of information through the educational system. Consider as an example the question of life on Mars.

Kindergarten students have simplistic ideas of how to search for life on Mars, yet these simple ideas have never been tried. The EMA mission plan is to permit kindergarten students to direct the Mars surface astronauts in the first search for life. After they complete their experiment, first grade students would be allowed to instruct the astronauts in conducting the first graders experiment in a search for life on Mars. This is to continue by grade level all the way through graduate school. Each grade builds upon the results of the lower grades, so that by the time the professional scientist becomes involved all of the obvious experiments will have been conducted, leaving the professional free to pursue the most subtle and esoteric of experimental designs. Due to the competitive atmosphere of this scheme students will tend to have their minds dragged to higher levels of sophisticated thinking as they strive to build on their underclassmates work and keep up with their upperclassmates experimental designs. Figure 5 illustrates sixth graders directing the astronauts in a search for life.

Figure 5 Students direct a search for life.

A Baseline Mission

Any serious program to explore Mars will entail at least 10 years of preparation time. During that time, a new kind of exploration astronaut could be created. We propose to establish a scholarship program for selecting preliminary astronaut candidates. Once each year, graduate students who have just earned a Masters Degree can take a battery of tests that we refer to as the Preliminary Candidacy Exam. The top scorers on those tests would receive scholarships to earn two additional Masters Degrees in two disparate fields at other universities. For example, one candidate may earn Masters Degrees in geology, meteorology, and electrical engineering. Even though this process will take years, we have the time to do it. From this expanding pool of preliminary candidates the pool of Mars surface exploration specialists will eventually be selected. The university coursework would be integrated with practical field experiences directed toward the Mars mission. As part of the EMA mission, the candidates would keep journals of their experiences and impressions.

As part of the training program, teams of preliminary candidate astronauts would be sent out to explore parts of Earth. These exploration trips would generally be low key, and conducted on a modest budget. For example, candidate students from Cornell, Northwestern, and MIT might be dispatched by train to spend a week examining a lava basin in eastern Washington. Only the last few exploration trips of the program would be the spotlight catching, expensive outings to exotic locales, such as Antarctica, a near Earth asteroid, or the Moon.

The EMA baseline mission calls for emplacing an unmanned base on the surface of Mars prior to dispatching any humans to Mars. The team that is sent to Mars is expected to spend nearly 700 days on the surface. We generally favor the hardware configuration and technical approach advocated by Robert Zubrin of Martin Marietta for our own baseline mission.

Upon returning to Earth, the critically important second exploration phase of Earth begins. Consider, for example, an exploration of the United States by the returning Mars astronauts. The US Congress is to select the landing site of the space shuttle carrying the returning astronauts. Congress is also to select the exploration route across the US to be followed by the astronauts. Our baseline mission calls for the astronauts to spend seven days exploring the US, and to travel around inside a specially prepared railroad car attached to a train. The astronauts would likely be quarantined, and not yet able to walk about in Earth gravity. Still, the passage of the train through many population centers would have a major impact on the residents of those cities. The astronauts would be recording in their journals their impressions of seeing those cities and those people and the countryside after having lived on another planet. The public would have a limited opportunity to examine some of the artifacts and specimens returned from Mars in an exhibit car to be attached to the train.

The EMA baseline mission terminates in the city of Washington, DC. The Mars astronauts are projected to first walk on Earth as they step from their train at Union Station in Washington. From there, they are to make their way a short distance up First Street to the Library of Congress. They will pass by the Supreme Court, congressional office buildings, and the Capitol building along the way. The EMA mission to Mars officially ends when the astronauts present their journals to the Library of Congress.

CONCLUSION

Project EMA involves many elements of our society. Students will be motivated to learn about other environments. Congress will select the Earth exploration routes symbolic of democracy. Environmentalists will prepare those routes for the post-Mars phase of the mission. Private firms will profit from the commercialization of the mission. Our project EMA provides a vision for space exploration by the United States.

REFERENCES

Stafford, Thomas P., "America at the Threshold: America's Space Exploration Initiative," Report to the Vice-President of the United States, June 1991.

"Final Report on Assessment of Technologies for the Space Exploration; Initiative (SEI)," AIAA, December 31, 1990.

Ride, Sally K., "Leadership and America's Future in Space," A Report to the NASA Administrator, August 1987.

Zubrin, R., Baker, D., and Gwynne, O., "Mars Direct: A Simple Robust, and Cost Effective Architecture for the Space Exploration Initiative," AIAA 91-0326, 29th Aerospace Sciences Meeting, Reno, NV, Jan. 7-10, 1991. Also published as AAS 90-168, in Proceedings of *The Case for Mars IV*, Boulder, CO, June 1990. ed. T. R. Meyer. American Astronautical Society, *Science and Technology Series*, Vol. 89, pp. 275-314, 1997.

CHAPTER 3:
HUMAN MISSIONS I:
TRANSPORTATION CONCEPTS

AAS 93-875

MAGNETOBRAKING FOR MARS RETURN VEHICLES

Geoffrey A. Landis*

It is proposed that magnetobraking may be used to dissipate hyperbolic excess velocity from a spacecraft returning from Mars to Earth's orbit. In magnetobraking, an electodynamic tether is deployed from the spacecraft. The Earth's magnetic field produces a force on electrical current in the tether, which can be used to either brake or accelerate the spacecraft without expenditure of reaction mass. The peak acceleration on the Mars return is 0.7 m/sec^2, and the amount of braking possible is dependent on the density and current-carrying capacity of the tether, but is independent of length.

Since energy is produced as the spacecraft velocity decreases, no on-board power source is required. A difficulty in magnetobraking is dissipating the energy produced. As the spacecraft approaches the Earth, the magnetic field increases and the power produced by the tether increases, reaching a maximum of about 80 kW per kilogram of spacecraft mass at closest approach. If the tether current is deceased during closest approach, this peak can be reduced, at the cost of reducing the braking effectiveness.

The energy produced as the spacecraft crosses the Earth's magnetic field can be used, for example, to drive an electric-propulsion engine such as an ion engine or a pulsed-inductive thruster. The power densities produced by an electrodynamic tether in the Earth's magnetic field can be orders of magnitude higher than power densities produced by other means, and hence extremely effective electric propulsion is possible. Used together, an electrodynamic tether and electric propulsion could be the most effective solution for braking a returning Mars vehicle.

INTRODUCTION

It has often been proposed that a vehicle returning from Mars will use aerobraking in the Earth's atmosphere to dissipate hyperbolic excess velocity to capture Earth orbit. Here a different system for dissipating excess velocity without expenditure of reaction mass, magnetobraking, is analyzed.

Magnetobraking uses the force on an electrodynamic tether in the Earth's magnetic field to produce thrust. An electrodynamic tether is deployed from the spacecraft as it approaches the Earth. The Earth's magnetic field produces a force on electrical current in the tether. If the tether is oriented perpendicularly to the Earth's magnetic field and to the direction of motion of the spacecraft, force produced by the Earth's magnetic field can be used to either brake or accelerate the spacecraft without expenditure of reaction mass.

---

* NASA Lewis Research Center, 302-1, 21000 Brookpark Road, Cleveland, Ohio 44135.

Since energy is produced as the spacecraft velocity decreases, no on-board power source is required. The magnetobraking concept is shown in schematic in Figure 1. (Regrettably, this approach cannot be used to brake a spacecraft arriving at Mars, since Mars has a very low or non-existent magnetic field.)

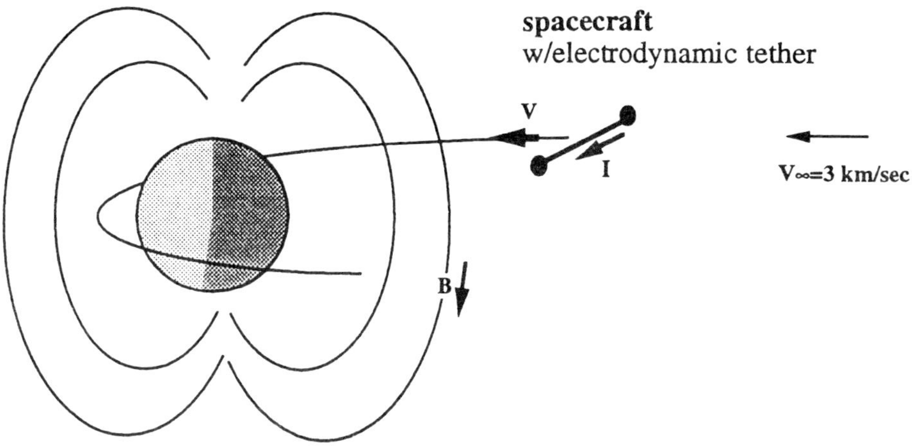

Figure 1 Schematic for magnetobraking a spacecraft in the Earth's magnetic field. The force produced by an electromagnetic tether is perpendicular to the current I and to the magnetic field B.

General principles of electrodynamic tethers are discussed in many references [1-4], and will not be reviewed here. There have been several demonstrations of tether operation in space. The experiment TSS-1 ("Tethered Satellite System-l), although it did not deploy to the full length, demonstrated controllability of a tether in space in the most critical regime of short tether lengths as well as verifying the V x B voltage expected by the conducting tether perpendicular to the Earth's magnetic field. The Delta-launched SEDS-l ("Small Expendable Deployer-l") demonstrated full deployment using a simple friction-braked reel. The follow-on experiment SEDS-2 demonstrated the plasma-contactor and the principle operational features of electrodynamic tether. Thus, all the required principles of tether technology have been demonstrated in orbit.

ANALYSIS

Can an electrodynamic tether be used to brake a vehicle coming in from Mars?

A schematic of the capture is shown in Figure 2. In order to capture into Earth orbit from a Mars return trajectory ("magnetocapture"), it is necessary that the tether remove the hyperbolic excess velocity V_∞ during the first pass past Earth. After the hyperbolic excess is removed, the resulting orbit can be reduced from the high elliptical orbit over as many subsequent passes as desired. Equivalently, it is necessary that the braking system remove an amount of energy E_∞ equal to the kinetic energy of the spacecraft before it enters the Earth's gravitational well.

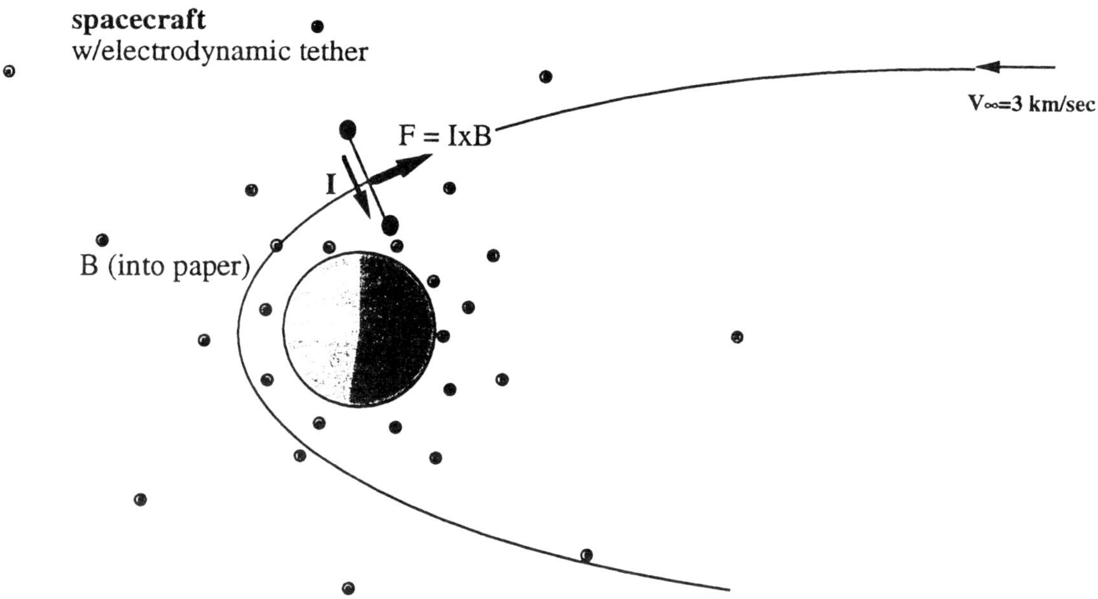

Figure 2 Magnetic force on a tether during magnetobraking (view from magnetic north).

The hyperbolic excess at arrival in a Hohmann transfer orbit from Mars to Earth ranges from 2.4 to 3.4 km/sec. For analysis of a typical case, a hyperbolic excess of 3 km/sec is assumed. Thus the energy to be dissipated is:

$$E_\infty = 1/2mV_\infty^2 = 4.5 \text{ Megajoules/kg} \quad (1)$$

The force produced on a tether by the Earth's magnetic field is:

$$F = IL \times B \quad (2)$$

where L is the length of the tether, I the current, and B the magnetic field. The Earth's magnetic field at a distance r from the surface can be expressed as:

$$B = B_0(R_e/r)^3 \quad (3)$$

R_e, the radius of the Earth, is 6.4×10^6 m. The surface field B o is:

$$B_0 = 35\mu\text{Tesla} \cos(\lambda) \quad (4)$$

where λ = latitude from magnetic equator. Here we will assume that the spacecraft comes in along the magnetic equator, and hence $\lambda = 0$. The tether is also assumed to be aligned perpendicular to the magnetic field.

The total energy dissipated by a force can be found by integrating the force over the distance:

$$E = \int F\,dx \quad (5)$$

In radial coordinates, dx = dr/cosϕ, where ϕ is the angle between the orbit direction and the radius vector. To simplify the analysis, we replace dr/cosϕ by dr. Since cosϕ is always greater than one, this will result in an underestimate of the amount of energy dissipated by the magnetic force, and hence be a conservative approximation.

This simplification is equivalent to assuming that the spacecraft trajectory is a radial line.

The total integral for flying in from infinity and flying out again is twice the integral from ∞ to R_e, so the energy is:

$$E = 2\int_{-\infty}^{R_e} LIB_o (R_e / r)^3 \, dr = LIB_o R_e \tag{6}$$

Inserting the values for the Earth's surface field and radius:

$$E = LI(3.5 \cdot 10^{-7})(6.4 \cdot 10^{6}) = 22LI \tag{7}$$

Setting this equal to the energy needed to be dissipated, 4.5 x 10^6 Joules/kg, gives the criterion for braking to be effective:

$$LI / m \geq 2 \cdot 10^{6} \; amp-meters / kg = 2 \cdot 10^{6} \; amp-km / ton \tag{8}$$

For example, one ton could be braked by a current of 1000 amps on a tether 2000 km long. Again, the approximations are conservative, and this result is an upper limit.

Note that the braking effectiveness is dependent only on the product of current and length, and not on the length of the tether. The product LI/m is referred to as the specific current density. For a tether with no payload, it is equal to the current density (amps per square meter) divided by the density of the material.

The maximum deceleration occurs at perigee,

$$F_{max} = LIB_o \tag{9}$$

Substituting the LI/m product of 2 x 10^6 amp-meters/kg, the maximum acceleration is 0.7 m/sec^2. This is a quite moderate acceleration, 1/14th of gee.

During this process, energy is being produced. The maximum energy production rate occurs at perigee, and at this point half of the hyperbolic excess velocity has been dissipated. The velocity is thus: $V_{perigee}$ = 11.3 km/sec. The rate of power production is:

$$P = FV = LIB_o V_{perigee} = 0.04LI \tag{10}$$

Substituting the LI/m product of 2 x 10^6 amp-meters/kg, P = 80 kW/kg.

The approximation used mean that this value is somewhat overestimated, and the actual peak value will be somewhat less. Nevertheless, this is extremely high by power system standards, although it is much lower than the rate of energy dissipation during an aerobraking maneuver. The dissipated power is lower because the Earth's magnetic field extends much further away from the surface than the Earth's atmosphere, and hence the same amount of energy is dissipated over a much longer time.

If the tether current is deceased during closest approach, this peak can be reduced, at the cost of reducing the braking effectiveness.

SUPERCONDUCTING TETHER

The values of current density and density will depend on the material used. The new "high temperature" (HT) superconductor materials are less well developed technically, but can be operated at temperatures of 77°K and higher, allowing the possibility of cooling by passive means in space. Superconductors are characterized by a maximum current density, known as the critical current. This is a function of the applied magnetic field and the temperature.

The most important value is the current density which has been obtained from material in the form of wires. $YBa_2Cu_3O_{7-x}$ (referred to as YBCO) is the most well-developed of the HT superconductor materials. In wire form, YBCO can be purchased with a critical current density (at 73°K) of 1000 A/cm^2. The density ρ of YBCO is 6.3 gr/cm^3. A somewhat less developed HT superconductor, Bi 2223 ($Bi_2Sr_2Ca_2$ Cu_3O), can be purchased in strip form with a critical current density of 3500 A/cm^2 (J = 3.5 x 10^7 amps/m^2) at 77°K.

For currently available Bi 2223 superconductor wire, the specific current density LI/m (=J/ρ) is 5.5 x 10^6 amps-meters/kg, twice the minimum specific current density needed for magnetobraking (i.e., a wire of this material could magnetobrake twice its own weight.).

However, both the achievable current densities and the technology for making HT superconductors in wire form is advancing rapidly. It is plausible that current densities now achieved in bulk material may in the near future be achievable in material manufactured into wires. YBCO has a current capability at 77°K on the order of 10^8 amps/m^2 in bulk form. This represents a specific current density of 1.6 x 10^7, eight times the minimum required for magnetobraking. Other superconductor materials have even higher critical currents. Critical current (at 77°K) for HoBaCuO is 3 x 10^{10} amps/m^2. If the HoBaCuO material could be made into wire form, it would have a specific current density of 4 x 10^9 amp-meter/kg, two thousand times higher than that required for magnetobraking. Hence, the mass of the conductor required for magnetobraking would be only 0.05% of the spacecraft mass. To this mass, of course, would have to be added other elements of the system, including a method of handling the power produced, plasma contactors, cooling systems, and structural elements.

In summary, the current densities required for magnetobraking are high, but within levels which are technologically achievable.

Can the tether radiate the power generated by braking and still remain superconducting? A superconducting tether will produce no power due to resistance. In the most desirable case, the power might be dissipated directly to space in the form of resistance in the space plasma or at the contact between the tether and the plasma. If there is insufficient resistance, however, it will be necessary to add resistance along the tether. Assume that this resistance is distributed along the tether, thermally in contact with the superconductor, and radiates directly to space.

Assume that the radiator area is equal to the tether area, and that the tether is shielded from the incident solar heat load. The tether is assumed to be of a rectangular cross-section of length L, width W, thickness t, and temperature T. The radiated power is then

$$\varepsilon\sigma T^4 = (P/m)t\rho \qquad (11)$$

Assuming that the emisivity $\varepsilon = 1$, and substituting in the values $\sigma = 5.67E\text{-}8$ $W/m^3 {}^\circ K$, P/m =80,000 W/kg, and (for a YBCO tether) $\rho = 6300\ kg/m^3$, then if t = 1 mm,

$$T = \sqrt[4]{\rho(P/m)t/2\sigma} = 1450\ K$$

This is hardly superconducting!

However, this temperature is purely a function of the mass to surface area ratio. If we now look at an alternative structure, a one-half micron thick film of superconductor deposited on a five-micron thick polymer substrate. The density of Kapton is roughly 1.4 gr/cm^3, and thus:

$$T = \sqrt[4]{\rho(P/m)t/\sigma} = 52\ K$$

In this configuration, the tether can radiate away the entire power generated by braking, and still remain superconducting (assuming other heat sources don't contribute).

Braking a mass of one ton on a 1000 km tether requires 2000 amps. Assume critical current density of 10^6 $amps/cm^2$ (i.e. values currently achieved by HoBaCuO at 77°K), the required tether width is 40 cm.

CONVENTIONAL TETHER

Is it possible to do magnetobraking using a non-superconducting tether? A non-superconducting tether must have some resistance. Using the notation R = resistivity (Ω) and R = resistance (blah),

$$\mathit{R} = RL/(tW) \tag{12}$$

then the power generated by resistance is

$$P = I^2\mathit{R} = I^2RL/(tW) \tag{13}$$

Using m = ρtWL and substituting values calculated earlier, the power dissipated by resistance is:

$$P/m\ (watts/gram) = 4\cdot10^{10}\ \rho(g/cm^3)\ R(\Omega\text{-}cm) \tag{14}$$

Assume an aluminum conductor, which has both low mass and low resistivity:

$$\rho = 2.7\ g/cm^3$$

$$R = 2.65\ \mu\Omega\text{-}cm$$

the power dissipated by resistance is 290 kW/gram! The power produced by braking is only 80 W/gr, so this power is far more than provided by the magnetic field. It is also far higher than the power level of any electrical power source that uses technology available now or in the foreseeable near-term future. Thus, it will not be possible to do magnetobraking in the Earth's magnetic field using a non-superconducting tether.

FURTHER ISSUES

This analysis has assumed that the tether can be kept straight, that a return loop through the plasma can be maintained, and that the energy produced can be dissipated, radiated, or absorbed.

The tether itself is only half of the electrical circuit; a return path for the current must be provided through the space plasma. This requires a "plasma contactor" at both ends of the tether to make the connection between the tether and the plasma. At the positive end (anode), the plasma contactor either collects electrons or emits ions; at the negative (cathode) end the contactor collects ions or emits electrons. Since electrons are lighter and hence more mobile than positive ions, most electrodynamic tether designs use an electron collector as a cathode and an electron gun as an anode. Many designs for plasma contactors have been proposed; several designs have been tested on the ground [5] and both electron guns and hollow-cathode plasma contactors have been tested in space. However, the typical currents of under 10 amps produced in today's designs are two orders of magnitude smaller than the currents required for magnetobraking even modest return payloads.

The material strength of the tether is important. This is most significant if the tether is of extremely high length. It is assumed that appropriate high-tensile strength fibers are incorporated as structural elements in the tether design. Since the force is uniformly distributed perpendicular to the tether direction, if the tether is left free, it will bend backward, rather than remain perpendicular to the magnetic field. Since a tether has negligible stiffness against bending, this requires a ship design which does not allow the tether to align with the magnetic field; for example, a design in which the returning ship is composed of two equal mass segments at each end of the tether.

Stability, guidance, and control are also significant issues which are not addressed here.

The question of space debris severing a tether is often brought up while discussing uses of tethers for critical missions. Fortunately, this problem has been recently addressed by Forward and Hoyt, who have shown that it is possible to design a multistrand tether which is tolerant to large numbers of debris impacts [6].

The energy produced as the spacecraft crosses the Earth's magnetic field need not be dissipated or radiated into space. It can be used, for example, to drive an electric-propulsion engine such as an ion engine or a pulsed-inductive thruster. The power densities produced by an electrodynamic tether in the Earth's magnetic field can be orders of magnitude higher than power densities produced by other means, and hence extremely effective electric propulsion is possible. Used together, an electrodynamic tether and electric propulsion could be the most effective solution for braking a returning Mars vehicle.

It has been assumed so far that the tether force has been entirely opposite to the spacecraft momentum. It will be more effective, however, if the thrust vector is oriented so that a component of the force is directed toward the Earth during closest approach. This prolongs the period of time that the tether spends in the highest magnetic field region (and also the highest plasma density region), hence increasing the total amount of braking possible and reducing the current density and the power dissipation required. In the limiting case, one could consider a tether that produces a radial inward force sufficient to circularize the trajectory; in this case the hyperbolic excess could be dissipated arbitrarily slowly. This would require, however, tethers with specific current density far in excess of those discussed so far.

OTHER APPLICATIONS

The concept of magnetobraking is not limited to Mars return applications. It can also be used during a gravity assist in the Earth's magnetic field, either to produce thrust, to change the orbital plane, or to add magnetic force to the gravity force in a magnetically-enhanced gravity assist.

Jupiter is a much better planet for electrodynamic tether use, with a stronger magnetic field ($12B_e$) and higher radius ($11.2R_e$). Magnetobraking could be used to capture into Jupiter orbit a probe to the Jovian moons; once captured, the electromagnetic tether could be used as a propulsion system throughout the Jupiter magnetosphere [7]. Alternatively, used in reverse, a magneto-gravity-assist could be used to propel a probe to Saturn or beyond.

Finally, a deep-space or interstellar probe could use a magneto-assist in the Sun's gravity field. By making a ΔV maneuver deep in the Sun's gravity well, considerable leverage of the ΔV is produced.

CONCLUSIONS

Magnetobraking in the Earth's magnetic field appears to be physically feasible. Acceleration and power dissipation are lower than for comparable aerobraking systems. Spacecraft design issues have not been studied. To reach the required current densities, superconducting tethers are needed. Passive dissipation of the generated energy appears to be feasible if the superconductor is in the form of a thin film. If the critical currents of superconducting thin-films can reach values achieved in the lab, in principle a magnetobrake could have a tether mass of less than 0.1% times the spacecraft mass.

A large number of issues remain to be addressed before it can be determined whether magnetobraking could actually be feasible for real space systems.

REFERENCES

1. P. A. Penzo and P. W. Ammann, eds., *Tethers in Space Handbook*, Second Edition, prepared by SRS Technologies for the NASA Office of Space Flight, under contract NASW-4341, May 1989.
2. *Tethers in Space: Toward Flight*; AIAA, Washington, D.C. May 1989.
3. G. von Tiesenhausen, ed., *The Role of Tethers on Space Stations*, NASA Technical Memorandum TM-86519, Oct. 1985.
4. L. Guerriero and I. Bekey, eds., *Space Tethers for Science in the Space Station Era*, proceedings of a conference in Venice, Italy, Oct. 1987, Societa Italiana di Fisica Conference Proceedings Volume 14, Bologna, Italy, 1990.
5. M. J. Gerver, D. E. Hastings, and M. Oberhardt, "Theory and Experimental Review of Plasma Contactors;" and J. D. Williams, "Plasma Contactor Research-1988" (NASA CR-182283), in Additional Presentations: 3rd International Conference on Tethers in Space Toward Flight, San Francisco, CA, May 1989, prepared by SRS Technologies, Washington DC, for NASA Office of Space Flight, 49-95.

6. R. L. Forward, "Failsafe Multistrand Tether Structures for Space Propulsion," paper AIAA 92-3214, presented at the 28th Joint Propulsion Conference, Nashville, TN, July 1992.

7. S. B. Gabriel, R. M. Jones and H. B. Garrett, "Alfven Propulsion at Jupiter," in L. Guerriero and I. Bekey, eds., *Space Tethers for Science in the Space Station Era*, 217-229.

AAS 93-876

EFFECT OF INERT PROPELLANT INJECTION ON MARS ASCENT VEHICLE PERFORMANCE

James E. Colvin* and Geoffrey A. Landis†

A Mars ascent vehicle is limited in performance by the propellant which can be brought from Earth. In some cases the vehicle performance can be improved by injecting inert gas into the engine, if the inert gas is available as an *in-situ* resource and does not have to be brought from Earth. Carbon dioxide, nitrogen, and argon are constituents of the Martian atmosphere which could be separated by compressing the atmosphere, without any chemical processing step.

The effect of inert gas injection on rocket engine performance was analyzed with a numerical combustion code that calculated chemical equilibrium for engines of varying combustion chamber pressure, expansion ratio, oxidizer/fuel ratio, and inert injection fraction. Results of this analysis were applied to several candidate missions to determine how the required mass of return propellant needed in low Earth orbit could be decreased using inert propellant injection.

NOMENCLATURE

AR - Area ratio of a nozzle
CH_4-O_2 - methane/oxygen fueled engine
CO-O_2 - carbon-monoxide/oxygen fueled engine
F - thrust (nt)
g - gravitational constant of Earth, 9.82 m/sec.
H_2-O_2 - hydrogen/oxygen fueled engine
I_{sp}(FO) - fuel+oxidizer specific impulse (sec); thrust per unit fuel and oxidizer mass flow
I_{sp}(P) - propellant specific impulse (sec); thrust per unit propellant mass flow
M_C - Earth return payload (kg), including crew capsule, crew, suits, consumables, and soil samples.
M_E - engine mass (kg)
$\dot{m}_{F+O}$ - mass flow of the energy-producing components of the propellant (kg/sec) (fuel and oxidizer without the inert).
M_f - burnout mass of stage (kg)
M_i - initial mass of stage (kg)
M_P - propellant mass (kg)
$\dot{m}_{total}$ - total mass flow of propellant (kg/sec)

* University of Arizona, Space Engineering Research Center, Tucson, Arizona.
† Sverdrup Technology, Inc., NASA Lewis Research Center Group, Brook Park, Ohio.

M_T - tank and structure mass (kg)
N_2 - nitrogen
O/F - ratio of mass flow rate of oxidizer to mass flow rate of fuel into engine
ΔV - velocity increment, km/sec

INTRODUCTION

It is possible for a return mission to the surface of Mars to be made significantly less costly if Mars derived resources can be used to provide some or all of the propellant required for the return to Earth. The mass of propellant that must be brought from low Earth orbit (LEO) to Mars for the return flight is significantly reduced by the use of local resources in propellant production.[1-5] The simplest resource available is the Martian atmosphere, which is present at every landing site and does not require prospecting or mining operations to acquire.

Most suggestions for utilization of the atmosphere of Mars involve gas separation and energy-intensive chemical processing into rocket fuel and oxidizer. However, it is worthwhile to consider uses of the Martian atmosphere which require no chemical processing.

One possible way to reduce the propellant mass brought from Earth would be to inject locally available "inert" gas into the propellant combustion flow of the return engines. (Here we use the word "inert" to designate a gas which adds mass to the exhaust but does not participate in chemical reactions with the propellant.) While this does not decrease fuel mass by as great a factor as chemical production of fuel or oxidizer, it has the advantage of being extremely simple: the process requires only compressing and liquefying the atmosphere.

Table 1 shows the composition of the Martian atmosphere.[6] The most easily available choices for inert gas available from the Mars atmosphere are carbon dioxide, nitrogen, and argon.

Table 1
COMPOSITION OF THE MARTIAN ATMOSPHERE (from Ref. 6)

Gas	Percent (by mass)
CO_2	96.8
N_2	1.7
Ar	1.4
O_2	0.1
CO	0.04
Ne	1.1 ppm
Kr	0.6 ppm
Xe	0.2 ppm

THEORY

Rocket propellant can be conceptually divided into two elements: energy source and reaction mass. For some rocket types, such as a nuclear rocket, these are separate items. In a conventional chemical rocket, the energy source and the reaction mass are the same. However, it is also possible to add additional reaction mass to the exhaust. In this case the energy source is the chemical fuel and oxidizer, and the reaction mass is this plus the added inert component.

In the idealized case where the inert mass does not affect the combustion energy release, the thrust (and hence the total impulse) increases as the square root of the total mass flow:

$$F(\text{with inert}) = F(\text{F+O only})\sqrt{\frac{\dot{m}_{total}}{\dot{m}_{F+O}}} \qquad (1)$$

Specific impulse (I_{sp}) is the thrust per unit of mass flow. When inert reaction mass is added into the propellant, it is possible to define specific impulse two different ways. We will define the *propellant* specific impulse $I_{sp}(P)$ as the thrust per unit *propellant* mass flow, where the propellant includes both the energy producing components (fuel and oxidizer) plus the additional inert mass. The propellant specific impulse thus decreases as the square root of the total mass flow:

$$I_{sp}(P) \text{ with inert injection} = I_{sp}(\text{no inert})\sqrt{\frac{\dot{m}_{F+O}}{\dot{m}_{total}}} \qquad (2)$$

Here I_{sp}(no inert) is the specific impulse of the fuel/oxidizer system without inert injection, $\dot{m}_{total}$ is the total mass flow of propellant, and $\dot{m}_{F+O}$ is the mass flow of the energy-producing components of the propellant (fuel and oxidizer with out the inert).

Alternately, we define the *fuel+oxidizer* specific impulse $I_{sp}(FO)$ as the thrust per unit fuel and oxidizer mass flow, not including the additional inert mass. This is related to the propellant specific impulse by:

$$I_{sp}(F+O) = I_{sp}(P)\left[\frac{\dot{m}_{fuel} + \dot{m}_{ox} + \dot{m}_{inert}}{\dot{m}_{ox} + \dot{m}_{inert}}\right] \qquad (3)$$

The fuel+oxidizer specific impulse *increases* as the inert flow increase. For a Mars-return vehicle where the fuel and oxidizer must be shipped from Earth, but inert gasses are locally available, the increase in fuel+oxidizer specific impulse may be more important than the decrease in propellant specific impulse.

For the case where the fuel and oxidizer are brought from Earth, it is desired to minimize the amount of fuel and oxidizer required per ton of payload delivered, or equivalently, to maximize the payload per unit amount of fuel and oxidizer. This is equivalent to optimizing the performance of a rocket where the energy source is fixed, but the specific impulse may be varied to maximize the payload. This can be calculated in closed form for the simple case where fixed mass such as tanks and engines are ignored[7,8]. If the (propellant) specific impulse is constant during thrust, the optimum is found when the specific impulse is proportional to the required mission delta-V.

$$gI_{sp} = 0.625\ \Delta V \qquad (4)$$

For low mission ΔV, the increase in thrust outweighs the loss of propellant specific impulse, and adding inert reaction mass is effective. For high mission ΔV, the additional launch mass required by the lower specific impulse means that adding inert mass is not effective.

It is also worth noting that if the specific impulse is allowed to vary during the flight (for example, if the inert mass flow is allowed to vary), the optimum occurs when the exhaust velocity [$gI_{sp}(t)$] is exactly equal to the mission velocity. This case was not examined in this study.

The present study addressed two questions:

1. How does the addition of inert mass to the combustion chamber affect the thrust and specific impulse of rocket engines when real-world loses due to thermodynamics and chemical reaction are included in the analysis?
2. Can the addition of inert reaction mass derived from the Martian atmosphere reduce the mission mass of a Mars return mission?

In addition to the possibility of increasing the amount of impulse produced by a fixed amount of fuel, it is also of interest to study inert injection for entirely Mars-derived fuel. There are two reasons for this. First, an actual operating plant to produce fuels from the Martian atmosphere will not produce fuels of 100% purity, and, in fact, it may add considerable mass and complexity to the system to remove the unnecessary ("inert") atmospheric components. Thus, it is of considerable interest to quantify how much degradation in performance results from residual inert composition of Mars-derived fuel or oxidizer. If, for example, an oxygen production plant still produces acceptable propellant if 50% nitrogen content remains in the fuel produced, the mass and complexity of the separation system may possibly be reduced.

The second reason is that propellant production on Mars is likely to be limited by the amount of energy available. Inert gas, on the other hand, will be produced by simple compression of the Martian atmosphere, a process requiring comparatively low energy consumption. Thus, like the case of Earth-derived fuels, it would be desirable to "stretch" the energy-intensive fuels with inert injection, if this is possible.

To study inert injection using Mars resources, a two phase parametric study was undertaken. In phase I, the computer code Complex Chemical Equilibrium (CEC)[9] was used to predict the effects on the vacuum specific impulse (I_{sp}) by varying the percent of inert loading levels in the total propellant combustion flowrate. This code calculates the equilibrium temperature, pressure, and chemical composition of the rocket combustion chamber and nozzle using known thermodynamic parameters of the propellant and exhaust products.

In phase II, the specific impulses calculated from CEC were used to examine the effect of inert propellant injection on several proposed mission scenarios. The desired conclusion from this study would be to show whether a mass savings from LEO could be realized by injecting inert propellant into the exhaust flow of the return engines.

EFFECT OF INERT INJECTION ON ROCKET EFFICIENCY

Three propellant combinations were studied. The hydrogen/oxygen (H_2-O_2) fuel combination is typical of the highest specific impulse system available. Methane/oxygen (CH_4-O_2) propellant has somewhat lower specific impulse, but the propellant is more compact, easier to store, and there is a possibility of manufacturing methane on Mars if hydrogen is brought from Earth.[3,4] Finally, the carbon monoxide/oxygen (CO-O_2) propellant system might be manufactured using only Mars resources.

Likewise, the three most common inert gasses in the Martian atmosphere, carbon dioxide, nitrogen, and argon, were considered.

Using CEC, each propellant was examined by investigating the addition of one inert at a time and then by varying the inert loading level to 0, 25, 50, 75, 90, and 95 percent of the total propellant mass flow-rate. The CEC code arbitrarily sets the total propellant mass flow-rate to be 1.0 kg/sec. The relative amounts of fuel and oxidizer is a variable. This is set as the mixture ratio (O/F) or the mass flow rate of oxidizer to the mass flow rate of fuel. The inert gas was, for convenience, assumed to be added in the same mass proportion to both the oxidizer and the fuel; this allows the combustion to have the same reaction stoichiometry independent of the inert injection. Practical engine design considerations will determine whether the inert is best added with the oxidizer, the fuel, or separately to the combustion chamber.

As an example calculation, consider the case of adding 25% inert to the baseline propellant H_2-O_2 with O/F equaling 6.0. For the specified propellant mass flow-rate of 1 kg/sec, with no inert the O_2 flow-rate is 0.857 kg/sec, and the H_2 flow-rate 0.143 kg/sec. To add 25% inert requires 0.2143 kg/sec of inert in the oxidizer and 0.0357 kg/sec of inert in the hydrogen flow. The flowrates of the original propellant components must then be reduced by 25%. Table 2 is a breakdown of the propellant flow-rate composition with the addition of 25% inert gas.

It can be seen from Table 2 that the total inert mass flow is 25% of the total, with the relative amounts of fuel and oxidizer still maintained at an O/F of 6.0. Similar calculations were used for other inert loading levels. The results of these calculations provided input for CEC and were used to investigate the three propellant mixtures.

Combustion chamber pressures considered were 1.38 MPa (200 PSI), representative of a pressure-fed engine, and 20.7 MPa (3000 PSI), representative of a pump-fed engine. The nozzle exit area ratio (AR) was 200. The engines were assumed to operate in vacuum, however, the atmospheric pressure on Mars is so low that there is negligible difference between operation in vacuum and on the Mars surface.

Table 2
EXAMPLE PROPELLANT FLOW-RATE

0.2143 kg/sec	:	inert in oxidizer
0.0357 kg/sec	:	inert in fuel
0.6428 kg/sec	:	oxidizer (O_2)
0.1072 kg/sec	:	fuel (H_2)
1.0000 kg/sec	:	Total Propellant

Figures 1-3 shows the change in propellant specific impulse with inert injection for hydrogen, methane, and carbon monoxide fueled engines. All of the inert gasses behave nearly the same, with a slightly worse performance from carbon dioxide injection compared to nitrogen or argon. The behavior follows almost exactly the ideal behavior (Equation 2). For the remainder of the study, nitrogen was chosen to be the inert added. Since there was little difference in performance between inert gasses, similar results will apply for all.

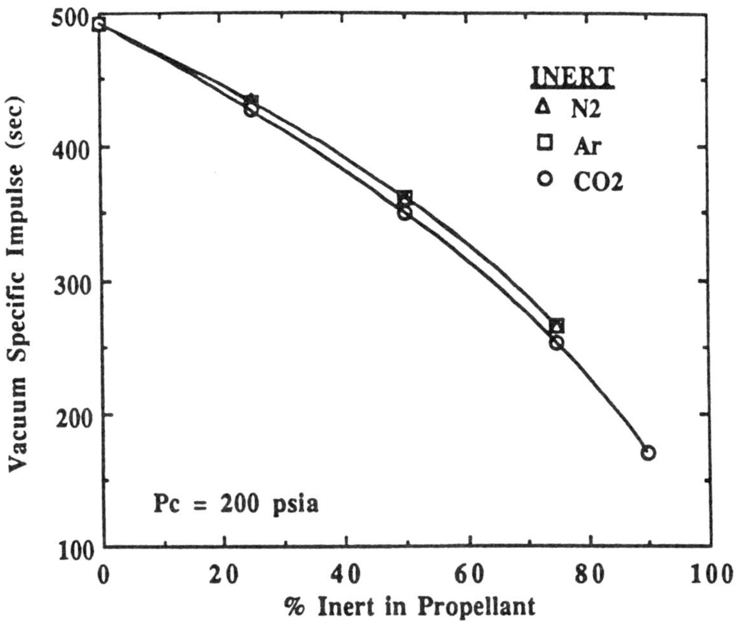

Figure 1 Effect on propellant specific impulse of addition of inert reaction mass to H_2-O_2 fueled engine.

Figure 4 shows the three fuels compared.

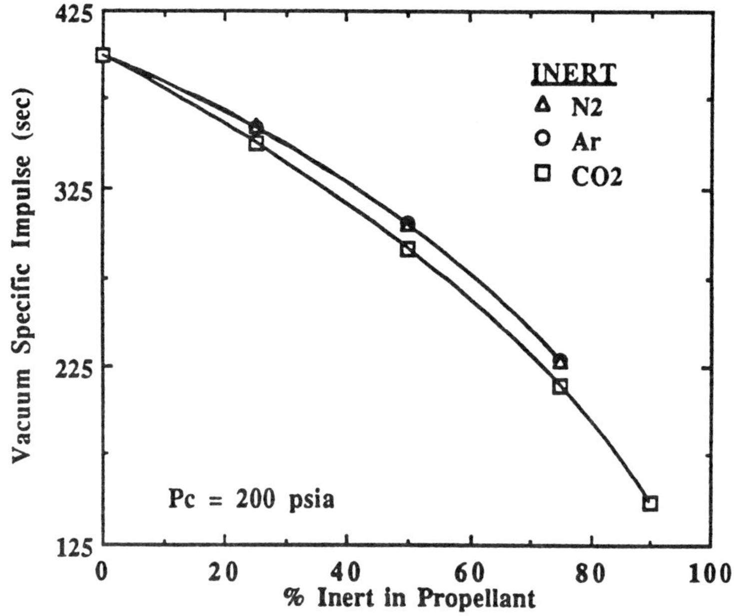

Figure 2 Effect on propellant specific impulse of addition of inert reaction mass to CH_4-O_2 fueled engine.

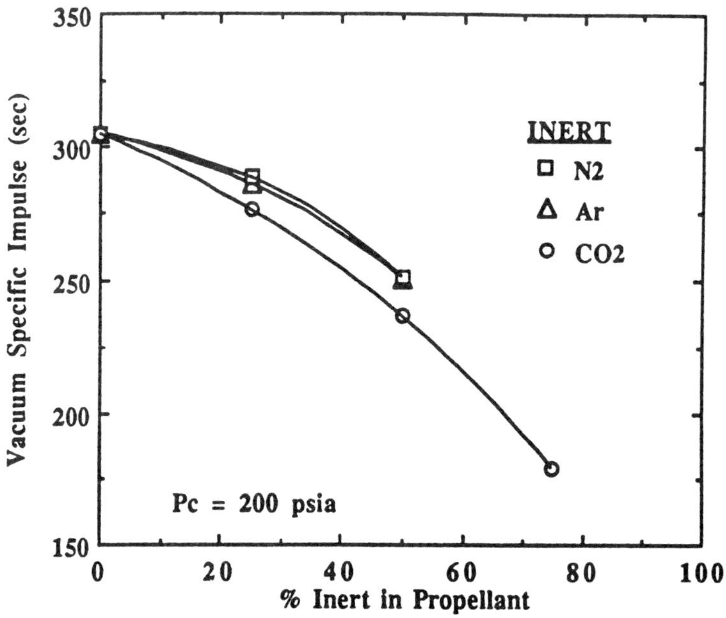

Figure 3 Effect on propellant specific impulse of addition of inert reaction mass to CO-O_2 fueled engine.

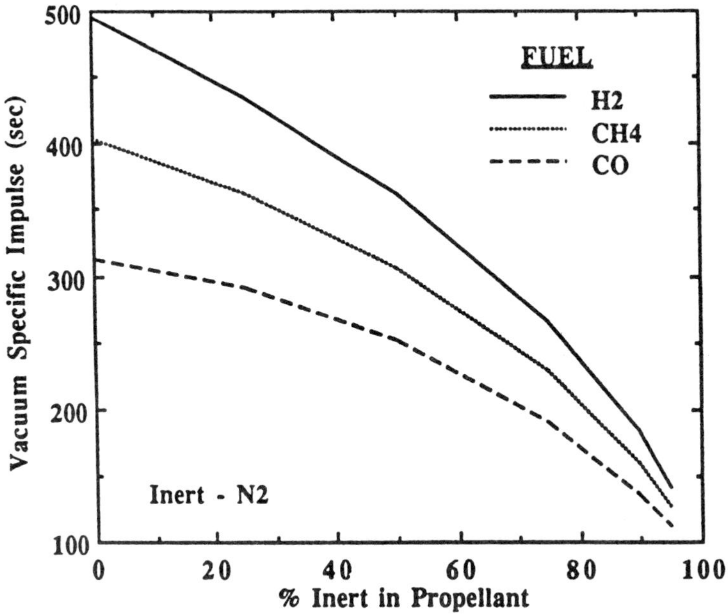

Figure 4 Comparison of effect on propellant specific impulse due to the addition of local inert to hydrogen, methane, and carbon monoxide fueled engines.

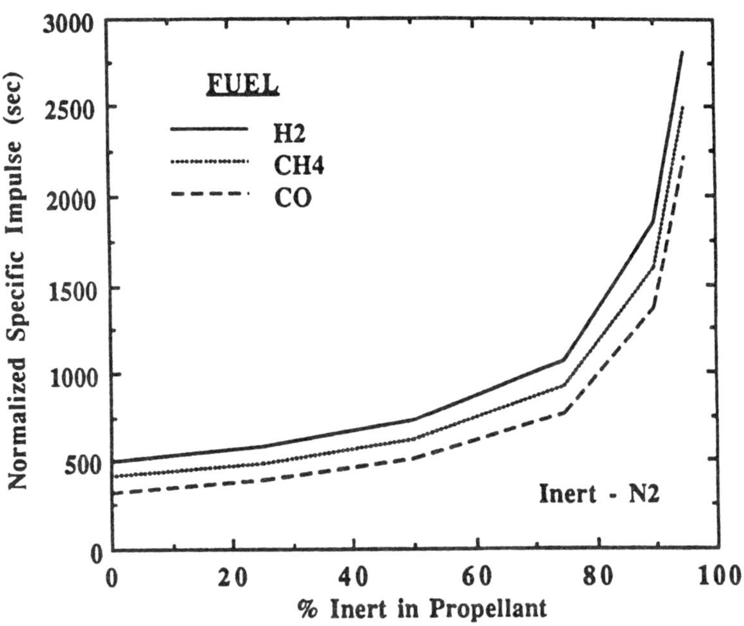

Figure 5 $I_{sp}(F+O)$ [Specific impulse normalized to fuel and oxidizer flow], showing effect of addition of inert reaction mass to hydrogen, methane, and carbon monoxide fueled engines.

Figure 5 shows the fuel+oxidizer I_{sp} (as defined above) plotted as a function of the percentage of inert in the propellant mass flow, for the same conditions as in Figure 4.

As expected from Equation 1, the fuel+oxidizer specific impulse increases as inert propellant is added to the mixture.

O/F mixture ratios used for the previous figures were: 6.0 for H_2-O_2, 3.4 for CH_4-O_2, and 0.55 for CO-O_2 fueled engines. These mixture ratios are nearly optimal for performance with no inert injection.

Similar results were obtained for the chamber pressure of 20.7 MPa (3000 PSI), shown in Figures 6-11. Here we show the effect of the inert loading on I_{sp}(P) and I_{sp}(F+O) for various mixture ratios, keeping the AR constant at 200. It is clear that the optimum mixture ratio does not depend on the inert injection.

Finally, the effect of inert loading is shown for varying nozzle area ratio, again for 20.7 MPa chamber pressure and O/F close to stoichiometric. Figures 12 through 17 show the results of this analysis. The effectiveness of inert injection does not significantly depend on the area ratio.

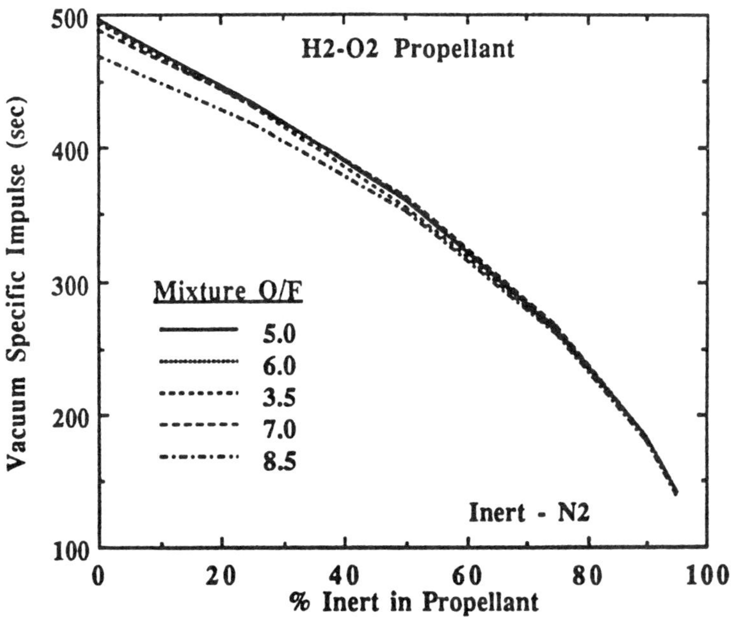

Figure 6 The effects on propellant specific impulse of addition of N_2 to H_2-O_2 fueled engine while varying the O/F mixture ratio (Exit Area Ratio Constant at 200).

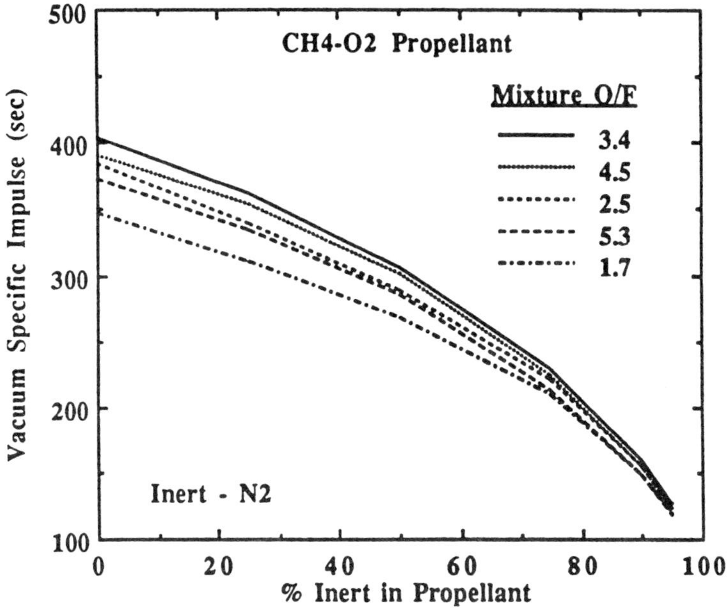

Figure 7 The effects on propellant specific impulse of addition of N_2 to CH_4-O_2 fueled engine while varying the O/F mixture ratio (Exit Area Ratio Constant at 200).

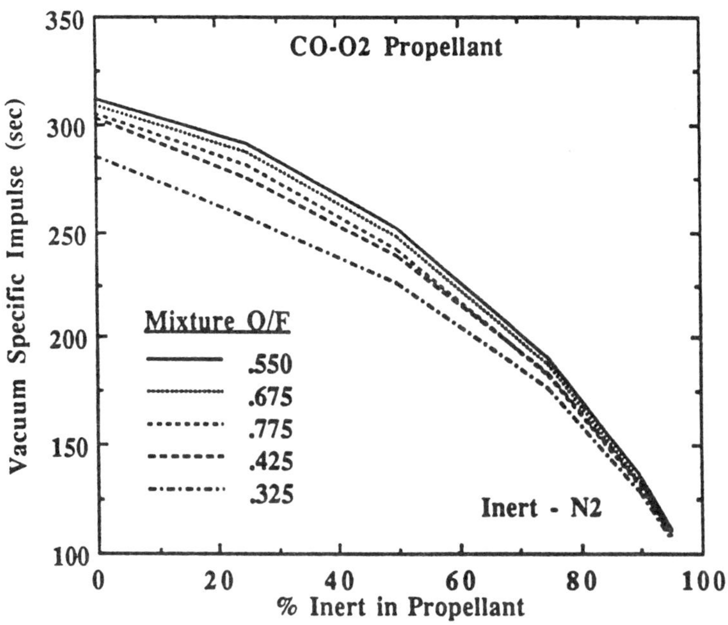

Figure 8 The effects on propellant specific impulse of addition of N_2 to CO-O_2 fueled engine while varying the O/F mixture ratio (Exit Area Ratio Constant at 200).

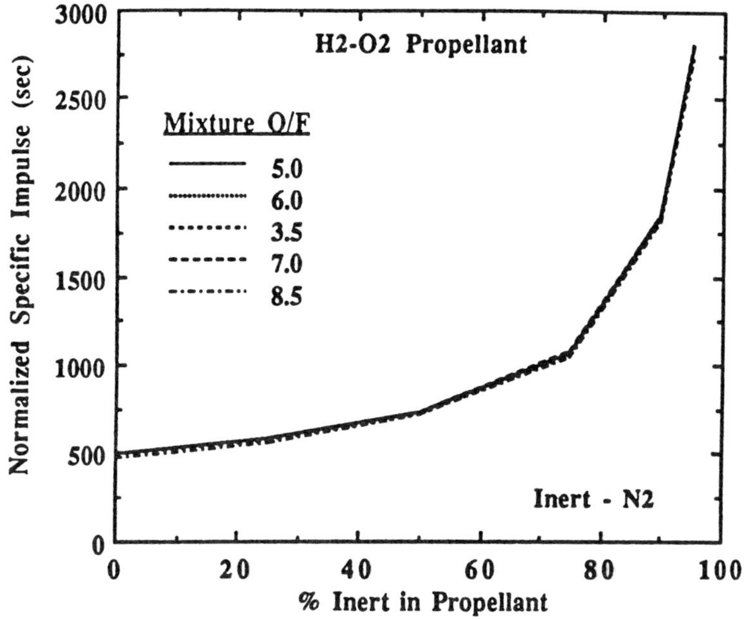

Figure 9 $I_{sp}(F+O)$ [Specific impulse normalized to fuel and oxidizer flow], showing effect of addition of inert reaction mass to hydrogen fueled engine while varying the mixture ratio (Exit Area Ratio Constant at 200).

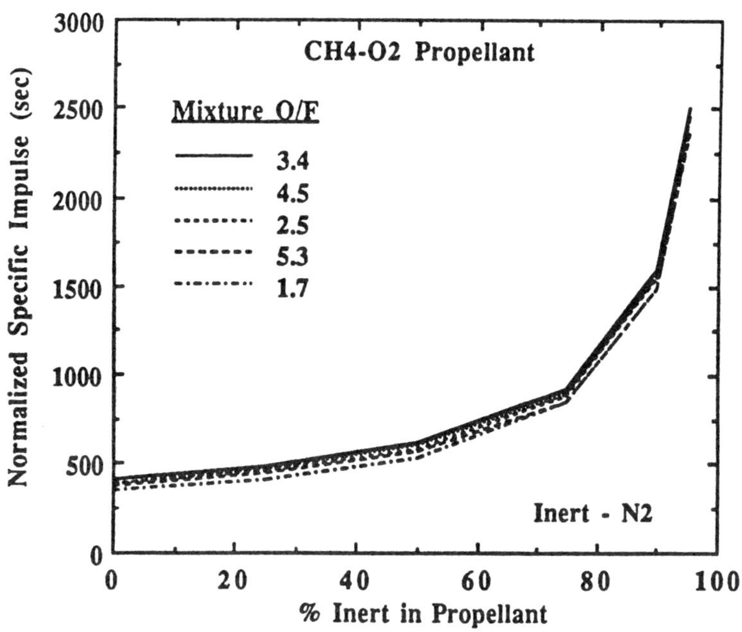

Figure 10 $I_{sp}(F+O)$ [Specific impulse normalized to fuel and oxidizer flow], showing effect of addition of inert reaction mass to methane fueled engine while varying the mixture ratio (Exit Area Ratio Constant at 200).

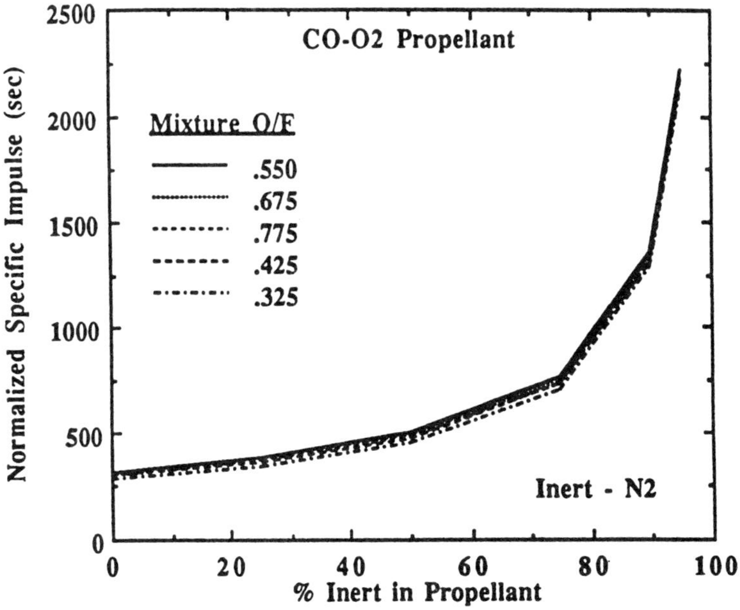

Figure 11 $I_{sp}(F+O)$ [Specific impulse normalized to fuel and oxidizer flow], showing effect of addition of inert reaction mass to a carbon-monoxide fueled engine while varying the mixture ratio (Exit Area Ratio Constant at 200).

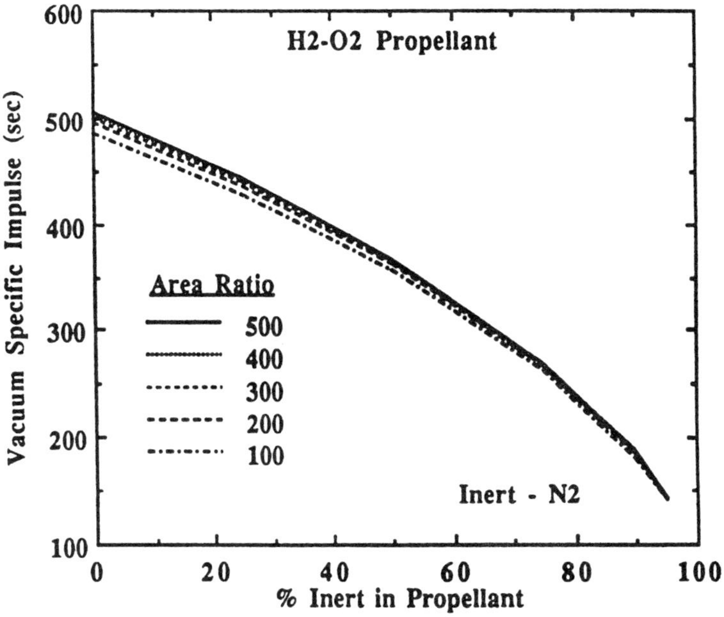

Figure 12 The effects on propellant specific impulse due to the addition of nitrogen to a hydrogen-fueled engine while varying the mixture ratio (Mixture Ratio - 6.0).

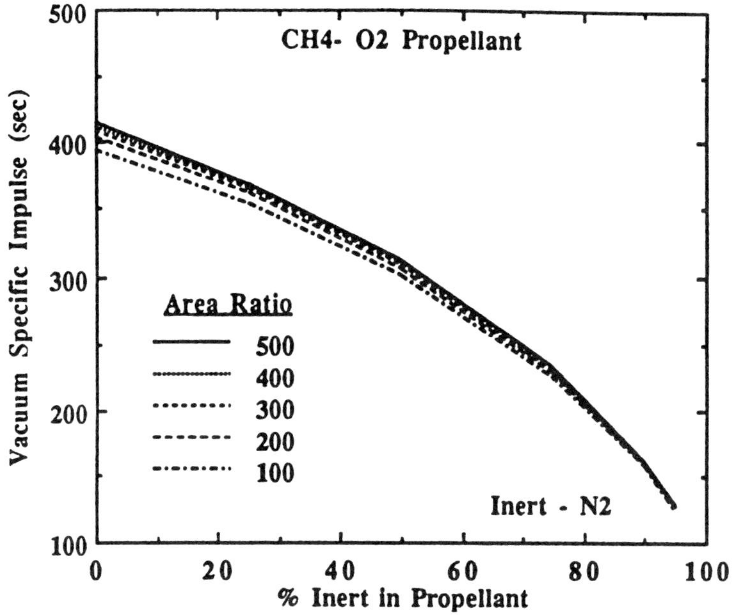

Figure 13 The effects on propellant due to the addition of motrpgem to a methane-fueled engine while varying the exit area ratio (Mixture Ratio - 3.4).

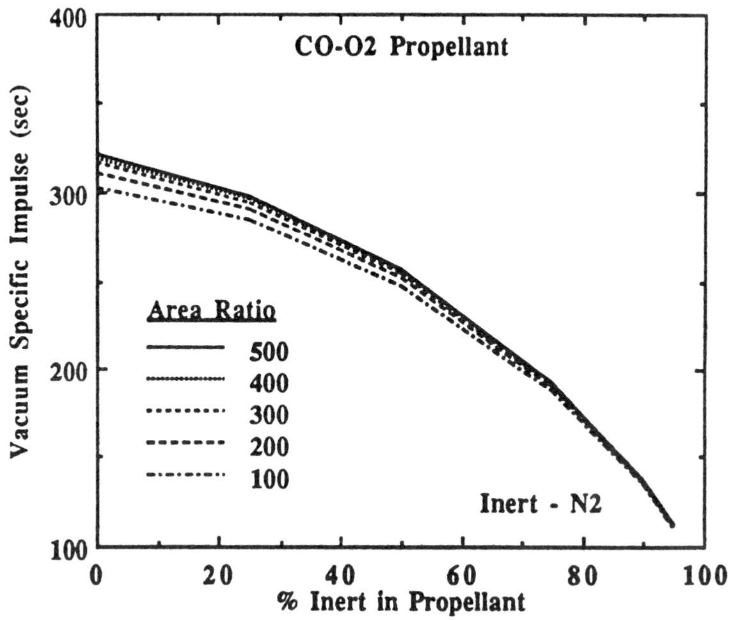

Figure 14 The effects on I_{sp} due to the addition of N_2 to a carbon monoxide fueled engine while varying the exit area ratio (Mixture Ratio - 0.55).

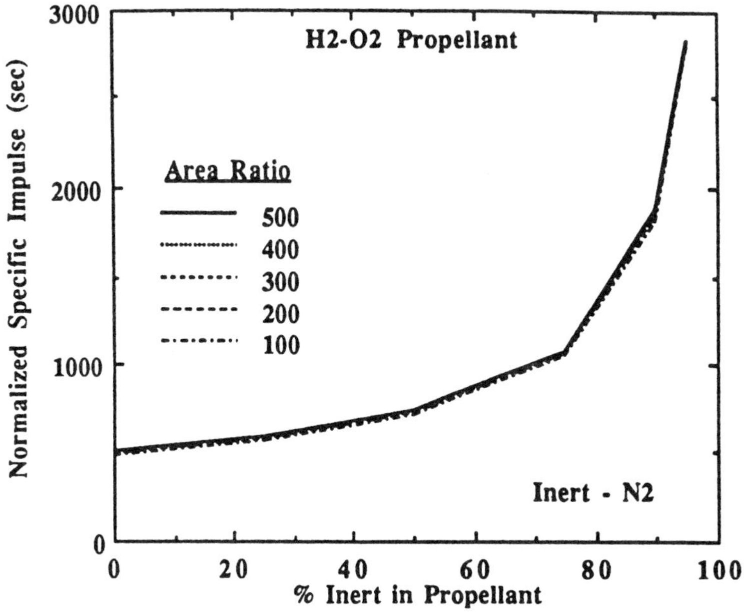

Figure 15 $I_{sp}(F+O)$ [Specific impulse normalized to fuel and oxidizer flow], showing effect of addition of inert reaction mass to a hydrogen fueled engine while varying the exit area ratio (Mixture Ratio - 6.0).

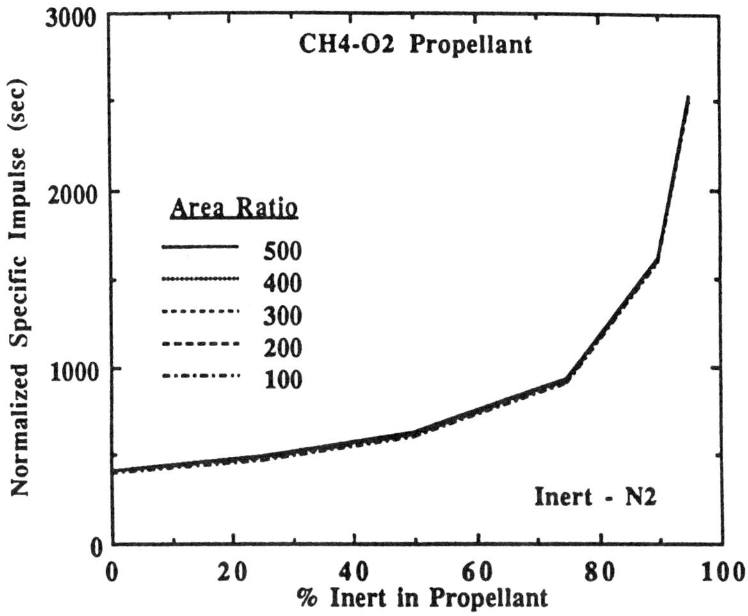

Figure 16 $I_{sp}(F+O)$ [Specific impulse normalized to fuel and oxidizer flow], showing effect of addition of inert reaction mass to a methane fueled engine (Mixture Ratio - 3.4).

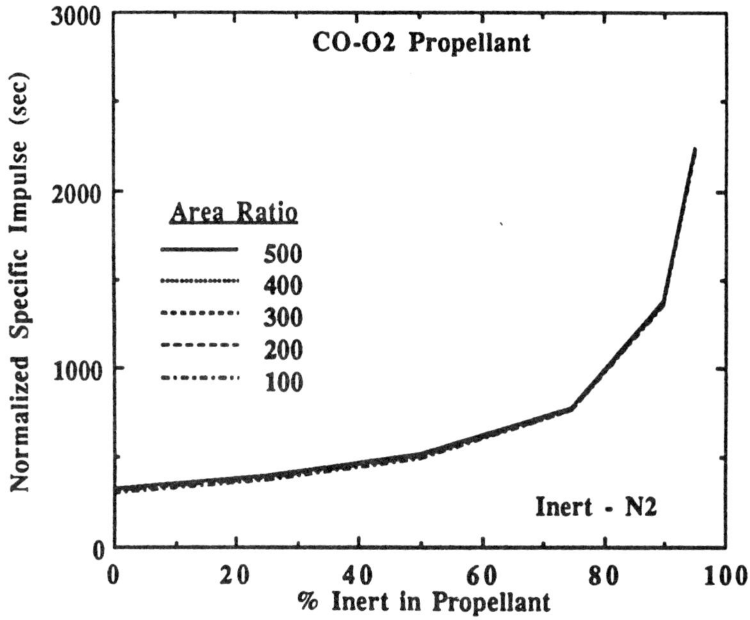

Figure 17 $I_{sp}(F+O)$ **[Specific impulse normalized to fuel and oxidizer flow], showing effect of addition of inert reaction mass to a carbon-monoxide fueled engine while varying the exit area ratio (Mixture Ratio - 0.55).**

EFFECT OF INERT INJECTION ON MISSION MASS

The data from Phase I show that as the inert loading level is increased, the propellant specific impulse decreases, but the fuel plus oxidizer specific impulse increases. The question of interest is, can the amount of H_2 brought from LEO be reduced by the addition of locally available inerts in spite of the decreasing propellant specific impulse? To answer this question, two proposed mission scenarios were examined parametrically. The first was a two-stage direct launch from the surface to an Earth transfer orbit, following the mission parameters used in the "Mars Direct" mission proposed by Zubrin and Baker.[3,4] The second scenario analyzed was a one-stage launch to Mars orbit for rendezvous with an orbiting Earth return vehicle.[10]

These two scenarios were chosen as representative of the range of values of Mars launch vehicle masses and their ΔV requirements for typical Mars return missions.

For the mission analyses, the inert used was nitrogen. Only the H_2-O_2 and CH_4-O_2 propellant combinations were considered, as the lower specific impulse of the carbon monoxide engine results in less advantage obtained from inert injection for the mission ΔVs considered.

"MARS DIRECT" MISSION

1. Engine:

For the purpose of this study, it was assumed that the engine would be very similar to that of the Advanced Space Engine designed by Rocketdyne. Mass and thrust levels were modified slightly. The assumed mass per engine was 181 kg. The assumed thrust per engine was fixed at 89 kN. In order to compare scenarios on the same basis, the thrust was assumed to be independent of the inert injection rate.

2. System Mass:

The two stage mission described by Zubrin and Baker has an Earth return payload of 12,250 kg (M_C). It was assumed for this parametric study that the tank and structure mass (M_T) would be 10% of the propellant mass per stage. The values of M_P, M_E, and M_T were determined for each level of inert injection from the rocket equation (5). Only the first stage had the inert added to its propellant mix.

3. Second Stage:

The calculation starts with the second (upper) stage, and then uses the second stage total mass as input to the calculation for the first (bottom) stage calculation. ΔV for the second stage bum was calculated from the quoted masses using the rocket equation:

$$\Delta V = gI_{sp} \ln\left[\frac{M_i}{M_f}\right] \tag{5}$$

where:

$$M_i = M_c + M_p + M_E + M_T \tag{6}$$

$$M_F = M_i - M_p \tag{7}$$

Using Zubrin and Baker's mass values for the second stage (M_p = 22,170 kg and M_E = 2,560 kg), a second-stage ΔV of 3,350 m/sec was calculated.

Injection of inert mass into the second stage engine does not improve performance, and for these calculations the second stage was assumed to have zero inert injection. The mass breakdown is therefore only dependent upon the type of propellant used. A baseline case of H_2-O_2 was compared with the CH_4-O_2 propellant assumed by Zubrin and Baker.

Table 3 shows the second-stage mass breakdown for H_2-O_2 propellant combination (I_{sp} = 495 sec.), and Table 4 shows the mass breakdown using CH_4-O_2 propellant (I_{sp} = 384 secretary).

Table 3
H_2-O_2 SECOND STAGE MASS BREAKDOWN

Propellant :	13,718 kg
Tank :	1,372 kg
Engine :	181 kg
Capsule and payload:	12,250 kg
TOTAL :	27,521 kg

Table 4
CH_4-O_2 SECOND STAGE MASS BREAKDOWN

Propellant :	20,784 kg
Tank :	2,078 kg
Engine :	181 kg
Capsule and payload:	12,250 kg
TOTAL :	35,294 kg

To verify that the assumed thrust levels and calculated masses are reasonable, the initial and final acceleration of the second stage bum were calculated, assuming the use of one 89 kN engine. Accelerations are shown in Table 5 as a fraction of Earth and Mars gravitational acceleration. The values calculated are well within tolerable ranges for both hardware and crew safety.

Table 5
SECOND STAGE ACCELERATIONS

Propellant	Initial Acceleration	Final Acceleration
H_2-O_2	$0.86g_{Mars}$	$1.72g_{Mars}$
	$0.33g_{Earth}$	$0.66g_{Earth}$
CH_4-O_2	$0.67g_{Mars}$	$1.64g_{Mars}$
	$0.26g_{Earth}$	$0.63g_{Earth}$

For the CH_4-O_2 propellant, the O/F ratio equals 3.4, and thus 0.2273 kg of CH_4 is needed per kg of propellant. The total amount of CH_4 needed is 4,724 kg. In the "Mars Direct" scenario, only the hydrogen is brought from Earth; methane is produced from it by a series of chemical processing steps. The relative masses of C and H in CH_4 is 3:1, and so the mass of hydrogen that must be brought from LEO for use in the second stage is 1,181 kg.

In the case of the H_2-O_2 propellant, the O/F ratio equals 6.0, resulting in the mass fraction of fuel in the propellant being 14.3%. The total amount of hydrogen that must be brought from LEO is 1,960 kg.

5. First Stage Accelerations:

A ΔV of 3,400 m/sec was used for the first stage bum. Mass values for the first stage for the CH_4-O_2 case (no inert) are M_p = 70,160 kg and M_E = 8,850 kg, plus the total second stage mass of 36,980 kg.

Table 6 and 7 show the first stage mass breakdown for the H_2-O_2 and CH_4-O_2 case.

Table 6

H_2-O_2 FIRST STAGE MASS BREAKDOWN

	0 % Inert	75 % Inert
Propellant	34,135 kg	110,838 kg
Tank and Structure	3,413 kg	11,084 kg
Engine	2,722 kg	2,722 kg
Payload	27,521 kg	27,521 kg
TOTAL	67,791 kg	152,164 kg

I_{sp}=495 sec (no inert); 266 sec (75% inert).

Table 7

CH_4-O_2 FIRST STAGE MASS BREAKDOWN

(ΔV = 3,400 m/s)

	0% Inert	75% Inert
Propellant	56,572 kg	196,900 kg
Tank and Structure	5,637 kg	1,969 kg
Engine	2,722 kg	2,722 kg
Payload	33,348 kg	33,348 kg
TOTAL	98,279 kg	252,660 kg

I_{sp}=404 sec (no inert); 229 sec (75% inert)

Again, acceleration was calculated to verify that the acceleration levels did not exceed reasonable limits. Consistent throughout these first stage calculations was the assumption of 15 engines with a total mass of 2,720 kg (M_E). These engines combine for a total thrust of 1,330 kN. The thrust and engine masses were kept constant to keep the number of variables to a minimum in this analysis. (For an actual mission, thrust levels would be optimized with the number of engines being adjusted, and the engines throttled as necessary). Accelerations were calculated for two cases for each propellant mix: 0% inert addition and 75% inert addition.

Calculated accelerations are shown in Table 8. The acceleration found is for all cases greater than one Mars gravity and within acceptable ranges. The purpose for using these two loading levels was to examine these values at two extremes. At an intermediate loading, the accelerations would be between these values.

Table 8

FIRST STAGE ACCELERATIONS (ΔV = 3,400 m/s)

	m/s^2	g_{Mars}	g_{Earth}
H_2-O_2	with 0% Inert		
A (initial)	19.68	5.26	2.01
A (final)	39.65	10.60	4.04
	with 75% Inert		
A (initial)	8.77	2.34	0.89
A (final)	32.29	8.63	3.29
CH_4-O_2	with 0% Inert		
A (initial)	13.58	3.63	1.38
A (final)	31.98	8.55	3.26
	with 75% Inert		
A (initial)	5.28	1.41	0.54
A (final)	23.93	6.40	2.44

6. Mass of Hydrogen Required

The mass of hydrogen brought from LEO for the first stage can now be calculated. We use the rocket equation, modified to account for tank and structure fraction (assumed to be 10% of propellant mass):

$$M_p = \frac{(\xi - 1)(M_c + M_E)}{1.1 - 0.1\xi} \tag{10}$$

where

$$\xi = \frac{M_i}{M_f} = \exp\left[\frac{\Delta V}{g I_{sp}}\right] \tag{11}$$

The mass of propellant necessary for the first stage engine can be calculated from these equations as a function of propellant specific impulse. To calculate the inert injection fraction which will give a specified propellant specific impulse, a fifth-order polynomial was fit to the calculated values of propellant specific impulse versus nitrogen addition shown in Figure 4. This was done using inert levels from 0% to approximately 90%.

The mass of hydrogen in the first stage propellant mass was calculated by a similar procedure to that used for the second stage calculations, modified to account for the inert loading level. The total mass of hydrogen to be brought from LEO is then just the sum of the amounts needed for the first and second stages.

Figures 18 and 19 show the required amount of hydrogen to be brought from Earth, for hydrogen and methane fueled engines respectively, as a function of the amount of inert added. In both cases, the amount of hydrogen required decreases with inert injection, up to an optimum inert fraction of 60% to 70%. This is consistent with the result estimated using equation (3), which predicts that the optimum propellant specific impulse for the 3400 m/sec first stage ΔV is about 215 sec., well below the specific impulse of either the hydrogen or methane propellants without inert injection. The actual optimum is slightly higher, as expected, since equation (3) does not account for engine and tank mass.

The values shown are hydrogen mass only, assuming that the rest of the propellant is manufactured locally on Mars. If both fuel and oxidizer are brought from Earth, and only inert production is done on Mars, these masses are higher by a factor of 7 (for H_2-O_2) and 17.6 (for CH_4-O_2). The location and relative value of the minimum does not change.

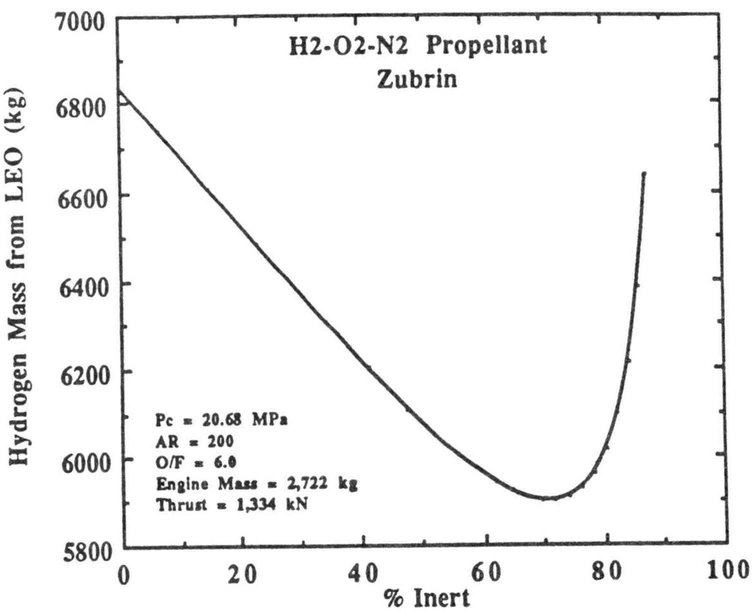

Figure 18 Hydrogen mass required to support the "Direct" return launch (two-stage launch from surface to Earth injection) using H_2-O_2-N_2 propellant mixture.

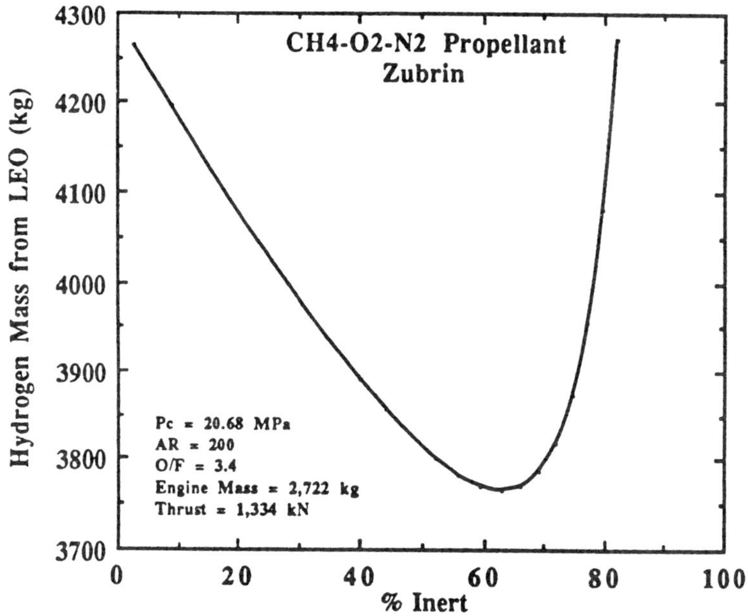

Figure 19 Hydrogen mass required to support the "Direct" return launch (two-stage launch from surface to Earth injection) using CH_4-O_2-N_2 propellant mixture.

2015 MISSION

The "2015 Mission" scenario describes a single-stage launch to a return vehicle waiting in low Martian orbit (LMO).[10] For the purpose of this study, it was assumed the engine mass was 600 kg, with a thrust level of 133.5 kN per engine. Three engines per mission were assumed. This results in a total engine mass of 1,800 kg and a total thrust of 400 kN. The payload mass is 5,400 kg, and the ΔV required is 5,320 m/s.

Table 9

H_2-O_2 PROPELLANT MISSION MASS BREAKDOWN

(Single Stage to Mars Orbit)

	0% Inert	50% Inert
Propellant	18,405 kg	40,430 kg
Tank and Structure	2,024 kg	4,447 kg
Engine	1,800 kg	1,800 kg
Payload	5,406 kg	5,406 kg
TOTAL	27,635 kg	52,083 kg

The propellant I_{sp} predicted by CEC for this engine with no inert injection was 495 seconds for the H_2-O_2 propellant and 404 seconds for the CH_4-O_2 propellant. With the addition of 50% nitrogen, these values were reduced to 363 sec. and 306 sec. respectively. Using these values for propellant specific impulse and Equations. 5-7, mass breakdowns for these missions were calculated. This analysis was performed using the assumption that the tank fraction is 11% of the propellant mass. Table 9 shows the mass breakdown for the two H_2-O_2 propellant missions, while Table 10 shows the mass breakdown for the two CH_4-O_2 propellant Missions.

Again, accelerations were calculated for the beginning and end of the engine bum to insure the feasibility of the scenario in regards to crew and equipment safety. This was performed using the assumed total thrust level of 400 kN with the derived system masses. The accelerations are shown in Table 11.

The mass of hydrogen which must be brought from LEO for the hydrogen- and the methane-fueled "2015" mission scenarios were calculated by the same method used earlier.

Table 10

CH_4-O_2 PROPELLANT MISSION MASS BREAKDOWN

(Single Stage to Mars Orbit)

	0% Inert	50% Inert
Propellant	29,553 kg	76,603 kg
Tank and Structure	3,251 kg	8,426 kg
Engine	1,800 kg	1,800 kg
Payload	5,406 kg	5,406 kg
TOTAL	40,010 kg	92,236 kg

Table 11

2015 MISSION ACCELERATIONS

	m/s^2	g_{Mars}	g_{Earth}
H_2-O_2 with 0% Inert			
A (initial)	14.49	3.87	1.48
A (final)	43.37	11.60	4.42
with 50% Inert			
A (initial)	7.69	2.06	0.78
A (final)	34.35	9.19	3.50

	m/s^2	g_{Mars}	g_{Earth}
CH_4-O_2 with 0% Inert			
A (initial)	10.01	2.68	1.02
A (final)	38.28	10.24	3.90
with 50% Inert			
A (initial)	4.34	1.16	0.44
A (final)	25.61	6.85	2.61

Figures 20 and 21 show the amount of hydrogen which must be brought from LEO as a function of the percentage of inert loading level. For this mission, there is no advantage to inert injection.

This is slightly surprising, since the simple estimation using equation (3) suggests that the optimum I_{sp} for this case should be about 340 sec. The difference shows that neglecting engine and tank mass does not result in an accurate estimate of whether inert injection will decrease the required hydrogen.

While there is no performance advantage of inert injection, the fact that inert addition has little effect on performance up to nearly 50% mass fraction of inert is itself a significant result. This indicates that high propellant purity is not a critical parameter in the design of manufacturing processes for Mars-manufactured propellant from a performance standpoint.

If the inert injection was done only at the beginning of the flight, when the required I_{sp} is low, inert injection would have also resulted in a decreased hydrogen requirement for this mission as well. However, this would have required the assumption of a more complex engine than the one assumed. While there is no reason why such an engine could not be made, this engine type is not currently under study.

CONCLUSIONS

A novel Mars mission strategy was studied, where liquefied inert gasses from the Martian atmosphere are used as inert reaction mass to inject into the combustion chamber of an Earth-return vehicle. This results in a greater mass of propellant used, but can result in a lower requirement for fuel brought from Earth. A standard rocket engine combustion code was run to analyze the effect of inert injection on engine performance.

Two missions and two propellant combinations were analyzed. For one case, a two-stage launch direct from the Mars surface to Earth, the inert injection was found to result in a reduced mass of fuel brought from Earth.

These missions were not designed to make optimum use of inert propellant injection, but were only representative cases of mission strategies currently being considered. The advantage of inert injection is greatest when inert injection is used on stages with low Δv; choosing the staging criterion carefully would clearly increase advantage. Design of new mission strategies specifically intended to take advantage of the increase in propellant leverage using inert gas injection may increase the amount gained by this strategy. For example, utilizing inert injection only during the beginning of the boost would allow the advantages of the injection but eliminate most of the penalty of higher propellant mass carried.

The mission analyses were all calculated using a fixed fraction of inert in each stage. It would be interesting to do the calculations assuming that the inert injection, and hence the I_{sp}, can be varied during the boost. This could result in additional mass savings.

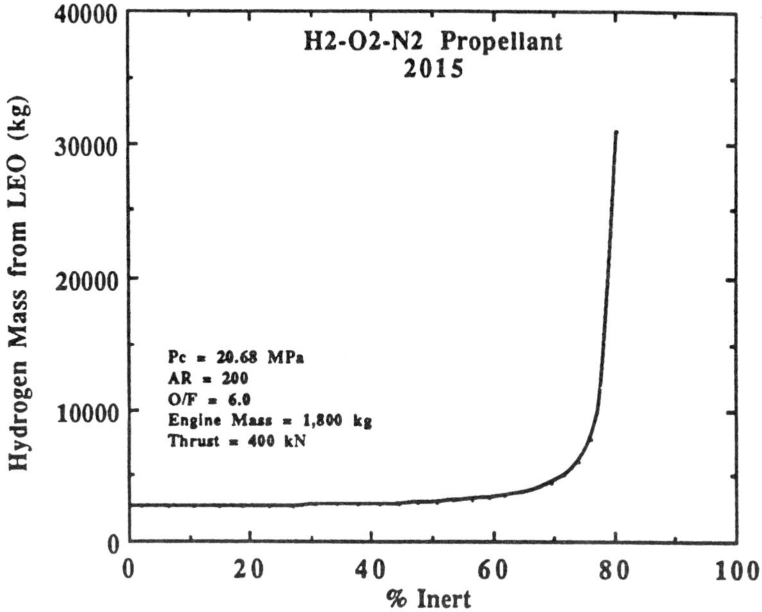

Figure 20 Hydrogen mass required to support the "2015 mission" return launch scenario (surface to Mars orbit) with H_2-O_2-N_2 propellant mixture.

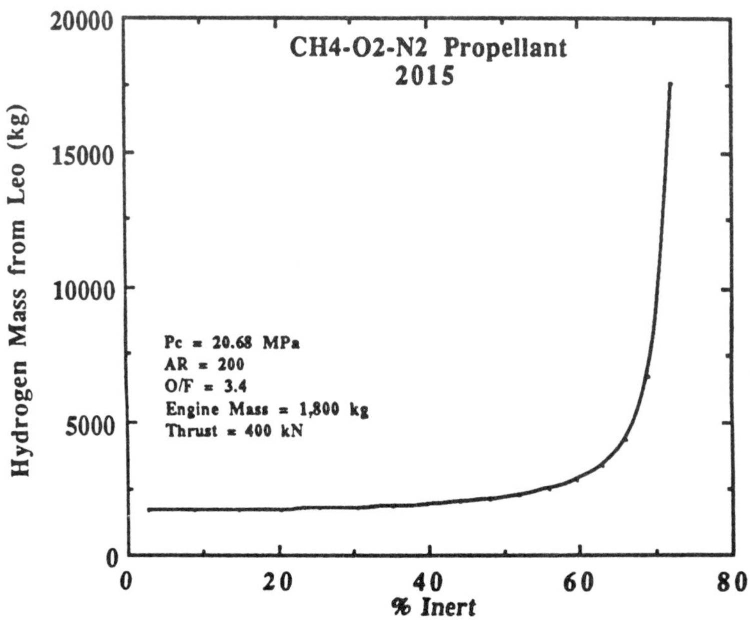

Figure 21 Hydrogen mass required to support the "2015 mission" return launch (surface to Mars orbit) with CH_4-O_2-N_2 propellant mixture.

REFERENCES

1. French, J. R., "Rocket Propellants from Martian Resources," *Journal of the British Interplanetary Society*, Vol. 42, pp. 167-170, 1989. Also published as AAS 97-375 in R. M. Zubrin (Ed.) *From Imagination to Reality: Mars Exploration Studies of the Journal of the British Interplanetary Society* (Part II: Base Building, Colonization and Terraformation), AAS *Science and Technology Series*, Vol. 92, pp. 43-51, (1997).
2. Stancati, M. L., Niehoff, J. C., and Wells, W. C., "Mars Automated Refueling System Final Report and Presentation," Report No. SAI 1-120-196-S7, October, 1978.
3. Zubrin, R. M., and Baker, D. A., "Humans to Mars in 1999!," in *Aerospace America*, August, 1990.
4. Baker, D. A., and Zubrin, R. M., "Mars Direct," *Journal of the British Interplanetary Soc.*, Vol. 42, No. 4, pp. 147-160, 1989.
5. Landis, G. A. and Linne, D. L., "Acetylene Fuel from Atmospheric CO_2 on Mars," *Journal of Spacecraft and Rockets*, Vol. 29, No. 2, pp. 294-296, 1992.
6. Kaplan, D., "Environment of Mars, 1988," NASA Technical Memorandum 100470, October, 1988.
7. Leitman, G., "Propulsive Efficiency of Rockets," 3rd International Symposium on Rockets and Astronautics, Tokyo, Japan, 1961; in expanded form in G. Leitman (ed.) *Optimization Techniques*, Chap. 13, pp. 377-388.
8. Zuppero, A., and Landis, G. A., "Optimum Rocket Propulsion for Energy-Limited Transfer," *1st Annual Symposium on Resources of Near-Earth Space*, 7-10 July, 1991, Tucson, AZ.
9. Gordon, S., and McBride, B. J., "Computer Program for Calculation of Complex Chemical Equilibrium Compositions, Rocket Performance, Incident and Reflected Shocks, and Chapman-Jouguet Detonations," NASA SP273, March, 1976.
10. Donahue, B.; Boeing Co., "Orientation/1st Intern Briefing," NASA Contract NAS8-37857, December 1989.

AAS 93-878

A STUDY OF POWERED ABORT TO EARTH ORBITS FROM EARTH TO MARS TRANSFER ORBITS

Stephen C. Merrihew*

As a significant fraction of the duration of a human exploration mission to Mars will be devoted to the interplanetary transfer orbit, in-flight failures leading to mission degradation or termination may occur. Additionally, the failure of other mission elements such as pre-supplied fuel caches or surface habitats may preclude landing on Mars. Therefore, emergency mission operations will depend upon the failure type and the mission phase. This paper investigates powered abort to Earth orbits as a mission abort mode for failures not affecting the vehicle's propulsion system. These powered abort to Earth orbits are analyzed for both high and low energy nominal Earth to Mars transfer orbits. This analysis includes contour plots of the required abort orbit change in velocity (delta-V) as a function of the time of abort orbit initiation and duration. These plots indicate regions of powered abort to Earth orbit opportunities for any given propulsive capability. Several proposed Mars exploration mission architectures are evaluated to determine their ability to perform such maneuvers. Finally, design guidelines are presented to enable abort to Earth capabilities in future Mars exploration mission architectures.

INTRODUCTION

Although Mars is the nearest hospitable planet for human exploration, the distances separating Earth and Mars are vast; 77,792,000 kilometers at the nearest and 376,992,000 kilometers at the furthest. It is these distances that make the exploration of Mars so difficult. Various mission designs, utilizing non-repeating Keplerian transfer orbit trajectories, have been proposed to place a crew of explorers on the Martian surface. These mission designs can be divided into two categories: long duration, minimum energy missions and short duration, high energy missions. Long duration, minimum energy designs aim to reduce the fuel requirements for a given mission by increasing the overall mission duration. Typically, minimum energy missions spend from 40% to 70% of the total mission duration in-flight to and from Mars. Short duration, high energy missions seek to reduce the total mission duration by increasing the energy requirements and hence the required fuel mass. These high energy, short duration missions typically require up to 90% of the total mission duration for the interplanetary flight phases.

As a significant fraction of the overall duration of a Mars exploration mission will be devoted to the interplanetary transfer orbit, care must be taken to ensure the safety of

* Ph.D. candidate, Aeronautics and Astronautics Department, Durand Building, Stanford University, Stanford, California 94305.

the crew in the event of an emergency. Emergency scenarios may include in-flight failure or degradation of primary life support, electrical power, thermal control or possibly command, control and communications systems. Potential in-flight emergencies may also arise due to crew injury or sickness. Alternatively, the failure may reside in a non-flight element; the failure of any pre-supplied systems at Mars such as habitats or fuel caches would threaten the success of the mission. Classical preparation for in-flight failures in Earth orbit and lunar missions have included abort to Earth capabilities for the crew, as demonstrated during the Apollo 13 mission. The importance of an abort to Earth capability in low Earth orbit is also indicated by the efforts to build an assured crew return vehicle for Space Station *Freedom*. In the case of Mars exploration missions the abort to Earth options are somewhat limited by the great distances between Earth and Mars. Mission abort options may be categorized according to failure type (life threatening and non-life threatening) and mission phase (Earth to Mars transfer, Mars surface and Mars to Earth transfer).

This paper investigates the powered abort to Earth orbit as an emergency response to life or mission threatening failures not affecting the propulsion system. The motivation for this work is the realization that unlike failures on the Martian surface, or even failures during the Mars to Earth transfer orbit, failures during the Earth to Mars transfer orbit may require immediate orbital maneuvers in order to ensure the safe return of the crew. The question then becomes what are the characteristics of these abort to Earth orbits? Once these characteristics are understood, mission architectures can be evaluated for abort to Earth capabilities. An analysis of free return abort to Earth trajectories (those requiring no propulsive maneuvers, hence providing an abort option in the case of propulsion system failure) is provided in wolf.[1]

The powered abort to Earth analysis was conducted in three phases. In the first phase, nominal Mars exploration transfer orbits for both high and low energy mission designs were calculated. These nominal transfer orbits constitute a double rendezvous, first the Earth to Mars leg, then the Mars to Earth leg. In the second phase, the energy requirements for the abort to Earth orbit are calculated for each combination of initiation time and abort orbit duration. Initiation time is defined as the number of days out from Earth in the nominal Earth to Mars transfer orbit while abort orbit duration is defined as the number of days to return to Earth from the point of abort orbit initiation. The change in velocity (delta-v) and atmospheric entry velocity (at Earth) are calculated and presented in contour and three dimensional plots. The resulting analysis mirrors the calculation of the nominal Mars to Earth transfer orbit as the abort orbit performs a one way transfer orbit from the nominal location in the Earth to Mars transfer orbit to Earth at the desired arrival time. With these delta-v and entry velocity values, various mission architectures are analyzed for abort to Earth capabilities in the third phase of the analysis. Mission design guidelines ensuring that abort to Earth capabilities are provided in future Mars exploration mission designs are then presented.

NOMINAL TRANSFER ORBITS

As stated previously, the Mars exploration transfer orbit is essentially a double rendezvous problem. The first rendezvous is with Mars on the outbound leg and the second with Earth on the inbound leg. The positions of Earth and Mars are functions of time, so the nominal transfer orbit is completely determined by defining the following: Date of Earth departure, date of Mars arrival, date of Mars departure, and finally date of Earth arrival.

It is important to consider the orbital geometry of Earth and Mars when calculating transfer orbit characteristics. As the orbits of both the Earth and Mars are elliptical (Mars eccentricity is 0.0934 while Earth's is 0.0167) the relative orientation of the planets varies considerably over time. The synodic period, the time required for any phase angle between the planets to repeat itself, is 2.135 Earth years. This synodic period determines the occurrence of opposition (closest approach) between the two planets. As the orbits of Earth and Mars are eccentric, non-coplanar (Mars in inclined 1.85 degrees from the elliptic) and their line of apsides do not coincide, sequential oppositions do not result in equal relative distances between the two planets. The period of minimum distance opposition is approximately fifteen years. Therefore, the relative orientation of Earth and Mars is characterized by two cycles, the 2.135 year synodic period superimposed on the approximately 15 year period of minimum distance opposition.

The energy requirements for a particular transfer orbit are expressed in terms of hyperbolic excess velocity (V-infinity) or the mass independent energy term C3 (V-infinity^2). The departure V-infinity is defined as the difference between the spacecraft heliocentric velocity at transfer orbit injection and the heliocentric velocity of the departure planet. The departure V-infinity is therefore determined solely by the departure planet and the desired transfer orbit (similarly for the arrival V-infinity).

The nominal transfer orbit calculations are performed as follows: the planetary positions are determined for a given date using an ephemeris, Van Flandern,[2] the heliocentric velocities (for either Earth or Mars) are determined using Gibb's method and finally the transfer orbit is determined given two position vectors (Earth at departure and Mars at arrival, for instance) and the time of flight using gauss's solution of Lambert's problem. Once the transfer orbit is determined the V-infinity terms are calculated as discussed above.

Many alternative nominal Mars transfer orbits have been presented by Hoffman, McAdams, and Niehoff.[3] This analysis focuses on two Mars exploration mission transfer orbit types; minimum energy-long duration mission transfer orbits and high energy-short duration mission transfer orbits (without a Venus swingby). These transfer orbit types are selected because they provide an approximate upper and lower bound for energy requirements of proposed mission designs.

Table 1 lists the V-infinity values for a minimum energy transfer orbit over a fifteen year Earth-Mars cycle. The departure date and durations listed in the table are drawn from the minimum energy transfer orbit data presented in Hoffman, McAdams and Niehoff,[3] while the V-infinity values are calculated by the author. A plot of a typical round trip minimum energy transfer orbit is presented in Figure 1.

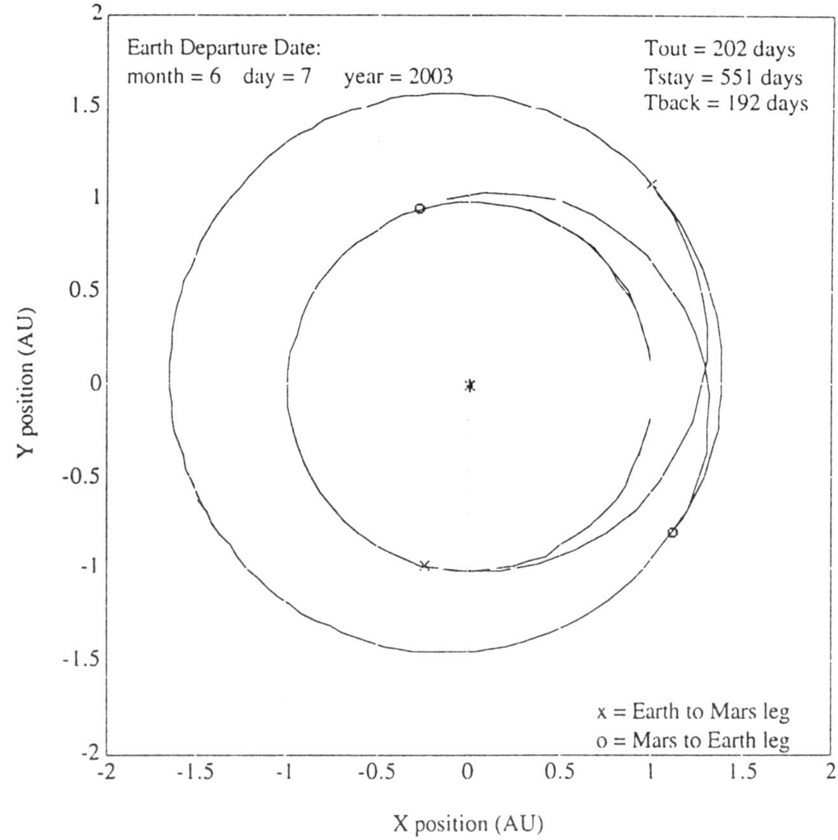

Figure 1 Typical Minimum Energy Class Transfer Orbit.

Table 1
MINIMUM ENERGY CLASS TRANSFER ORBIT V-INFINITY VALUES

Dep. Date (M,D,Y)	T out (days)	V_∞ Earth Dep. (km/s)	V_∞ Mars Arrival (km/s)	T stay (days)	T back (days)	V_∞ Mars Dep. (km/s)	V_∞ Earth Arrival (km/s)
6/7/2003	202	2.96	2.71	551	192	3.68	3.71
9/1/2005	402	3.95	3.55	286	283	3.19	2.93
9/22/2007	369	3.58	2.85	308	297	2.80	2.84
10/14/2009	328	3.21	2.46	340	333	2.59	3.54
11/8/2011	298	2.99	2.75	392	335	2.41	4.79
1/1/2014	329	2.97	4.45	383	300	2.32	5.49
3/21/2016	305	2.84	5.45	418	212	2.46	3.20
5/17/2018	235	2.78	3.26	515	191	3.39	3.33

Note: V_∞ equals V-infinity

Table 2

HIGH ENERGY CLASS TRANSFER ORBIT V-INFINITY VALUES

Dep. Date (M,D,Y)	T out (days)	V_∞ Earth Dep. (km/s)	V_∞ Mars Arrival (km/s)	T stay (days)	T back (days)	V_∞ Mars Dep. (km/s)	V_∞ Earth Arrival (km/s)
6/7/2003	120	4.10	6.38	30	120	16.69	12.51
8/15/2005	100	6.25	8.90	30	120	19.85	14.39
9/22/2007	150	5.46	7.71	30	150	18.86	19.82
10/1/2009	130	9.66	13.53	30	130	18.12	15.44
11/8/2011	140	8.62	13.73	30	140	15.98	16.37
1/1/2014	130	6.14	12.87	30	130	15.01	15.73
3/1/2016	130	3.97	9.04	30	130	12.49	13.05
5/1/2018	120	4.06	7.63	30	120	14.79	11.84

Note: V_∞ equals V-infinity

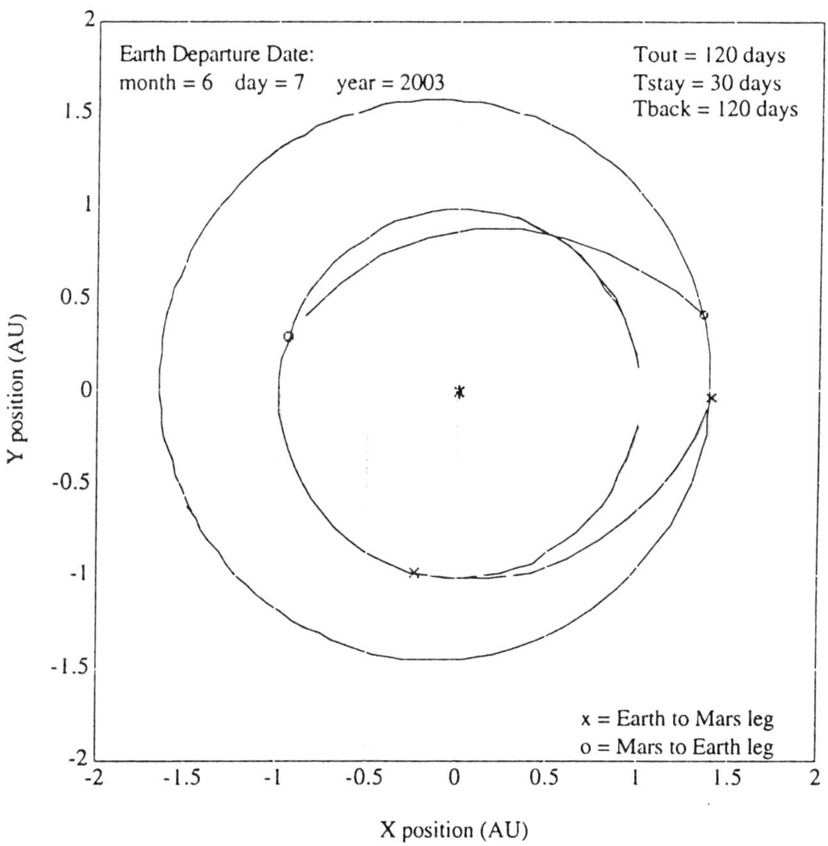

Figure 2 Typical High Energy Class Transfer Orbit.

Table 2 lists the energy requirements for a sampling of high energy, short duration transfer orbits for the same fifteen year period presented in Table 1. For the purpose of this paper, the Mars surface stay time is selected to be 30 days. The duration of the Mars surface stay time can be varied, yet in general stays greater than 30 days imply higher energy requirements. A plot of a typical round trip high energy transfer orbit is presented in Figure 2.

ABORT ORBIT ANALYSIS

Once the nominal Earth to Mars transfer orbits are determined, the analysis of the abort to Earth orbits can proceed. The abort orbit initiation time determines the position of the spacecraft along the nominal transfer orbit while the abort orbit duration determines the location of Earth for rendezvous at the completion of the abort orbit. Abort orbit initiation and duration can be varied to create a grid of potential abort orbits.

The goal of this analysis is the determination of the required spacecraft performance for the abort orbits, so the delta-V and atmospheric entry velocities are calculated. The delta-V required to inject the spacecraft into an abort orbit from a nominal transfer orbit is termed the initiation delta-V. This value is determined from the nominal outbound trajectory and the required abort orbit heliocentric velocities.

The V-infinity terms defined in the previous section determine the magnitude of the Earth arrival delta-V. The Earth arrival delta-V is determined from the V-infinity of the inbound abort orbit and the subsequent Earth orbital velocity of the spacecraft (the Earth orbital attitude is assumed to be 400 kilometers). The atmospheric entry velocity is determined by the abort orbit V-infinity value at Earth arrival for an atmospheric entry altitude of 120 kilometers.

ABORT ORBITS FROM NOMINAL MINIMUM ENERGY TRANSFER ORBITS

The dashed line shown in Figure 3 represents a typical abort orbit from the minimum energy transfer orbit shown in Figure 1. The nominal outbound leg is drawn as a solid line while the location of Earth at departure and arrival is displayed by an "x" and an "o" respectively. As stated in Figure 3, the abort orbit initiation is 90 days from the nominal Earth departure and the abort orbit duration is 125 days. A qualitative evaluation of this figure would indicate that the delta-V at abort orbit initiation is greater in magnitude than that at Earth arrival. This qualitative assessment is supported by Figures 4 and 6. Figures 4 and 6 present a contour plot of the required delta-V at abort orbit initiation and arrival, respectively. A range of orbit initiation times from 50 to 125 days and duration of 50 to 200 days is provided. The delta-V values indicated by the contour lines in Figure 4 (abort orbit initiation) range from 6 to 25 km/s while those found in Figure 6 (abort orbit arrival) range from 4 to 10 km/s.

The dramatic difference in the delta-V values for abort orbit initiation and Earth arrival are due to the large angle of incidence between the nominal and abort orbits at the time of initiation. The large delta-V required to re-orient the spacecraft trajectory from the nominal orbit to the abort orbit results in the higher delta-V values found in Figure 4.

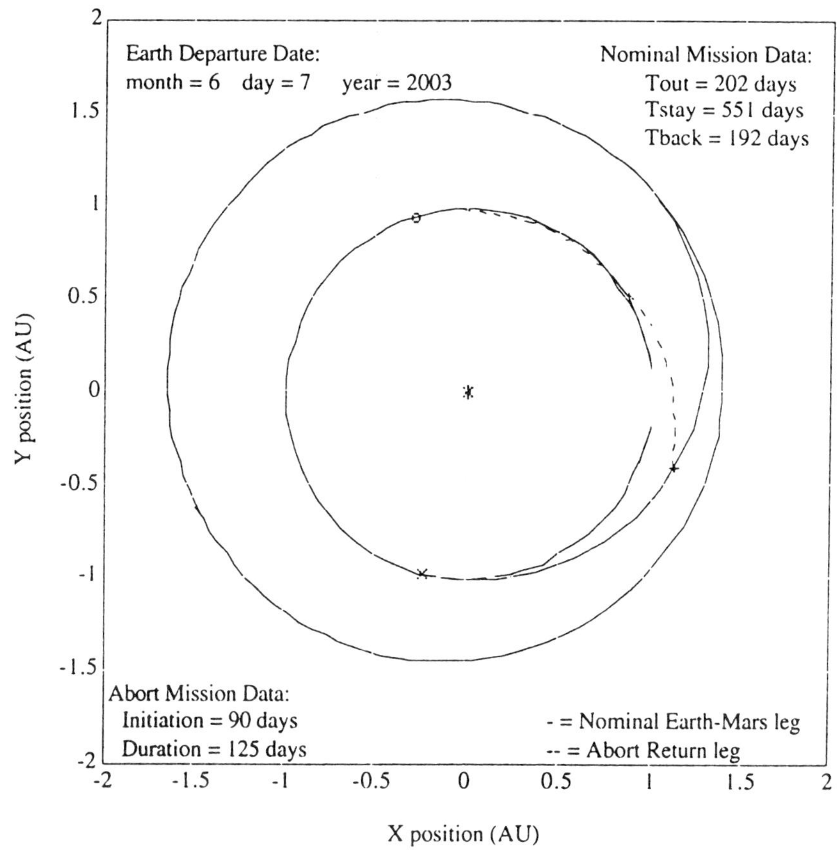

Figure 3 Typical Minimum Energy Class Abort Orbit Geometry.

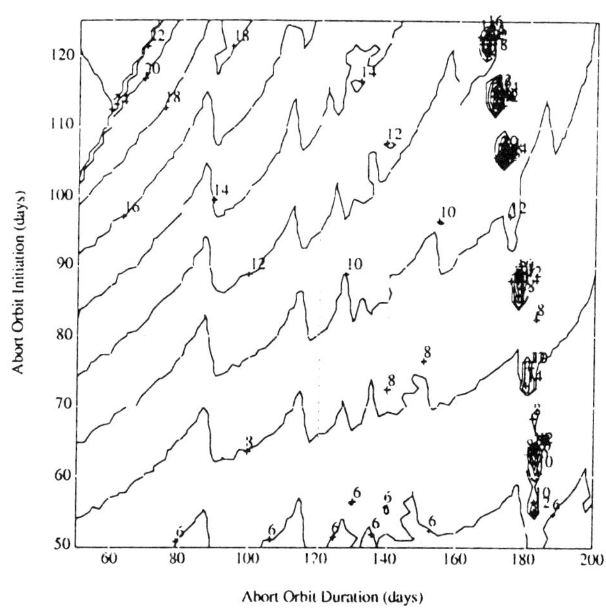

Figure 4 Minimum Energy Abort Orbit Initiation Delta-V, Contour Plot (km/s).

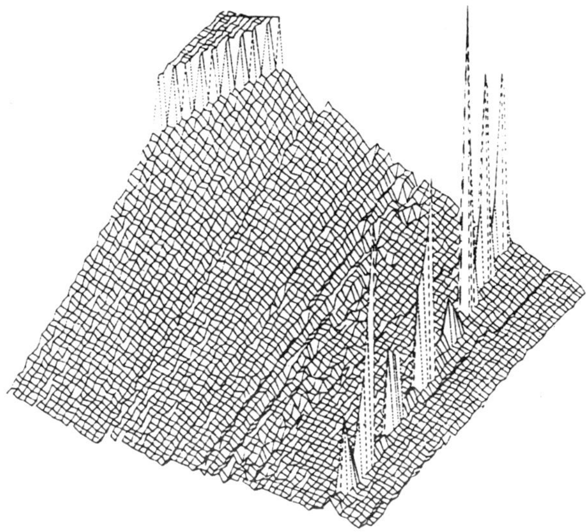

Figure 5 Minimum Energy Abort Orbit Initiation Delta-V, 3-D Mesh Plot.

Figures 5 and 7 present three dimensional mesh plots of delta-V values for abort orbit initiation and Earth arrival. The viewer is located 30 degrees above the X-Y plane with the plot rotated –30 degrees about the Z axis. Hence the 3-D mesh is viewed from 30 degrees above the 200 day duration, 50 day initiation grid point.

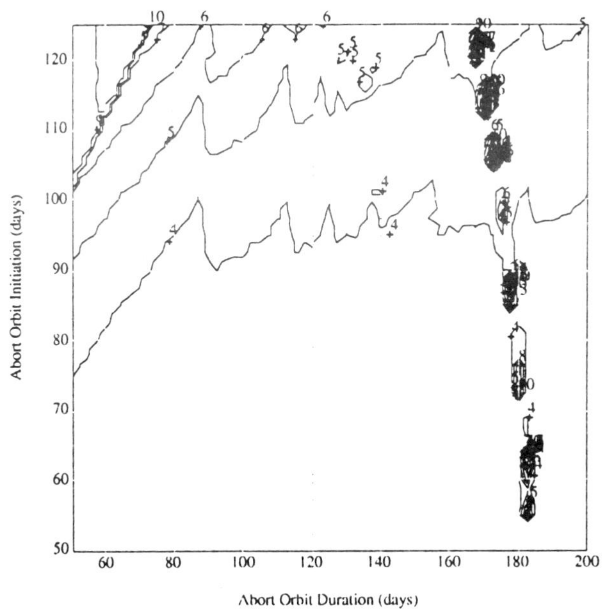

Figure 6 Minimum Energy Abort Orbit Arrival Delta-V, Contour Plot (km/s).

The abrupt peaks in Figures 5 and 7 correspond to the "islands" found in Figures 4 and 6 (approximately along the 180 day duration line). The cause of these abrupt increases in delta-V can be ascertained from Figures 8 and 9. Figure 8 shows an X-Y ecliptic plane view of an abort orbit with an initiation of 115 days out and a duration of

170 days. In this case, the result is an abort to Earth orbit with a transfer angle of 180 degrees (the position of abort orbit initiation and Earth arrival are directly opposite the Sun). As described earlier, the orbit of Mars is inclined to the ecliptic plane by 1.85 degrees. As the plane of the abort orbit must contain the Earth, Sun, and Mars, the orbital plane must rotate about the Earth-Sun-Mars line in the 180 degree transfer orbit case. This rotation produces a large inclination relative to the ecliptic plane. This inclination, when viewed in the X-Y plane causes the abort obit to appear as highly eccentric. However, when viewed in the X-Z plane, the abort orbit is found to be highly inclined to the ecliptic instead. This highly inclined characteristic is true of the abort orbits with excessively large delta-V values presented in this paper. It should be noted that the appearance of peaks is an artifact of the coarseness of the grid. If the grid fineness was on the order of 0.5 days, the peaks would merge into a solid wall. The large delta-V values found for the 180 degree transfer orbit cases do not impose any severe restrictions on the abort orbit operations. As can be seen in both the contour plots and the 3-D mesh plots, the neighboring V-infinity values are consistent with the delta-V ranges mentioned above.

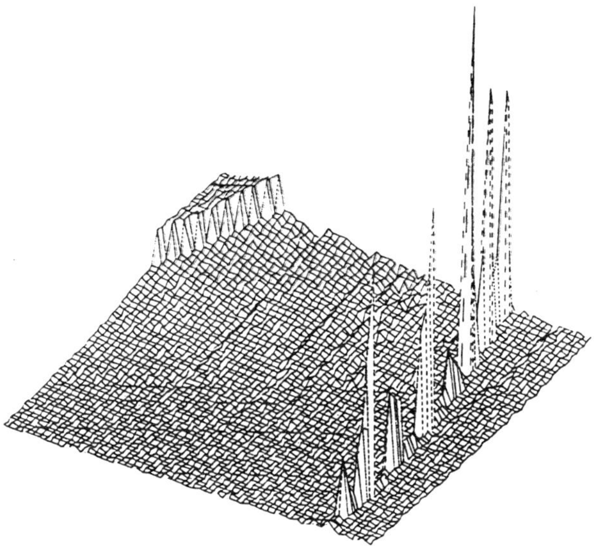

Figure 7 Minimum Energy Abort Orbit Arrival Delta-V, 3-D Mesh Plot.

The Earth atmospheric entry velocity for the abort orbit is presented in Figures 10 and 11. The entry velocity contours found in Figure 10 range from 12 to 17 km/s. Both the contour and 3-D mesh plots exhibit the 180 degree transfer angle peaks discussed above.

Minimum Energy Class Abort Orbits, Interpretation

In order to interpret the delta-V and atmospheric entry velocity values presented in Figures 4 through 7 and 10 through 11, the nominal Earth-Mars-Earth transfer orbit delta-V and Earth atmospheric entry values are required (see Table 3). These values are calculated assuming a 500 km circular orbit about Mars, a 400 km circular orbit above Earth and an atmospheric entry altitude of 120 km.

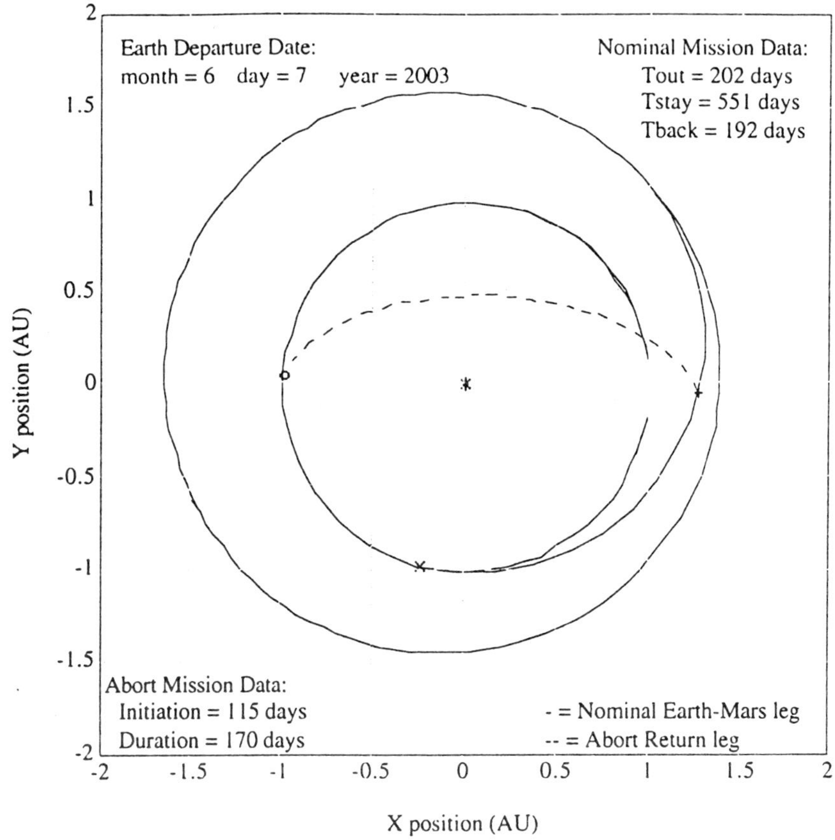

Figure 8 Minimum Energy Abort Orbit, 180 Degree Transfer Angle Case, X-Y Ecliptic Plane View.

The minimum abort orbit initiation delta-V value presented in Figure 4 is found to be 6 km/s. This minimum is nearly three times the nominal Mars arrival delta-V of 2.10 km/s and more than twice the nominal Mars departure delta-V of 2.65 km/s. If the spacecraft is to perform any of the abort to Earth orbits within the range presented in Figure 4, additional delta-V capabilities must be provided. Assuming both the Mars arrival and departure delta-V capabilities are available (as would be the case for an all propulsive mission architecture resulting in a total capability of 4.75 km/s), an additional 1.25 km/s of propulsive delta-V must be provided.

The 6.0 km/s contour may be deemed overly restrictive as the only abort to Earth orbits available are in the range of 50 days initiation and 110 to 200 days of duration. If the mission design requirements specify greater powered abort capabilities—providing up to 10 km/s of abort orbit delta-V, for example—then abort orbits are available for any duration if initiated prior to 65 days out in the nominal transfer orbit. Following the 10 km/s delta-V contour in Figure 4 shows initiation and duration combinations within the 10 km/s capability. The contour plots therefore provide a map of abort orbit opportunities for any given delta-V capability.

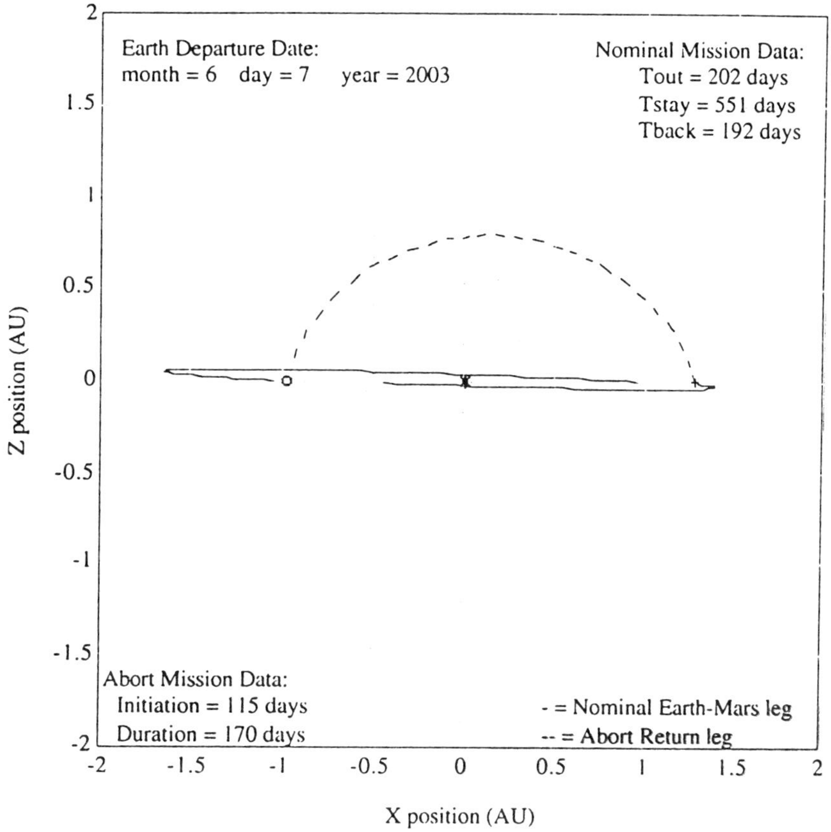

Figure 9 Minimum Energy Abort Orbit, 180 Degree Transfer Angle Case, X-Z Ecliptic Plane View.

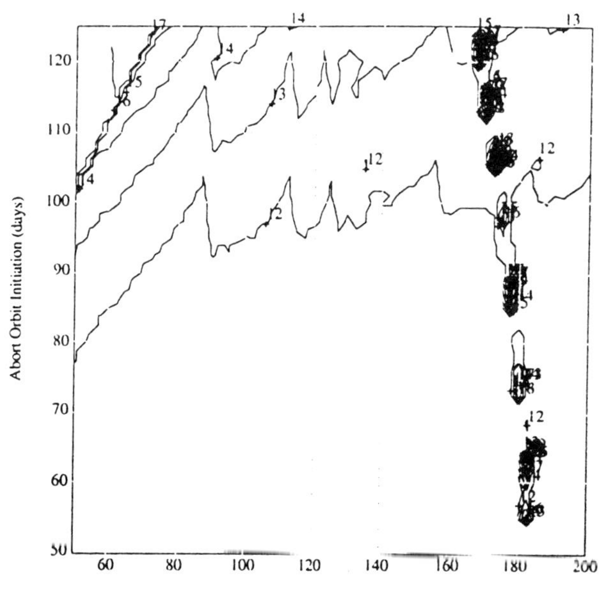

Figure 10 Minimum Energy Abort Orbit Atmospheric Entry Velocity, Contour Plot (km/s).

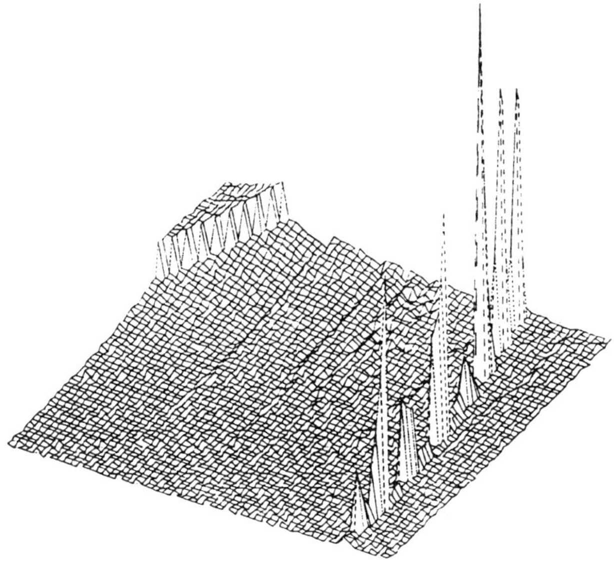

Figure 11 Minimum Energy Abort Orbit Atmospheric Entry Velocity, 3-D Mesh Plot.

Table 3
NOMINAL MINIMUM ENERGY CLASS TRANSFER ORBIT DATA

Mission Definition:	
Earth Departure Date	= 6/7/2003
Outbound Flight Time	= 212 days
Mars Stay Time	= 551 days
Inbound Flight Time	= 192 days
Performance Requirements:	
Delta-V, Earth Departure	= 3.71 km/s
Delta-V, Mars Arrival	= 2.10 km/s
Delta-V, Mars Departure	= 2.65 km/s
Delta-V, Earth Arrival	= 3.79 km/s
Earth Atmospheric Entry Velocity	= 11.68 km/s

The preceding discussion assumes that the mission architecture is all propulsive. If the mission design incorporates an aeroshell for aerocapture at Mars, the nominal propulsive delta-V available for abort orbit initiation is limited to that required for nominal Mars departure (2.65 km/s, this assumes that an aerocapture maneuver is to be performed at Earth as well as at Mars). If it is desired to retain the 10 km/s delta-V capability discussed above, a proportionately greater impact on the mission architecture of 7.35 km/s (versus 5.25 km/s for the all-propulsive mission) must be incorporated. It should be noted that the total abort orbit delta-V capability is the same (10 km/s) but the relative impacts differ.

As shown in Figure 6, the minimum abort arrival delta-V is 4 km/s. This minimum delta-V provides abort arrival opportunities for all durations if the initiation time is less than approximately 80 days out from Earth. The minimum value of 4.0 km/s is very close to the nominal transfer orbit Earth arrival delta-V of 3.79 km/s. As a result, the abort arrival capability can be greatly enhanced for a relatively minor mission architecture impact of approximately 200 m/s delta-V. The atmospheric entry velocities found in Figure 10 are very nearly the same as the nominal transfer orbit Earth atmospheric entry velocity of 11.68 km/s. Virtually the entire abort initiation-duration space is available for atmospheric entry velocities less than 14 km/s.

Minimum Energy Class Abort Orbits, Summary

The previous analysis supports the qualitative assessment drawn from Figure 3. Simply stated, the abort orbit initiation delta-V is more restrictive than either the Earth arrival delta-V or the atmospheric entry velocity. The analysis also indicates that in order to satisfy the minimum abort orbit initiation delta-V of 6 km/s, mission architectures incorporating a propulsive braking maneuver at Mars require a smaller impact on the spacecraft design (1.25 km/s of delta-V) as opposed to a Martian aerobrake mission architecture (3.35 km/s) An analysis of the October 2009 minimum energy transfer orbit presented in Table 2 results in findings similar to those discussed above. Note that the 2003 and 2009 transfer orbit opportunities are equivalent to the beginning and middle of the fifteen year Earth-Mars cycle discussed previously.

ABORT ORBITS FROM HIGH ENERGY, SPRINT CLASS TRANSFER ORBITS

The abort orbit from the nominal high energy transfer orbit for June 7, 2003 shown in Table 2 is similar to that shown in Figure 3. Figures 12 and 13 provide contour plots of the abort orbit initiation and Earth arrival delta-V values respectively (the 3-D mesh plots are not included here for the sake of brevity). The Earth atmospheric entry velocity contour plot is provided in Figure 14. As discussed earlier, the peaks apparent along the right hand edge of the contour plots are the result of 180 degree transfer angle abort orbits. Again, these 180 degree transfer angle abort orbits are special cases that can be avoided operationally by varying either the abort orbit initiation, or duration, or both.

High Energy Class Abort Orbits, Interpretation

The nominal high energy transfer orbit delta-V values are presented in Table 4. These delta-V values are calculated for the baseline Earth and Mars orbit parameters as defined for Table 3.

Due to the high energy requirements of the nominal sprint class transfer orbit mission, significantly greater delta-V capabilities are available for the abort orbit initiation than was the case for the minimum energy transfer orbit analysis discussed above. Up to 18.62 km/s of delta-V are available from an all propulsive mission architecture designed for the 2003 high energy transfer orbit defined in Table 4. This large available delta-V more than offsets the greater abort orbit initiation delta-V imposed by the faster nominal transfer orbit. As can be seen in Figure 12, an 18 km/s delta-V capability enables virtually any abort orbit initiation. If the mission architecture incorporates an aerobrake for Mars aerocapture, 14.02 km/s will be available for abort orbit initiation.

The 14 km/s contour line, while more restrictive than the 18 km/s contour, indicates that nearly any abort orbit duration is available if initiated prior to approximately 65 days out from Earth.

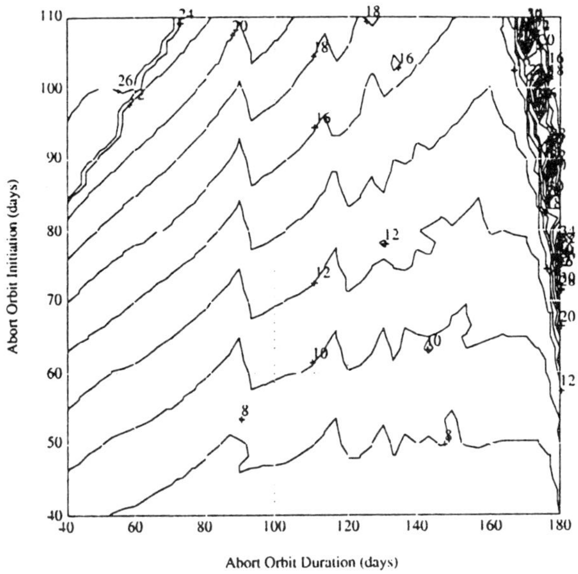

Figure 12 High Energy Abort Orbit Initiation Delta-V, Contour Plot (km/s).

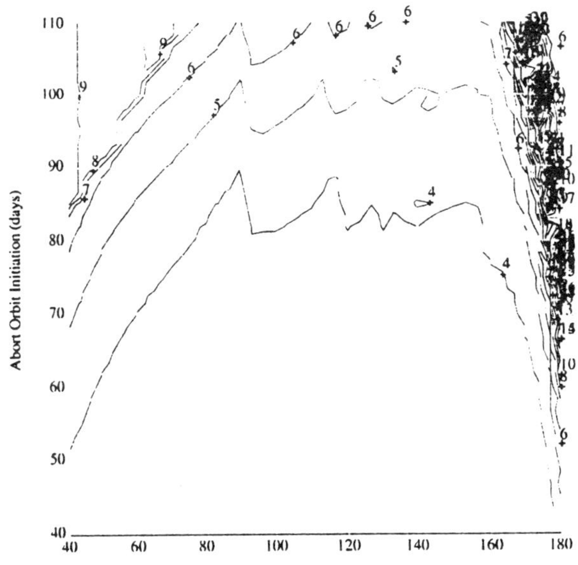

Figure 13 High Energy Abort Orbit Arrival Delta-V, Contour Plot (km/s).

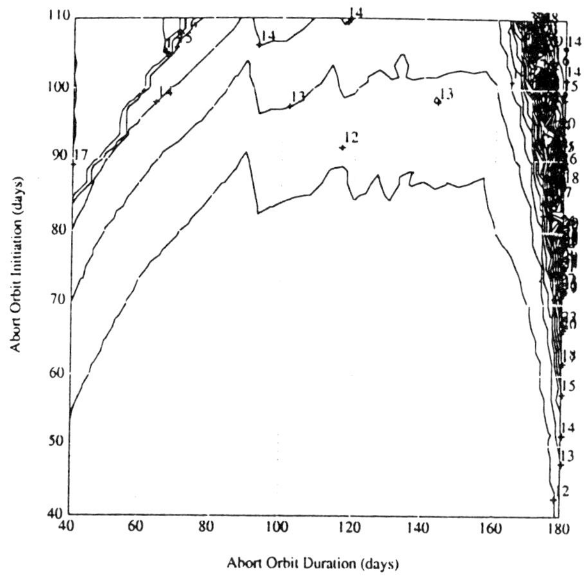

Figure 14 High Energy Abort Orbit Atmospheric Entry Velocity, Contour Plot (km/s).

Table 4
NOMINAL HIGH ENERGY CLASS TRANSFER ORBIT DATA

Mission Definition:	
Earth Departure Date	= 6/7/2003
Outbound Flight Time	= 120 days
Mars Stay Time	= 30 days
Inbound Flight Time	= 120 days
Performance Requirements:	
Delta-V, Earth Departure	= 3.82 km/s
Delta-V, Mars Arrival	= 4.60 km/s
Delta-V, Mars Departure	= 14.02 km/s
Delta-V, Earth Arrival	= 8.89 km/s
Earth Atmospheric Entry Velocity	= 16.71 km/s

The Earth arrival delta-V and atmospheric entry velocity analysis also reveals that most of the analyzed abort orbit initiation-duration space is available for the nominal mission design. If the mission architecture incorporates a propulsive braking maneuver at Earth arrival, the nominal 8.89 km/s delta-V capability enables any abort orbit duration if initiated before approximately 85 days out from Earth. If the Earth arrival maneuver is to incorporate an aerobrake, the results are virtually the same as most of the abort orbit opportunities that are available for atmospheric entry velocities below 16.71 km/s. It should be noted that most references recommend a maximum atmospheric entry velocity of 14 km/s, hence the nominal mission architecture would be required to perform a propulsive braking maneuver to reduce the atmospheric entry velocity. Such a maneuver would also be required for any abort orbit entry velocities greater than 14 km/s.

High Energy Class Abort Orbits, Summary

This analysis indicates that for the high energy transfer orbit discussed here, most of the abort orbit opportunities are available to an all propulsive mission architecture. Incorporation of an aerobrake into the design reduces the abort opportunities, but no additional propulsive capability is required for satisfactory abort orbit performance. An analysis of the October 2009 high energy transfer orbit mission defined in Table 2 indicates that similar abort orbit opportunities exist roughly half way into the 15 year Earth-Mars cycle discussed previously.

It is important to recognize that, unlike minimum energy transfer orbits, there is great freedom in the design of high energy transfer orbits. Potential high energy transfer orbits may be designed to different mission criteria and therefore not result in such robust abort orbit capabilities as those discussed above.

MISSION ARCHITECTURE REVIEW AND DESIGN GUIDELINES

Various mission architectures, defined here by transfer orbit type (minimum and high energy) and planetary braking technique (aerobraking or propulsive braking), have been analyzed for abort orbit performance. In general, minimum energy transfer orbit mission architectures incorporating aerobraking at Mars suffered the greatest penalty (3.35 km/s for the 2003 mission) in order to satisfy the minimal abort orbit initiation requirement of 6.0 km/s. The propulsive braking mission architecture required an additional 1.25 km/s over the nominal orbit requirement in order to satisfy the same 6.0 km/s abort orbit initiation requirement. High energy transfer orbit mission architectures were found to provide robust abort orbit capabilities with either all propulsive or aerobraking designs. No additional propulsive capability was required for satisfactory abort orbit performance.

This analysis supports the common sense conclusion that given greater propulsive capacity (as with the high energy transfer orbit mission architectures) mission abort orbit performance is improved. This conclusion is found to be true even though the high energy transfer orbits require greater abort orbit delta-V's and atmospheric entry velocities than the minimum energy transfer orbits.

Recommendations that may be drawn from this analysis to improve the abort orbit capabilities for any Mars exploration mission include the use of aerobraking at Earth combined with propulsive braking at Mars. Additionally, high energy transfer orbits (with their attendant robust propulsive capabilities) greatly enhance the abort orbit performance of any mission. The abort orbit capabilities of a minimum energy mission architecture can be significantly enhanced by the addition of abort orbit propulsive capacity. The design of split crew and cargo mission architectures, and the resulting potential for a reserve fuel supply enroute to Mars along with the crew transfer vehicle, may also improve mission abort orbit performance. While these abort orbit mission architecture design recommendations must, of course, be weighed against other mission design criteria, careful consideration must be made of the abort to Earth opportunities of any proposed mission architecture.

REFERENCES

1. Wolf, A. A., "Free Return Trajectories for Mars Missions," AAS Paper No. 91-123, in *Spaceflight Mechanics 1991*, J. K. Soldner *et al.*, eds., *Advances in the Astronautical Sciences*, Vol. 75-I, pp.317-329, 1991.
2. Van Flandern, T. C., Pulkkinen, K. F., "Low Precision Formulae for Planetary Positions," *Astrophysical Journal Supplement Series*, pp. 391-411, 1979.
3. Hoffman, S. J., McAdams, J. V., Niehoff, J. C., "Round Trip Trajectory Options for Human Exploration of Mars," AAS Paper No. 89-201, in *Orbital Mechanics and Mission Design*, J. Teles, ed., *Advances in the Astronautical Sciences*, Vol. 69, pp. 663-679, 1989.

ZEPHYR: A MICROSAT TRANSFER ORBIT STAGE

George William Herbert*

This paper discusses the design of "Zephyr," a transfer stage for microsat-class vehicles. Zephyr is capable of propelling 50 kg class payloads on Mars Transfer Orbits from the Pegasus launch vehicle, for total projected launch costs of under $20 million.

INTRODUCTION

Microsatellites (sub-ton class satellites) have proved their merit in near-earth applications. The potential for light planetary explorers of similar mass has become evident; small rovers, small science stations, and other applications have appeared in literature (for example, the JPL mini rover series). This paper presents an orbital transfer vehicle, the "Zephyr," suitable for sending such planetary microsats to Mars.

The vehicle that was to become the Zephyr transfer-vehicle/upper-stage was conceived during a lecture on robotics in planetary exploration at International Space University's summer 1992 program. The concept began after a presentation on the JPL "Rocky" minirover. The author wondered if these small payloads could be sent using dedicated small launch vehicles. The obvious benefits are cheaper project costs, faster, more frequent missions, and increased access due to lower cost.

The author had just come across the technical specifications on the engine that Zephyr now uses,[1] put two and two together, and the first preliminary design was completed by that afternoon using basic design guidelines.[2] Since then, it has been refined and developed as a low-priority project at Retro Aerospace.

REQUIREMENTS

The first step undertaken in this project was identification of a complete, if not rigid, set of mission requirements. These are:

- The transfer vehicle must be compatible with the existing Pegasus vehicle.
- The vehicle must be able to propel useful payloads onto a Hohmann transfer orbit to Mars.
- The vehicle must be priced affordably for Microsat/Explorer class missions.
- The vehicle must use existing components as systems allow.

* President, Retro Aerospace, 2240 Blake Street, #101, Berkeley, California 94704. E-mail: gwh@retro.com.

Pegasus Compatibility

The Pegasus launch vehicle is currently the standard microsatellite launch vehicle. It can place a payload of approximately 350 kg in a 150 km circular equatorial orbit, or lesser payloads to higher altitude and/or inclination. Its payload fairing has a useful envelope measuring 116 cm (46 inches) in diameter, with a maximum length of 182 cm (72 in.). Pegasus has a relatively benign launch environment.

Useful Payload to Mars Transfer Orbit

The author felt that the minimum acceptable payload for a mission would be on the order of 50 kg placed into MTO on an average alignment launch window. If a package of that mass were aerocaptured at Mars, the resulting payload to Mars surface would be on the order of 15 kg., enough for a basic minimal rover or immobile basic science station.

Affordable Pricing

An initial guess at reasonable pricing for microexplorer payloads was that launch costs should not exceed twenty million dollars. As the Pegasus vehicle procurement costs are $112-15 million, this limits transfer stage costs to about $5 million.

Existing Components

This final constraint was imposed to reduce development and certification costs and to reduce program risk. A lengthy, expensive development program for a cheap stage for cheap missions makes no sense.

OVERALL CONFIGURATION

The vehicle that evolved within the stated constraints is named the Zephyr transfer stage. Its core is the SEP 8kn engine, around which are positioned 12 small spherical fuel tanks. Dry mass of the stage is approximately 55 kg, and it can hold up to 2990 kg of Nitrogen Tetroxide and Monomethyl Hydrazine propellants. The whole vehicle fits within the 116 cm diameter Pegasus shroud, though volume and shape constraints have forced an unusual aft-facing configuration during launch. (See Figure 1). It is assumed that the Pegasus third stages' maneuver capability will be used to orient the Zephyr vehicle and payload in a forwards facing alignment before it separates and the Zephyr fires.

The overall configuration of the Zephyr vehicle was influenced by two tight constraints; the outer envelope of the Pegasus shroud, and the volume and shape of the relatively massive engine. Of the 182 cm length of useable volume within the shroud, the engine takes up 145 cm, and much of the diameter for that length. Around the front of the motor, where it is thinnest, are the 12 propellant tanks, each 34 cm (13 in.) in diameter. Between the tanks and the engine are a sheet aluminum heat shield and some limited additional thermal insulation. The vehicles' systems are mounted on its front and between the rows of propellant tanks. (Figure 2).

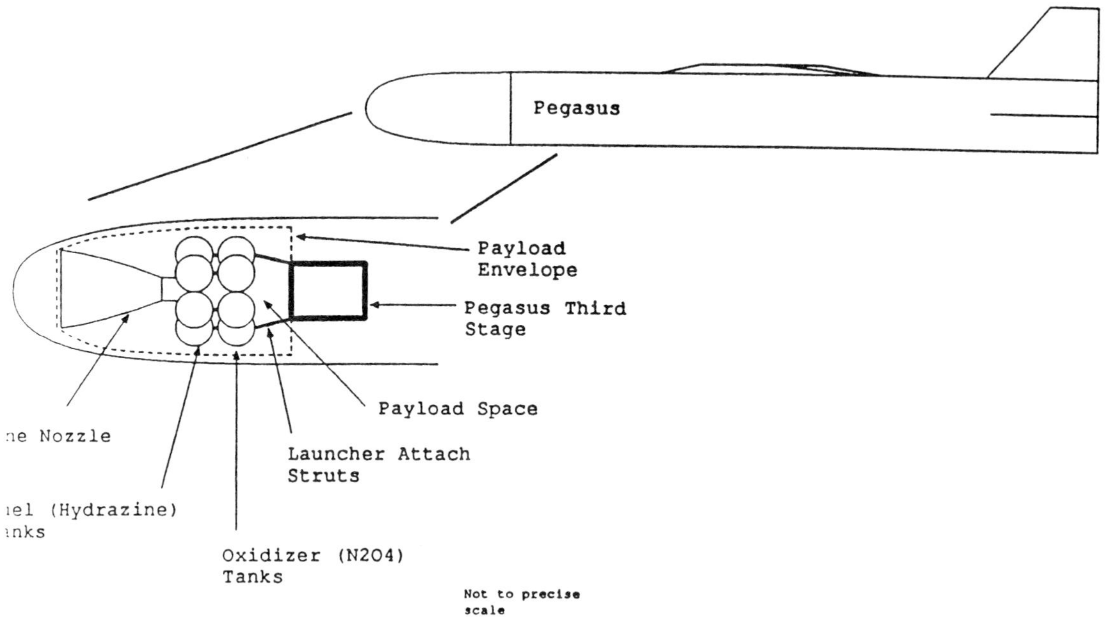

Figure 1 Zephyr/Pegasus Configuration.

Mission Profile

Zephyr is launched into an equatorial 150 km orbit by the Pegasus launch vehicle. Following third-stage burnout, the Zephyr stage remains attached to the Pegasus' third stage. The Pegasus third stage then uses its cold-gas control system to flip the two connected vehicles around, so that the Zephyr is oriented nozzle-aft relative to its orbital motion, and spins the two vehicles up to 30 RPM. Following spinup, the Pegasus third stage separates from the Zephyr vehicle and uses its remaining propellant to leave the immediate vicinity of the Zephyr. (Figure 3).

Once the Pegasus stage is clear, the Zephyr then begins its firing sequence. First, the pressurization system begins operation and brings the propellant tanks up to operating pressure. The propellant is already settled by the vehicles' 30 RPM rotation. The pressurization breaks rupture plates, allowing the propellants to flow into the motor.

Once pressure is established and stable, the motor is commanded to fire. Its internal valves operate to allow propellants into the combustion chamber. The motor is swiveled to control vehicle orientation during the burn. The motor burns for approximately 100 seconds, until a predefined final velocity vector is achieved, and then shuts down.

The initial 30 RPM spin is sufficient to stabilize the vehicle with a sufficient margin that increased or decreased roll rate due to nozzle irregularities or other unexpected performance is negligible.

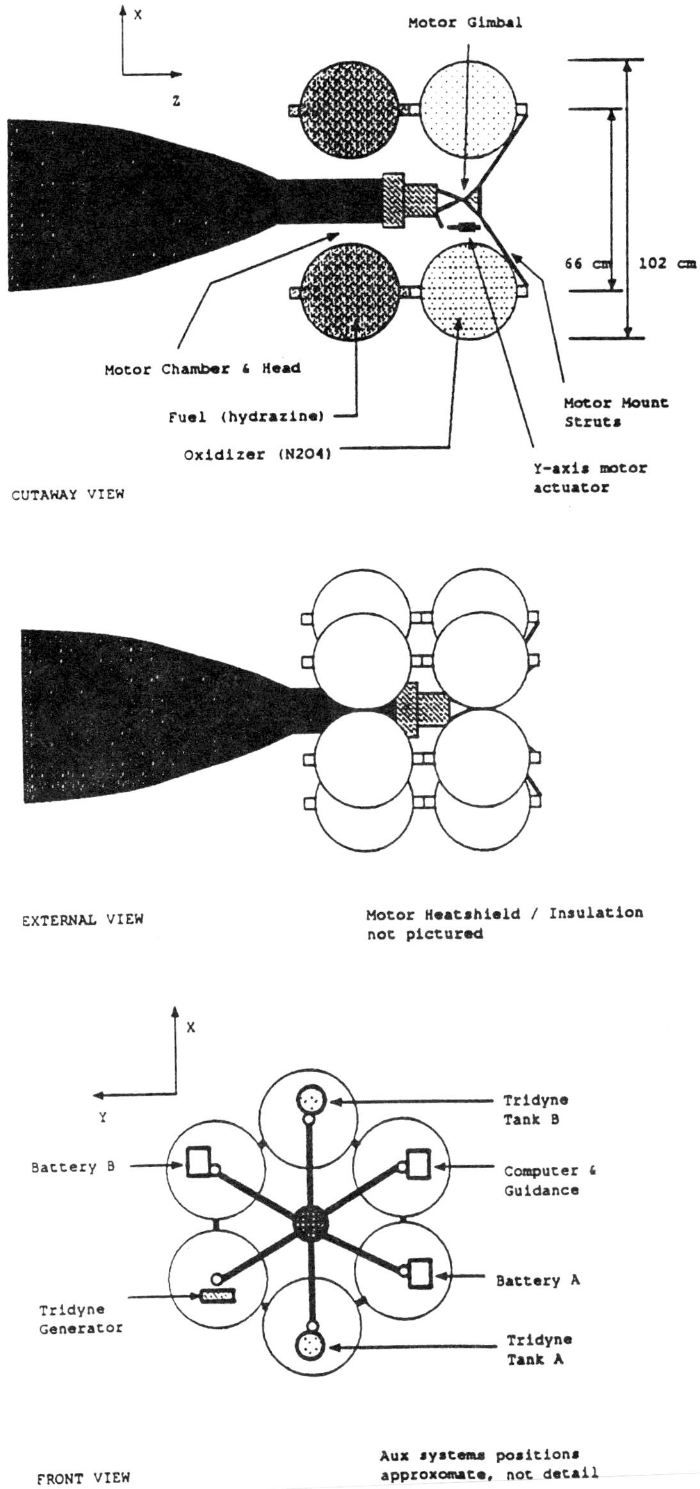

Figure 2 Zephyr Vehicle Configuration and Details.

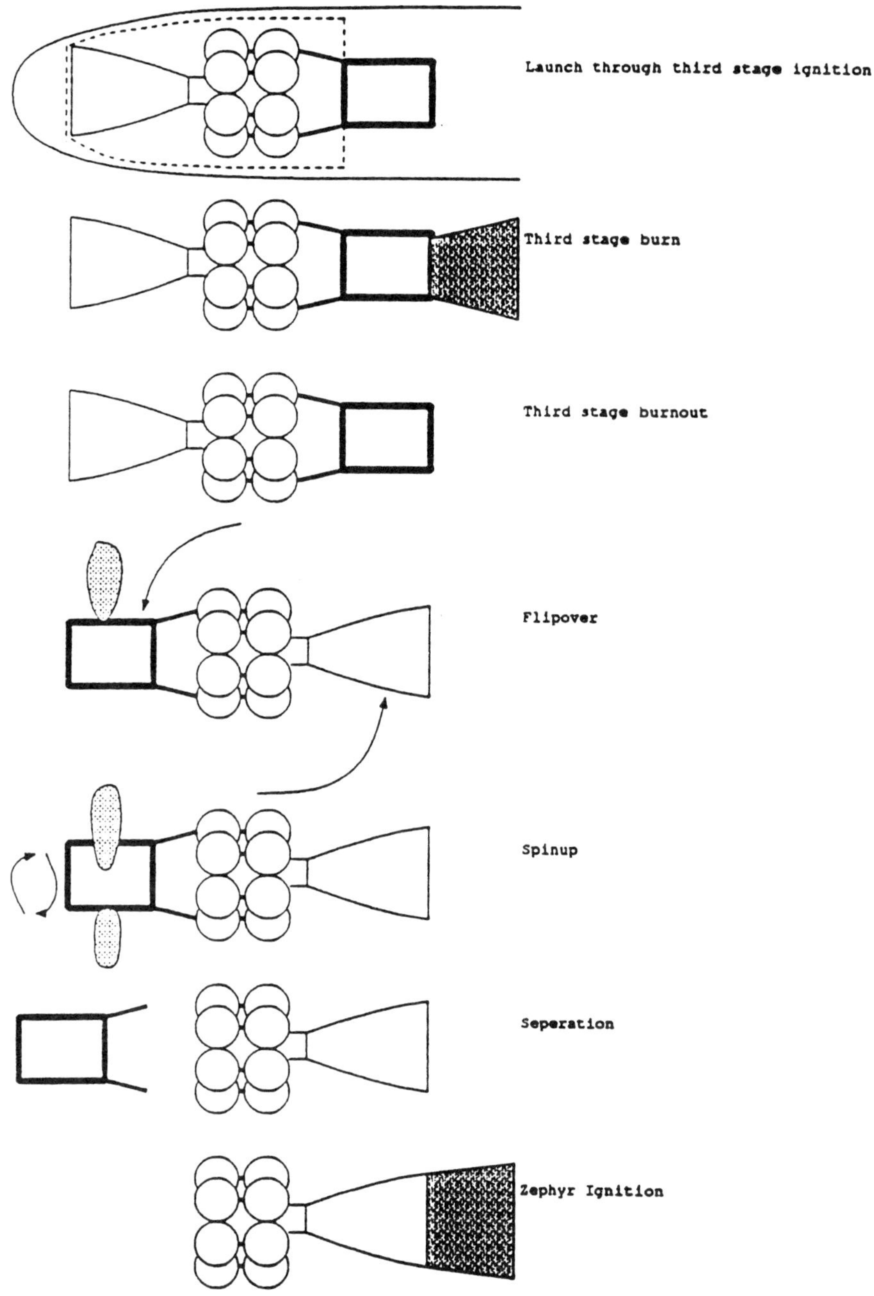

Figure 3 Zephyr/Pegasus Flight Profile.

As Zephyr is merely a transfer stage, despin, midcourse trajectory corrections and similar requirements are offloaded to the payload items. Little of Zephyr's hardware would be of use to a payload during the nine months of a transfer, and its added mass would significantly affect midcourse correction fuel requirements. Retro Aerospace is investigating a bus design to handle midcourse corrections, keep alive power etc., during the transfer, but this has not yet been completed.

Overall Mass Budget

Any transfer stage (indeed, any space vehicle)[3] is a highly mass-critical design. The following table summarizes the current mass budget for the Zephyr vehicle.

Table 1
MASS BUDGET AND BREAKDOWN

System	Level	Mass (kg)	Totals
Engine Mounts	Detail	1.2	
Propellant Tanks	Detail	12.0	
Motor	Built	21.0	
Battery	Detail	4.0	
Propellant Pressure	Det. Est.	4.5	
Other Structure	Det. Est.	0.8	
Margin & subtotal	5%	2.2	45.7
Insulation	Est	3.0	
Motor Actuators	Est.	3.0	
Guidance & Control	Est.	3.0	
Communications	Est.	1.0	
Margin & subtotal	15%	1.5	11.5
Total			57.2

Those systems which have been through at least one detail design examination are described as "designed." Detail estimates are those for which parametric modeling was used, based on other existing systems, in which the author has high confidence. Estimates are systems where parametric modeling had less precise background information and a correspondingly higher margin is used.

The mass margin is about 10% if the motor, which has been built, is excluded from the margin calculation. This is reasonable for a design at this stage. Alternate payloads are listed in section 6.0 for a overweight condition (20% margin) should this prove inadequate.

Cost Estimation

The per-flight costs of the Zephyr are estimated at $3.625 million. The largest single portion of this is the purchase of the main motor, at $1.0 million dollars each in small quantities. Other costs are approximately $150,000 for the control and guidance system, $50,000 for the pressurization system, $225,000 for the tankage, $250,000 for other systems and hardware, and 10,000 man-hours of assembly, testing, and operational preparation, including margins in each systems' cost.

It is hoped that the development cost for the vehicle, estimated at $5-7-5 million, can be absorbed by commercial interest in the stage. However, it is possible that this will not occur. Amortizing the maximum estimate for development cost over a five-vehicle buy gives a total program cost of $5.125 million per flight ($25.6 million total). The target cost was $5.0 million per flight. Zephyr therefore meets its price goal.

STRUCTURAL DESIGN

The structure of the Zephyr vehicle is broken up into a number of subsystems: the fuel tank structure, the engine mount structure, secondary structures, and launcher interface structure.

Fuel Tank Structure

Fuel is contained in 12 spherical fuel tanks, pressurized to 330 PSI in operation. An early analysis of the vehicle indicated that using the tanks as their own support structure was the most efficient design solution: in addition to pressure loads they are strengthened to carry their own inertial loads. Each tank is a spherical carbon-fiber/epoxy structure,[4] 34 cm in diameter, with a stiffening cylinder about 4 cm diameter running through it along the Z axis. The stiffening cylinder also provides convenient connection points at the top and bottom and a handle during fabrication. A thin (0.1 mm or thinner) aluminum liner ensures propellant isolation.

Minimum composite thickness, determined by the pressurization loading and a reasonable safety factor, is 1.5 mm.[5] The tanks are reinforced where other loads are significant. Detail design of the tanks is being delayed until a finite element stress analyses can be performed, but estimated mass per tank is 1.00 kg based on the basic tube & sphere structure and a conservative margin.

The tanks are arranged in 2 rows of 6, hexagonally situated around the motor. The front row is the oxidizer tankage, the back row fuel. Each tank is attached to the two neighboring tanks in its own row and to the corresponding tank in the other row. The resulting interconnected structure of tanks is self-supporting.

Connections between tanks in a given row are via flanges formed integrally to the tanks, which are riveted together. Between the oxidizer and fuel tanks a cylindrical coupler fits over the two stiffener cylinders and is riveted to each. At the top of the oxidizer tanks where the tank stiffener cylinders meet the motor mounts, payload mounts, and launcher interface brackets, a more complex fitting is riveted via an integral coupler to the oxidizer tank.

Engine Mounts

The engine is mounted at the axis of the vehicle. It is attached to the ring of oxidizer tanks (the front ring) by six stainless steel thin-wall tubes, with appropriate end hardware for the connections. The mounts mass 0.20 kg each.

Secondary Structures

There are a number of secondary structures on the Zephyr vehicle; a thermal shield and insulation blanket separates the engine from the fuel tanks; this is massed at 3.0 kg.

The payload mass is supported directly from the top of the forward six propellant tanks, at the same positions that the motor mounts tie in. The vehicle's other systems, such as control, propellant pressurization are mounted on struts between existing primary structure nodes.

The insulation may be reduced based on detailed heat-transfer calculations and thermal stability of the tank materials. The insulation listed amount is a worst-case value.

Launcher Interface Structure

The Zephyr vehicle is connected to the Pegasus launch vehicle's third stage by three or more brackets. They are bolted to the Pegasus payload adapter ring and attached using pyro bolts to the Zephyr vehicle's structure at the top of the oxidizer tanks, at the same point the main motor is attached to them. This reduces the Zephyr vehicle's dry mass.

PROPULSION

The Zephyr vehicle has a single motor, used for both propulsion and directional control. It is used to provide any needed velocity changes.

Motor

Primary propulsion is provided by a SEP 8000 N (1750 lbt) storable bipropellant motor, which utilizes a ceramic matrix composite combustion chamber and nozzle. The engine burns 2.5 kg of nitrogen tetroxide and monomethyl hydrazine propellant per second, and develops a specific impulse of 320 sec. Its mass is 21 kg, including its mounting bearing but not including actuator masses.

This engine has been developed and test-fired extensively by SEP. This application will require no modification of the engine, and has burn times of 1/3 of the demonstrated life.

The SEP motor is the best choice for the Zephyr vehicle due to a combination of thrust level and specific impulse. Ideally, a motor with about half the mass and thrust (but similar specific impulse) would be a better choice, but there is no such motor. A lower specific impulse has a significant negative impact on payload: the Space Shuttle RCS thrusters have a maximum Isp of 304, and even this reduction reduces payload significantly. An examination of the Royal Ordinance 500 N motor is underway, but is not yet complete. This motor reportedly achieves a Isp of 325 (maximum) and in its standard form achieves Isp 314 with a dry mass of 5 kg. This would represent a slight improvement over the SEP motor.

Any motor change will have to balance additional considerations such as commercial use of the stage. The SEP motor version can place 70 kg in Geosynchronous Transfer Orbit from a standard Pegasus, or 106 kg from a Pegasus XL vehicle. A vehicle with much lower thrust ratings would have degraded performance due to longer burn time and gravitational losses. It may be beneficial to keep the SEP motor for its dual-use flexibility.

Propellant Feed

Propellants are pressure-fed into the engine's combustion chamber, requiring a pressurized storage and feed system. The feed system is fairly simple, with pipes from the 12 tanks joining the two main feed lines and then connecting to the motor, where an internal cutoff valve controls operation. The feed system operates at 2.3 MPa (330 PSI), with a 0.1 MPa drop from tank pressure to the engine manifolds.

Pressure within the propellant tanks is maintained by a small hot-gas helium (Tridyne) generator system. The design of this system was scaled from existing testbed systems.

The Tridyne system produces hot helium with a small percentage of water vapor mixed in. The total amount of water is on the order of 300 grams, and little of it will end up condensing into the propellants. Slight contamination of the Hydrazine fuel by that water has no effect. Water in the nitrogen tetroxide oxidizer may produce some nitric acid and the propellant tank lining and feed tubing must be somewhat resistant to acid conditions. This is easily accomplished, and the amount of acid formed by water condensation is very small (roughly 0.003 molar absolute worst case).

SYSTEMS

This section describes other systems not yet covered.

Guidance

Position and velocity vector information for the Zephyr vehicle are generated by a missile-type ring laser inertial system. Initial position and velocity are uplinked from the ground, and the guidance system merely has to provide a delta from those values over the roughly hundred-second burn time, for direction vector and burnout velocity control.

An alternative GPS guidance solution is being investigated as promising, but is not yet baselined. Further flight data on GPS guidance is desirable.

Control

The Zephyr vehicle is controlled by an on-board microcomputer of as yet undetermined type. Retro Aerospace is investigating using a ruggedized PC-type computer with existing realtime operating systems and appropriate input/output and control capabilities, at great savings compared to highly customized control computer design and programming techniques.

Directional Control

Directional control of the Zephyr is provided by main engine pointing. Two electromechanical actuators (X,Y planes) provide control motions. Use of electromechanical actuators allows the vehicle to have no hydraulic or pneumatic systems, decreasing complexity. No secondary directional control is provided, although an add-on system would be easy (at some loss of payload capacity) if a particular mission demanded it.

Electrical Power

The Zephyr vehicle uses electrical power for three purposes: computer and guidance system power, main engine pointing actuation, and valve operation. By far the highest peak loads are from main engine pointing actuation, estimated at 1.4 kW averaged over the 100 second burn. Total power requirements include computer operation for three hours after reaching orbit, in case checkout procedures or problems delay orbital insertion. Power is provided by a primary (non-rechargeable) battery.

Three-Axis Stabilization

The Zephyr vehicle's baseline configuration is a spin-stabilized flight mode. A preliminary design was done on a version using 3-axis stabilization, indicating that the changes would add about $1.2 million dollars in price and a loss of 5-10 kg of payload capacity. Analysis on this is at a preliminary level. Customer input may increase emphasis on this mode, but at this time it is still an undeveloped option.

The three-axis system was baselined as a series of cold-gas thrusters at the base of the vehicle, and a nitrogen tank placed above the engine mounts (reducing some payload volume).

PERFORMANCE

This section discussed the payload mass and volume of the Zephyr vehicle.

Payload Mass

The Zephyr vehicle as designed in detail exceeded its performance requirements to a large degree. The following table summarizes the vehicle's performance when launched by the Pegasus vehicle and if the Pegasus XL enhanced-performance vehicle is available.

Table 2

ZEPHYR MARS TRANSFER ORBIT PAYLOAD SUMMARY

Specified Mass Margin case

Launch Vehicle	Best Case Payload	Worst Case
Pegasus	51 kg	43 kg
Pegasus XL	71 kg	57 kg

Overweight stage case

Launch Vehicle	Best Case Payload	Worst Case
Pegasus	47 kg	39 kg
Pegasus XL	67 kg	53 kg

In each case, the best and worst case payloads were derived from constructive and destructive pairings of aphelion and perihelion for Mars and Earth.[6] Actual launch velocities for upcoming windows have not been calculated for more detailed payload in year/window figures.

The payloads which Zephyr can accelerate to transfer orbits are quite reasonable for achieving scientific objectives on or around Mars.

The average transfer time is nine months for these orbits.

Payload Envelope

The Zephyr vehicle's maximum payload envelope is a disk 30 cm (12 inches) deep and 116 cm (46 inches) in diameter. At the center of the disk, about 5 cm more depth is permissible. Through this volume must run at least three brackets to connect the Zephyr vehicle to the Pegasus third stage. The nominal space inside the brackets is a disk 30 cm deep and 60 cm in diameter. (Figure 4).

DISCUSSION

The Zephyr vehicle as described in this paper has fulfilled the primary design requirements quite well. From the Pegasus vehicle, injection payloads of 45 kg are easily attainable, which should allow an on-the-ground payload of perhaps 15 kg at Mars if aerocapture deceleration is used. This is plenty enough for a miniature rover, science station, or for a single-experiment more detailed mission.

As the total launch is low, and probe cost should be low due to very low mass, missions for under $30 million are quite possible. The political attractiveness of small, fast to develop and fly missions is also not to be discounted in a fluctuating budget situation. The Zephyr vehicle allows relatively good performance potential within this range, using conventional and low-risk technology. Additionally, potential near-Earth applications may make developing Zephyr on a commercial basis feasible, unburdening those costs from scientific programs.

Analysis of services required by scientific payloads may suggest that some of the Zephyr's services may be more optimally moved to the payload (main computer, guidance) or the proposed light payload bus, so that duplication of service is avoided. Further analyses and input from potential customers will help optimize this tradeoff.

ACKNOWLEDGEMENTS

I would like to thank Henry Spencer, Paul Dietz, Gordon D. Pusch and Graham O'Neill for technical and other feedback which has greatly improved the final version of this paper. I would also like to thank Kevin William Ryan and Dominic Herrity for questions which noted weak areas and Paul Carr for pointing me towards alternative propulsion options.

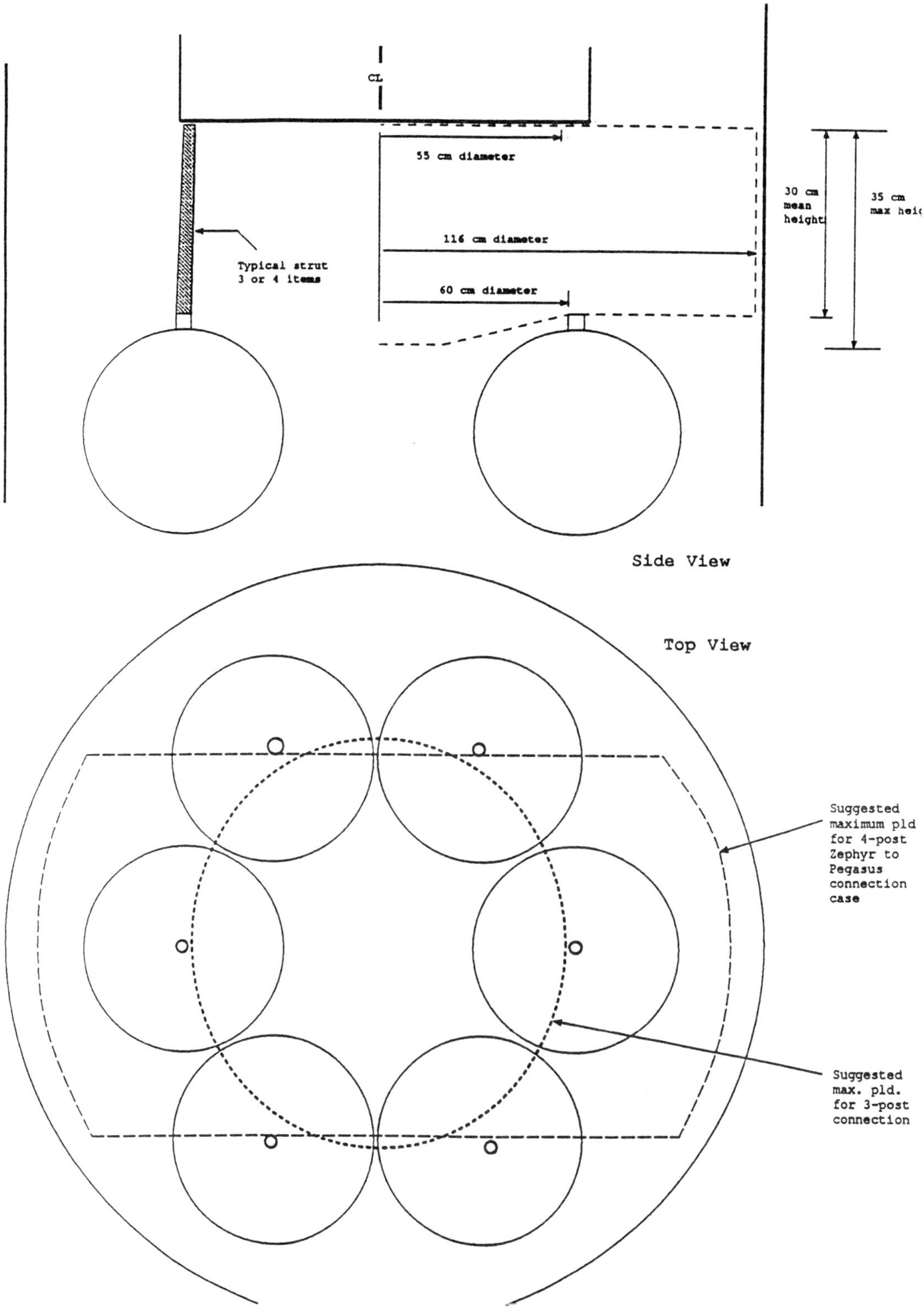

Figure 4 Zephyr Payload Volume Desription.

REFERENCES

1. *8000-Newton Engine data sheet*, SEP, 1992, Paris.
2. *Space Handbook*, Doc. No. AU-18, ed. Cochran, C. D. *et al.*, 1985, Air University Press, Maxwell Air Force Base, Alabama.
3. "Multi-Role Capsule System Description," Hempsell, C. M. and Hannigan, R. J., in *JBIS* (full cit. unknown).
4. *Design with Advanced Composite Materials*, ed. Phillips, Leslie N., 1989, Springer-Verlag, London.
5. *Introduction to Mechanics of Solids*, Popov, Egor P., 1969, Prentice-Hall, Englewood Cliffs, NJ.
6. Hohmann Ellipse Transfer Data, Armento W. J., in *Handbook of Chemistry and Physics*, 65th edition, ed. Weast, R. C., 1985, CRC Press, Boca Raton, Fla.

AAS 97-370

PRACTICAL METHODS FOR NEAR-TERM HUMAN EXPLORATION OF MARS†

Robert M. Zubrin and David B. Weaver

This paper investigates means for achieving human expeditions to Mars utilizing existing or near-term technology. Both mission plans described, Mars Direct and Semi-Direct, are accomplished with tandem direct launches of payloads to Mars using the upper stages of the heavy lift booster used to lift the payloads to orbit. No in-orbit assembly of large interplanetary spacecraft is required. *In situ* propellant production of CH_4/O_2 and H_2O on the Martian surface is used to reduce return propellant and surface consumable requirements and thus total mission mass and cost. Chemical combustion powered ground vehicles are employed to afford the surface mission the high degree of mobility required for an effective exploration program. Data is presented showing why medium-energy conjunction class trajectories are optimal for piloted missions. Mission analysis is given showing what technologies are optimal for each of the missions primary maneuvers. The optimal crew size and composition for initial piloted Mars missions is presented, along with a proposed surface systems payload manifest. The back-up plans and abort philosophy of the mission plans are described. An end to end point design for the Semi-Direct mission using either the Russian Energiya B or a U.S. Saturn VII launch vehicle is presented and options for further evolution at the point design are discussed. It is concluded that both the Mars Direct and Semi-Direct plans offer viable options for robust piloted Mars missions employing near-term technology.

† This article appears in full in *From Imagination to Reality: Mars Exploration Studies of the Journal of the British Interplanetary Society* (Part I: Precursors and Early Piloted Exploration Missions), R. M. Zubrin, Ed., AAS *Science and Technology Series*, Vol. 91, pp. 289-315, 1997. The article was first published in the *Journal of the British Interplanetary Society*, Vol. 48, No. 7, pp. 287-300, July 1995.

CHAPTER 4:
HUMAN MISSIONS II:
POWER AND FUEL SYSTEMS ON MARS

AAS 93-833

THE VIABILITY OF PHOTOVOLTAICS ON THE MARTIAN SURFACE

James R. Gaier and Marla E. Perez-Davis*

The viability of photovoltaics (PV) on the Martian surface may be determined by their ability to withstand significant degradation in the Martian environment. Probably the greatest threat is posed by fine dust particles which are continually blown about the surface of the planet. In an effort to determine the extent of the threat, and to investigate some abatement strategies, a series of experiments were conducted in the Martian Surface Wind Tunnel (MARSWIT) at NASA Ames Research Center. The effects of dust composition, particle size, wind velocity, angle of attack, and protective coatings on the transmittance of light through PV coverslips were determined. Both initially clear and initially dusted samples were subjected both to clear winds and simulated dust storms in the MARSWIT. It was found that wind velocity, particle size, and angle of attack are important parameters affecting occlusion of PV surfaces, while dust composition and protective coatings were not. Neither induced turbulence nor direct current biasing up to 200 volts were effective abatement techniques. Abrasion diffused the light impinging on the PV cells, but did not reduce total coverslip transmittance by more than a few percent.

INTRODUCTION

On the twentieth anniversary of Apollo 11's historic achievement of the first manned landing on the moon, then President Bush proposed, "...a journey into tomorrow, a journey to another planet, a manned mission to Mars." NASA Lewis Research Center was charged with developing plans to provide power on the Martian surface for such a mission. One of the leading candidates as a source of power is photovoltaic (PV) arrays. Calculations indicate that the insolation at the Martian surface is similar to that on the terrestrial surface.[1] A question which, however, has not been as straightforward is the effect of the fine dust which covers the Martian surface. Ultimately, it may be the dust effects which determine whether PV arrays are a viable power source for the Martian surface.

It was against this background that a series of studies was initiated designed to provide experimental evidence on the effects of dust on PV performance on Mars. It was decided that the only practical way to achieve this goal was terrestrial simulation of Martian dust under simulated Martian conditions. The key to these experiments was the existence and availability of a facility in which the effects of Martian winds could be simulated, the Martian Surface Wind Tunnel (MARSWIT) at NASA Ames Research

* NASA Lewis Research Center, 21000 Brookpark Road, Cleveland, Ohio 44135.

Center. This paper summarizes the results obtained from several suites of experiments run over a four year period.

METHODS AND MATERIALS

There are a variety of conditions which could affect dust sticking and removal from a photovoltaic surface on Mars. The conditions which were included in these studies were dust composition, particle size, wind velocity, the angle between the wind velocity vector and the array (angle of attack), height above the ground, composition of the PV surface, whether or not there was dust initially on the surface, and whether the wind was "clear" or dust laden. In addition, two dust abatement strategies were tried, erection of a "snow fence" in front of the samples, and direct current biasing the samples.

Samples

No decision has yet been made about the type of PV cell which will be used in the early Mars exploration missions. It was assumed the PV cells will probably be protected by a coverslip and that SiO_2 is a likely coverslip material. Hence, most of the tests were done with 2.5 cm square, 0.13 mm thick glass coverslips. In order to determine if there might be a better choice to minimize dust effects, some of the coverslips were ion beam sputter coated with candidate protective coatings including SiO_2, polytetrafluoroethane (PTFE), a 50% mixture of SiO_2 and PTFE, indium tin oxide (ITO), or diamond-like carbon (DLC). Table 1 summarizes the coatings.

Table 1
PHOTOVOLTAIC CELL COATINGS TESTED

COATING†	THICKNESS	DEPOSITION	SUBSTRATE
SiO_2	650 Å	ion beam	glass
PTFE	1000 Å	ion beam	glass
PTFE/SiO_2	1000 Å	ion beam	glass
ITO	1000 Å	ion beam	glass
DLC	1000 Å	ion beam	glass

†abbreviations defined in text

The samples were mounted in specially designed sample holders by means of foil tabs which stretch across two corners, and by a third foil tab attached to a removable pin (Figure 1). The idea was to try to maintain the flat surface as much as possible. Samples were held at angles of 0° (horizontal), 22.5°, 45°, 67.5°, or 90° (vertical) from the wind tunnel floor. The holders could also be held horizontally for initial dust deposition or for optical transmittance measurements.

Figure 1 Sample Holder of the Type Used to Test Photovoltaic Coverslips.

Optical transmittance measurements were made by sliding the sample into a transmittance measuring device (TMD) shown in Figure 2. In the TMD an incandescent light source was suspended above the sample, and the sensing head of a Coherent Model 212 Power Meter was beneath the sample. Absolute transmittance measurements were converted into percent transmittance measurements.

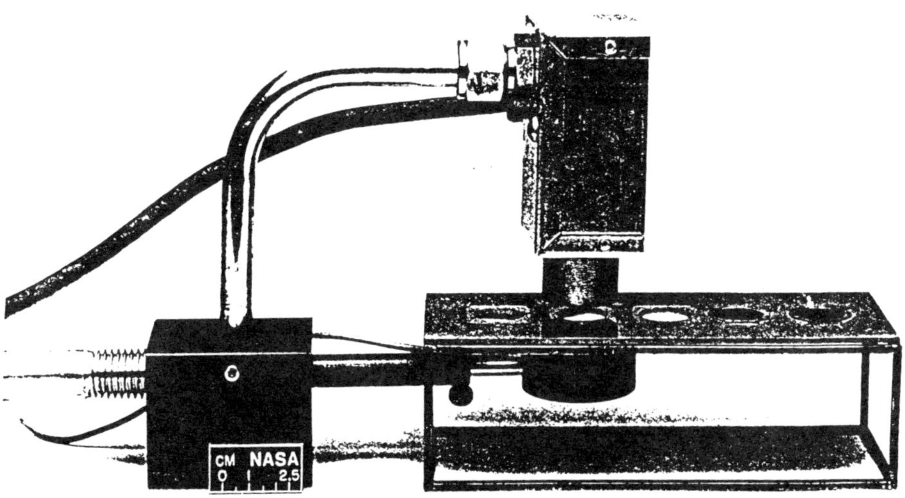

Figure 2 Transmission Measuring Device (TMD) Used to Test Transmission of Light Through Photovoltaic Coverslips.

In the case of the initially clear samples the ratio of the final to initial transmittance ($T_f - T_o$) was calculated. Note that this function ranges from zero (when so much dust has accumulated that no light can be detected) to one (when no dust has been deposited.).

In the case of initially dusted samples, the situation is more complex. Measurements were made before and after the samples were dusted (T_o and T_d respectively), and after the dusted samples were subjected to the MARSWIT tests (T_f). The amount of dust cleared from the sample was evaluated using a dust clearing parameter, which was defined as the ratio of the transmittance change on exposure of the dusted samples ($T_f - T_d$) to that of the transmittance change upon dusting ($T_o - T_d$). This function is constrained to vary from zero (no clearing of the dust by the wind) to one (complete clearing of the dust by the wind.). If there is a net accumulation the dust clearing parameter becomes negative. Note then that the values of zero and one have different meaning for the initially dusted and initial clear case.

In order to verify that our coverslip measurements were valid tests, conventional SiO_2 covered and uncovered N/P silicon PV cells (Spectrolab) were affixed to the sample holders described above in place of one of the coverslips. All PV cells were mounted on 45° sample holders. Damage was assessed by inspection using optical and scanning electron microscopes. The illuminated (AMO) cell characteristics were also measured using a Spectrolab X-25L xenon arc solar simulator calibrated by an aircraft calibrated silicon reference standard.

Martian Dust Simulants

The composition of the Martian dust is not well understood. The elemental composition was determined by the Viking Landers,[2] and based on the optical properties developed from terrestrial materials analogs have been proposed.[3] The Viking biology experiments, however, dramatically showed that the chemistry of the dust is unique to Mars. It should be noted that the purpose of these experiments was not to try to accurately simulate the Martian soil, but to try to determine how sensitive dust clearing is to dust composition. Thus, a variety of dust compositions were used to simulate various aspects of the Martian dust. The chemical composition of these simulants, as well as the elemental composition determined from the Viking Landers is found in Table 2.

The distribution of particles sizes in the Martian atmosphere and on the loose surface has not been narrowly constrained by experiments to date. Best estimates of particle size distributions suggest 0.1 to 2000 μm diameter particles at the Viking Lander sites.[4] The smallest particles (0.1 to 10 μm) remain suspended in the atmosphere indefinitely, and the sand size particles (60 to 2000 μm) will not rise more than a few meters from the surface. Particles in the intermediate sizes (5 to 100 μm) are expected to be elevated during dust storms and quickly settle out onto, among other things, PV surfaces.

An 1800 grit aluminum oxide based optical polishing grit (American Optical Co.) was the first test dust. The important property of the optical polishing grit is its low tendency to agglomerate with moisture in the air. The particle sizes, ranging from 7 to 25 μm, were chosen to be in the range where the particles would be elevated on Mars during times of high wind speed, but which would settle out on the surface after the winds subsided.

Table 2
COMPOSITION OF TEST DUSTS

MINERAL	VIKING LANDERS	OPTICAL POLISH	TRAP ROCK	IRON OXIDE	ARTIFICIAL GLASS
SiO_2	44.6 %	6.6 %	46.6 %	0.0 %	53.5 %
Fe_2O_3	18.1	0.6	13.0	100.0	21.7
MgO	8.3	0.0	6.1	0.0	9.9
Al_2O_3	5.7	89.0	16.6	0.0	6.8
CaO	5.6	0.0	11.1	0.0	0.0
TiO_2	0.9	3.0	2.0	0.0	1.1
Na_2O	?	0.0	2.3	0.0	6.7
Cr_2O_3	0.0	0.6	0.0	0.0	0.0
MnO	0.0	0.0	0.3	0.0	0.0
P_2O_5	0.0	0.0	0.1	0.0	0.0
K_2O	0.0	0.0	1.1	0.0	0.3
CO_2	?	0.0	0.1	0.0	0.0
TOTAL	83.3 %	99.8 %	99.3 %	100.0 %	100.0 %
SIZE	.1-2000 μm	7 - 25 μm	5 - 20 μm	.5 - 2.5 μm	5 - 100 μm

A finely ground basalt called "trap rock" was the second simulant. The Martian soil appears to be basaltic in nature, so a typical basalt was thought to be appropriate. Its grey-green color is an indicator, however, that it differs from the orange dust of the Martian surface. The particles sizes, 5 to 20 μm were similar to the optical polishing grit.

The third simulant was finely ground iron (III) oxide (Fe_2O_3). The Martian soil is known to have a high iron content in high oxidation states. Its particle size was an order of magnitude smaller than the first two dusts.

The final simulant was a glass manufactured (Ferro Corp, Independence, OH) to approximate the composition of the Martian dust as closely as possible. This was sifted into several different size ranges and served to enable study of the effects of particle size over a range of about 5 to 100 μm. The larger particles (greater than about 75 μm) might better be described as sand.

The dust was applied to the samples by elevating it within an enclosure by a stream of dry air, and allowing it to settle out on the sample surfaces. The technique is described in detail elsewhere.[5] The uniformity and extent of the dust deposition were monitored optically. For most runs, the ratio of the transmittance after dusting to before (Td) was between 0.4 and 0.7. The specular transmittance was used as a probe of the extent of occlusion and abrasion. Power is also generated in PV cells from diffuse light, and so the specular transmittance cannot be converted directly to PV performance. However, Katzan *et al.* have related coverglass transmittance in the TMD to N/P silicon PC cell performance at AM0.[6] Their data, reformatted in Figure 3, indicate that the normalized transmittance correlates with PV power (calculated from the I-V curves) to within the error of the measurement.

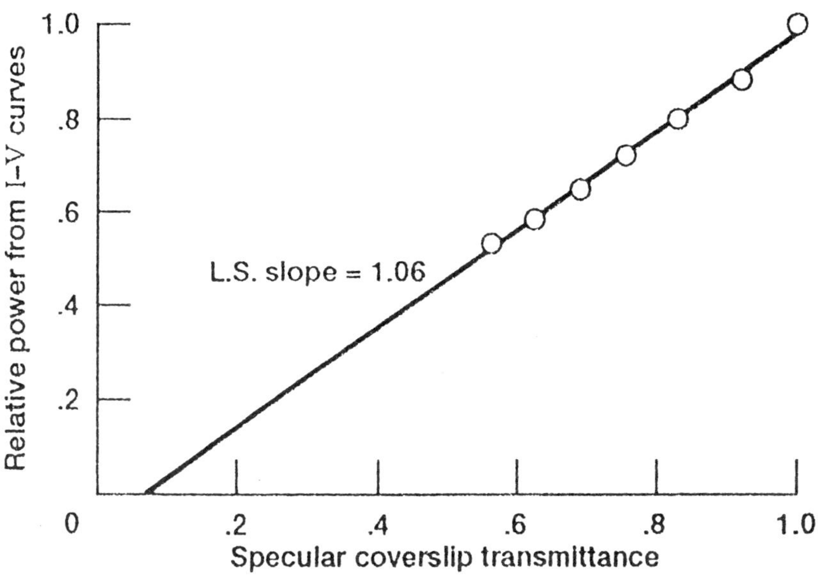

Figure 3 Relationship Between Coverslip Transmittance as Measured in This Study and PV Power Calculated From I-V Curves on N-Doped on P-Doped Silicon Cells at AM0.

Experimental Conditions

The first suite of experiments had as its goal to determine whether dust which would settle out on photovoltaic surfaces would be blown away by the tenuous Martian winds. The effect of angle of attack, height from the surface, and various coatings on cover glasses were also investigated. Optical polishing grit was used as the test dust. Wind tunnel velocities varied from 10 to 124 m/s (22-277 mph).

A principal focus of the second suite of experiments was the effect of different dust types. Thus, many of the experiments from the first suite of experiments were carried out using trap rock or iron (III) oxide instead of the optical polishing grit. Glass and plastic cylindrical sample holders were also introduced at this stage. A horizontal "snow fence" was tried as an abatement technique. Wind velocities varied from 30 to 95 m/s (67-213 mph).

The third set of experiments investigated the effects of basaltic dust being blown onto the surfaces. Thus a series of experiments using initially clear and initially dusted samples were subjected to simulated Martian dust storms. Dust blown wind velocities varied from 19 to 97 m/s (43-217 mph).

The fourth set of experiments investigated the effects of particle size on the dust clearing and depositing properties. For these experiments the Ferro glass Martian simulant was sifted so that a dust which had a composition relevant to the Martian surface and available in different particle sizes was used. Wind velocities varied from 23 to 116 m/s (51-259 mph).

In the fifth set of experiments complete PV cells were tested (as opposed to coverslips) and dome geometry was tested. The PV cells were run neutral, and also biased up to ± 200 V dc.

The winds on Mars were simulated using the Martian Surface Wind Tunnel (MARSWIT) at the NASA Ames Research Center. The MARSWIT is a flow through wind tunnel 14 m in length with a 1 by 1.1 by 1.1 m test section located 5 m from the tunnel entrance. It is located within a 3000 m^3 vacuum chamber which can be pumped down to a few mbar. It is capable of developing wind speeds in excess of 100 m/s at 10 mbar. Its characteristics are described in detail elsewhere.[7]

In order to simulate a Martian dust storm, dust was fed through a hopper from the top of the MARSWIT, near its entrance (Figure 4). First the wind was generated in the MARSWIT at a velocity well below that which would clear dust off of predusted samples. Then the hopper feed was started, dropping the dust into the air stream. Immediately thereafter the wind velocity was increased to the test conditions. The larger particles dropped to the floor of the MARSWIT, while the fine particles were carried along the wind stream and struck the particles, much as would happen during a dust storm on Mars. The MARSWIT was shut down before the hopper was turned off, so there was no time when high velocity clear air hit the samples.

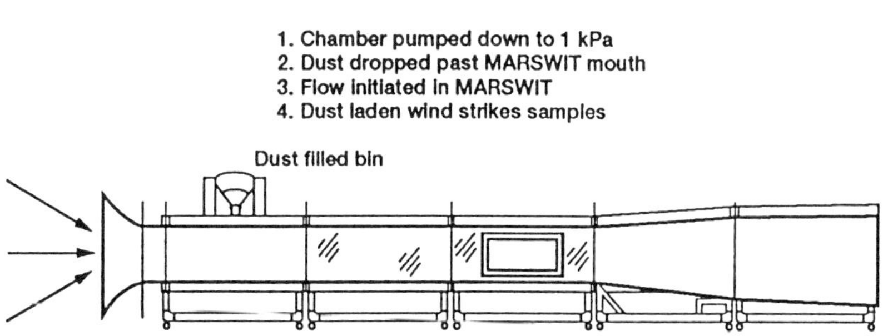

Figure 4 Schematic Drawing of the Martian Surface Wind Tunnel (MARSWIT).

Some of the samples were gold coated and imaged in a Cambridge Model 200 Scanning Electron Microscope (SEM). Photomicrographs were made of samples both before and after dust was cleaned from the samples. Particles size distributions were determined using a Cambridge Quantimet 900 image analyzing computer.

RESULTS AND DISCUSSION

Dust Clearing from Initially Dusted Samples

The first issue to be addressed revolved around the ease of clearing dust from a PV surface by the winds on Mars. Although there is evidence of high wind velocities on Mars,[8] the atmospheric pressure is only about one percent that of Earth. Thus force per unit area exerted by the wind, the dynamic pressure ($1/2\rho v^2$, where ρ is the gas density, and v is the wind velocity) is only about one percent that of comparable wind velocities on Earth. Since the velocity term is squared, a Martian wind velocity ten times that of Earth's will give a similar dynamic pressure. For example, a 100 mph wind on Mars is comparable to a 10 mph wind on Earth.

As might be expected, wind velocity was the most important variable determining whether dust was cleared from the samples. At wind velocities of 95 m/s and higher, dust was effectively cleared (dust clearing parameter greater than 0.90 from all surfaces (Figure 5c). The maximum wind velocity recorded by the Viking Landers was a bit over 30 m/s.[9] It should be noted, however, that the Viking Lander data are averaged over 1 minute periods, so gusts of wind could well have exceeded 30 m/s.

Another important variable for dust clearing efficiency was found to be angle of attack. If the samples were tilted with respect to the wind (that is, not horizontal), the threshold clearing velocity drops dramatically, to 30 to 40 m/s. This is illustrated in Figure 6, which shows the threshold clearing velocity for samples held at a 45° angle. When the angle of attack reaches 90° (vertical) the threshold velocity increases again, to 55 to 85 m/s.

The third important variable was dust particle size. Smaller particles have greater surface to volume ratios, so it is expected that surface interactions, such as adhesion, would be more important for small particles. Experiments bear out that the threshold clearing velocity is dependent on the particle size. Small particles (10 μm and below) did not clear efficiently at 40 m/s, whereas larger particles (30 μm) did. This is clearly shown in Figure 5 where the threshold clearing velocity of particles of the same composition under the same conditions, differing only in their size, are compared.

The largest particle sizes used, being larger than 75 μm, might be better described as sand than dust. Particles smaller than this, because of their high adhesion to the sample plates could be tipped to 90° or even inverted with no significant loss of particles. Particles larger than 75 μm however, would slide off of the plates when tipped to a high angle. Our measurements indicate that the angle of repose for these particles was about 27°. However, these measurements were done in a gravity field of 1.0 g. On Mars, with a gravity of 0.377 g. a sliding force equal to that created by 27' in 1.0 g could not be developed at any angle, so this material would be expected to exhibit dust-like behavior. In fact, any particles with an angle of repose greater than about 22° would be expected to exhibit dust-like behavior. Therefore, particles larger than 75 μm were not used to dust sample plates, but were used in the wind stream to examine the effects of particle size on clearing, deposition, and abrasion. The deposition data for angles greater than 27° is not expected to be representative however, of deposition on the Martian surface since the particles would slide off.

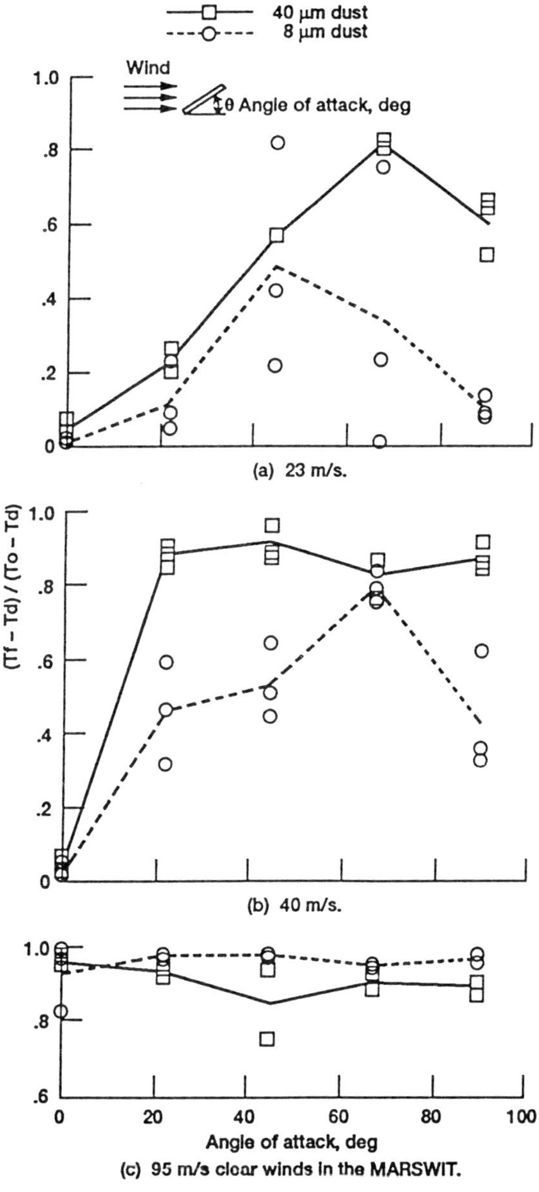

Figure 5 Dust Clearing Parameter as a Function of Angle of Attack Initially Dusted Samples in Clear Winds in the MARSWIT.

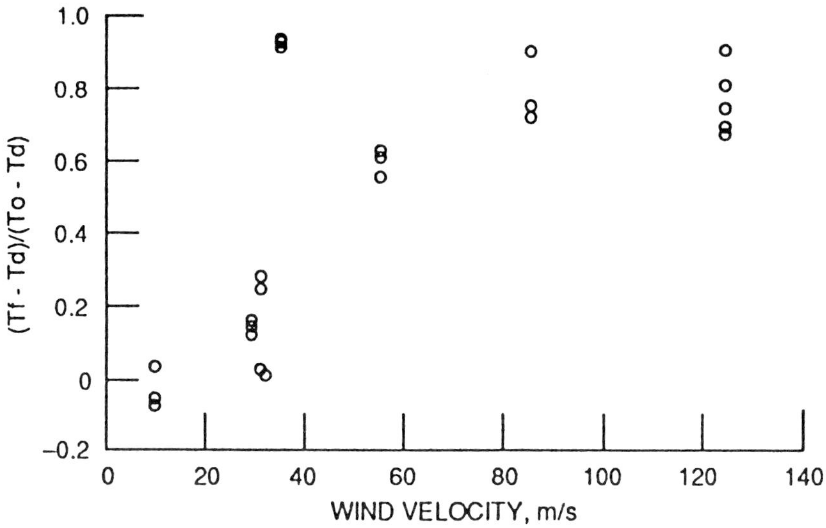

Figure 6 Dust Clearing Parameter in Clear Wind as a Function of Wind Velocity for 45° Angle of Attack in the MARSWIT.

The threshold clearing velocity for initially dusted samples in clear Martian wind is dependent on the angle of attack. (Figure 5). Horizontal samples require very high velocities (85 m/s) whereas those cocked at a 45° angle require only about 35 m/s. The actual angular dependence is somewhat broad, with 22.5° and 67.5° sample clearing to a similar degree. At 90° (vertical) however, the threshold clearing velocity is somewhat higher than at the intermediate angles.

Given the angular dependence of the dust clearing, one might suspect that the mechanism of detachment would involve the rolling or sliding of dust particles, followed by their becoming airborne. For the most part, however, this did not appear to be the case. Photomicrographs of the dust layer remaining on the dusted glass surfaces subjected to 35 m/s winds at different attack angles showed no directionality to the dust removal. Only on the samples with an attack angle of 22.5° could it be discerned from the photographs the direction of wind arrival. This was further confirmed by photographs of the cylindrical samples subjected to the same conditions,. Only when both the wind velocity and the attack angle were very low was there any appreciable rolling or sliding of the particles. Thus, turbulence at the surface must act to aerodynamically lift the particles out in a direction which is approximately normal to the surface. This view is supported by the classical models of Bagnold in which aerodynamic lift plays a key role in particle motion from a surface at the threshold velocity.[10]

This lead to the investigation of the relative clearing ability of laminar and turbulent flow over the samples. Laminar flow gives higher wind velocities which are fairly uniform across the surface. Turbulent flow results in somewhat lower average velocities, but perhaps higher local velocities. In the first experiment the samples were placed at about 3, 6 and 9 cm from the wind tunnel floor, all within the boundary layer. Although the clearing was slightly higher as height increased, the effect was small. Samples which were tested at about 5 cm off the floor of the MARSWIT behaved similarly to those held 50 cm off the floor.

Turbulence was also induced by placing a "fence" of cylindrical rods in front of the samples at a wind speed near the threshold, in an attempt to determine whether the turbulence would lower the threshold clearing velocity. The fence was found to raise rather than lower the threshold velocity.

A wide variety of photovoltaic cell coatings were tested to determine of coatings could be used to lower the threshold clearing velocity. Because of probable differences between the simulant and actual Martian dust this test may have limited applicability, but it was hoped that trends might be observed. The results are presented in detail elsewhere, but the general result is that the coatings did not change the threshold clearing velocity appreciably.

Dust Adhesion to Initially Clear Samples

Dust was found to stick on initially clear samples which were exposed to simulated Martian dust storms. Once again the important variables were wind velocity and angle of attack (Figure 7). Transmittance decreases as angle of attack increases from 0° (horizontal) to 90° (vertical). There was less than 5 percent degradation in the transmittance of samples when the attack angle was 0° over a wide range of wind velocities. In contrast, those with an attack angle of 90° suffered a 17 to 36 percent degradation. Note that higher velocity winds produced larger degradation.

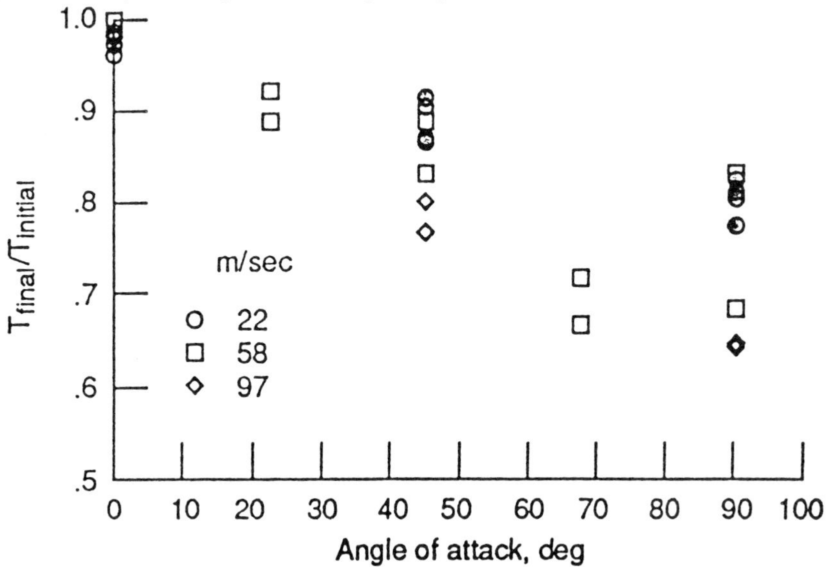

Figure 7 Change in Coverslip Transmittance as a Function of Angle of Attack for Initially Clear Coverslips Subjected to Dust Laden Wind.

Transmittance of the coverglass can actually be degraded by two mechanisms, occlusion and abrasion. Abrasion was not found to be significant until the wind velocity was about 85 m/s. As is apparent in Figure 8, abrasion increased with angle of attack. At 97 m/s wind driven dust horizontal samples were essentially undamaged while vertical samples sustained substantial damage. However, even the highly damaged samples lost little of their transmission when the occluding dust was removed. None of the samples lost more than about 3 percent (Table 3). In fact, in a study of PV coverglass pitting by micrometeoroids it was found that a totally pitted coverglass (not original surface remaining) that the transmission only drops by about 5 percent. The transmitted light does go from specular to diffuse. This will not be a problem in conventional PV cells, but this could be a serious difficulty for solar concentrator cells.

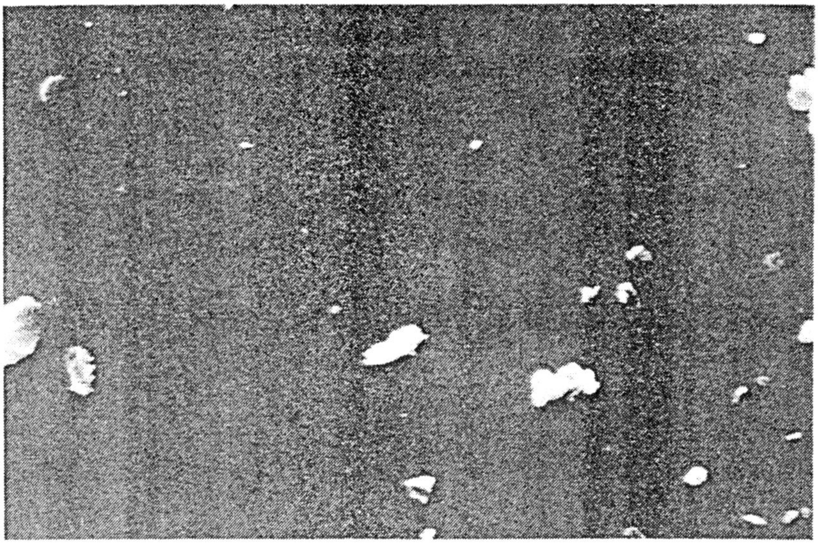

(a) Angle of attack = 0°.

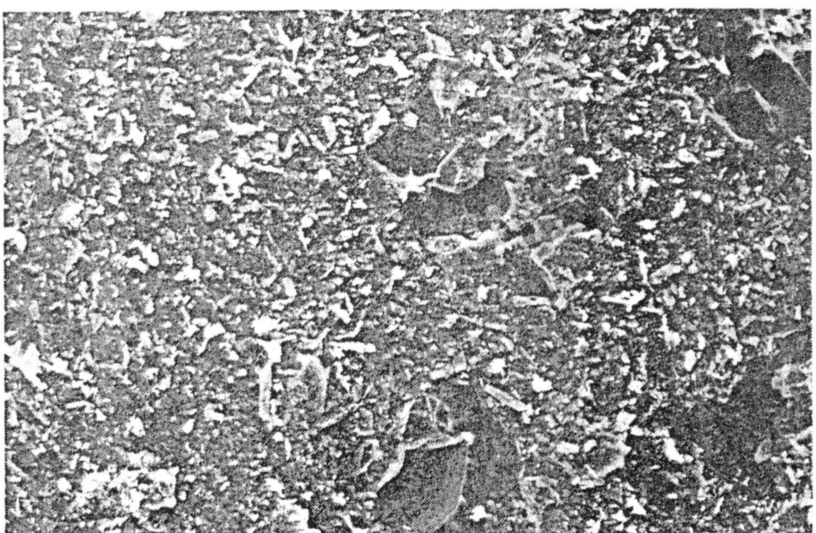

(b) Angle of attack = 90°.

Figure 8 Scanning Electron Micrographs of Coverslip Surfaces Subjected to 97 m/s Dust Laden Wind.

Table 3
PERFORMANCE OF PV CELLS

	Cover Glass	Initial Dust μm	Wind Dust μm	Wind Speed m/s	Pitted by SEM?	AMO Eff %	Tf/To 45° Cover
PV-1	none	30	none	23	no	9.6	1.00
PV-2	none	10	none	32	no	9.1	1.00
PV-3	SiO_2	30	none	95	yes	9.6	0.97
PV-4	SiO_2	10	none	96	yes	9.5	0.97
PV-5	none	10	15	23	no	---	1.00
PV-6	none	none	15	23	no	9.3	1.00
PV-7	SiO_2	10	15	89	yes	8.5	0.98
PV-8	SiO_2	none	15	89	yes	9.4	0.95

Dust Adhesion to Initially Dusted Samples

The most likely scenario for photovoltaic cells which are emplaced on the surface for long periods of time is that they will have a coating of dust, and will be occasionally swept by dust laden wind. All samples experienced a net dust clearing when predusted samples were exposed to dust laden wind. At low velocity (19 m/s) the net clearing increased with angle from horizontal through vertical (Figure 9). Even though this was well below the threshold clearing velocity for clear air, all samples had a net clearing. The horizontal samples cleared by somewhat less than 20 percent, but the vertical samples cleared by about 90 percent. Thus, as wind driven particles collide with the surface, they will more likely knock particles off of the existing surface than stick themselves. This will lead to a dynamic equilibrium in which dust thickness will vary with wind velocity and entrained particle concentration. Particle size distributions also support the dynamic equilibrium model. Although the dust storm experiments only lasted about 15 seconds it was found that the proportion of small particles was larger in the predusted samples exposed to dust laden wind. Larger particles which were initially on the sample surface were blown off, and smaller particles from the wind were blown on.

At high velocity the low angle clearing become more efficient. The clearing of horizontal plates improved from less than 20 percent for 24 m/s to about 80 percent at 100 m/s. At 45° there was an improvement from about 50 percent to about 70 percent. The vertical plates showed little, if any, increase in dust clearing, probably because the deposition rate increased also.

The effects of entrained sand size particles (75 μm) were somewhat different. There was a net clearing to about 80 percent of the clean plate value during both high and low velocity tests. Further, this value was found to be nearly independent of attack angle.

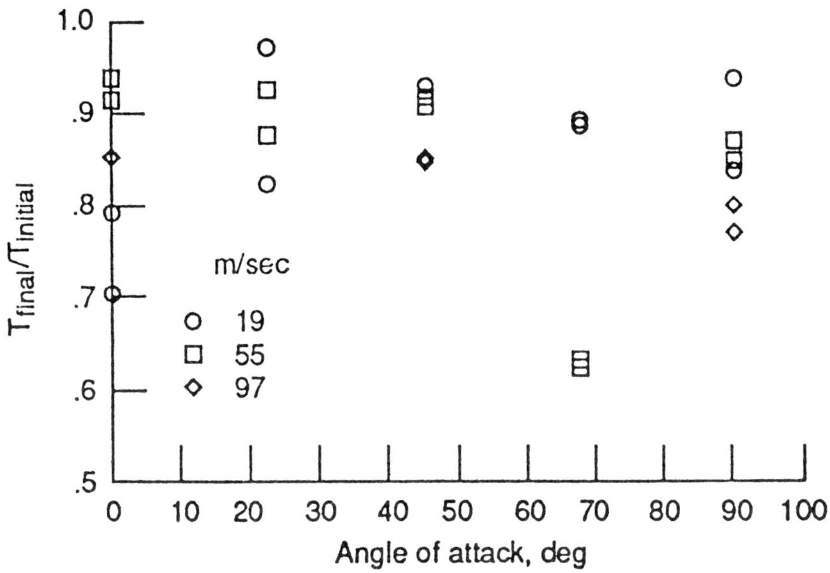

Figure 9 Change in Coverslip Transmittance as a Function of Angle of Attack for Initially Dusted Coverslips Subjected to Dust Laden Wind.

PV Cell Performance

Coverglass transmittance, though thought to be a good predictor of PV cell performance, is not synonymous with it. The degradation of transmittance due to occlusion is thought to reasonable reflect the degradation in PV power. This is what was measured by Katzan, *et al.*, as noted earlier. Abrasion, however, might not be accurately reflected in these results. Thus, PV cell characteristics were actually measured in a few cases. Abrasion of PV cell surfaces was only observed in those samples subjected to 89 m/s winds and higher. The abrasion pits in the SiO_2 coverglass looked like those in coverglasses when imaged under the SEM. The efficiencies calculated from the current-voltage characteristics of the PV cells subjected to the MARSWIT were measured after these tests (Table 3). Although the efficiencies were not determined before the tests, the sample subjected to 23 m/s winds showed no visible signs of damage, and thus were thought to reflect the initial conversion efficiencies. Given that assumption, there was no significant loss in the efficiencies of any of the PV cells. Bowman, *et al.*, have reported degradation in PV cell performance from coverglass abrasion caused by hypervelocity impact of about 5 percent.[12] These tests had very high impact densities which perhaps set an upper bound for the abrasive component of cell degradation. Wade reported that erosional effects on the contacts could be an important factor degradation[13] but, with covered cells and not inter-cell contacts, such effects were not observed in this study.

Cells without coverglasses were also tested in similar conditions (see Table 3). No significant differences were observed between those cells with coverglasses and those without. It should be noted, however, that none of the unprotected cells were tested at high wind velocity.

During use conditions PV arrays often develop a voltage bias with respect to their surroundings. Since electrostatic interactions are expected to be an important dust adhesion mechanism, it was thought that this bias might affect the threshold clearing velocity. Thus, PV cells were biased up to + 200 V or – 200V dc in an attempt to determine this effect. No differences between the biased and unbiased cells.

CONCLUSIONS

The degradation of PV coverglasses on the Martian surface due to dust was investigated by a series of experiments conducted in the MARSWIT. The effects of dust composition, particle size, wind velocity, angle of attack and protective coatings on the transmittance of light through PV coverglasses were determined. Both initially clear and initially dusted samples were subjected both to clear winds and simulated dust storms in the MARSWIT.

It was found that the ability of clear winds to remove dust from surfaces increases as wind velocity and particle size increase, and when the angle of attack is roughly 45° (± 25°). Threshold clearing values for samples held at a 45° angle were about 35 m/s when the particle size was in the 7 to 25 μm range. This is about the same as the highest winds recorded by the Viking landers. Dust composition and protective coatings had little effect on the threshold velocity.

The amount of dust deposited in artificial wind storms also varied with particle size, wind velocity and angle of attack. Smaller particles were more likely than larger to be deposited, or more probably, less likely to be removed once deposited. Similarly, lower wind speeds resulted in more net deposition. Unlike the dust clearing behavior, more dust was deposited as the angle of attack increased from horizontal to vertical. Occlusion was not sensitive to dust composition.

It was also found that wind velocity, particle size, and angle of attack are important parameters affecting occlusion of PV surfaces which were initially dusted and then subjected to dust laden wind. The higher the wind velocity and higher the angle of attack the more net clearing of dust from the surfaces was achieved. The threshold clearing velocity was much lower with dust laden wind, occurring below 22 m/s in all but the horizontal sample cases. A dynamic model of smaller particles being attached while larger are removed was supported by a net decrease in the particle sizes which remained on the samples.

The brief search for effective abatement techniques was unsuccessful with neither induced turbulence nor direct current biasing up to ± 200 volts being effective. Perhaps the best abatement is control over the attack angle. Horizontal deployment of PV arrays which are expected to operate over the long term should be avoided. Tilting the arrays into the wind even a relatively shallow angle (20°) could pay large dividends.

Significant abrasion was only seen in sample subjected to high velocity (> 85 m/s) winds. This is much higher than the highest winds recorded by the Viking landers. Abrasion diffused the light impinging on the PV cells, but did not reduce total coverglass transmittance by more than a few percent. Concentrator PV cells and mirrors for solar dynamic power systems would be much more vulnerable to the effects of abrasion.

PV cells were substituted for coverglass in a few of the experiments with the general result being to validate the coverglass data.

ACKNOWLEDGEMENTS

The authors wish to express appreciation to R. Leach and R. Greeley (Arizona State University) and the MARSWIT support staff (NASA Ames Research Center) for their assistance in the wind tunnel tests, the Space Instrumentation and Studies Branch of NASA Ames Research Center for their financial assistance, M. Marabito (Cleveland State University) for his technical assistance, Alia Moinuddin (Cleveland State University) and R. Martinau (NASA Lewis Research Center) for particle size analysis, D. Scheiman of Sverdrup Technologies for I-V measurement on the PV cells, and the Photovoltaic Branch of NASA Lewis Research Center for their technical assistance.

REFERENCES

1. J. Applebaum and G. A. Landis, "Solar Power on Mars: Update 1991," NASA Technical Memorandum 105216 (1991).
2. P. Toulmin III, A. K. Baird, B. C. Clark, K. Keil, H. J. Rose, Jr., R. P. Christian, P. H. Evans, and W. C. Kelliher, "Geochemical and Mineralogical Interpretation of the Viking Inorganic Chemical Results for Martian Surface Materials," *J. Geophysical Research* 82, (1977) pp. 4625-34.
3. R. E. Arvidson, E. A. Guinness, H. J. Moore, J. Tillman, and S. D. Wall, "Three Mars Years, Viking Landerl Imaging Observations," *Science* 222 (1983) 463-8.
4. H. J. Moore, "Martian Surface Material," *Sand and Dust on Mars*, NASA Conference Publication 10074 (1991) p. 30.
5. M. E. Perez-Davis, J. R. Gaier, R. Kress, and J. P. Grimalda, "Simulation of Martian Dust Accumulation on Surfaces," NASA Conference Publication 3096, 16th Space Simulation Conference (1990) pp.447-56.
6. C. M. Katzan, D .J. Brinker, and R. Kress, (1991), "The Effects of Lunar Dust Accumulation on the Performance of Photovoltaic Arrays," Space Photovoltaic Research and Technology Conference (SPRAT XI), NASA Conference Publication, p. 35-1.
7. R. Greeley, B. R. White, J. B. Pollack, J. D. Iverson, and R. N. Leach, "Dust Storms on Mars: Considerations and Simulations," NASA Technical Memorandum 78423, (1977).
8. A. R. Peterfreund and H. H. Kieffer, "Thermal Infrared Properties of the Martian Atmosphere III - Local Dust Clouds," *J. Geopohys. Res.* 84 (1979),pp.2853-63.
9. J. B. Pollack, C. B. Leovy, Y. H. Mintz, and W. Vancamp, "Winds on Mars During the Viking Season - Predictions Based on a General Circulation Model with Topography," *Geophys. Res. Let.* 3 (1976) pp. 479-482.
10. R. A. Bagnold, "The Flow of Cohesionless Grains in Fluids," *Philos. Trans. R. Soc. Ser.* A29 (1) (1956) pp. 239-97.

11. J. R. Gaier, M. E. Perez-Davis, and M. Marabito, "Aeolian Removal of Dust Types from Photovoltaic Surfaces on Mars," NASA Conference Publication 3096, 16th Space Simulation Conference (1990) p. 379.

12. R. L Bowman, M. J. Mirtich, and J. R. Jack, "Performance of N/P Silicon; and Cadmium Sulfide Solar Cells as Affected by Hypervelocity Impact," NASA Technical Note D-5133 (1969).

13. F. A. Wade, "Martian Environmental Effects on Solar Cells and Solar Cell Coverglasses," Jet Propulsion Laboratory Contract 952582, TTU-3101 (1971).

AAS 93-836

SOLAR ARRAY CORRECTIONS FOR MARS SURFACE

Dale R. Burger*

Power generation of a solar array on Mars is affected by a number of environmental conditions. Corrections for these conditions are necessary to provide an accurate estimate of available power. A computer model has been constructed to make these corrections. The quantities that must be calculated include: optical depth and surface albedo; sun declination angle and number of daylight hours; zenith angle; normalized net flux function; global, direct beam, and diffuse insolation; effect of array tilt and azimuth angles; efficiency correction for spectrum red-shift; Fresnel reflectivity correction; shadow losses; array temperature; Voc and Isc temperature corrections; array losses from packing factor, radiation, UV, and cell mismatch; and Mars to Earth hours correction. The paper presents hourly and total insolation and power levels, and temperature and efficiency corrections for the cells of a baseline solar array.

INTRODUCTION

The Mars Environmental Survey Mission (MESUR) is a low cost, near term mission to study the Martian surface, atmosphere, and interior. To meet the science objectives of the mission, a network of 16 stations will be placed at various latitudes and longitudes on Mars. A Pathfinder mission is being planned to demonstrate the entry and landing as well as some of the engineering subsystems of the Network landers. MESUR Pathfinder will be launched in 1996.

In order to survey the surface of Mars with inexpensive landers an affordable source of power is required. Mars atmospheric models have been published by Anon. [8] and more recently by Landis and Applebaum [1] and Haberle, *et al.* [2]. The recent models are based upon analyses of Viking lander data. The models show that, even during severe dust storms with high optical depth, there is quite a bit of diffuse light present at the planetary surface. This availability of solar energy on the surface throws a new light upon the possible use of solar arrays for powering Mars survey landers and rovers.

BACKGROUND

The amount of light available at any one place on the surface of Mars varies from the following causes: changes in the Mars-Sun distance from orbital eccentricity, obliquity (axial tilt of Mars to the orbital plane), solar zenith angle, and optical depth. The changes in available light from orbital eccentricity, obliquity, and zenith angle are easily calculated [1,2]. However, the optical depth is not so easily modeled due to variable

* Mr. Burger is affiliated with the Jet Propulsion Laboratory, California Institute of Technology, 4800 Oak Grove Drive, Pasadena, California 91109-8099.

amounts of atmospheric water and dust. Mars has light clouds and, near the north polar region, occasional fog. Since there is little water, the optical depth for fog and clouds only ranges from 0.01 to 0.5 and thus is not a serious problem. Dust, however, is a major concern.

There are long periods when the atmosphere is relatively clear of dust - an optical depth of 0.5 or less. When there is dust in the atmosphere, it changes with local, regional (long axis of affected area > 2000 km), and occasional planet-encircling (one hemisphere) or global dust storms [3]. About 100 local storms occur each Mars year all over the planet with higher probability near polar caps when the caps are evaporating. These local dust storms last only a few days and have an optical depth ranging from thin haze ($\tau < 1$) through thick haze ($\tau > 1$). Global dust storms (estimated to reach an optical depth of 6 in southern latitudes with lower optical depths elsewhere) generally start in the southern tropical latitudes (0 to –30) during southern summer which occurs near perihelion (at an areocentric longitude of about 249 degrees). Areocentric longitude is measured from 0 at the northern latitudes vernal equinox.

Since the occurrence of dust storms at any one location is based on probabilities it was decided to fully model the effects of optical depth on solar array operation. This model had to include the effects discussed above as well as the intensity, temperature and spectral changes due to atmospheric dust.

COMPUTER MODEL

The computer model for available Mars surface solar irradiance was based upon equations presented by Applebaum and Landis [1] and Haberle, *et al.* [2]. In particular, there are two optical depth models presented in Ref. 1 which use different assumed starting points for the two 1977 global dust storms. The first optical depth model assumes that the two dust storms seen by the Viking landers started at latitude –30 and areocentric longitudes 215 and 295 respectively:

$$\tau(\phi, L_s) = \max\left\{0.5, \left[\frac{16787\left(1+\frac{\phi}{150}\right)}{1917+(\phi+38.27)^2}\left[0.779e^{-\frac{(L_s-215)^2}{730}}+e^{-\frac{(L_s-295)^2}{730}}\right]\right]\right\}$$

where Φ is lander latitude and L_S, is areocentric longitude. The second optical depth model assumes that the first dust storm started at latitude –30, longitude 215 and the second started at latitude –10, longitude 295:

$$\tau(\phi, L_s) = \max\left\{0.5, \left[\frac{19500\left(1+\frac{\phi}{150}\right)}{4000+(\phi+48.1)^2}e^{-\frac{(L_s-215)^2}{730}}+\frac{12700\left(1+\frac{\phi}{410}\right)}{2465+(\phi+13.1)^2}e^{-\frac{(L_s-295)^2}{730}}\right]\right\}$$

Either of the above models can be selected in the program along with the capability to manually enter values for optical depth and surface albedo. When the optical depth is manually entered the value for albedo is usually determined from a relationship [1]:

$$al = \max\left\{al_T, \min(0.18\tau, 0.4)\right\}$$

where *al* is albedo, τ is optical depth, and al_τ, is minimum albedo of 0.1.

Using input values for areocentric longitude, L_s, and lander latitude, Φ, the program first calculates or uses given optical depth, and albedo values. The declination angle is calculated next using the following relationship:

$$\sin\delta = \sin\delta_o \sin L_s$$

where δ_o is the obliquity.

. Once the declination angle is known the number of daylight hours can be calculated from:

$$T_d = \frac{2}{15}\cos^{-1}\left(-\tan\phi\tan\delta\right)$$

Next the program calculates the hourly zenith angle of the Sun using variable AM and PM hour limits as determined above. Note that a 24-hour Martian day is used (a Martian day is 24.65 Earth hours). Tilt and azimuth of the array or angle and attitude of a facet of an array are then taken into account during the calculation of the cosine of the angle of the Sun to the array normal:

$$\cos\alpha = \sin z\cos(H - A)\sin T + \cos z\cos T$$

where H is hour angle, T is tilt angle, and A is azimuth angle.

The areocentric longitude and orbital eccentricity is used to calculate the amount of light available above the atmosphere and Beer's law [1] is used to determine the direct beam light reaching the surface:

$$G_b = G_{ob}\exp\left[-\tau m(z)\right]$$

where G_b is direct beam irradiance on Mars surface and G_{ob} is the irradiance at the top of the Mars atmosphere. The air mass function, m(z), is usually approximated by 1/cos z, where z is the zenith angle.

The total global insolation reaching the surface is taken from analyses by Pollock, *et al.* [9] based upon data taken from the Viking 1 and 2 landers and is calculated by a polynomial function of the optical depth, albedo, and zenith angle [1]:

$$f(z,\tau,al) = \left[\sum_{i=0}^{5}\sum_{j=0}^{5}\sum_{k=0}^{1} p(i,j,k)\cdot\tau^i\cdot\left(\frac{z}{100}\right)^j\cdot(al)^k\right](1-al)$$

The p coefficients are listed in Table 1. The value for the diffuse light is taken as the difference between the global and the direct beam values. In cases of high optical depth, the direct beam value may be only a few percent of the total global value. The direct beam value is then modified to account for Fresnel reflection losses at the solar array coverglass surface.

Table 1
NORMALIZED NET FLUX FUNCTION COEFFICIENTS

k = 0

j/i	0	1	2	3	4	5
0	1.0028	- 0.228681	0.019613	0.000231	-0.00013	0.000003
1	-0.450073	1.335955	-1.131691	0.402126	-0.063967	0.003758
2	5.566705	-16.912405	13.739701	-4.756079	0.74374	-0.043159
3	22.471579	64.909973	-52.50947	17.997548	-2.786548	0.16034
4	36.334497	-101.800319	79.895539	-26.762885	4.074117	-0.231476
5	-20.42049	53.207148	-39.949537	12.977108	-1.931169	0.107837

k = 1

j/i	0	1	2	3	4	5
0	0.009814	0.226139	-0.117733	0.030579	-0.004090	0.000218
1	-0.156701	0.396821	-0.313648	0.099227	-0.013508	0.000651
2	1.361122	-3.758111	3.007907	-0.987457	0.141693	-0.00732
3	-4.365924	12.539251	-10.394165	3.486452	-0.513123	0.027401
4	5.991693	-17.498138	14.29137	-4.765323	0.703675	-0.03796
5	-2.915099	8.275686	-6.593125	2.173999	-0.320308	0.017335

ARRAY CONSIDERATIONS

In order to calculate the array power output, a temperature calculation is made for the array surface of interest each hour using the Fresnel reflectivity modified total global insolation value for heat input and an estimated value for convection cooling. The actual calculated temperatures fall within a range of a few degrees above zero to as low as –113°C for dawn and dusk. These values are comparable to Mars surface temperature values since the Martian surface absorptance and emissivity ratio is about 1 - the same ratio as a silicon solar array. Constants in the program can be adjusted to give different temperature values for different array surfaces and cell optical properties. Calculation of array power uses separate algorithms for the direct beam, diffuse and albedo power contributions.

Corrections are made to the calculated power value to allow for the non-linearity of Voc and Isc due to low temperatures and intensities. A fifth order polynomial function [4] is used to make temperature corrections:

$V_{oc}(T,H) = A(T) + B(T)\log_{10}H$, where
$A(T) = a_0 + a_1T + a_2T^2 + a_3T^3 + a_4T^4 + a_5T^5$
$B(T) = (b_0 + b_1T + b_2T^2 + b_3T^3 + b_42^4 + b_5T^5)$, and

$I_{sc}(T,H) = C(T)H$, where
$C(T) = c_0 + c_1T + c_2T^2 + c_3T^3 + c_4T^4 + c_5T^5$

The polynomial coefficients for silicon and gallium arsenide are given in Table 2.

Table 2
POLYNOMIAL THERMAL COEFFICIENTS FOR SOLAR CELLS

	sub	a	b	c
2 Ω-cm Silicon	0	.480653E+03	.801129E+O2	.914727E+00
	1	-.241702E+01	.118309E+00	.108713E-02
	2	.294965E-02	-.224011E-02	-.695706E-05
	3	-.206159E-04	.140478E-04	.226603E-07
	4	-.184560E-06	.816163E-07	.171090E-09
	5	.114636E-08	-.548025E-09	-.144039E-11
10 Ω-cm Silicon	0	.507642E+03	.480243E+02	.100581E+01
	1	-.300212E+01	.455891E+00	.143787E-02
	2	.407518E-02	-.470785E-02	-.315471E-05
	3	.459916E-04	-.909273E-05	-.121720E-06
	4	-.323551E-06	.205860E-06	.538927E-10
	5	-.164756E-09	-.323078E-09	.462234E-11
GaAs/Ge	0	.861295E+03	.897263E+02	.891758E+00
	1	-.191649E+01	.680781E-02	.513316E-03
	2	-.517813E-02	.208990E-02	.265402E-05
	3	.191307E-04	-.759904E-05	-.121918E-08
	4	.173634E-07	-.975039E-08	-.501139E-10
	5	-.256290E-09	.863732E-10	-.116910E-12

A correction is also made which accounts for the spectral shift due to different levels of atmospheric dust. This correction uses the delta-Eddington approximation [5,6] and makes use of light transmittance versus wave length equations and data from Ref. 2 and solar AM0 spectrum values from Ref. 7:

$$T(\mu,\tau,A) = c_1 e^{-k\tau}(1+P) + c_2 e^{+k\tau}(1-P) - (\alpha+\beta-1)e^{-\frac{\tau}{\mu}}, \text{ where}$$

$$c_1 = -\frac{(1-P)e^{-\frac{\tau}{\mu}}\left[A+(1-A)\alpha-(1+A)\beta\right]-(\alpha+\beta)e^{+k\tau}\left[(1-A)+P(1+A)\right]}{(1+P)e^{+k\tau}\left[(1-A)+P(1+A)\right]-(1-P)e^{-k\tau}\left[(1-A)-P(1+A)\right]}$$

$$c_2 = +\frac{(1+P)e^{-\frac{\tau}{\mu}}\left[A+(1-A)\alpha-(1+A)\beta\right]-(\alpha+\beta)e^{-k\tau}\left[(1-A)-P(1+A)\right]}{(1+P)e^{+k\tau}\left[(1-A)+P(1+A)\right]-(1-P)e^{-k\tau}\left[(1-A)-P(1+A)\right]}$$

where μ is the cosine of the zenith angle and A is the surface albedo.

The optical constants for the Martian atmosphere were determined from data measured by photo-diodes located on the Viking landers. This data was accumulated during dust storms where the optical depth reached as high as 3.6 so there is good reason to use the data up to an optical depth of 4 which is the upper operating limit for Pathfinder. Values for use in the above equations are:

$$k = \left[3(1-\omega_0)(1-g\omega_0)\right]^{\frac{1}{2}}$$

$$P = \frac{2}{3}\left[\frac{3(1-\omega_0)}{(1-g\omega_0)}\right]$$

$$\alpha = \frac{3}{4}\mu\omega_0\left[\frac{1+g(1-\omega_0)}{1-(\mu k)^2}\right]$$

$$\beta = \frac{1}{2}\mu\omega_0\left[\frac{\left(\frac{1}{\mu}\right)+3g\mu(1-\omega_0)}{1-(\mu k)^2}\right]$$

where ω_0 is the single scattering albedo and g is the asymmetry factor as shown in Table 3 (also taken from Ref. 2).

Table 3
MARS DUST PARTICLE SINGLE SCATTERING PROPERTIES

Wavelength, μm	ω_0	g
0.35	0.598	0.736
0.40	0.627	0.682
0.45	0.659	0.634
0.50	0.695	0.592
0.556	0.735	0.556
0.669	0.840	0.497
0.80	0.901	0.492
1.0	0.739	0.585

The product of the light transmittance function and the AM0 spectrum gives a spectrum which is increasingly red-shifted with increasing atmospheric dust and zenith angle. Wave lengths around 800 nm are favored in the transmittance function. The red-shifted spectrum is then multiplied by the ratio of the total flux in the unmodified spectrum divided by the total flux in the modified spectrum. This rescaling cancels out any problems with intensity changes. The rescaled, red-shifted AM0 spectrum is then multiplied by the spectral response curve of the selected cell material to obtain the expected current output. A correction factor can then be obtained by comparing the integrated total current from the rescaled, red-shifted spectrum with the current from the standard spectrum. Correction factors for silicon and gallium arsenide are presented in

Table 4 for various optical depths and zenith angles. Finally, corrections are made for array packing factor, radiation damage, UV darkening, and assembly mismatch. Mars hours are converted to Earth hours for correct energy results.

Table 4
RED-SHIFT SPECTRUM CORRECTION FACTORS

Optical		Zenith Angle				
Depth	0	15	30	45	60	75
0.25	1.012620	1.012775	1.013302	1.014451	1.017070	1.025377
0.50	1.019673	1.020004	1.021129	1.023580	1.029076	1.044619
1.00	1.036765	1.037495	1.039955	1.045205	1.056145	1.078059
2.00	1.078247	1.079879	1.085257	1.095953	1.114097	1.135195
3.00	1.119917	1.122489	1.130708	1.145717	1.166621	1.185343
4.00	1.161785	1.165211	1.75781	1.193442	1.214383	1.231021
5.00	1.203856	1.207932	1.220041	1.238593	1.258014	1.272542
6.00	1.244658	1.249097	1.261786	1.279717	1.296808	1.309167

FUTURE WORK

While the models used are believed to be the best presently available, there is additional work needed on the distribution of diffuse light and atmospheric refraction effects at high zenith angles. This work will be incorporated as it becomes available. Additional thought is also needed on issues such as dust deposition, chemical weathering, and electrostatic charging. Any findings in these areas are not considered to be appropriate for this program and will be covered elsewhere.

ACKNOWLEDGMENT

The work described in this paper was performed by the Jet Propulsion Laboratory, California Institute of Technology, under a contract with the National Aeronautics and Space Administration.

REFERENCES

1. J. Applebaum and G. Landis, "Solar Radiation on Mars - Update 1991," Technical Memorandum NASA TM-105216, September 1991.
2. R. M. Haberle, *et al.*, "Atmospheric Effects on the Utility of Solar Power on Mars," Draft of a revised chapter for the book, *Resources for Near-Earth Space*, May 1992.
3. L. J. Martin and R. W. Zurek, "An Analysis of the History of Dust Activity on Mars," *J. Geophys. Res.*, v98, E2, pp. 3221-3246, Feb. 25, 1993.
4. R. E. Patterson and R. K. Yasui, "Parametric Performance Characteristics and Treatment of Temperature Coefficients of Silicon Solar Cells for Space Application," JPL Tech. Report 32-1582, May 15, 1973.
5. M. D. King and Harshvardhan, "Comparative Accuracy of the Albedo, Transmission, and Absorption for Selected Radiative Transfer Approximations," NASA RP-1160, January 1986.

6. J. H. Joseph, W. J. Wiscombe, and J. A. Weinman, "The Delta-Eddington Approximation for Radiative Flux Transfer," *Jour. of the Atmos. Sci.*, v 33, December 1976, pp. 2452-2459.

7. Tada, H. Y., *et al.*, *Solar Cell Radiation Handbook*, 3rd Ed., JPL Pub. 82-69, November 1, 1982.

8. Anon., "Summary Report: Computer Program for Planetary Solar Array Development," EOS Report 4042-SR, NASA CR-107014 (N70-11912), 10 Nov. 1969.

9. J. B. Pollock, *et al.*, "Properties and Effects of Dust Particles Suspended in Martian Atmosphere," *J. Geophys. Res.* 84(B6), pp. 2929-2945, 1979.

AAS 93-880

METHANOL: A FUEL FOR EARTH AND MARS

Kelly R. McMillen,* Thomas R. Meyer* and Benton C. Clark†

To sustain a high quality human presence on Earth as well as to facilitate the human exploration and habitation of Mars, the development of renewable fuel technologies for both planets will be essential. Methanol is an outstanding candidate for the fuel because on Earth, it can aid the transition from fossil fuels to renewable technology while providing cost-effective and sustainable opportunities to meet world energy demands during the next century. It can be produced from varied sources such as natural gas, coal, biomass, and industrial carbon dioxide waste. Methanol could play a key role in reducing global carbon dioxide emissions while greatly improving the efficiency of fuel resource utilization. Methanol can be used in automobiles with only minor conversion and it is the leading choice for fuel cell electric vehicles now being developed by most car manufacturers. It can be used to produce octane-boosting additives that reduce emissions, it is a feedstock for producing many industrial products, and it is a convenient currency for the storage and transportation of natural gas. It is a particularly attractive alternative for resource poor nations that are restricted to more efficient energy strategies. On Mars, methanol can be produced from atmospheric carbon dioxide and a source of hydrogen. It is a low fire-risk fuel that can be stored as a liquid at Martian ambient temperatures and pressures. It can be used in fuel cells to produce electricity for Martian habitats, in Martian surface vehicles and for ascent rockets. More importantly, a mission strategy that employs an *in situ* fuel/oxygen production system on the surface of Mars significantly reduces payload mass requirements and mission costs. Thus, in the waning glory of a petroleum-driven era and in the dawning age of Mars exploration, the development of *in situ* fuel/propellant production systems for Mars promotes the human exploration of Mars, and furthers the development of important renewable energy technologies needed for Earth. The purpose of this analysis is to provide a rationale for the use of methanol on Earth and Mars and to review relevant technologies currently used on Earth.

INTRODUCTION

Methanol has the potential to play a major role in the future of both Earth and Mars. On Earth methanol could serve as the bridge to a renewable energy future allowing an easy transition away from fossil fuels. On Mars it could power future exploration outposts and colonies and become the currency of future Martian energy reserves.

One of the crowning achievements of our technological civilization is the capability of space flight. This has opened new vistas of opportunity that have never before been available to mankind. Beginning with the Viking mission in 1976, a continuing se-

* Boulder Center for Science and Policy, P.O. Box 4877, Boulder, Colorado 80306. Authors to whom correspondance should be addressed.

† Lockheed Martin Astronautics, P.O. Box 179, Denver, Colorado 80201.

ries of probes has stimulated a growing interest in Mars. Mars is compelling not only because it offers a wealth of scientific knowledge including possible answers to the riddle of life, but also because of its similarity and proximity to Earth and its apparent potential as a territory where humans may some day live. We already know that Mars has all of the resources necessary to sustain life in some accessible form on the surface. However to truly embrace this opportunity requires that we assess its risks and resources, assess its habitability, conquer its harsh environment, and most importantly, develop energy technologies for Mars that will enable human civilization to thrive in that new world.

A key difference between Earth and Mars is that Mars has no fossil resources from which to produce chemical fuels such as methanol or other energy products. On Mars chemical fuels must either be imported or synthesized from indigenous materials such as water or carbon dioxide, materials that do not have energy value as do fossil fuels. Therefore synthesis requires an expenditure of the same amount of energy as would later be obtained from combustion plus that due to the processing inefficiency. Sources for this energy on Mars could be solar, wind or possibly nuclear power.

By contrast, the Earth has been endowed with large deposits of natural gas, coal and oil that together with atmospheric oxygen represent an enormous cache of stored energy. On Earth most energy products are merely refined from fossil fuels, their energy value having already been imparted by photosynthesis at the time of their formation millions of years ago. Thus chemical fuels, electric power and innumerable petrochemical products are produced at relatively little additional energy cost except that required for refining and conversion.

The Earth's fossil fuel resources have gone largely unnoticed and untapped for millions of years until starting in the 20th century fossil fuels began to be exploited on a massive scale for industrialization. The speed with which this non-renewable energy has been able to be extracted and utilized has vastly accelerated productivity and the rate of economic development over what was possible when only renewable energy resources were available prior to the discovery of oil. In the industrialized nations, cheap energy has been the enabling factor in the growth of many basic industries including mining, manufacturing, agriculture and transportation. In this era, science and engineering also made great strides that helped utilize these energy resources to produce countless material goods, labor saving devices, and public utilities. On the one hand this has resulted in making possible an unprecedented standard of living, but on the other, it has produced global environmental impacts, and created an enormous dependency on fossil fuel resources that has yet to be reconciled with their finite nature.

In the space of just a few generations, Western society, and particularly the United States, has been dramatically transformed from the farm where renewable energy was the principal source of fuel, to a highly material and energy intensive urban lifestyle that is heavily dependent on fossil fuels. In 1998, the United States, with less than 5 percent of the world's population, consumed 25% of the world's energy production, of which 85% was derived from fossil fuels. A second wave of even greater demand is now building, driven by the global population explosion. By 2020 world energy demands are expected to rise 65% over 1996 levels as the developing nations overtake the industrialized nations as the primary consumers of energy [EIA 1999].

Annual world oil consumption is now within two orders of magnitude of the total amount of conventional oil resources estimated to remain on the planet. Assays of remaining oil reserves suggest that we are approaching a time early in the 21st century when petroleum production on Earth will peak and begin to decline as reserves become depleted [MacKenzie, 1996, Bernabé 1998]. Thus market forces can be expected to raise the price of energy and thereby increase the attractiveness of alternative fuels and renewable energy technologies. They are also likely to stimulate technology improvements that will open less economical sources and increase extraction efficiencies that will ameliorate the decline [Taylor 1997]. But eventually a full transition to recycling and renewable energy will become necessary if we are to achieve a sustainable energy policy that will not compromise the ability of future generations to meet their own needs and that will maintain the long-term integrity of the Earth's environment and biosphere.

One of the major environmental concerns is from carbon emissions resulting from the massive consumption of fossil fuels. Total world carbon emissions are expected to reach 8 billion metric tons by 2010. Efforts to reduce emissions related to energy consumption have become the collective concern of the international community through efforts such as the reduction scenarios proposed under the Kyoto Protocols [EIA 1999].

Methanol has the potential to play an important role in all of these major concerns. As a transition fuel for Earth it could serve as the common fuel during a shift from petroleum to natural gas to renewable energy and it could facilitate a graceful conversion of major end use applications in transportation and electric power generation to more efficient technologies. It could replace gasoline without requiring major alterations to current internal combustion engines and it has the advantage of lower emissions. It could serve as the fuel for electric vehicles resulting in a 50%-85% reduction in total petroleum use. Though currently produced from natural gas, it could also be produced from biomass or directly from carbon dioxide and hydrogen. When recycled or natural sources of carbon dioxide are used, and the hydrogen is produced from renewable energy, the production of an equivalent amount of fossil fuel is fully displaced. This same technology is also applicable to Mars where the atmosphere is a rich source of carbon dioxide. The physical properties of methanol make it attractive for use as an energy storage medium on both Earth and Mars. On Earth it has the potential to be used to improve the efficiency of electric power generation systems. Methanol fuel cells are scalable all the way from public utility applications to use in personal portable and handheld equipment. Methanol could also facilitate a transition to a fuel cell based distributed power generation infrastructure supplied by locally produced renewable energy products.

As has been our experience on Earth, our ability to adapt to the Martian environment, utilize resources, and expand human opportunity will ultimately depend on the amount of energy available. Thus one of the most important challenges of the exploration program will be to develop technologies for energy production and storage on

Mars. Just as it is on Earth, wealth on Mars will be measured in terms of the amount of stored energy that is available. To achieve this on Mars, power systems must be put in place and energy gathered and stored. As we can learn from our experience on Earth, Mars will have no future as an abode for humans without having substantial energy resources. In contrast to Earth today, the lack of energy resources on Mars represents the same constraint that the Earth will face when its fossil fuel reserves become depleted. Thus research devoted to developing energy technology for Mars serves a dual purpose since similar renewable energy technology solutions will ultimately be needed on Earth.

The purpose of this analysis is to discuss the rationale for the use of methanol on Earth and Mars, to outline methods for the production of methanol on Mars based on current industrial techniques, and to highlight some relevant technologies and applications. The present study does not go into detail about the specific design and yields from such systems.

ROLE OF METHANOL ON EARTH AND MARS

Methanol is the leading candidate to be the principal fuel and energy storage medium on Mars as well as the transition fuel to a renewable energy future on Earth.

The physical and chemical properties of methanol make it attractive for both Earth and Mars. Its low freezing temperature and low vapor pressure allow it to be easily stored as a liquid at most times and in most locations on Mars without heating or expensive pressurized containment. Because of its very low vapor pressure, methanol will not boil even at the nominal 6 mb ambient pressures on Mars at temperatures below about –25°C. This makes it convenient to use as a liquid fuel for vehicles and equipment on the Martian surface (Figures 1 and 5), and gives it an advantage over other easily synthesized fuels such as methane, acetylene, carbon monoxide or hydrogen. Methanol is equally storable on Earth where its boiling point at 1 atmosphere is about 65°C.

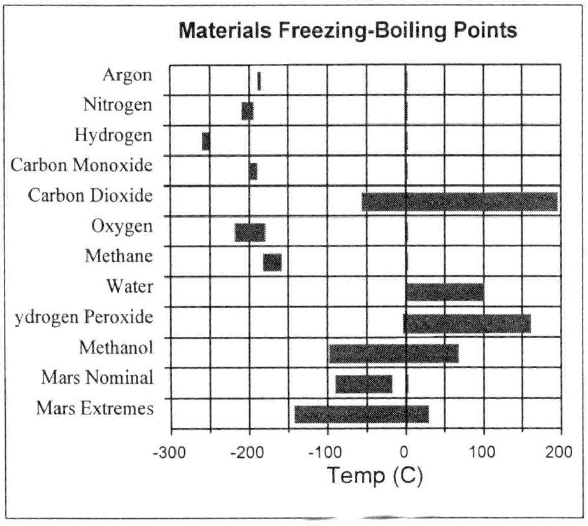

Figure 1 Materials freezing and boiling points versus Mars nominal and extreme temperatures [Clark, 1991].

Using Methanol on Earth

The critical issues to be solved on Earth for which methanol could facilitate solutions are, (1) the need to transition society away from its dependence on fossil fuels, and (2) the need to limit the emission of carbon dioxide and other pollutants that result from the combustion of fuels.

Ninety percent of the methanol produced today comes from natural gas, but it can also be produced from a variety of renewable resources such as biomass and from carbon dioxide waste, thus enabling methanol to serve as a bridge to renewable technology. It has been estimated that bio- and chemical-energy resources such as biomass residues, organic waste, woody crops, natural gas (methane), and carbon dioxide could be converted into alcohol fuels and potentially supply up to half of the world energy demands during the twenty-first century [EIA, 1997, IEA 1978, OTA 1993].

One of the most important virtues of methanol is its flexibility to be manufactured from many different types of feedstock and to serve many applications and niches. This affords the opportunity to implement incremental changes in energy policy to orchestrate changes, spread costs and minimize impacts to the existing industry infrastructures during the transition to renewable energy technology.

On Earth methanol is the leading alternative to gasoline and could be used to facilitate a global transition from fossil fuels to renewable fuels. Transportation energy use is projected to constitute more than half of the World's oil consumption by 2020 [EIA 1999]. In the near term, internal combustion engines could be adapted to use methanol with relatively little impact on the automotive industry while at the same time reducing emissions. The principal changes required for conversion of conventional internal combustion vehicles is to change the material used in fuel wetted lines and gaskets to a material compatible with methanol, and to increase the size of the fuel tank (by a factor of 1.7) to maintain the same cruising range.

Methanol is also a top contender for use in fuel cell electric vehicles currently being developed by most major automobile manufacturers because of their potential for high fuel efficiency and low emissions. The most mature fuel cell technology available today is based on hydrogen and oxygen. However conversion to hydrogen as a fuel has drawbacks because of the lack of production and distribution infrastructures, the impact it would have on the existing distribution industry, and the difficulty of handling and containment of hydrogen at refueling stations and on board vehicles. Methanol on the other hand is very similar to gasoline and could easily be adopted. Furthermore, methanol, gasoline and a variety of other hydrocarbon fuels can be used with hydrogen/oxygen fuel cells by incorporating a reformer to convert the fuel to hydrogen and carbon dioxide [Oniciu, 1976].

The first application of fuel cells was for the Gemini spacecraft that used a solid polymer electrolyte fed by hydrogen and oxygen. Presently five types of hydrogen/oxygen fuel cells have been developed that are distinguished by their electrolyte type. Listed in order of increasing operating temperature, these are the polymer electrolyte membrane (60-100°C), alkaline (90-100°C), phosphoric acid (175-200°C), molten carbonate (600-1000°C), and solid zirconium oxide (600-1000°C) fuel cells.

The most attractive hydrogen/oxygen fuel cell technology for transportation use is the polymer electrolyte membrane (PEM) fuel cell. These cells use a solid organic polymer that must contain water to support the transport of hydrogen ions. Their typical operating temperatures are 60-100°C. Since their operation is limited to temperatures at which water is a liquid, thus imposes constraints on their potential applications in the cold environment on Mars. The electrochemistry of a hydrogen/oxygen fuel cell can be described in terms of the process occurring at each electrode,

Oxidation (anode)	$2\,H_2$	$\rightarrow$	$4\,H^+ + 4\,e^-$
Reduction (cathode)	$O_2 + 4\,H^+ + 4\,e^-$	$\rightarrow$	$2\,H_2O$.
Total Reaction	$2\,H_2 + O_2$	$\rightarrow$	$2\,H_2O$ (1)

Methanol can be steam reformed at 260(C for use with PEM fuel cells according to the following overall reaction,

$$CH_3OH + H_2O \rightarrow CO2 + 3\,H_2 \qquad \Delta H = 49.38\ \text{kJ/mol} \qquad (2)$$

Alternatively, the direct methanol fuel cell is a new and promising technology that is unique as a low temperature fuel cell system that does not require hydrogen. It burns methanol and oxygen directly without the need for a reformer, one of the highest cost items in a fuel cell engine. As an alternative to hydrogen, the direct methanol fuel cell would eliminate the heavy and expensive pressurized hydrogen fuel tank needed in each vehicle, methanol being relatively similar to gasoline in its storage and handling requirements [Surampudi 1993].

The direct methanol fuel cell consists of a catalyst-impregnated membrane that separates a negatively charged chamber containing a methanol/water solution, and a positively charged chamber containing air or oxygen. The hydrogen ions from the methanol pass through the membrane from the negative to the positive side thus producing an electric current. Efficiencies for conversion of methanol to electrical power have been reported to exceed 34%, about twice the efficiency of current gasoline engines [AMI 1998].

There are two critical problems associated with direct methanol fuel cells that must be solved before they can be commercialized. In order to achieve sufficiently high current, a much larger amount of platinum catalyst is presently needed compared to polymer electrolyte membrane fuel cells. Secondly, membrane technology must be improved to prevent methanol from crossing over from the anode to the cathode and thereby wasting fuel and decreasing performance.

In a direct methanol fuel cell the reactions are:

Oxidation (anode)	$CH_3OH + H_2O$	$\rightarrow$	$CO_2 + 6\,H^+ + 6\,e^-$
Reduction (cathode)	$3/2\,O_2 + 6\,H^+ + 6\,e^-$	$\rightarrow$	$3\,H_2O$.
Total Reaction	$CH_3OH + 3/2\,O_2$	$\rightarrow$	$CO_2 + 2\,H_2O$ (3)

Using Methanol on Mars

Methanol is an attractive candidate to be included in the initial planning phases of a Mars exploration program because it can be produced from Martian atmospheric carbon dioxide and thus reduce the high cost of importing fuel from Earth.

As human exploration of Mars progresses, the development of a Mars synthetic fuel industry based on the *in situ* production of methanol and oxygen from Martian atmospheric carbon dioxide is both feasible and highly desirable. This will create a safer, more robust and self-sufficient infrastructure, increase the exploratory capability, and lay the foundation for eventual permanent settlements on Mars.

Initially the hydrogen required to drive the methanol synthesis reaction may be obtained by importing it from Earth; later it might be obtained by splitting water obtained locally on Mars. The energy required for processing may be obtained from solar, wind or nuclear power systems, and by recovering excess heat given off from associated materials processing reactions. Methanol together with oxygen produced from Martian resources is a potential energy storage medium and it has substantial applications for support of a Martian outpost. It can be used for the production of electricity from fuel cells, it is well suited for use in internal combustion engines in the cold Martian climate, and it is a viable propellant for ascent vehicles. It can also be reformed to make syngas, a feedstock for synthesis of a wide range of organic chemicals.

Rocket Performance

As a rocket propellant methanol compares well with other candidates. In a 500 psi thrust chamber with a 100:1 nozzle, methanol provides 340 sec of ISP by comparison to carbon monoxide (285 sec), kerosene (348 sec) and methane (358 sec). Although methane is also easily prepared from atmospheric carbon dioxide, methanol has a higher density than methane, thus making the engineering of feed systems easier. In contrast to methane, methanol has a lower vapor pressure and does not require cryogenic storage therefore simplifying tank design. The engineering of rocket engines based on methanol is aided by the fact that methanol can be decomposed to carbon monoxide (CO) and hydrogen (H_2). This thermochemical process enables a simpler expander cycle engine to be both feasible and more cost-effective (Zubrin 1997b).

Rover Performance

A highly capable roving vehicle will be needed on Mars for long distance traverses as well as for base activities. Whether based on a fuel cell electric or internal combustion engine design, methanol is an attractive fuel choice. Its low freezing point enables it to remain liquid even at night at most mid-latitude locations throughout the year, and its low vapor pressure allows the convenience of a simple fuel tank system.

The polymer electrolyte membrane fuel cell is attractive for Mars applications because it has the lowest operating temperature but its operational requirements are that the fuel and oxygen stream must be humidified and its polymer electrolyte membrane must be kept hydrated by liquid water. Likewise, the direct methanol fuel cell technology also uses a methanol/water solution with a membrane. These aspects present challenges for applications that require on demand power and zero maintenance energy during shut down. Therefore fuel cells may be better suited for semi-stationary applications in heated environments while internal combustion engines may be superior for rovers that can quick start in the cold Martian environment.

Because methanol has a low vapor pressure and high heat of vaporization, conventional methanol fueled internal combustion engines may have difficulty in cold start conditions, at temperatures below 10°C. One possible solution is to use on board dimethyl ether (DME) generation to aid Methanol fueled engine cold starting [Karpuk 1988]. Methanol will exothermically dehydrate to form DME according to the following equation:

$$2\,CH_3OH \rightarrow CH_3OCH_{3)} + H_2O \tag{4}$$

In an earlier study by Clark [1991], a methanol internal combustion engine/generator for use in Mars surface rover operations was shown to be a mass efficient method of providing on-demand capability. The use of a hot surface ignition system would enable it to cold-start in the Arctic-like temperatures of Mars. Hilger *et al.* [1990] have developed a system that consists of a modified hot surface igniter glow plug in a direct injection stratified charge diesel engine that has diesel like efficiencies with Otto-like emissions, thus making it of potential use on Earth.

Agricultural Applications

Methanol has been shown to significantly improve plant growth in certain species while reducing water consumption thus providing an alternative plant growth medium in water limited environments such as Mars, and in arid regions on Earth (Arizona, Mexico, Egypt, Israel, North Africa, and Brazil) [Nonomura 1992]. Both the EPA and the Arizona Department of Agriculture classify methanol as a fertilizer. While it has little effect on C4 plants such as corn sorghum and many noxious weeds, it shows distinct advantages when used with C3 plants such as cotton, wheat, strawberries and melons that are known to be less efficient users of photosynthesis. By spraying plants with a highly dilute solution of methanol during peak sunlight, it is believed that additional carbon is made available thus inhibiting photo-respiration where plants reject energy collected during times of peak photosynthesis if there is insufficient natural carbon dioxide available. Although plant toxicity has been observed for concentrations exceeding 30%, none has been observed when used in highly dilute solutions. When experimental plants growing under dry conditions were sprayed with dilute methanol, wilting was greatly reduced during the hot part of the day and plants were shown to grow larger and about twice as quickly thus halving their total water requirement. Although studies of the process and its effects are continuing, methanol shows great promise for increasing crop yields and reducing agricultural water requirements in arid regions on Earth. The same benefits may also hold true on Mars where water resources are limited.

Industrial Uses of Methanol

About three-quarters of all methanol produced in the United States is used for the production of three main industrial products, MTBE (methyl tertiary butyl ether), formaldehyde, and acetic acid. Nearly half of all methanol production is used to make MTBE, a high-octane oxygenate used in reformulated gasoline to reduce emissions. In second place, formaldehyde resins are used to make engineered wood products such as particleboard made from waste wood, and for the production of fabrics including Spandex fibers. Methanol-based acetic acid is used in making PET plastic bottles and polyester fiber used in clothing and carpets. Methanol is also the principal ingredient used in chlorine-free bleaches, paints, solvents, refrigerants, and disinfectants [AMI 1998b].

On a mature Martian base, the production of formaldehyde resins and acetic acid could be applied to the manufacture of construction materials, plastics, fabrics and other bulk materials necessary for the expansion of the base. Formaldehyde resins could be used to process cellulose waste products from Martian greenhouses to manufacture furnishings and construction materials. Although the optimal design for Martian greenhouses has not been resolved, a requirement for their construction would imply potentially large material requirements. Products derived from methanol could address the need for fabrics for inflatable or semi-rigid inflatable structures, insulation, windows and water containment systems [Boston 1981].

Environmental Aspects

Methanol offers a number of important environmental benefits. It is currently used to produce the oxygenate MTBE that is added to approximately 25% of the gasoline sold in the United States to promote cleaner burning fuels and reduce emissions. Methanol is produced from natural gas with 60% total energy efficiency and most methanol plants supplement their production with purchased carbon dioxide that would otherwise have been released to the atmosphere [AMI 1998].

Studies of the fate and transport of methanol in the environment show that, relative to gasoline that has constituents such as benzene, methanol has fewer adverse impacts on the environment [Pirnie 1999]. Methanol is capable of completely mixing with water, degrades quickly in the atmosphere, and will rapidly biodegrade in surface waters underground. Methanol spills to the soil, groundwater, and surface water will quickly biodegrade under both aerobic and anaerobic conditions and, consequently, methanol is not expected to persist. Methanol exhibits a lower toxicity to both humans and indigenous microbes than conventional gasoline and is neither mutagenic nor carcinogenic.

OBTAINING ENERGY AND RESOURCES

Atmospheric Resources on Mars and Earth

The atmosphere of the Earth which contains 20.9% oxygen both supports life and is indicative of a legacy of past life. Earth has large deposits of petroleum and natural gas, whereas Mars contains no known energy resources. Viking data shows that the atmosphere of Mars consists almost entirely of carbon dioxide (95.3%), a compound of carbon that is in its lowest energy state. The balance of the atmosphere consists mostly of inert gases (N_2 2.7% and Ar 1.6%), a small amount of water, plus minute traces of oxygen and carbon monoxide that result from photo-dissociation of carbon dioxide in the atmosphere.

Viking data indicates that throughout its history, the conditions on Mars have not been suitable to support substantial biological processes that could have generated fossil fuel resources. Therefore we should not expect to find deposits such as petroleum and natural gas, or wood as we do on Earth. Oxygen, the other essential component of combustion, is generally taken for granted on Earth. Its presence is the direct result of the Earth's vigorous and thriving life forms that continuously produce and maintain the atmospheric oxygen at its present level through photosynthesis. By comparison, the Martian atmosphere is impoverished of free oxygen. Thus on Mars, all supplies of chemical fuels and oxidants must either be imported or manufactured from indigenous Martian resources such as water and carbon dioxide, materials that are fully oxidized and have no energy value. In marked contrast to the situation on Earth, processing these resources will require as much energy input as will be derived from their ultimate combustion, plus the cost of manufacturing inefficiencies. Additional energy may also be required for mining and benefaction of the necessary feedstock.

Energy Options for Mars

The only known energy sources on Mars are solar energy and wind power. Possible technologies for their utilization include photovoltaics, solar thermal systems, windmills, and biomass production. Energy might also be produced using nuclear reactors. The solar constant at Mars' orbital mean is 590 W/m^2 as compared to 1371 W/m^2 for Earth [Meyer & McKay 1989]. Recent advances in monolithic solar cells have achieved efficiencies of 32% using a x47 concentrator at standard Earth solar intensity [Kurtz 1999] thus significantly improving their competitiveness. Wind speeds at the Viking Lander were not chosen to be optimal for wind but were observed to exceed 10 m/sec. Even though its atmosphere is 100 times thinner than Earth's, estimates of the power that could potentially be extracted from Martian wind if a location were found with wind speeds of 14 m/sec is 30 kW/200 m^2 windmill surface area [Haslasch 1989]. In principle, bioenergy crops could also be grown on Mars. Since the ambient pressure and temperature on Mars is too low to support liquid water which is essential for plant growth, heated, pressurized greenhouses would be required [McKay 1993]. Bioenergy crops may not be directly competitive with other forms of energy production. However, biomass that is normally produced as a by-product of food production and from habitat wastes would be a valuable source of methane or synthesis gas (CO, H_2).

Since the energy cost for producing chemical fuels and oxidants on Mars is high, the rationale for their manufacture will be based on factors such as necessity, safety, convenience, and practicality. For example, chemical fuels are necessary for rockets and may be the most practical option for surface vehicles. Chemical fuels could be used for heating remote or temporary facilities, and they could be used to provide convenient, portable, on-demand electrical power when used in fuel cells [Clark 1991].

Obtaining Hydrogen

On Earth as well as on Mars, hydrogen is the largest cost element in the production of methanol from CO_2 and H_2. Extracting hydrogen from fossil fuels through methane reforming, oxidation of oils and coal gasification has been the predominant method of obtaining hydrogen on Earth. Non-fossil fuel production of hydrogen can be achieved by electrolysis, thermochemical splitting of water, or photolysis [Hepp 1991], all methods that may also be used on Mars.

The primary source of hydrogen on Mars is indigenous water. Water is believed to exist in the Martian regolith and can be extracted by heating the soil. Estimates of the energy cost of extraction vary from 0.9 kW/kg for polar ice, to 9 kW/kg for water obtained by heating bulk soil [Clark 1989]. The atmosphere of Mars is also a ubiquitous and inexhaustible source of water. Because the partial pressure of water vapor in the Martian atmosphere is very small, water extraction by condensation and cooling techniques would be very expensive [Meyer & McKay 1981], but water can potentially be economically extracted by temperature swing water vapor adsorption [Finn 1996; Williams 1995].

During the early phases of Mars exploration, until such time as ready sources of water are developed, it may be more practical to import hydrogen from Earth to use for the production of chemical fuels. Hydrogen is a powerful reducing agent that can easily be reacted with atmospheric carbon dioxide along with various catalysts to produce fuels such as methane and methanol plus oxygen (as water). A strategy for fuel production whereby only the hydrogen is imported is advantageous since hydrogen represents only 5% of the total mass of the constituents of the fuel [Zubrin 1991], with the balance of the mass being obtainable from the Martian atmospheric carbon dioxide (Eqn. 9).

For a hypothetical crew return spacecraft that required 100 metric tons of methanol/oxygen fuel on the Martian surface, 5 tons of this mass (5%) in the form of hydrogen would need to be imported to Mars. At an estimated shipping cost of $50,000 per kilogram landed on the surface of Mars, the cost to ship all 100 tons of fuel and oxidant would be $250 million, compared to $12.5 million for the hydrogen alone. To produce 5 tons of hydrogen by electrolysis, 45 tons of water would be required and approximately 200 megawatt-hours of electricity, or the equivalent of a 100 kWh power supply operating for 2000 hours (83 days). Although the water might be obtained on Mars, the mass of the extraction and electrolysis equipment must be added to the shipping cost of items brought from Earth, though possibly they can be amortized over multiple missions.

The cost to transport a power supply, such as a nuclear or photovoltaic system, is also significant. A power supply would be needed regardless of whether the oxidant component of the fuel was produced from water or from the reduction of carbon dioxide since the bonding energies involved are approximately the same. However, one benefit of producing hydrogen by electrolysis from a local source of water is that all of the oxygen needed for combustion is produced directly and no cryogenic storage of hydrogen is needed. This is inherently more efficient than producing the oxygen separately when the fuel is produced from imported hydrogen.

Conservation of H_2

On Mars hydrogen or water is expected to be a rare and expensive commodity, therefore conservation and recycling will be necessary. An important conservation measure is to capture and recycle the water by-product from methanol combustion. This may be done wherever methanol fuel is used on the surface such as in habitat heating, surface vehicles, and fuel cell power systems.

Recovery or optimal utilization of hydrogen and water from *in situ* synthesis processes will also be important. Hydrogen can be separated from waste gas using, for example, hollow-core polymeric fiber or palladium [Peachey 1995] metal membranes, and water can be recovered by methods such as cooling and condensation, or with cold fingers.

In methanol production, where the reverse water gas shift reaction is used to reduce carbon dioxide to carbon monoxide, the water by-product can be removed before the methanol synthesis, thus methanol produced from low-CO_2 content syngas may have a water content below 1% and require little purification [Unnasch 1991]. Thus, one of the most valuable Martian resources, water, may be recovered and recycled.

Sources of H_2 for Producing Methanol on Earth

Ninety percent of the methanol produced on Earth today is produced directly from natural gas with up to 60% energy conversion efficiency [AMI 1998]. Through the process of steam reforming or partial oxidation, natural gas is converted to syngas (CO, CO_2 and H_2) that forms the basic feed stock for methanol synthesis.

From an environmental perspective it is desirable to produce methanol from non-fossil sources of hydrogen and carbon dioxide since this will displace the production of an equivalent amount of fossil fuel and thus reduce the overall emissions of CO_2 into the Earth's atmosphere. It is not sufficient to simply use a non-fossil source of carbon dioxide and an industrial source of hydrogen, because nearly all hydrogen produced today is also obtained from fossil fuels, by methane reforming, oxidation of oils, or coal gasification, all of which produce carbon dioxide as a by-product. Likewise, hydrogen produced by electrolysis of water using power obtained from the electric power grid is also likely to have its origin in a coal or natural gas fired power plants. Over half of all the electricity produced in the United States today is produced from coal.

Carbon dioxide itself has no energy value since the carbon is fully oxidized. In methanol produced from carbon dioxide and hydrogen, it is the hydrogen that brings the energy value to the methanol. The production of hydrogen from fossil fuels bears only the cost of conversion of an existing stored energy resource, whereas production of hydrogen from renewable resources means that all of the energy that will ultimately be obtained by burning the methanol must first be obtained from a primary energy source. For this reason methanol produced in this way is not likely to be competitive with methanol produced from natural gas unless the price of the renewable energy source is especially low, but these economics may change as natural gas sources become depleted in the years ahead.

The cost of electricity produced from biomass is said to be competitive with production from coal or natural gas, but there is no substantial infrastructure in place for biofuel production at this time [Sloan 1997]. On Earth where we have long been committed to the use of fossil fuels, the capital investment needed to build an alternate infrastructure of renewable energy systems is also a barrier. However, as global fossil fuel reserves continue to dwindle, their cost will inevitably increase, and renewable energy systems will one day become competitive. On Mars where there are no fossil fuel resources, this is the case today, and renewable energy technology is the only option.

Obtaining Carbon Dioxide

Mars has abundant carbon resources, but unlike Earth, it has no known fossil fuels. The presence of carbon dioxide on Mars was first identified spectroscopically by Kuiper in 1948 [Kuiper 1949], and its atmospheric abundance (95.3%) was determined by the Viking mission in 1976 [Owen 1977]. Carbon dioxide together with a source of hydrogen or water makes an ideal feedstock for the production of oxygen and fuel, two of the most important bulk consumables needed for a Martian research outpost.

Acquisition of CO_2 from the Martian atmosphere can potentially be accomplished at very low energy cost using an adsorption-based separator operating as a temperature-swing adsorption pump. Zeolites and carbon have been successfully tested in this application [Finn 1996b]. At the low temperatures of the Martian night, CO_2, loading can approach theoretical limits for certain adsorbents. Operating on the change in temperature from day to night, this diurnal cycle provides an effective optimization for use of adsorbent mass. Additionally, under the low-pressure conditions of Mars, adsorption pumps are often able to produce pressure ratios much larger than single-stage mechanical compressors. They use fewer moving parts and are more easily scaled to small dimensions than are mechanical methods [Finn 1996; Finn 1996b].

The abundance of carbon dioxide in the Martian atmosphere (95.3%) is an ample feedstock that can be used directly for the production of methanol and oxygen. Its ubiquity makes it highly attractive for use in methanol/oxygen propellant plants that can be located wherever needed.

Carbon dioxide is also one of the most abundant carbon resources on Earth. All fossil fuels presently in use come from carbon dioxide via photosynthetic processes. With the generation of atmospheric carbon dioxide steadily increasing from combustion of fossil fuels, it is desirable to develop techniques whereby it can be diverted from direct release into the atmosphere and converted to valuable hydrocarbons for use as fuels or industrial feed stocks [Kuei 1991; Halmann 1993].

Carbon dioxide is an essential substance in the life cycle of plants and animals where it is continually exchanged via the atmosphere. The production of biomass is an effective method of extracting carbon from the Earth's atmosphere even though its concentration is only 0.03%. The most attractive sources of CO_2 for industrial applications are those that produce it in the highest concentration. Power plants are by far the largest producers of CO_2, their volume exceeding most other anthropogenic sources by a factor of 10 to 100. But the concentration of power plant emissions is only in the range of 13-18% and therefore this source of CO_2 is not used for chemical synthesis. Some of the highest purity CO_2 comes from ethanol plants (~99%), followed by ammonia, hydrogen, ethylene oxide, and natural gas processing plants (~97%). There are also quite significant natural sources of high purity CO_2. Diverting CO_2 from any of these sources diverts it from entering the atmosphere and displaces the production of an equivalent amount of fossil fuel [Unnasch 1991]. One advantageous way to divert it is to co-locate natural gas based methanol production plants near these sources of CO_2 since steam reforming of methane to produce methanol results in excess hydrogen that can utilize the outside source of CO_2 to produce additional methanol.

Other low-grade sources of CO_2 include dissolved carbon dioxide and calcium carbonate in seawater, and calcium carbonate sands. To be competitive and environmentally beneficial, methanol produced from carbon dioxide requires a low cost source of renewable electrical power to produce hydrogen from water.

Producing Oxygen on Mars

On Earth, atmospheric oxygen (21%) is freely available for use as an oxidant but Mars' atmosphere contains only trace amounts (0.13%), therefore *in situ* production of oxygen is necessary to supply the needs for life support, surface vehicles and especially for ascent spacecraft. *In situ* production of oxygen will enhance crew safety and is a key component of the strategy to minimize payload cost.

There are several approaches for production of oxygen by reduction of atmospheric carbon dioxide. These include the electrochemical reduction using a zirconia membrane [Sridhar 1993; Ramohalli 1993; Colvin 1992] use of an electrochemical pump [Gwynne 1993; Ash 1978] photochemical reduction [Hepp 1991], and the reverse water gas shift process. Each carbon dioxide reduction method has advantages and disadvantages [Stoker 1993]. The zirconia membrane method involves high temperatures and fragile components, and its practical efficiency has not been proven. The photoelectrochemical reduction of carbon dioxide method is theoretically attractive because it could use the solar UV incident on Mars, but this has not yet been demonstrated. The reverse water-gas shift process normally does not go to completion and its equilibrium constant is only about 0.1 at practical catalyst operating temperatures (400°C). However it can be made to achieve high utilization of its hydrogen and carbon dioxide feedstock by the use of a condenser to recover the water by-product and a membrane to recover and recycle the unused feedstock [Zubrin 1997].

Occasions where oxygen is an available by-product are perhaps rare, but one example would be from the electrolysis of water to prepare hydrogen for use in dirigibles or balloons.

H_2O_2 as an Oxidant

Only a trace amount of hydrogen peroxide (~ 1 ppm) is believed to exist in the Martian soil and does not occur in sufficient concentration to be used as a resource [Zent 1994]. However H_2O_2 could be produced on Mars and used both as an energy storage medium as well as an oxidant for the combustion of methanol. It can also be used as a monopropellant and for a myriad of other applications. Preparation of H_2O_2 would require a source of indigenous water and electrical power [Clark 1987]. A comparison of the combustion reactions of methanol with oxygen and hydrogen peroxide is given in Equations 5 and 6:

$$CH_3OH(g) + 1.5\ O_2 \rightarrow CO_2(g) + H_2 \qquad \Delta H = -675.72\ kJ/mol \qquad (5)$$

$$CH_3OH(g) + 2\ H_2O_2 \rightarrow CO(g) + 4\ H_2O \qquad \Delta H = -609.76\ kJ/mol \qquad (6)$$

In a scenario in which 500 kg of fuel/oxidant provides 250 km of rover travel using H_2O_2 as the oxidant, 180 kg of the H_2O_2 is oxygen and 200 kg water. Of the 200 kg of water, 130 kg can be recovered from the exhaust gas. Other candidate oxidizers compatible for combustion with methanol fuel include: nitrogen tetroxide (N_2O_4), and nitric acid (HNO_3) [Clark 1991]. Since oxygen and hydrogen peroxide can both serve as oxidants and consumables (with H_2O_2 requiring oxidation), they are likely candidates for early *in situ* production on Mars.

MANUFACTURE OF METHANOL

Synthesis Scenarios on Mars

The first major application for *in situ* propellant production on Mars could be for Earth-return spacecraft for robotic or human missions. Because of the anticipated difficulty of obtaining water on Mars during the early phases of exploration, hydrogen can be brought from Earth at a relatively modest cost compared to importing the entire mass of fuel. The combustion equation for methanol shows that hydrogen represents only 5% of the total mass of the fuel/oxidant materials by weight, the balance being carbon and oxygen that can be obtained on Mars.

$$CH_3OH(g) + 1.5\ O_2 \Leftrightarrow CO_2(g) + H_2O(g) \qquad \Delta H = -676.81\ kJ/mol \qquad (7)$$

Efficiency considerations for the production of large quantities of fuel and oxidant dictate that these be produced in the correct proportions* required for combustion (Eqn. 9). The amount and type of feedstock needed for *in situ* propellant production will depend on whether the hydrogen requirement is met by using imported hydrogen or by splitting (indigenous) water. Figures 2a and 2b illustrate these two synthesis scenarios. The left side of each figure shows the amount of feedstock needed by each reaction and the right side shows the net amount of propellant, oxidant and waste by-product pro-

* The actual proportions depend on the application. When used as a bi-propellant for rocket engines the optimum mixture ratio may be slightly fuel-rich, fuel cell applications may require stoichiometric proportions.

duced. The methanol processes are simplified to show only the net reaction. Companion processes for carbon dioxide reduction using the reverse water-gas shift reaction and water electrolysis to produce oxygen are also shown.

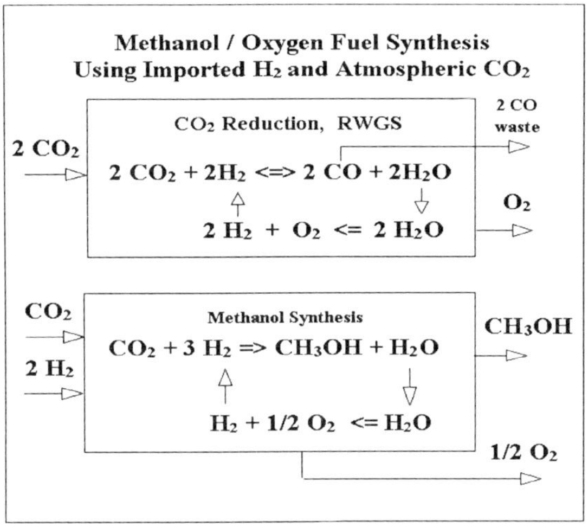

Figure 2a. Ideal mass flow requirements for methanol-oxygen fuel production using imported hydrogen and Mars atmospheric carbon dioxide.

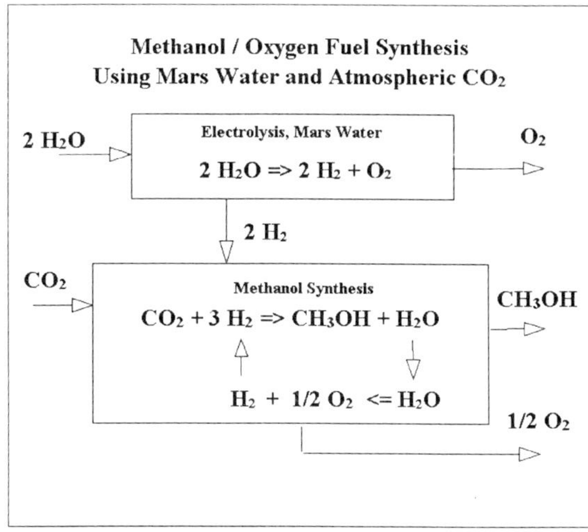

Figure 2b. Ideal mass flow requirements for methanol-oxygen fuel production using water and carbon dioxide obtained on Mars.

When imported hydrogen is the feedstock for the methanol/oxygen plant, the oxygen requirement must be met by reducing additional carbon dioxide. In this case two moles of CO waste are generated for each mole of methanol produced (Figure 2a.). Alternatively, if water is used as the feedstock, then the amount of oxygen resulting from the electrolysis of the water (and water by-products) will be in the correct proportion for combustion. In this case only one-third as much carbon dioxide is required, and there is no waste product (Figure 2b).

Note that the largest energy cost for running these reactions is for water electrolysis and that the amount of electrolysis required by the two scenarios is the same. However, in the case where imported hydrogen is used, the additional cost of operating the reverse water-gas shift reaction for producing oxygen must be added to the electrolysis cost.

Methanol Synthesis

Methanol can be produced from nearly any compound containing carbon. Presently, nearly all methanol made on Earth is produced from natural gas by steam reforming or partial oxidation of methane to produce syngas (H_2, CO, and CO_2). One exception is methanol produced from synthesis gas made by coal gasification. On Mars these resources are not available but methanol can be manufactured from atmospheric CO_2 and a source of hydrogen or water.

Methanol is produced by catalyzing carbon dioxide and carbon monoxide with hydrogen according to the mechanism,

$$CO + 2\,H_2 \Leftrightarrow CH_3OH \qquad \Delta H = -90.68 \text{ kJ/mol} \tag{8}$$

and

$$CO_2 + 3\,H_2 \Leftrightarrow CH_3OH + H_2O \qquad \Delta H = -49.38 \text{ kJ/mol} \tag{9}$$

These reactions are exothermic and proceed under volume contraction and therefore are favored by low temperature and high pressure. Early processes for the production of methanol operated at high temperatures (300-400°C) and pressures (275-360 atm). Today most industrial producers use the ICI or Lurgi processes that operate at lower temperatures and pressures (250°C, 50 bar), with a $CuO/ZnO/Al_2O_3$ catalyst. Synthesis of methanol using Cu^{n+} as the pivotal catalytic species is a well-known fact [Nakamura 1996].

All conventional methanol catalysts preferentially convert CO rather than CO_2; therefore nearly all CO_2 must be converted to CO. This can be done using the water-gas shift that also serves to couple the methanol reactions (Equations 8 and 9). However a small amount of CO_2 (2.5 - 3.5 percent) must be present to facilitate conversion [Supp 1990].

$$CO + H_2O \Leftrightarrow CO_2 + H_2 \qquad \Delta H = -41.3 \text{ kJ/mol} \tag{10}$$

In the ideal case the stoichiometry number of the synthesis gas, SN = $(H_2 - CO_2)/CO - CO_2)$, would be 2.0, the ratio of H_2 to CO in methanol. In practice however, using a small excess of hydrogen has been found to improve catalyst performance. A stoichiometry number of 2.05 has been shown to produce methanol most efficiently. Syngas gas produced from fossil fuels generally needs to be adjusted to this value. Syngas from coal gasification has a stoichiometry number of about 0.5, but this can be raised if some of the excess CO is reacted with steam to produce additional H_2. Syngas produced from natural gas by steam reforming has a stoichiometry number of about 2.8 [Unnasch 1991]. This can be reduced by a combination of steam reforming and partial oxidation of the syngas [Liebner 1988]. Thus the optimal stoichiometry number for efficient methanol production can be approached from either side.

Methanol can also be produced from hydrogen and carbon dioxide. On Mars, carbon dioxide can be filtered and compressed from the CO_2-rich atmosphere at little energy cost by using a temperature-swing adsorption pump [Finn 1996]. Most of the CO_2 must be converted to CO using the reverse water-gas shift in order to use it in the conventional methanol synthesis process. Since hydrogen will be supplied from an external source, the CO_2 content and stoichiometry number can be adjusted to 2.05 to maximize the methanol yield.

Generally, side reactions also occur during methanol production resulting in small amounts of methane and other light hydrocarbons. These can be separated and reformed to CO and H_2 and recycled back into the shift reactor. Figure 3 shows a schematic for methanol production from CO_2 and H_2 based on conventional technology for natural gas-based production.

Recent studies have shown improved yields and selectivity for methanol production over copper catalyst enhanced by a pyrochlore support, $La_2Zr_2O_7$. Using this catalyst, which has good thermal and chemical stability, high selectivity for methanol formation (vs. the competing water-gas-shift reaction) was obtained by operating at a low reaction temperature, high reaction pressure, high H_2 partial pressure and high H_2/CO_2 ratio [Kieffer 1991]. Alternatively, the effect of additives such as ZrO_2, MgO, Al_2O_3, or Cr_2O_3 on the activity of CuO-ZnO catalysts improves both yield and selectivity. For example, with CuO-ZnO-ZrO_2 (molar ratio 42/47/11) operating at 13 bar pressure and 160-220°C, the yield was 17.4% at 220°C with a selectivity of 94.8% for methanol at 160°C. Optimal conditions for methanol synthesis on a CuO-ZnO-Al_2O_3-Cr_2O_3 catalyst were obtained at 7-mPa pressure, 250°C with an H_2/CO_2 ratio of 3/1 and space velocity of GHSV = 1800 h^{-1}. Selectivity was 79% with a methanol yield of 20% [Halmann 1993].

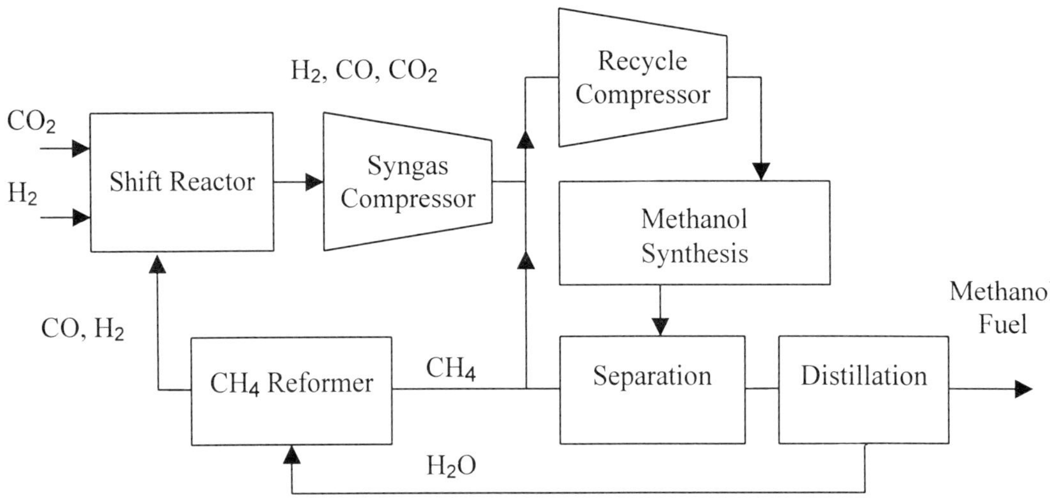

Figure 3 Schematic for Methanol Production from Carbon Dioxide and Hydrogen [Unnasch 1991].

Role of Water

Water plays a very important role in modifying the reaction environment and consequently the rate of methanol synthesis [Parameswaran 1987] through the water-gas shift and reverse water-gas shift reactions. Methanol synthesis rates are maximized at a certain optimal water concentration in the reactor [Klier 1982]. However, an accumulation of water in the catalytic system may cause damage by leaching of catalyst components and by promoting crystallite growth. These problems are considered to be more severe in the liquid phase process than in the vapor phase process because hydrocarbon oil in which the catalyst is suspended poses additional resistance to the effective removal of water from the catalyst surface. The optimal concentration of water within the reactor is determined by a balance between the rates of CO_2-hydrogenation and water-gas shift reactions. This optimum is dependent to a large extent on the composition of the syngas in the reactor and to a lesser extent on the reaction temperature [Parameswaran 1989]. Water that may evolve in this system can be extracted using a cold finger or condenser and subsequently recycled using electrolysis (Eqn. 11, 12).

$$2\,H_2O(l) \Leftrightarrow 2\,H_2 + O_2(g) \qquad \Delta H = +571.40 \text{ kJ/mol} \quad \text{(Electrolysis)} \quad (11)$$

or

$$2\,H_2O(g) \Leftrightarrow 2\,H_2 + O_2(g) \qquad \Delta H = +483.40 \text{ kJ/mol} \quad \text{(Electrolysis)} \quad (12)$$

Recent progress in sensor technology has made it possible to improve the techniques for process control systems involving methanol. A sensor for monitoring methanol concentration in water [Narayanan 1999] was developed at NASA's Jet Propulsion Laboratory that resembles a small liquid-feed, direct-oxidation fuel cell. It is a polymer-electrolyte membrane coated on one side with a catalytic electrode comprised of Pt/Ru-alloy powder and on the other with a catalytic electrode made of Pt black. The methanol sensor is based on an electrochemical oxidation of methanol coupled with the electrochemical reduction of water via the following reactions:

$$CH_3OH + H_2O \rightarrow CO_2 + 6\,H^+ + 6\,e^- \quad \text{at the Pt/Ru electrode and} \quad (13)$$

$$6\,H^+ + 6\,e^- \rightarrow 3\,H_2 \quad \text{at the Pt electrode.} \quad (14)$$

Previously developed spectroscopic methanol sensors responded too slowly to change and were insensitive at high concentrations. By contrast, this sensor responds rapidly over a wide range of concentrations (0.01 M to 5 M), temperatures (0 to 100°C), and sustains a larger transport-limited current (current increases with concentration of methanol).

Methanol Production from Methane

On Earth the conversion of natural gas (methane) into liquid methanol is advantageous because methanol is less volatile and less toxic in the event of leakage or spills during manufacture and transport. Conversion of natural gas to methanol that is currently flared on offshore oil wells or that is associated with ocean bottom methane hydrate deposits may be the only economical means to recover and transport this valuable resource. Preliminary estimates of methane hydrate ice deposits and trapped free methane indicate that these deposits may are comparable to the combined world reserves of oil, gas and coal [Fouda 1998]. With current technology, production of methanol from methane is a fairly efficient process. The catalytic conversion of methane into syngas components using oxygen over a Rhodium surface [Hickman 1993] is given below:

Methane to Syngas:

$$CH_4(g) + 1/2\,O_2 \Leftrightarrow CO(g) + 2\,H_2(g) \quad \Delta H = -35.57 \text{ kJ/mol } CH_4 \quad (15)$$

Or, the direct production of methane to methanol [Unnasch 1991]:

$$CH_4(g) + CO_2(g) \Leftrightarrow CH_3OH(g) + CO(g) \quad \Delta H = +119.73 \text{ kJ/mol } CH_3OH \quad (16)$$

More recently the development of a homogeneous liquid catalyst that converts methane to methanol at temperatures below 100°C has achieved yields as high as 70 percent [Periana 1993; Periana 1998; Fouda 1998]. During this process the liquid catalyst creates a chemical intermediate, methyl ester, that is protected from oxidation and is relatively stable.

Produced *in situ* from atmospheric carbon dioxide and used as a propellant for the Earth Return Vehicle [Zubrin 1995; Zubrin 1997], methane is likely to be available on Mars and if so, can be utilized to produce synthesis gas (CO, H_2) for the production of methanol (Eqn. 8, 9). In this scenario, methane and oxidant are produced *in situ* from atmospheric carbon dioxide and imported hydrogen via the "Sabatier/Electrolysis" or "SE" cycle. Methane and water are produced in a 1:2 molar ratio. The methane is liquefied and stored and the water is condensed and split by electrolysis yielding oxygen which is also stored, and hydrogen that is recycled to produce additional methane. Equations 17 and 18 demonstrate this process.

Sabatier:

$$CO_2(g) + 4\,H_2(g) \Leftrightarrow CH_4 + 2\,H_2O \qquad \Delta H = -41.21 \text{ kJ/mol} \qquad (17)$$

Electrolysis:

$$2\,H_2O(l) \Leftrightarrow 2\,H_2(g) + O_2(g) \qquad \Delta H = +571.40 \text{ kJ/mol} \qquad (18)$$

In order to fully utilize the hydrogen, if the water is continuously recycled all of the hydrogen can be shifted into methane, but the quantity of oxygen produced will be only half of the stoichiometric requirement for combustion. If the hydrogen for this process was imported, then it has no oxygen counterpart that could be used for combustion of the excess methane. One solution would be to use the reverse water gas shift to reduce atmospheric CO_2 to produce the additional O_2. The hydrogen involved in the reverse water gas shift reaction is largely conserved and two moles of waste CO will be generated for each mole of O_2 produced.

Methanol Production from CO

A waste product such as carbon monoxide represents valuable stored energy (283 kJ/mol when burned with oxygen) if it can be effectively utilized. Carbon monoxide evolves as a waste from carbon dioxide reduction systems used for air recycling in habitats and from oxygen production for rocket propellant. It is also a by-product of the direct production of methanol from methane (Eqn. 16) and is a product in steam reforming or the partial oxidation of methane to make syngas.

It may be tempting to try to utilize the "waste CO" by combustion, either directly or after it has been transformed into a hydrocarbon such as methanol (Eqn. 8). However on Mars the oxidizing component of combustion is not available for free in the atmosphere. Thus new oxygen would need to be produced from indigenous raw material, the principal resource options being CO_2 and H_2O. Since these materials are in their lowest energy state, the energy cost to produce the requisite amount of oxygen will be at least as great as that obtainable from the original "waste CO". For example, if the amount of oxygen needed were obtained by reducing more CO_2, then again as much CO waste would be produced at an energy cost equal to the energy value of the original CO plus processing inefficiencies, thus resulting in no net gain. Had the original CO been first transformed to methanol, a similar analysis would apply.

A better way to utilize waste CO is to find applications where its reducing power can be applied directly to a utilitarian purpose. As operations on Mars advance, we can begin to utilize carbon monoxide as a reducing agent in iron ore processing thus turning it back to CO_2 [Stoker 1996]. Carbon monoxide is used for all three stages of iron ore processing in a modern blast furnace.

Energy Storage and Conversion on Mars

For unmanned sample return missions, propellants produced for an Earth return spacecraft would most likely take place over an extended period of time and therefore would require energy efficient storage for methanol and oxygen. For a manned outpost, methanol/oxygen fuel production would be a flexible and efficient means of energy storage that would serve a variety of applications including surface and space transportation and electric power production.

The maintenance of a large cache of chemical fuel would increase the robustness of the base, play a vital role in assuring crew safety in the event of the failure of the primary power system, assure continuity during maintenance, and increase the overall exploratory capability of the base. Storage systems can be coupled directly to rocket propellant transport depots and vehicle-refueling stations, as well as to fuel cell based electric power generation plants. They can provide buffering for production so that the fuel plant can operate at a constant throughput for maximum efficiency as well as load leveling so that demand fluctuations are handled by inventory changes in the storage system.

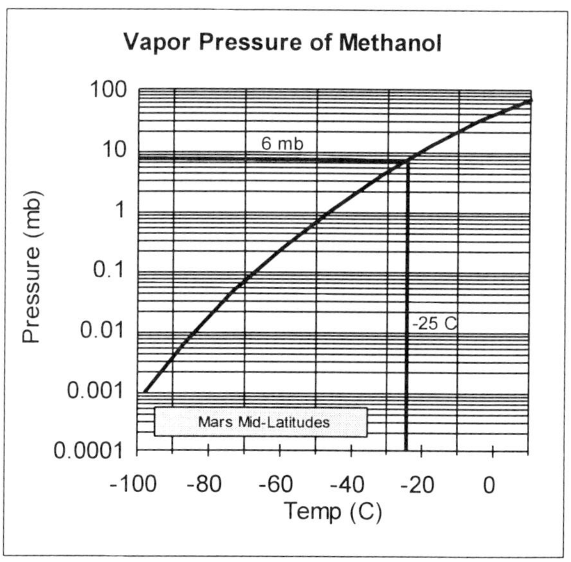

Figure 4 Vapor pressure of methanol compared to Mars mid-latitude average daily surface temperatures. At 6 millibars, the nominal surface pressure on Mars, methanol boils at –25°C. The left endpoint is the methanol freezing point, –97.68°C.

Because of its low vapor pressure and low freezing point (–97.68°C) methanol can be easily stored throughout the year at most locations within about +/–33° latitude without pressurized containment or cryogenic cooling and with little or no thermal support to prevent freezing (Figure 4). Based on the Mars General Circulation Model [Haberle

1993, Lewis 1999] the average daily temperatures range from a low of about –95°C in the Northern winter to a high of –35°C in the Southern summer. Although the diurnal temperature variations may be as much as +/–40°C, insulated storage tanks and a large thermal mass of stored fuel would significantly moderate this effect thus allowing methanol to remain liquid at all times.

Materials Cryogenic Storage

For oxygen storage on Mars as well as other liquid consumables such as methane and hydrogen, a Pulse Tube Refrigeration (PTR) system (Figure 5), could be very efficient [Lee 1995]. A pulse tube cooler is a device for cooling to low temperatures. It is a closed system that uses an oscillating pressure at one end (typically produced by a compressor) to generate an oscillating gas flow in the rest of the system. This gas flow (usually helium gas) can carry heat away from a low temperature point (cold heat exchanger) if the conditions are right. An orifice controlling the flow at the other end of the cooler can provide the appropriate condition for cooling to occur. The amount of heat a pulse tube cooler can remove is only limited by its size and the power used to drive it. Its efficiency is comparable to other systems such as Stirling coolers. The primary advantage of pulse tube coolers over Stirling coolers is that they have no moving parts in the low-temperature region. This means that there is no friction, no wear and essentially no vibration, so the low-temperature sections have an indefinite lifetime. Oxygen can also be stored in either a gaseous (GOX) or liquid (LOX) form under Mars conditions. Compression of carbon dioxide for use as a cryogen may also be feasible for use on Mars.

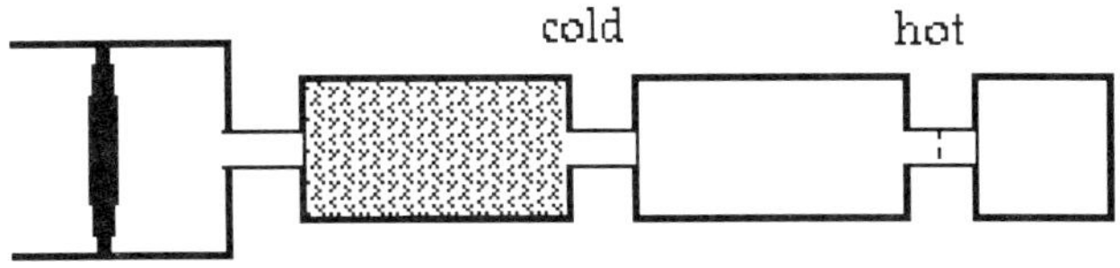

Figure 5 Basic components of a pulse tube cooler [Lee & Kittel *et al.*, 1995].

Oxygen Storage in Hydrogen Peroxide

An alternative method of oxygen storage would be to incorporate it into hydrogen peroxide. While additional energy is required for its production, this could be returned as heat upon controlled catalytic decomposition.

$$2\,H_2O_2(aq) \rightarrow 2\,H_2O(l) + O_2(g) \qquad \Delta H = -196\text{ kJ/mol }O_2 \qquad (19)$$

Hydrogen peroxide freezes at –2°C and can be stored frozen indefinitely without cryogenic measures. One cubic meter of hydrogen peroxide stored at –55°C has a mass of 1.25 tons. It contains 661 kilograms of water and 588 kilograms of oxygen, and would yield 996 kWh of energy upon decomposition. It can be used as a mono-propellant or in combination with a fuel such as methane as a bi-propellant [Clark 1987].

Some commercial production of hydrogen peroxide is done by electrolysis of concentrated sulfuric acid to produce oxydisulfuric acid that can be hydrolyzed to form sulfuric acid and hydrogen peroxide. Another process uses cold sulfuric acid and hydrated barium peroxide to produce barium sulphate and hydrogen peroxide. Hydrogen peroxide being less volatile than water can be concentrated by evaporation under reduced pressure.

DISCUSSION AND CONCLUSION

In summary, methanol is an excellent candidate fuel for both Earth and Mars because it can be produced from atmospheric carbon dioxide on Mars, industrial carbon dioxide waste on Earth, biomass and methane. Its manufacture and applications represent important dual-use technologies whose advancement will hasten vital objectives for both planets. On Earth, methanol is an excellent alternative fuel that could aid the transition from fossil fuels to renewable energy technology. Its similarity to gasoline, its versatility to be produced from many different materials, and its ability to be used by many different applications, allows it to be used to advantage to minimize the economic impact of a transition. As an alternative fuel, methanol gives high performance and has a higher octane rating than gasoline. It is also an efficient hydrogen carrier that may open the door to practical electric vehicles using direct methanol fuel cells. Moreover methanol could play a key role in reducing global carbon dioxide emissions while greatly improving the efficiency of fuel resource utilization.

On Mars methanol and oxygen can be produced by autonomous propellant plants to provide the fuel needed for robotic spacecraft returning samples from Mars as well as for crew return vehicles during human exploration. If methanol is produced from materials indigenous to Mars, this offers a significant reduction in mission cost by reducing payload requirements from Earth. Methanol has excellent storage and handling properties that make it practical for many applications in the cold environment of Mars including rockets, surface vehicles, and electric power production using fuel cells. Its low vapor pressure and low freezing point also means that it can be stored as a liquid without heating or cryogenic measures at mid-latitudes throughout the year thus making it an ideal energy storage compound when used in conjunction with primary power systems on Mars.

Methanol is less toxic to humans than gasoline and is neither mutagenic nor carcinogenic. It is capable of completely mixing with water; degrades quickly in the atmosphere; and will rapidly biodegrade in surface waters and underground and therefore is likely to be less dangerous to humans and the environment than conventional gasoline under equivalent release or spill scenarios. Moreover, methanol has been shown to enhance C3 crop growth while reducing water consumption, thus reducing agricultural water utilization in arid regions on Earth [Nonomura 1992]. Additionally, it serves as a chemical feedstock for compounds such as acetic acid and formaldehyde that can contribute to the establishment of a human presence on Mars.

Table 1: SUMMARY OF REACTIONS IN METHANOL SYNTHESIS

SYNTHESIS PROCESS	REACTION	ΔH (kJ mol^{-1})	CATALYST/ Process CuO/ZnO/ Al_2O_3	Temp/Press (° K/ bar)
Syngas	$CO_{2(g)} + H_{2(g)} \Leftrightarrow CO_{(g)} + H_2O$ $CO_{(g)} + 2\,H_{2(g)} \Leftrightarrow CH_3OH_{(l)}$ $CO_{2(g)} + 3H_{2(g)} \Leftrightarrow CH_3OH_{(l)} + H_2O$	+41.30 -90.68 -87.33	1. ICI or Lurgi (Unnasch) 2. Liquid phase (Lee)	523 K/50 bar 510 K/64 bar
Carbon Dioxide Hydrogenation	Hydrogenation: $CO_{2(g)} + 3\,H_{2(g)} \Leftrightarrow CH_3OH_{(l)} + H_2O$ The reaction could be assumed to proceed in two stages, as follows: $CO_2\,(g) + H_{2(g)} \Leftrightarrow CO_{(g)} + H_2O$ $CO + 2\,H_2 \Leftrightarrow CH_3OH_{(l)}$	- 87.33 + 41.30 --90.68	1. Vapor Phase	A. Nair 600K, 50-100 bar B. Chinchen uses 503-573K, 50-100 bar
Carbon Monoxide Reduction	$CO_{(g)} + 2\,H_{2(g)} \Leftrightarrow CH_3OH_{(l)}$	-90.68	or -128.63 for liquid	

Table 2: METHANE TO METHANOL SYNTHESIS

SYNTHESIS PROCESS	REACTION	ΔH (kJ mol^{-1})
Methane	$4H_{2(g)} + CO_{2(g)} \Leftrightarrow 2H_2O + CH_4$ (Sabatier)	-41.21
	$CO_{2(g)} \Leftrightarrow CO_{(g)} + 1/2\,O_{2(g)}$	+283.00
	$CH_4 + 1/2\,O_{2(g)} \Leftrightarrow CH_3OH_{(g)}$	-126.25
	$CH_4 + CO_{2(g)} \Leftrightarrow CH_3OH_{(g)} + CO_{(g)}$	+156.75
Partial Oxidation of Methane, no catalyst, 1500-1800K, 30-80 Bar	$CH_{4(g)} + 5\,O_{2(g)} \Leftrightarrow CO + 2H_2O\,(g)$	-518.97
Steam Reforming of Methane using Ni^+ K_2O; 1000-1100 K, 15-40 Bar	$CH_{4(g)} + 2H_2O\,(g) \Leftrightarrow CO + 3H_{2s}$	+206.13

REFERENCES

1. American Methanol Institute (AMI), Global Climate Benefits of Methanol. 1998.
2. American Methanol Institute (AMI), Methanol: North America's Clean Fuel and Chemical Building Block. 1998.
3. Ash, R. L., Dowler, W. L., Varsi, G. 1978. "Feasibility of Rocket Propellant Production on Mars." *Acta Astronautica*. 5: 705-724.
4. Bernabé, Franco. Chief Executive of the Italian oil company ENI SpA cited in "The End of Cheap Oil", by Howard Banks; *Forbes Magazine*, June 1998.
5. Boston, P. J. 1981. "Low pressure greenhouses and plants for a manned research station on Mars." *J. British Interplanetary Soc.* 34:189-192.
6. Clark, B. C., J. L. Kalkstein, S. Meyer. 1991. "The Case for the Methanol Powered Planetary Rover." 42nd Congress of the Intl Astronautical Federation, Oct. 5-11, 1991, Montreal Canada.
7. Clark, B. C. 1989. "Survival and prosperity using regolith resources on Mars." J.B.I.S. 42(4):161-166. Also published as AAS 97-374 in R. M. Zubrin (Ed.) *From Imagination to Reality: Mars Exploration Studies of the Journal of the British Interplanetary Society* (Part II: Base Building, Colonization and Terraformation), AAS *Science and Technology Series*, Vol. 92, pp. 31-41, 1997.
8. Clark, B. C. and D. R. Pettit. 1987. "The Hydrogen Peroxide Economy On Mars." in *The Case for Mars III*, ed. C. R. Stoker. AAS 87-212, American Astronautical Society, *Science and Tech. Series*, Vol. 74, pp. 551-557.
9. Colvin, J., Schallhorn, P. Ramohalli, K. 1992. "Fuel Systems Engineering Design and Operation of an O_2 Plant." *J. of Propul. & Power* 8(5): 1103-1108.
10. Finn, J. E., McKay, C. P., Sridhar, K. R. 1996b. "Martian Atmospheric Utilization by Temperature-Swing Adsorption." 26th Intl Conf. on Environmental Systems, Monterey, CA July 8-11, 1996. SAE #961597.
11. Finn, J. E., Sridhar, K. R., McKay, C. P. 1996. "Utilization of Martian Atmospheric Constituents by Temperature-Swing Adsorption." *J.B.I.S.* 49: 423-430. Also published as AAS 97-360 in R. M. Zubrin (Ed.) *From Imagination to Reality: Mars Exploration Studies of the Journal of the British Interplanetary Society* (Part I: Precursors and Early Piloted Exploration Missions), AAS *Science and Technology Series*, Vol. 91, pp. 127-142, 1997.
12. Fouda, S. A. 1998. "Liquid Fuels from Natural Gas." *Scientific American*, March, 92-95.
13. Gwynne, O. 1993. "Handbook for a Mars Base," in the Proceedings for the Case for Mars V, June 11-16, 1993, Boulder, CO, ed. P.J. Boston.
14. Haberle, R. M., Pollack, J. B., Barnes, J. R., Zurek, R. W., Leovy, C. B., Murphy, J. R., Lee, H. and Schaeffer, J. (1993) "Mars atmospheric dynamics as simulated by the NASA/Ames General Circulation Model I. The zonal-mean circulation," *J. Geophys. Res.* 98, 3093-3124.
15. Halmann, M. M. 1993. *Chemical Fixation of Carbon Dioxide: Methods for Recycling CO_2 into Useful Products*. CRC Press, Boca Raton, FL.
16. Haslach, H. W., Jr. 1989. "Wind energy: A resource for a human mission to Mars," *J. Brit. Interplant. Society* 42(4), 147-160. Also published as AAS 97-376 in R. M. Zubrin (Ed.) *From Imagination to Reality: Mars Exploration Studies of the Journal of the British Interplanetary Society* (Part II: Base Building, Colonization and Terraformation), AAS *Science and Technology Series*, Vol. 92, pp. 53-70, 1997.
17. Hepp, A. F., G. Landis and D. Linne. 1991. Material processing with hydrogen and carbon monoxide on Mars. NASA Tech Memorandum 104405.
18. Hickman, D. A. and L. D. Schmidt. 1993. "Production of Syngas by Direct Catalytic Oxidation of Methane." *Science* 259:343-346.

19. Energy Information Administration (EIA), U.S. Department of Energy, International Energy Outlook March 1997.

20. Energy Information Administration (EIA), U.S. Department of Energy, International Energy Outlook March 1999.

21. Kieffer, R. 1991. "Methanol Synthesis from CO_2 + H_2." *Proc. Int. Symp. Chemical Fixation of Carbon Dioxide*, Nagoya, Japan, Dec. 2-4, 1991. 151-158.

22. Klier, K., Chatikvanij, V., Herman, R. G., Simmons, G. W. 1982. "Catalytic synthesis of methanol from CO/H_2: IV. The effects of Carbon Dioxide." *J. Catalysis*. 74: 343-360.

23. Kuei, C.-K., Chen, W.-S., Lee, M.-D. 1991. "Hydrogenation of Carbon Dioxide on Iron Manganese Catalysts." *J. of the Chinese Chem. Soc.* 38, 127-137.

24. Kuiper, G. P. 1949. Survey of Planetary Atmospheres. In *The Atmospheres of the Earth and Planets*. Univ. of Chicago Press. 304-345.

25. Kurtz, S. R.; Olson, J. M.; Friedman, D. J.; Geisz, J. F.; Kibbler, A. E.; Bertness, K. A. (1999). *Passivation of Interfaces in High-Efficiency Photovoltaic Devices*. 18 pp. NICH Report No. CP-520-26494.

26. Lee, J. M., Kittel, P., Timmerhaus, K. D., Radebaugh, R. 1995. Steady Secondary Momentum and Enthalpy Streaming in the Pulse Tube Refrigerator. In *Cryocoolers 8*. R. G. Ross, Jr., Ed. Plenum Press, NY. 359-369.

27. Lewis, S. R., Collins, M., Read, P. L., Forget, F., Hourdin, F., Fournier, R., Hourdin, C., Talagrand, O., Huot, J-P. (1999) "A climate database for Mars," *J. Geophys. Res.* 104, 24,177.

28. Liebner, W. And Supp, E. (Lurgi GmbH), "Combined Reforming: A Most Economical Way from Natural Gas to Alcohols and Synfuels," presented at the VIII International Symposium on Alcohol Fuels, Tokyo, Japan, November 13-16, 1988.

29. MacKenzie, James J., *Oil as a Finite Resource*. World Resources Institute, 1996.

30. McKay, C. P., Meyer, T. R., Boston, P. J., Nelson, M., MacCallum, T., Gwynne, O. 1993. Utilizing Martian Resources for Life Support. In *Resources of Near-Earth Space*. Lewis, J., Matthews, M. S., Guerrieri, M. L., Eds. Univ. Arizona Press, Tucson, AZ. 819-843.

31. Meyer, T. R. and C. P. McKay. 1989. "The resources of Mars for human settlement." *J. Brit. Interplanetary Soc.* 42(4):147-160. Also published as AAS 97-373 in R. M. Zubrin (Ed.) *From Imagination to Reality: Mars Exploration Studies of the Journal of the British Interplanetary Society* (Part II: Base Building, Colonization and Terraformation), AAS *Science and Technology Series*, Vol. 92, pp. 3-29, 1997.

32. Meyer, T. R. and C. P. McKay. 1981. "The atmosphere of Mars: resources for the exploration and settlement of Mars." In: *The Case for Mars I* Proceedings, Ed. P. J. Boston. AAS 81-244, Amer. Astronautical Soc. *Sci. and Tech. Series*, Vol 57. pp. 209-232.

33. Nakamura, I.., Fujitani, T., Uchijima, T., Nakamura, J. 1996. "A Model Catalyst for Methanol Synthesis: Zn-deposited and Zn-free Cu surfaces." *J. Vac. Sci. Technol.* A 14 (3, Pt. 2) 1464-1468.

34. Narayanan, S., W. Chun, T. Valdez. 1999. Sensor for Monitoring Concentration of Methanol in Water. NPO-20125, NASA Tech Briefs 22(4): 46-47.

35. Nonomura, A. M., Benson, A. A. 1992. "The Path of Carbon in Photosynthesis: Improved Crop Yields with Methanol." *Proc. Natl. Acad. Sci.* 89, 9794-9798.

36. Oniciu, Liviu. 1976. *Fuel Cells*. Abacus Press, Tunbridge Wells, Kent, United Kingdom.

37. Owen, T., Biemann, K., Rushnek, D. R., Biller, J. E., Howarth, D. W., and Lafleur, A. L. 1977. "The composition of the atmosphere at the surface of Mars," *J. Geophys. Res.* 82: 4635-4639.

38. Parameswaran, V. R., Lee, S. 1987. "Pore Diffusional Limitations in the Liquid-Phase Methanol Synthesis Process." *Energy & Fuels* 1: 217-222.

39. Parameswaran, V. R., Lee, S., Wender, I. 1989. "The Role of Water In Methanol Synthesis." *Fuel Science & Technology Int'l* 7(7): 899-918.

40. Peachey, N. M., Snow, R. C., Dye, R. C. 1995. "Composite Pd/Ta Metal Membranes for Hydrogen Separation." *J. of Membrane Sci.* 3:123-133.

41. Pirnie, M., *Evaluation of the Fate and Transport of Methanol in the Environment*, Prepared by Malcolm Pirnie, Inc., 180 Grand Avenue, Suite 1000, Oakland, California 94612, for American Methanol Institute, 800 Connecticut Avenue, NW, Suite 620, Washington, DC 20006. Jan 1999.

42. Periana, R. A., D. J. Taube. 1993. "A Mercury-Catalyzed, High-Yield System for the Oxidation of Methane to Methanol." *Science* 259:340-343.

43. Periana, R. A., D. J. Taube. 1998. "A Mercury-Catalyzed, High-Yield System for the Oxidation of Methane to Methanol." *Science*.

44. Ramohalli, K. N. 1993. "Fueling a revolution in space processing." *Aerospace America*. November, 38-41.

45. Sloan, M. 1997. Supporting Documentation for the Texas Renewable Energy Poster, Virtus Energy Research Associates.

46. Sridhar, K. R. 1993. "Mars Sample Return Mission With ISRU." 44th Congress of the IAF, Graz, Austria, Oct. 16-22, 1993. IAF-93-U.4.567.

47. Stoker, C. R. , Emmart, C., Editors. 1996. *Strategies for Mars: A Guide to Human Exploration. Science & Technology Series*, Vol. 86. AAS, San Diego, CA.

48. Stoker, C. R., Gooding, J. L., Roush, T., Banin, A., Burt, D., Clark, B. C., Flynn, G., Gwynne, O. 1993. The Physical and Chemical Properties of and Resource Potential of Martian Surface Soils. In *Resources of Near-Earth Space*. Lewis, J., Matthews, M. S., Guerrieri, M. L., Eds. Univ. Arizona Press, Tucson, AZ.

49. Supp, Emil. 1990. *How to Produce Methanol from Coal*. Springer-Verlag, New York.

50. Surampudi, S., Narayanan, S. R., Vamos, E., Frank, H., Halpert, G. "Advances in Direct Methanol Fuel Cells at JPl." *Proceedings of the Eighth Annual Battery Conference on Applications and Advances*. Jan. 12-14, 1993.

51. Taylor, P. J., "Discussion of Modeling the U.S. Oil Industry; How Much Oil is Left?" *Journal of Petroleum Technology*, paper SPE 526000 (May 1997).

52. The International Energy Agency (IEA), 1978 Bioenergy Agreement. 1997. Wood Supply Research Group, Univ. of Aberdeen, Dept. of Forestry, U.K.

53. Unnasch, S., Luscher, D. 1991. Evaluation of methanol production from hydrogen and waste carbon dioxide. Acurex Corp (Mtn View, CA) Contract # 68-02-5285 for US EPA, Air and Energy Engineering Research Laboratory, Research Triangle Park, N.C.

54. US Congress, Office of Technology Assessment (OTA). 1993. Energy Efficiency Technologies for Central and Eastern Europe. OTA-E-562, Washington DC United States Govt. Printing Office.

55. Williams, J. D., Coons, S. C., Bruckner, A. P. 1995. "Design of a Water Vapor Adsorption Reactor for Martian In Situ Resource Utilization." *J. Brit. Interplan. Soc.* 48(8):347-354. Also published as AAS 97-355 in R. M. Zubrin (Ed.) *From Imagination to Reality: Mars Exploration Studies of the Journal of the British Interplanetary Society* (Part I: Precursors and Early Piloted Exploration Missions), AAS *Science and Technology Series*, Vol. 91, pp. 59-73, 1997.

56. Zent, A. P., McKay, C. P. 1994. "The Chemical Reactivity of the Martian Soil and Implications for Future Missions." *Icarus* 108:146-157.

57. Zubrin, R. M. Baker, D. A. , Gwynne, O. 1991. "Mars Direct: A Simple, Robust, and Cost Effective Architecture for the Space Exploration Initiative." AIAA 91-0326, 29th Aerospace Sciences Mtg, Reno, NV. Also published as AAS 90-168, in Proceedings of *The Case for Mars IV*, Boulder, CO, June 1990. ed. T. R. Meyer. American Astronautical Society, *Science and Tech. Series*, Vol. 89, pp. 275-314, 1997.

58. Zubrin, R. M., Price, S., Mason, L., Clark, L. "Mars Sample Return with In-Situ Resource Utilization." Contract NAS 9-19145, Martin Marietta Final Report to Johnson Space Center, Jan. 13, 1995.

59. Zubrin, R., Meyer, T. R., McMillen, K. R. 1997. Mars In-Situ Resource Utilization Research (MIRUR): Design of a RWGS System. NASA JSC Contract NCC9-45.

60. Zubrin, R. M., Burnside-Clapp, M., Meyer, T. R. 1997b. "New Approaches for Mars In-Situ Resource Utilization Based on the Reverse Water Gas Shift." 35th Aerospace Sciences Meeting, AIAA, Jan. 6-10, 1997, Reno, NV. AIAA# 97-0895.

CHAPTER 5:
HUMAN MISSIONS III:
TERRAFORMING

AAS 93-845

THE LONG-TERM HABITATION OF MARS

Martyn J. Fogg*

One follow-on objective of the human exploration of Mars might be the permanent settlement of the planet. This entails examining issues considered by previous CFM conferences over much longer timescales.

In this paper a systems approach is adopted to consider the acquisition of energy and material resources on Mars and their organization within life-support systems. Since such resources will have to be initially manufactured on Earth, an early settlement will represent a particularly expensive example of fossil fuel subsidized infrastructure. Thus, growth of a Martian civilization cannot be realistically expected unless near complete independence is attained. A crucial advance early in any settlement scheme will be the bootstrapping of all the industrial capacity required for survival and growth from a minimum inventory of terrestrial goods.

Similarly, whilst it seems possible to fabricate closed loop ecological life-support systems in which colonists might live, such energy-intensive and constrained habitats appear a far less attractive option for a large population and a long-term future. The need to perpetually maintain the integrity of habitat structure, mechanically recycle water and wastes and monitor and control the balance of gases cycling through the biota compares poorly with the benefits of the autonomous and fully bioregenerative life-support system that is the Earth's biosphere. Since Mars may be the only planet in the Solar System where the creation of an unconstrained planetary biosphere may be a realistic proposition, it seems likely that Martian colonists will be attracted to the possibility of terraforming their adopted world. Recent scientific progress on the subject of terraforming Mars is briefly reviewed.

It is suggested that the permanent settlement of Mars represents a thought experiment, largely unclouded by non-scientific details, of great relevance to the present day Earth. Like future Martian colonists, the people of Earth are faced with the increasing need to find an alternative to fossil fuels and an ultimate requirement to maintain and manage a global life-support system.

1. INTRODUCTION

The short-range goals of Mars exploration are basically intellectual and scientific [1,2]. To understand the evolution of a planet, from a time when its environment was much more similar to that of Earth [3], into its present cold and desiccated state will be of immense value to planetary science. However, Mars has traditionally stood for being much more than a proving ground for theory. It holds a unique place in human culture as somewhere far away, but real, to which the imagination is irresistibly drawn. Despite

* Probability Research Group, c/o 44 Hogarth Court, Fountain Drive, London SE19 1UY, United Kingdon.

the ongoing debate as to the proper route to space settlement, and some claims that planetary surfaces might be by-passed altogether [4], a fascination with Mars has continued to increase. Mars has become a metaphor for a future rich with magnificent possibility.

The permanent settlement of the Red Planet has been a consistent theme of *The Case for Mars* conferences, positioned as a backdrop to the more prominent and immediate issues of travelling to Mars and establishing an initial presence. Amongst some of the reasons for going to Mars discussed by the "grass roots" participants of the first *CFM* gathering [5] are such notions as establishing an independent branch of humanity, pushing human intelligence into the unknown and creating a "new Earth". The increasing respectability of settling Mars as a long-range goal of the U.S. civilian space program was emphasized by the National Commission on Space [6] who expressed the purpose of establishing free societies on new worlds, *"..from the highlands of the Moon, to the plains of Mars."*

Public interest is also on the increase as evidenced by private funding and media coverage. The Biosphere 2 project represents a bold, but possibly premature, attempt to step beyond on-paper designs and to construct a pilot model of an enclosed biospheric habitat, with well publicized applications to living on Mars [7,8]. The plans for the "Hyperion Project"—a Japanese Martian city theme park [9] are similarly touted as being *"... a model for a future, self-contained, biosphere colony on Mars."* Whilst such claims contain an element of exaggeration, they serve a purpose beyond that of science and engineering—their value being to provide inspiring prototype visions of tomorrow.

The consummate goal of settling Mars, and the most long-range vision of all, is the possibility of engineering the environment of the planet to resemble that of Earth. This concept of "terraforming" has received intermittent scientific attention for twenty years and is currently experiencing its most vigorous period of interest to date. A number of technical journals have published recent papers dealing with prospects for planetary engineering on Mars [e.g. 10-13]: sessions have been held on the subject at a number of academic symposia and its exciting and controversial nature has not been lost on the media and popular press.

If the colonization of Mars does happen, then the population of settlers will eventually grow to the point where any mass exodus back to Earth in the face of discomfort or disaster will be impracticable. The new Martians will have to like or lump their adopted world—there will be no turning back. Bearing this in mind, what *The Case for Mars* is ultimately about is *the creation of a new civilization*. It is appropriate therefore to examine this long-range view over the relevant timescales, which history tells us ranges from decades to millennia. The options for Martian settlers over this span of time, with respect to their access to planetary resources and choice of life-support systems, are likely to be wider than those immediately available to early explorers. Whilst no forecasts concerning the eventuality and progress of settling Mars can be made, it is at least possible to qualitatively analyze certain aspects of the problem.

This paper examines the issues of acquisition of matter and energy on Mars and their organization in life-support systems. An attempt is made to compare possible solutions suitable for *self-sufficiency* and *long-term habitability*. Irrespective of whether humanity ever leaves its home planet, the thought experiment represented by the permanent settlement of Mars is shown to have a surprising relevance to the persistence of our high-tech civilization on Earth.

2. SELF-SUFFICIENCY: MATTER AND ENERGY

Whilst early Martian settlements will have to be parasitic on the resources and industry of the Earth, this will be undesirable and impractical in the long-term. Independence for Mars will involve an almost complete self-sufficiency in matter, energy and manufactured goods, and deficit being made up, not by an umbilical cord to Earth, but by the two-way traffic of interplanetary trade. Like terrestrial civilization, its Martian counterpart will have to rely on the resources of its own environment for survival and growth. This ability, not just to "live off the land" but to prosper from it, will be the deciding factor determining whether pioneering settlements can grow into a planetary civilization.

Two essential pre-requisites for growth can be identified:

1. Raw materials essential for biological and technological systems must be available in forms that are practical and economic to exploit.
2. It must be possible to generate primary energy with a high net energy ratio. (The ratio of energy yielded to the economy to that required for its processing.)

It is simple to see why these requirements are fundamental. If raw materials are difficult to locate and extract, infrastructure expensive to maintain, and the quality of energy resources is poor, the net energy ratio of power production could fall to ≤ 1 and no growth would be possible.

It is the high net energy provided by an abundance of fossil fuels that has driven the rapid growth of terrestrial civilization since the industrial revolution. Such a concentrated energy source in unlikely to exist on Mars and even if it did, oxygen would have to be manufactured for combustion, wiping out any net energy gain. Mars is thus impoverished in at least one resource with respect to the Earth—but then so is any other potential colonization site where there has not been a biosphere working for millions of years to sequester chemical potential energy. Apart from this however, the combined quantity and quality of a wide range of other Martian resources may be good.

2.1 Matter

The material resources of Mars that we know of, and deposits analogous to those on Earth that are likely or unlikely to be present are shown in Table 1.

Our ignorance of the true picture is extensive. The Martian surface has only been sampled superficially at two sites, data indicating the presence of silicate rocks, low in Al and K, and a soil thought to consist of iron-rich clays contaminated with sulphate salts [14]. However, unlike the moon, volatiles are present both in the atmosphere and as ices exposed in the polar regions. At middle to high latitudes, water ice may also exist underground in a layer of permafrost [15]. The presence of volatiles has three important implications for colonization:

1. water may have played an important role in geological processes that create mineral deposits;
2. volatile reagents for industrial mineral processing will be already present or might by manufactured on site;
3. life support systems can be opened to easy re-supply of biogenic compounds and are thus freed from the requirement for perfect balance and recycling.

Table 1: Mars - Matter Resources.

Known.
Atmosphere — CO_2, N_2, Ar, H_2O
Ices — H_2O, CO_2
Silicate Rocks
Clays
Salts
Possible Mineral Concentrations.
Hydrothermal deposits
Evaporites
Sedimentary deposits
Layered igneous intrusions
Unlikely Mineral Concentrations.
Organic deposits
Orogenic deposits

Remote sensing data suggests that the constitutions of the two Viking sites are not typical of most of the Martian surface [16]. However, even their basic inventories are of biological and technological use [17,18]. The Martian atmosphere can be processed into breathable air and water [19] and Martian soil, with perhaps the addition of K and N fertilizers, might be a suitable medium for plant growth [20]. Manufacture of functional bulk materials also appears possible. Examples include construction materials, such as cements, ceramics and glass [17], "duricrete" [21], and cellulose (wood) [22]; fuels, such as CO and CH_4 [23]; and an assortment of other chemicals such as reagents oxidizers, explosives and fertilizers [17]. However, production of such "low-tech" materials locally would only reduce, but not eliminate, re-supply requirements from Earth. A long-term commitment to be self-sufficient in the wide range of products characteristic of a technological civilization demands that Martian geology has concentrated rarer substances in high grade mineral deposits.

The global tectonic style of Mars differs fundamentally to that of Earth, it appearing to be a one-plate planet, its ancient crust showing no sign of the large-scale orogenic movements characteristic of plate tectonics. Thus its crust may not be as differentiated as the Earth's and one might conclude from this that the prospects of locating rich mineral deposits are poor [24]. However, a more upbeat assessment of Martian resources that draws an analogy between Mars and Africa [25] has identified at least four ore generating geologic processes that have probably occurred (see Table 1) and concludes that Mars might be very well endowed with minerals. The key to this attractive possibility lies in the planet's past when the climate and geological activity may have been more earth-like [3]. There is surface evidence in several areas of the interaction between lava and ground ice [26,27]. At depths, hydrothermal circulation around cooling magma bodies could have precipitated large quantities of metallic ores arranged zonally about igneous centers. What with the substantial magmatic activity in the planet's past and the common occurrence of sub-surface water, such localities might be commonplace, especially if local melting and mineralization can also be produced by large impacts. Layering produced by differentiation within the magma bodies themselves might also produce rich ores, accessible where erosion has brought them close to the surface. Erosion and deposition on a warm, watery Mars would have brought about the formation of sedimentary rocks [28,29] giving rise to the possibility of mineral concentrations in residual or detrital deposits. Lastly as the transient seas and lakes of Mars dried up, evaporites would have formed. Sulphates and chlorides are hinted at in the Martian soils of the Viking sites, [14] but of greater interest to the long-term settlement of Mars would be the discovery of nitrates which only precipitate in the driest of locales, such as the Atacama desert in Chile. Since the Martian "waters" dried up completely, substantial nitrate deposits may await discovery.

This claimed similarity between Mars and Africa may not be fully appropriate [30] and so one can only be cautiously optimistic about the prospects for mineral recovery on Mars. Although Martian ores may have formed in a comparatively short space of time, in the first quarter of the planet's history, the lack of a tectonic recycling mechanism may have preserved them in abundance to the present day. However, counterbalancing this possible advantage is the probability that the lesser erosion that has occurred on Mars may have left many ores deeply buried. Optimistically, and excepting organic deposits, (assuming the planet has always been lifeless, there may be areas of Mars as rich as modern Earth, maybe more so since the best deposits in many countries were exhausted in the days of pre-industrial civilization. It is possible therefore that the long-term growth of Martian civilization may not be significantly thwarted by a deficiency of raw materials.

2.2 Energy

Such a potential abundance of raw materials will only be of use if energy is available to prospect, extract and process them into useful products. Self sufficiency for a Martian civilization therefore involves procuring an independent, reliable and long-term energy supply. Since Mars in unlikely to be in a position to pay for such a supply by exporting products in bulk that are unobtainable on Earth or elsewhere in the solar system, then its energy must stem from sources on or near Mars and must be processed and distributed by an infrastructure built from and maintained by Martian raw materials. The energy the economy must feed back into enabling and maintenance of its flow must be less that that yielded to its economy in the first place. Ideally, this net energy ratio must be as high as possible to promote rapid initial growth.

Table 2: Options for Power Production on Mars.

System	Problems
Surface Solar	¤ Latitude dependency ¤ Diurnal dependency ¤ Half power of Earth equivalent ¤ Dust storms and precipitation
Wind	¤ Tenuous atmosphere ¤ Wind speed fluctuations ¤ Low power rating per unit
Nuclear	¤ Complex and expensive infrastructure ¤ Mining and processing of expensive fuels ¤ Safety requirements ¤ Waste disposal
SPS & Laser	¤ Reliance on space-based economy ¤ Untested
Biomass	¤ Reliant on artificial life-support ¤ Needs large surface areas ¤ Lower insolation ¤ Large water requirement
Nuclear Geothermal	¤ Needs specific geological conditions ¤ Pollution hazard ¤ Untested ¤ Politically incorrect
Fusion	¤ Unproven

Options for power production that might be available are shown in Table 2 along with comments on specific disadvantages in each case. The first five systems in the list have already been suggested as methods of powering initial settlements [31-35], however in these cases the energy processing infrastructure will have been fabricated on Earth and transported to Mars. In the case of nuclear power [33] the fuel will have come from Earth too, both fuel and reactor being shipped in one package. Self-sufficiency requires Martian civilization to close the energy supply/feedback loop both internally and profitably.

Solar and Wind Power. Solar and wind energy is readily available on Mars and preliminary studies [31,32] have suggested that these renewable resources could be readily tapped. However, the dependency of solar power on time of day, the intermittent nature of wind power and the expensive manufacture of components for both make these systems low net energy providers on Earth [36], although technical progress and experience is continuing to improve the picture [37]. On Mars, the relative efficiency of such systems would be diminished further by the half strength of sunlight and the low density of the atmosphere. It is likely therefore that a growing Martian civilization may use solar and wind power as secondary auxiliary sources.

Nuclear Power. Nuclear power is an obvious primary energy source for Mars, but only if accessible uranium deposits are located. However, the mining and processing of rare fuels and the construction and maintenance of the complex machinery of a nuclear power plant consumes a great deal of power that has to be fed back into the energy processing system. Net energy is generated, but not in the bonanza quantities of fossil fuels. Proper waste disposal requires more power, unless the early Martians are prepared to be more lax over safety standards. This will no doubt be tempting on a planet that is initially sparsely populated and sterile, but may be reserving significant problems for the future.

Solar Power Satellites. If the materials for solar power satellites (SPS) can be obtained economically from the Martian moons or perhaps from the Mars-crossing Amor asteroids, then these systems might prove a good choice for primary power production, despite the weaker Martian sunlight. The SPS has the advantage of non-stop illumination and good transmission of microwaves through the dry Martian atmosphere. The potential high power rating, simplicity and reliability of the SPS system was the reason for its choice as the principal power source for a Mars base designed by the Ohbayashi Corporation to house 150 explorers [34]. This establishment is clearly envisioned as being a precursor step to the immigration of ordinary people and thus demands a large excess of energy for growth. (It is interesting to note that the planners also chose to incorporate windmills in the design of their settlement, but only as auxiliary sources of power.) Enthusiasm for SPS should however be tempered by the fact that it remains an unproven, conceptual system and it would mean that a Martian economy would be reliant, possibly to an unacceptable degree, on the economics of space-based manufacturing—something which itself is highly uncertain. These reservations perhaps apply even more strongly to the other concept of beamed power from orbit that has been proposed for Mars—the laser [35]. Here the technology on the scale envisaged is more speculative and, although a laser beam might be more flexible in its range of applications, such problems as long-term reliability and extremely fine pointing accuracy would have to be solved first.

Biomass Energy. Biomass (plant matter) accounts for 15% of world energy use, although much of it is converted very inefficiently with primitive technologies that have been available throughout history. However, prospects of modernizing bioenergy appear to be promising [37]. It has been claimed that bioenergy might be made competitive with fossil fuels by adapting coal gasification technology to accept the yield from dedicated energy farms and plant residues from agriculture and forestry. Since colonizing Mars will involve growing lots of plants as part of life-support systems (see Section 3.2), then one can envisage a settlement deliberately designing ecosystems where production exceeds consumption and the biomass and oxygen excess can be used to generate power. If net energy is produced then it is possible to imagine life generating the power to construct new habitats and to fuel its own expansion across the surface of the planet. The problem with this idea is that energy plantations need large surface areas to be cost effective, a good supply of water and quite precise environmental conditions to maximize yield. On Mars, plants also require a life-support system of their own, with energy being fed back into temperature regulation, artificial hydrological and nutrient cycles and maintenance of the habitable enclosure. With these extra subsidies that must be fed back into energy processing, and taking into account the weaker Martian sunlight, it not clear that biomass production on present day Mars will be capable of generating significant net energy. It is likely however that incineration of biomass might be used as a method of balancing any tendency to oxygen excess in closed ecological life-support systems (as in the Bios 3 experiment) and so bioenergy may be available as a useful, small-scale, auxiliary source of power, even if large-scale production turns out to be uneconomic.

Nuclear Geothermal Power. It is probably too much to hope that large and exploitable reserves of geothermal energy are available on Mars. On Earth, profitable sources of geothermal energy require the fortuitous combination of regions of high heat flow and favorable groundwater conditions occurring close to the surface. Although areas of "recent" volcanic activity on Mars have been proposed, [38-40], perhaps involving the liquefaction and release of ground ice [38] it seems likely that such localities will be much rarer than on Earth and probably absent altogether. However it might be possible to create Martian geothermal energy by the artificial injection of heat underground into selected frozen aquifers. Nuclear stimulation of geothermal energy has been studied for use on Earth—both for the storage and safe recovery of the energy released by thermonuclear explosives [41] and for the improving of the efficiency of energy extraction from impermeable rock [42]. Hot rock at ~350°C in the rubble chimney created by a one megaton explosion would provide ~4 times as much energy as that released by the device itself [42] but this optimum output could only be achieved in regions with a pre-existing high heat flow. Even though such localities may not be present on Mars, the process becomes increasingly economic as the power of the device is increased, since cost of nuclear explosives is a very weak function of yield. On a very sparsely populated planet, it is possible to envisage setting off very powerful explosions deep under remote areas which would create large volumes of hot rock and circulating meltwater from which geothermal power might be extracted. Scaled up sufficiently, nuclear geothermal energy could compete favorably with conventional nuclear power stations, since the infrastructure is much less complicated and the mass of fission products to be worried about is far smaller. Ideally, if a way could be found to dispose of the fission thermonuclear trigger, possible radioactive contamination of the Martian environment (unlikely from a completely contained explosion hundreds of meters underground) would be limited to short-lived neutron activation products and residual tritium. Once again however, nuclear geothermal energy is an unproven concept which remains only a speculative possibility for powering a Martian civilization and which requires the model of a porous ice-rich regolith to be correct. Moreover, even if Martian conditions are found to be ideal for the emplacement and extraction of this kind of energy, the fact that "nukes" are involved as part of the process would likely result in any program being buried in controversy.

Controlled Nuclear Fusion. The above proposal basically involves the use of the only net thermonuclear energy we currently know how to generate—the explosive kind. The controlled generation of fusion energy is hoped for but the experimental reactors presently being applied to the problem still have to exceed energy "breakeven" let alone produce net energy after accounting for all the subsidy embodied in structure and maintenance. Optimists who believe that these problems can be overcome have stated that economical fusion reactors may be only decades away from realization. Since this is the sort of timescale commonly quoted for getting Martian colonization started and since fuels such as deuterium or lithium should be obtainable on Mars. Then perhaps controlled fusion might have an important and ultimately crucial role. However, one can rely on no more than the vaguest of projections that controlled nuclear fusion will become the major primary power source of the 21st century.

Given the choice in Table 2 of power generation methods, and another fifty years advance on the current state of the art, it seems possible that a Martian civilization will be able to generate sufficient primary energy for growth. Although all the methods outlined have their problems and unknowns, especially in an extraterrestrial setting, an intelligent combination of alternatives may be the approach that is ultimately followed. However, the large net energy ratios of fossil fuels (up to 50:1 [36]) are unlikely to be achievable. Thus whilst it has been stated [17] that energy resources will be a less critical limiting factor than material resources on Mars, this may only apply initially. If the thorough exploration of the Red Planet turns up the wide range of mineral resources that some suspect may be there [25] then growth can be expected to be proportional to the quantity of net energy fed into the economy. Matter will need energy for its organization and energy will need organized matter for its flow. A self-sufficient Martian civilization will have to integrate its matter and energy supply into one system.

2.3 The Martian Matter-Energy System

The "metabolism" of a Martian civilization, and how it must evolve to self-sufficiency, can be encapsulated in simple system diagrams. Figure 1 depicts three distinct stages in the settlement of Mars using the "energy language" symbolic contention of Odum [36]. Energy flows and the embodied energy in matter flows are combined and are shown with countervailing flows of money.

Stage One is that of an Apollo-type mission to Mars that might occur as an initial manned visit. *Everything* required for the astronauts' stay is transported from Earth, this including energy sources, all the infrastructure, life-support systems and consumables such as gases, water and food. This temporary "Mars Colony" is thus totally subsidized from the Earth. Energy flow is one way and cash flow the opposite—the space program has a large bill to pay.

Stage Two shows the system that might be expected to describe early settlements and depicts the state that most proposals for "living off the land" have reached [e.g. 34]. Martian resources are now being tapped to provide energy from the sun [31] and wind [32] and to manufacture air, food and build "low-tech" materials [17,18]. All the complicated infrastructure however—nuclear reactors, solar cells and windmills; life-support machinery; exploration vehicles; mining equipment; manufacturing apparatus; and resident's accoutrements etc., are still supplied from Earth. Thus it is important not to be naive about what a colony at this developmental state really represents. Such a settlement is *not* yet "living off the land" in a full sense, neither is it generating net energy from Martian resources. The enormous energy subsidy embodied within its Earth-made infrastructure makes it in reality a *fossil fuel* based establishment—in fact probably the most expensive example of its kind, since all the components need to be shipped over interplanetary distances from their site of manufacture. Dollar flow is still one way; from the space program to the industries providing the goods and services. Given unlimited cash, reaching Stage Two and maintaining an arbitrary population on Mars is *easy* to envisage—we understand resource utilization and manufacturing on our energy-rich world and the principles of travelling through space very well. However, any realistic program to colonize Mars is likely to be cash limited, both in the total amount of money available and the time over which it is to be spent. If Mars colonies are to survive and prosper, they must *rapidly* develop an indigenous manufacturing industry capable of providing the essentials for survival and growth before the gravy train from Earth is halted.

Stage Three shows the ultimate goal of *The Case for Mars*—an independent and self-sufficient Martian civilization. A system has been created like on Earth that is internally maintained by *domestic* feedback of work into resource acquisition and processing. Martian energy sources and infrastructure are home-grown, the planet has become a trading economy, exporting as well as importing goods. Interplanetary cash flow is thus two way, eliminating the continuing cost of Mars to the citizens of Earth. Shown in the form of a system diagram, the transition from Stage Two to Stage Three looks deceptively simple, merely the disconnection and adjustment of a couple of flows. However, it represents a transition of considerable conceptual *difficulty*. This is not just because of our ignorance of the inventory and physical state of Martian resources—more fundamentally than this it is because there is simply *no historical precedent for it*. We have no experience of building an industrial civilization on anything other than the high net energy provided by fossil fuels. Yet the necessity to effect this "extraterrestrial transition" is an absolute pre-requisite for the success of colonization, not just of Mars, but of any locale in space.

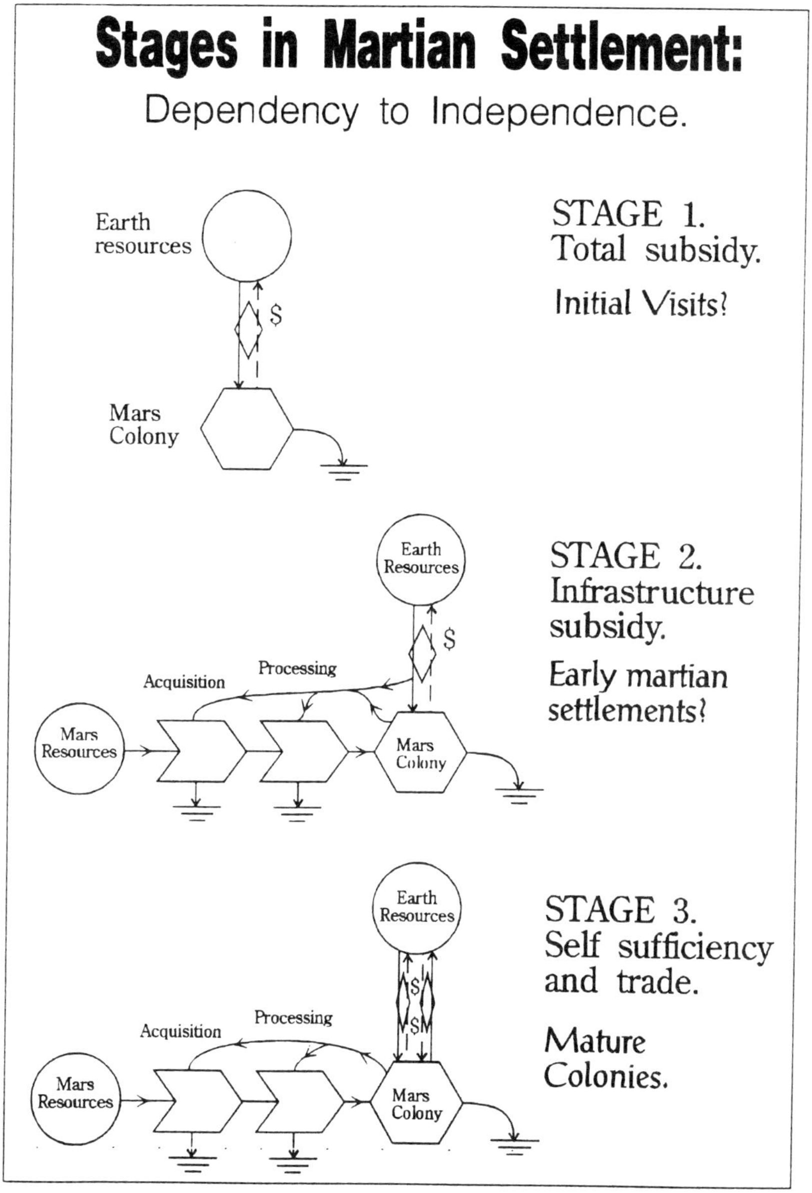

Figure 1 Systems diagram of stages in Martian settlement using the symbolic convention of Odum [36].

The essentials of the transition are illustrated in Figure 2. A Mars infrastructure is shown growing at the expense of Earth. However, the Earth is represented by the "passive storage" symbol, indicating that the quantity of resources it is prepared to supply to Mars is finite; once the supply ceases, the Martian infrastructure will depreciate and the colony will perish. The crucial question is; *can net energy be generated from Martian resources whilst subsidy is still being provided from Earth?* This cannot yet be answered but if it turns out to be yes, then the implications are far reaching. If the titles of Figure 2 are modified; "Earth Resources" being replaced by "Fossil Fuels"; "Mars Infrastructure" being replaced by "Earth Infrastructure" and "Mars Resources" by " Renewable Resources," the diagram now sums up the dilemma facing the future of industrial civilization *on the Earth.* Thus, terrestrial and extraterrestrial transitions are seen to be essentially *the same*—a change from a reliance on fossil fuels for *net energy* to some other source. It is not surprising therefore that the contents of Table 2 look similar to a list of power generation options for the Earth in a post-fossil fuel era (excepting natural geothermal power and systems involving the flow of water.).

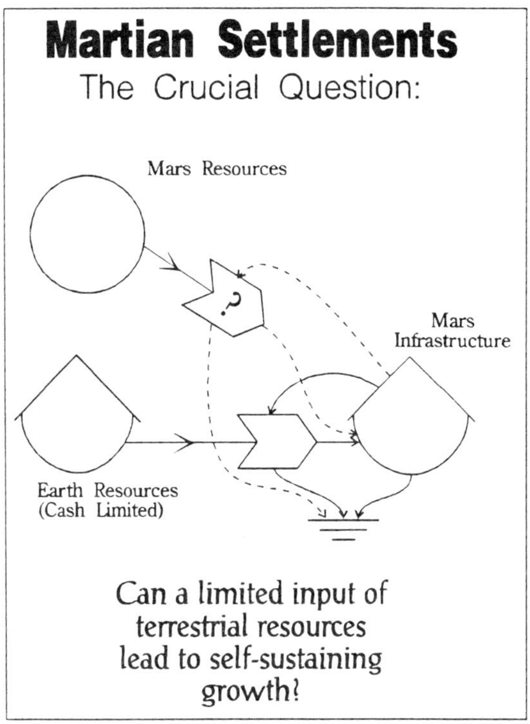

Figure 2 Summary of the crucial question of Martian colonization.

3. LONG-TERM HABITABILITY: LIFE SUPPORT SYSTEMS

A more basic question than that of the capacity of Mars to support technology, is its capacity to support life. Eventual independence from the Earth will bring the issue of life-support systems to a new focus. Concerns will change from day to day survival, propped up by an external supply of life-support system components, to that of creating a permanent future for life on Mars. Pioneers who look forward to such a prospect will ask; *can life be sustained on Mars, not just temporarily, but indefinitely, and if so, what is the best way of tackling an indeterminate need for life-support on a presently lifeless planet?*

At its most fundamental, a life-support system involves a flow of energy through a space that drives an internal cycling of matter into which the specific cycles characteristic of life can be integrated. The net effect is a controlled conversion of energy from a low to a high entropy state, thereby maintaining the order and information content of the system's living components. The goal of long-term habitability will entail the creation of a Martian life-support system that is stable, reliable, and inexpensive in terms of energy and conscious effort to maintain. More specifically, a system with the following desirable attributes is called for:

1. It must not require ongoing re-supply of energy and matter from beyond Mars.
2. It must rely principally on utilization of *gratis* energy, such as that of Martian sunlight and natural, solar-powered, matter cycles. Thus, it must minimize dependence on technologically processed energy, such as electrical power and the energy embodied in manufactured life-support infrastructure.
3. Its matter cycling must be *regenerative*—production and consumption must either be balanced, or capable of easy compensation from the exterior Martian environment.
4. Its components should have the properties of *self-replication* and *self-maintenance*, so as to reduce the human role in manufacture and repair.
5. Its internal dynamics must be *self-stabilizing* and *autonomous*—as safe as possible from dangerous perturbations and keeping essential human monitoring and control to a minimum.

The types of life-support system (hereafter LSS) , of which we have direct experience, or which have been proposed, are shown in Table 3. Only two are in reliable operation at the present time—the Earth's biosphere and the "open loop" LSS of spacecraft. The fully autonomous, pre-agricultural, biosphere of the Earth is no more and the other systems listed remain unperfected or conjectural. Looking at Table 3 from left to right, it is seen that the LSS characteristics vary in their degree of *closure* (the extent the system is independent of material resupply); the balance of mechanical and biological sub-systems involved in recycling and whether or not it is structurally constrained.

In choosing an ideal for the long-term habitation of Mars we can rule out open-loop LSS or variants of the EC/LSS (*Environmental Control/Life-Support System*). Both are of more use for spacecraft or surface exploration vehicles than for extended and comfortable occupancy. The former does not recycle anything and the latter (proposed for the "Freedom" Space Station) is only partially closed, relying on machinery to revitalize air and purify water, but still needing an external food supply and a repository for unprocessed solid wastes [43]. Such systems may be used for initial visits to Mars but not as the backbone supporting a permanent human presence. *Bioregenerative* LSS are the obvious choice for this purpose since they can exploit the self-reproducing, self-maintaining and self-stabilizing properties of life. They can also provide an internal source of food and can operate, to a varying degree, on the power of sunlight. The choice therefore boils down to three options from Table 3: a *Controlled Ecological Life-Support System* (CELSS - sometimes referred to in terms such as *"enclosed biosphere," "biosphere habitat"* etc.) [7-9,44-46]; a paraterraformed planet (basically a much more voluminous, planet-wide CELSS) [47]; or a *terraformed* planet (endowing Mars with a biosphere that mimics that of the Earth) [10-13].

Any initial life-support system for Martian colonists is going to have to be small and enclosed, irrespective of whether this is an ideal long range solution. Thus the first few generations of settlers, at the very least, will be living within some form of CELSS.

Table 3

CATEGORIES OF LIFE-SUPPORT SYSTEM

	Open Loop	**EC/LSS**	**CELSS**	**Para-terraformed Mars**	**Present day Earth or terraformed Mars**	**Pre-civilised Earth**
Air revitalization	Stored supply & waste	Mechanically recycled	Biologically recycled	Biologically recycled	Biologically recycled	Biologically recycled
Water reclaimation	Stored supply & waste	Mechanically recycled	Biologically & mechanically recycled	Quasi bio-hydrological cycle	Natural bio-hydrological cycle	Natural bio-hydrological cycle
Solid waste management	Treated & stored	Treated & stored	Biologically & mechanically recycled	Biologically recycled?	Biologically recycled	Biologically recycled
Food service	Stored	Stored	Horticulture	Agro-ecosystems	Agro-ecosystems	Natural ecosystems
Essential monitoring and control	Continuous	Continuous	Continuous	Continuous	Partial	Uneccessary
Containing structure	Present	Present	Present	Present	None	None
External resupply	Total	Partial	Minimal	Minimal	None	None

3.1 Constrained Biospheres

The concept of a CELSS can be easily understood by considering the attachment of a horticultural greenhouse to a human habitat—the purpose of this arrangement being to create a simple artificial ecosystem, with the greenhouse biota being producers and people being consumers. Not only would the greenhouse plants produce a supply of food, but they would also revitalize air by photosynthesis and purify water via transpiration. Outline designs for such life-support systems have been published for a number of years [44-46] and there have been two attempts at building working models capable of supporting a small human crew—Bios 3 in Siberia and Biosphere 2 in Arizona [7,8].

To best fulfill the criteria for long-term habitability discussed above, the ultimate CELSS would aspire to emulate all the vital functions of the Earth, containing varied ecosystems and having no need for complex life-support machinery. The biota would operate all necessary cycles and would have the capacity to maintain their enclosed environment at a stable equilibrium. The basic human quality of life should be effortless—the essential processes of life-support becoming imperceptible and merging with the background of "nature" with which one is surrounded. Unfortunately, however, we currently do not have the knowledge to perfectly scale down all of the functions of the Earth's biosphere. In fact, there are reasons to believe that mini-biospheres capable of supporting man may always need the assistance of human tended machinery to keep them going. Perusal of the literature on constrained biospheres reveals quite a long list of essential aspects of ecological life-support that may require a technological solution (shown in Table 4).

Table 4: Enclosed Martian Biospheres: Aspects Requiring a Technological Solution.

- Maintenance of habitable enclosure
- Temperature regulation
- Dehumidification of air
- Solid waste recycling
- System stabilisation
- Management of horticulture
- Trace contaminant control
- Replacement of leaked gases
- Pest and disease control
- Radiation protection
- Emergency life-support

The need for maintenance of the habitat's containing structure is the most obvious, being an activity of far greater importance than the upkeep of ordinary buildings on Earth. The provision of shelter on Mars involves having to hermetically seal a high pressure habitable interior against the lethal near-vacuum outside and to continually compensate for leakages (gradual and catastrophic). Comparing the drama involved in coping with a broken greenhouse window on Mars with a similar incident on Earth gives some idea of the extra day to day work Martians will have to put into offsetting the fragility of their life-support surroundings.

The other items in Table 4 are perhaps less evident, but mostly result as consequences of the small volume of a constrained biosphere with its relatively small mass of atmosphere, held down onto the biota by an impermeable roof just a short distance above. The absence of terrestrial type weather, means that Martian CELSS will probably need to rely on mechanical air conditioning systems to circulate air and for temperature regulation. Since rain will not be able to form naturally, air will have to be artificially dehumidified to close the habitat's hydrological cycle and to prevent condensation in areas where it is unwanted.

Biospheric containment also has an effect of seriously reducing ecological stability. This is because the ratio of biomass to atmospheric mass is enormously increased requiring biogenic materials to be cycled through the biota very rapidly. For example, Biosphere 2, which has a $m_{bio}/m_{atm} > 300$ greater than Earth, has to recycle all the CO_2 in its atmosphere *in just a few days*, as opposed to Earth's eight years [7]. Any habitable equilibrium must therefore be extremely precise as imbalances will run away very rapidly. The less-than-total predictability of the dynamics of ecological systems may make achievement of such a fine balance very difficult, especially as natural stabilizing biotic responses that function within the vast and massive biosphere of the Earth, may be too slow or inoperable in a confined space. Thus, the system will be sensitive to destabilization by minor perturbations, the speed of subsequent environmental change overwhelming the ability of ecosystems to compensate. Examples of events that could precipitate a sudden environmental crisis include a reduction in the amount of sunlight due to season or a dust storm; a need to take in extra colonists because of an emergency elsewhere; and an attack of plant pests or disease.

Such difficulties have been predicted in the technical literature and learnt by experience. Bios 3, for instance, did not achieve balance and had to cope with an oxygen excess by incinerating indigestible plant biomass. Much worse is the situation afflicting Biosphere 2, where consumption has been in excess of production, apparently because of an underestimate of the amount of sunlight available and a plague of crop-eating spider mites [48]. Fresh air has had to be let into the habitat to compensate for dangerously declining oxygen levels. This would not be an option on Mars; although the presence of a Martian atmosphere can buffer an oxygen excess (CO_2 is imported and excess oxygen and biomass is stored, dumped or used for growth), overcoming an oxygen deficit would require using technological systems for manufacturing oxygen.

The inevitable solution to the problem of the viability of small enclosed habitats therefore seems to be the back-up of ecological life-support with technology and conscious monitoring and control of the entire system. Machinery capable of manufacturing precise amounts of O_2 and CO_2 could compensate for unwanted deviations in the balance of these gases. The most indigestible of solid wastes might be broken down chemically and reprocessed into nutrients or ejected to the exterior. Similarly, a mechanical solution to cleansing the atmosphere from the build up of toxic trace contaminants might also be required. Uncertainties in agricultural production could also be reduced by exploiting the wide range of technologies involved in horticulture to deliver precise fluxes of light and flows of water, nutrients, pesticides, etc. Protection from complete failure of the habitat's ecological life-support systems would require a comprehensive back-up mechanical life-support system capable of functioning in a last resort, and for some considerable time, following a serious emergency. *Sophisticated and reliable machinery is therefore the key to survival on Mars.* Whilst this has been recognized by the designers of Biosphere 2, the fact that the project's publicity has tended to emphasize the role of its living components has tended to overshadow just how much it is reliant on technology. Table 5 shows a list of Biosphere 2's mechanical sub-systems—it is interesting to note just how similar this catalogue is to that in Table 4.

Table 5: Biosphere 2: Mechanical Life-Support Systems.

[List compiled from Allen(1991) — may not be exhaustive.]

- Fans to circulate air
- Pumps to circulate water
- Wave machine to oxygenate "ocean"
- Desalination systems to maintain fresh water reservoir
- Misting machine for "rainforest" cloud
- Dehumidifyers in basement air coolers
- Temperature regulation system
- Algae scrubbers to transfer water between "salt marsh" and "ocean"
- Compost machine for rapid recycling of waste
- Soil bed reactor to neutralize trace contaminants in air
- External structural maintenance teams
- Monitoring systems
- CO_2 scrubbers (not included in Allen, 1991).

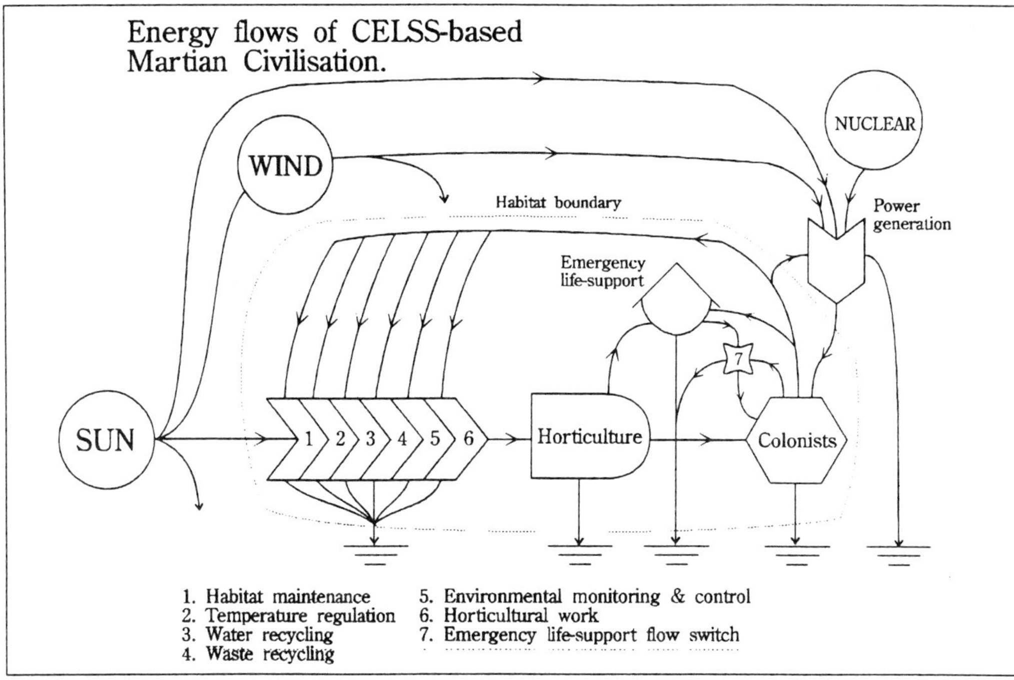

Figure 3 Systems diagram of energy flows of a CELSS-based Martian civilization.

Operating a CELSS will therefore require the feedback of a lot of high quality energy into its sub-systems. This is illustrated as an energy flow diagram in Figure 3. One can get a feeling for the magnitude of this power requirement by comparing the assessments for per capita usage in Table 6. The average use of primary power on Earth, where life-support services of the biosphere come "free," are ~2 kW/person and consumption of electric power are ~0.2 kW_e/person; these averages in the industrial nations are ~6 kW/person and ~0.7 kW_e/person respectively [49]. Estimates of power requirements in space for *life-support alone* exceed these values. The consumption of an EC/LSS, such as that designed for the "Freedom" Space Station, will be ~2 kW_e/person [43], a figure which does not include the substantial costs of resupplying food and spare parts. Since food-producing organisms also need life-support, the power consumption involved in operating a CELSS would be greater; one concept of a CELSS intended for a lunar base, cultivating a highly productive strain of wheat, quotes an estimate of ~5.5 kW_e/person [46]. It would seem therefore that the per capita power requirements for overall human activities on Mars will be significantly greater that on Earth and probably much higher. The large Ohbayashi base, designed to be self-sufficient in food and with enough power to support exploration, growth, and to cope with emergencies, has a calculated demand of ~50 kW_e/person [34]. There is an urgent need to determine how realistic these on-paper estimates are likely to be, as Biosphere 2, the only CELSS currently operating in practice, apparently requires ~100 kW_e/person to function! (Biosphere 2 has a power supply of 1 MW and has 8 crew [7]. In a context where fossil fuels are not present and net primary energy is less, the much greater per capita power requirements for basic survival on Mars will be another factor restricting the growth of a self-sustaining civilization.

Table 6: Human Power Consumption: Earth-Mars Comparison.

Earth (Gratis Life-Support).	
World Average:	~ 2 kW/person (Ref 49) ~ 0.2 kW_e/person
Industrial Nations:	~6 kW_e/person (Ref 49) ~ 0.7 kW_e/person
Extraterrestrial Estimates.	
EC/LSS only:	~ 2 kW_e/person (Ref 43)
Lunar CELSS only:	~ 5.5 kW_e/person (Ref 45)
Ohbayashi Mars Colony 2057:	~ 50 kW_e/person (Ref 34)
Biosphere 2:	~ 100 kW_e/person (Ref 7)

An integrated ecological and technological LSS can probably be made to work on Mars, nonetheless conventional CELSS provide only a poor match for the desirable attributes of a truly long-term habitable environment. They are expensive in terms of quality energy and will not be fully bioregenerative, self-maintaining and autonomous. Scaling up CELSS to enclose a much larger volume, such as in the scenario of paraterraforming [47], could permit the operation of quasi-natural matter cycles and, with an m_{bio}/m_{atm} only ~10 times that of Earth, internal stability would be improved (see Table 3). However, these benefits would have been bought at the price of a massive and vulnerable containing structure, the non-stop maintenance of which would be both essential and expensive.

It is difficult to imagine a civilization which looks forward to an indefinite future being content with relying on the eternal function of machines for the fundamentals of survival. This is especially when there is an alternative that we at least know works very well, even if we do not yet fully understand how. As has been recognized [50], *unconstrained* habitats offer such advantages for a civilization which aims to survive millennia that the notion of eventually terraforming Mars, even if only for the benefit of their remote descendants, will be debated by the earliest settlers.

3.2 Unconstrained Biospheres

An unconstrained biosphere is defined as one that has no artificial structural containment and which is merely restricted in its extent by gravity. Thus, such a biosphere would naturally englobe the entire planet on which it is situated and would extend vertically as far up into the atmosphere and as far down into the crust as life can survive. This description applies to the biosphere of the Earth and would also apply to a terraformed Martian biosphere. Possible methods of terraforming are reviewed later in Section 4; here, we consider the life-support properties of a fully terraformed Mars.

In Section 2.2, Martian solar power was considered to be a low net energy provider when processed by technological means. Similarly, because of the large subsidy of quality energy needed to operated a CELSS, the net photosynthetic energy yield of the interior biota would also be poor. However, by terraforming Mars, the efficient use of solar energy intercepted *by the entire planet* can be maximized (50). Energy flows on a terraformed Mars are shown in Figure 4 and it is instructive to compare them with those of a CELSS in Figure 3. It illustrates that Martian civilization would now be unrestricted on a fully habitable planetary surface; hermetically sealed, radiation-proof structures are no longer needed and life-support would be subsidized by two powerful and *gratis* energy flows. Normal weather now drives the circulation of air and the "uphill" branch of the hydrological cycle. The provision of fresh air, clean water, and the recycling of most wastes are services also provided for "free" by the natural "wilderness" biota. Feedback of human work will still be needed for agriculture, but not in the quantity needed by the intensive, sealed, horticulture of a CELSS.

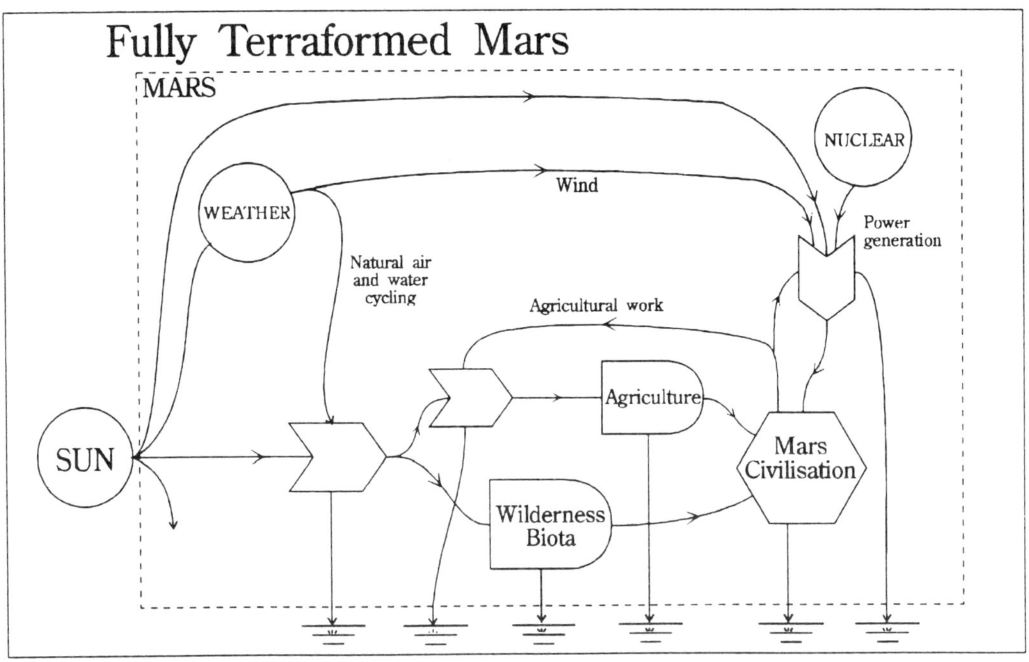

Figure 4 Systems diagram of energy flow on a fully terraformed Mars.

The enormous volume and consequent slower pace of change within an unconstrained Martian biosphere would render the environment more stable, especially if, on a global scale as on Earth, the biota could be brought to an overall ecological climax. The planetary LSS would therefore not be as prone to catastrophic change and would be in less need of continuous monitoring and control. True, human beings would now be subject to the annoyances of unpredictable weather, but this will be more than compensated for by the provision of an aesthetic and challenging environment in which to live, along with the freedom from relying on nearby machinery for the basic essentials of life.

The magnitude of these *gratis* life-support subsidies can be estimated by comparing those that benefit terrestrial civilization (see Table 7). The present world primary power production is ~10 TW and human "control" of net primary productivity (NPP), because of the large areas reserved for agriculture, amounts to ~20 TW. The NPP of the global biota is however ~90 TW and the power embodied in weather is a huge ~2500 TW [51]. Thus humanity, being inseparable from the biosphere and a consumer at the top of the ecological hierarchy, receives as a whole about 2600 TW of *gratis* life-support, equivalent to ~500 kW per 5 x 10^9 people. Seen in from this perspective, the 100 kW_e/person power consumption of Biosphere 2 does not appear so bizarre and unreasonable, except for the fact that it has to be provided by electrical generators. A Martian civilization capable of providing 10 TW of primary power production (the same as presently on Earth) might therefore only be able to support a maximum of 2 x 10^7 people. In reality however, it would have stopped growing long before the population reached this level as available net energy would have fallen to zero because of the excessive demands of life support.

Table 7: Terrestrial Energy Flows.

Natural.	
Total Insolation (top of atmosphere) :	**~ 175 000 TW.**
'Weather' :	**~ 2 500 TW.**
Net Primary Productivity (NPP) :	**~ 90 TW.**
Geothermal Heat Flow :	**~ 40 TW.**
Anthropogenic.	
Human 'control' of NPP :	**~ 20 TW.**
Primary Energy Production :	**~ 10 TW.**
Hypothetical (hopefully).	
Twelve Hour Nuclear War :	**~ 1000 TW.**

The ideal requirements of an indefinite future for life and civilization on Mars are therefore met by an earthlike, terraformed, biosphere. Such a creation on Mars might match well those desirable attributes, listed above, of a truly long-term life support system—self-contained, producing net energy from ordinary sunlight, bio-regenerative, self-maintaining, self-stabilizing and autonomous.

4. TERRAFORMING MARS

The assessment in the above paragraph assumes that it is possible to create a duplicate of the Earth's biosphere on Mars. Is this assumption reasonable?

The previous sections of this paper have considered the issues involved in establishing the seed of a self-sufficient Martian civilization. By comparison, the task of engineering Mars on a *global scale* may appear too enormous to take seriously. Yet this is not the case. Purposeful modification of the environment is a quintessential human activity and has been since prehistoric times. Recently, anthropogenic changes have grown from those with local or regional effects, to alterations in global parameters, such as the composition of the atmosphere and the average surface temperature. In some instances human influence outpaces that of "nature," such as in the case of mineral extraction in which up to two orders of magnitude more rock is shifted annually than by natural erosion [52]. Admittedly, these alterations of the Earth's environment are currently a haphazard side effect of the routine fluxes of energy and matter involved in the economy of modern technological civilization. However they demonstrate that planetary engineering, of a sort, is already possible. From this realization, it is not such a great step to envisage an organization and scaling up of such a capacity in order to purposefully alter a planetary environment [13]. Thus the future ability to terraform Mars depends on whether the planet can successfully support a technological civilization—this, as outlined above, depends on whether net energy is available to the economy of such a civilization. *The basic assumption underlying whether terraforming Mars is possible is therefore identical to that determining whether man can settle there in the first place.*

There has been intermittent scientific attention paid to the possibility of terraforming Mars since the first papers appeared in 1973 [53,54]. Ensuing highlights in the progress of the subject have included several workshops held at the Ames Research Center, a published official NASA study [55] and a particularly strong recent surge of interest since 1989, both in the USA and UK [10-13, 47, 50, 56-59]. The principal modifications required of planetary parameters are shown in Table 8. The column under the title *Minimal Ecopoiesis* shows those changes needed for anaerobic life to be established, whilst that under *Terraforming* displays the necessities of an environment habitable for man. It is evident that five main alterations to Mars will be essential: the atmospheric pressure and surface temperature must be increased: the surface flux of UV radiation must be reduced, the surface must be made wet; and the chemical composition of the atmosphere must be changed.

Table 8: Mars - Terraforming Requirements.

Parameter	Present value	Modification required for: Minimal ecopoiesis	Terraforming
Surface Gravity	0.38g	Not possible	
Rotation Period	24.6 h	Unnecessary	
Axial Inclination	25°12'	Unnecessary	
Insolation	593 W m^{-2}	Increase up to 1370 W m^{-2}	
Albedo	0.17	Low as possible	
Mean Surface Temperature	~ 220 K	~60 K Increase	
Surface atmospheric pressure	~ 7 mb	> 10 mb	500 - 5000 mb
CO_2 Partial Pressure	~ 7 mb	> 0.15 mb	< 10 mb
O_2 Partial Pressure	~ 0.006 mb	> 1 mb	130 - 300 mb
N_2 Partial Pressure	~ 0.12 mb	> 1 - 10 mb	> 300 mb
Hydrosphere	0%	> 0%	>> 0%
UV Flux 0.2-0.3 μm	~ 7 W m^{-2}	Reduction	Zero

Two distinct approaches to the problem of creating a Martian biosphere have emerged, which have been categorized as *ecocentric* and *technocentric* [50]. The former viewpoint [10,12,57,58] regards Mars from the broad perspective of the totality of life and is mainly concerned with using the mildest possible planetary engineering techniques consistent with the creation of an environment that can support some form of biology—however primitive. Ecocentric scenarios usually reduce human control of the engineering process to a minimum and invoke a purely biological, "Gaian," mechanism for subsequent stabilization of the global environment. A human-habitable biosphere is not the specific goal of ecocentrism [57] and is only considered as a remote possibility and in the vaguest of terms. In contrast the technocentric stance emphasizes the technological aspect of terraforming, arguing that since many planetary engineering tasks will require the work of humans and their machines, then full consideration of human motivations is appropriate [11, 47,50, 59]. Thus the primary goal of technocentric scenarios is to create a fully habitable Martian biosphere—as close a facsimile as possible to that of the Earth. Technocentrists are therefore less inhibited about proposing much more radical and speculative planetary engineering solutions.

A first step to terraforming Mars is commonly considered to be to thicken the atmosphere and to use its increased greenhouse effect to warm the planet's surface. Speculations on how this might be done require an assumption as to which Martian climate model is correct—i.e., the quantities of volatiles available, their physical state, location and how easy they are to mobilize. The nature of this assumption explains a lot about the bifurcation between ecocentric and technocentric approaches and is summed up on the catastrophe graph in Figure 5.

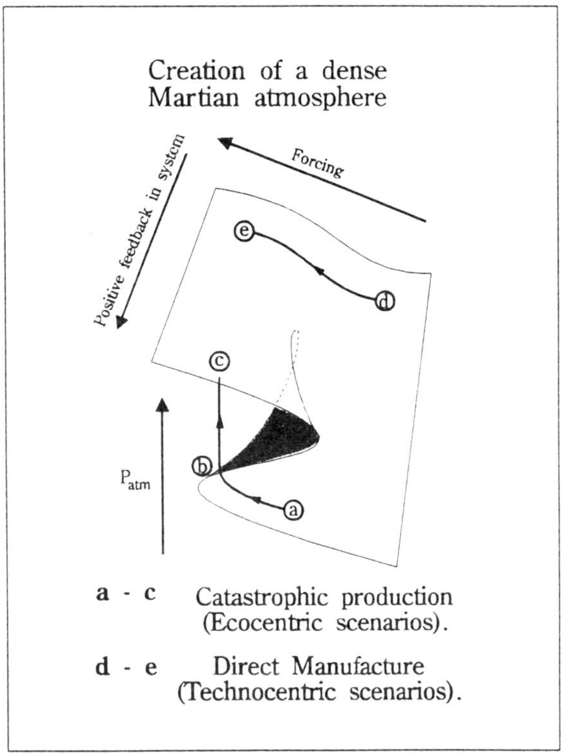

Figure 5 Contrast of terraforming scenarios.

Ecocentric scenarios are able to minimize the planetary engineering involved, (the degree to which the environment must be artificially "forced" into a biocompatible state.) by postulating a Martian climate that is *highly unstable*. Positive feedback between surface temperature and atmospheric pressure is assumed to be very strong—large excursions being possible as a result of a minor initial perturbation. This is represented on Figure 5 by the trajectory **a-c**. The Mars of today is at the location **a**, and planetary engineering takes it to point **b**, whereupon the climate becomes unstable and rapidly undergoes a spontaneous transition to a new high-pressure state at **c**. *Most of the* *work therefore to manufacture a dense Martian atmosphere comes from the innate amplification within the system and only a modest amount of planetary engineering is needed to set the process going.* This notion of "tweaking" Mars to life lends natural support to minimalist approaches that distance themselves from human activities and motivations. The opposite assumption, that the climate of Mars is now grounded in a highly stable equilibrium state, requires terraforming to follow the route **d-e**. This time there is little positive feedback assisting the production of an atmosphere and all of it must be *directly manufactured.* In such a situation, which obviously calls for a much greater degree of planetary engineering, it is more difficult to foresee terraforming getting underway, but once it does it is easier to imagine it being done with a commitment to "go the whole way," creating a fully earth-like environment.

4.1 Terraforming Mars—The Standard Paradigm

It is fair to say that path **a-c** on Figure 5 represents what might be described as the "standard paradigm" for terraforming Mars. It is also the oldest, being based originally on the now obsolete "Long Winter Model" of the Martian climate [60]. This proposed that the planet's polar caps contained an atmospheric pressure equivalent of ~1 bar CO_2 in the form of dry ice and that the partitioning of carbon dioxide between solid and gaseous phases was very sensitive to surface temperature. It was thought that a minor warming of the polar caps, perhaps by reducing their albedo by a few percent with dark dust or plant growth [54,55] would be sufficient to cause some of the CO_2 to sublime and add to the mass of the atmosphere. This would, in turn, increase both the atmospheric greenhouse effect and the strength of advective heat transfer from the equatorial regions, thereby warming the caps still further, causing more CO_2 to vaporize and so on... with sufficient initial forcing, the Martian environment would experience a rapid ~100 year change from a situation where atmospheric gases are frozen into a cold surface, to one where they are gaseous, blanketing a warm surface. Solar energy would drive most of this change (Figure 5; **b-c**) and the new climatic regime, with a CO_2 pressure of ~1 bar and average surface temperature ~273 K, would be stable. Whilst not habitable for higher forms of life, Martian ecosystems of primitive and hardy organisms, such as algae and lichens, might be established [55].

The "Long Winter Model" was discarded when it was realized that the Martian polar caps were likely to contain much less than the requisite amount of carbon dioxide. However, the terraforming model that was based upon it survived, its basic mechanism unchanged, by switching its CO_2 source to a proposed reservoir of gas adsorbed on mineral grains at shallow depths in the regolith. Again, modest warming by such techniques as albedo reduction, orbiting mirrors, or the manufacture of artificial trace greenhouse gases (such as CFC's) eventually triggers a runaway release of CO_2, the bulk of it coming from the regolith, as opposed to the poles. The most recent modeling of this process [10] suggests that, if circumstances are favorable, an artificial warming of ~25 K would set off a transition to the high-pressure state, ~2 bars of CO_2 being needed to maintain Mars permanently at ~273 K in the absence of continued warming by other methods. It is proposed at this stage to seed Mars with plants and then to sit back for

~100,000 years and wait for the oxygen partial pressure to build up to a breathable level.

However, as has been pointed out [11,56] the standard paradigm is afflicted by non-trivial difficulties. Studies that have calculated just how much CO_2 gas the regolith could store, even if ideally absorbent, suggest that much less than the ideal ~2 bars is likely to be present in this easily releasable form [61,62]. It seems more probable, especially if Mars was once a warmer and wetter planet, that the chemical weathering of silicate rocks, which would have occurred inevitably under such conditions, would have resulted in most of the planet's dense CO_2 paleo-atmosphere turning into carbonate rock [3] (see Table 9).

Table 9: Martian Carbon Dioxide Reservoirs.

Constrained		
Atmosphere	~7 mb	
Polar Caps	< a few mb	Ref 61
Adsorbed in regolith	< 280 mb	Ref 61
	< 190 mb	Ref 62
Loss to space over past 3 Gyr	< 10 mb	Ref 66
Unconstrained		
Carbonate formation	several bars	Ref 3
Early impact erosion.	> 1 bar	Ref 67

Carbonates cannot be tweaked into giving up their volatiles, rather they need applied temperatures of ~1000 K or shock pressures of ~10 5 bars. Another problem is that, even if a substantial amount of adsorbed regolith CO_2 is present, heat conduction through the regolith, down from the surface where artificial warming would be applied, would be very slow. Thus, instead of a rapid thickening of the atmosphere over a period as short as a century, it might build up with a trickle, to a pressure perhaps only a tenth of that wanted, after several thousand years. Equally problematic would be the release of water. Aquifers at higher elevations would need to be thawed and drained so that water might pool at the surface in low lying natural basins. Again, passive downwards heat conduction would only melt sub-surface ice very gradually—one estimate [11] suggesting that after 100 years a globally averaged depth of only ~1 m of meltwater will have been produced, rising to just ~3 m after 1000 years.

The most excessive claims for the standard terraforming paradigm are therefore questionable. Perhaps the likely end result is a planet resembling a particularly cold and dry Precambrian Earth, where only the most hardy and primitive life forms can eke out a sparse existence. Not only would such an environment be of little use as a human life-support system, it would also have a limited relevance to the human future [50]. With no seas or oceans in which to bury organic carbon, the hoped for build up of atmospheric oxygen is unlikely. The bacterial biota would probable reach some ecological climax at conditions suitable for themselves and far away from the tolerance range of higher organisms.

4.2 Terraforming Mars—An Alternative

A number of proposals have been published detailing more technologically intense solutions to Martian terraforming [11,47,59]. Only one of these [11] is briefly reviewed here.

Possibly a more realistic starting point is to choose a conservative estimate of the amount of adsorbed regolith CO_2. If only ~190 mb is available [62], then Mars' initial position on Figure 5 is about half way between **a** and **d**. Planetary engineers would still be able to exploit a small amount of climatic instability, but with the disadvantage that the greenhouse effect of this quantity of CO_2 would not warm Mars above freezing. Some atmosphere might have to be manufactured from carbonate deposits: if the ~30mb CO_2 proposed to be contained within the layered deposits in the Valles Marineris was to be liberated (the easiest method being explosive devolatilization [56], then an atmosphere of ~227 mb would be created. This would generate an extra greenhouse warming of ΔT ~16 K and would be adequate for later processing into 160 mb of O_2—equivalent to the partial pressure on a Colorado ski slope. Additional warming increments would be provided by other planetary engineering methods acting in concert with the CO_2 greenhouse. Orbiting mirrors that increase the Martian insolation by 30% would give ΔT ~15 K and artificial trace greenhouse gases might provide ΔT ~30 K. None of these three methods of warming Mars seems capable of providing the full ΔT ~60 K, but the job might be possible by combining them and others, such as albedo reduction and biological production of greenhouse gases.

Unless a Martian hydrosphere is to take many thousands of years to accumulate, the problem of melting of sub-surface ice deposits would have to be overcome. *This could only be achieved by injecting large quantities of heat underground*, directly into the frozen aquifers, so as to rapidly melt and mobilize water. According to one model of Mars' early history [63], this process may have happened naturally several times in succession—pulses of magmatism in the Tharsis region triggering catastrophic outburst flooding. The large discharge rates indicated by the geomorphology of the outflow channels suggest that a globally averaged depth of 70-450 m of meltwater would have drained into a temporary Boreal Ocean in just a few years. Mars would have enjoyed quite long periods of "maritime" climate, an active hydrological cycle, before returning to its previous frozen condition. If this model is correct, then the mechanism presumable weakened as Martian igneous activity declined. Emulation of this extinct natural process would require planetary engineers to use an alternative heat source. Assuming that 10% of the thermal energy is converted into latent heat of fusion of ice, then ~3.4 x 10^{25} J would be needed to discharge 70 m of water, enough to fill a small Boreal Ocean to the –2 km contour. Over a 200 year project timescale, this energy release entails a power production of ~5000 TW—500 times the output of terrestrial civilization! It is difficult to conceive of any Martian civilization having such a huge quantity of primary power to spare for making oceans; the only sufficiently concentrated source they might use would be explosive thermonuclear fusion.

The concept of mining Mars for its volatiles by using buried thermonuclear explosives usually provokes two kinds of critical responses; the first being that it is technically unfeasible and the second expressing concerns over environmental ethics. Arguments about the technical aspects usually make the point that the total energy release required would entail the detonation of up to a billion devices of megaton yield and, even if such a large number could be manufactured and emplaced, Mars would be left dangerously radioactive [64]. This however rests on the false contention that military-type devices, optimized for war, are also the kind that will be suitable for planetary engineering. This is not the case. Since there appears to be no theoretical upper limit to the yield of a thermonuclear explosive [65], and since their cost is only a very shallow function of yield, scaling up the individual explosive units hugely increases net energy return and substantially reduces cost. Residual radioactivity also becomes less of a problem since the fraction of the yield represented by the contribution of the fission triggers can be reduced to a negligible quantity. If fission triggers can be eliminated altogether then the radioactive isotopes remaining would be neutron activation products from dountry rock, which mostly have half lives of a few days or less, and residual tritium, with its half life of 12.3 years. Preliminary studies as to the feasibility of the thermonuclear mining scenario have identified nothing that render it technically impossible [11,56]; for instance, the requisite quantities of deuterium and lithium from which to make the thermonuclear fuel should be obtainable on Mars and the amount of rock needed to be shifted to emplace the devices is already within the capability of terrestrial civilization. The only other idea which addresses the need to rapidly melt Martian permafrost at depth—that of impacting the planet with asteroids [64]—compares poorly with the thermonuclear mining scenario since it is much more wasteful of resources and far less controllable. As to the ethics of the situation—currently this issue mostly boils down to personal opinion, not to any inherent environmental "law". Do Martian rocks have rights? The issue is a complex and unresolvable one and has been discussed at length elsewhere [57,58]. However, in the opinion of this author, terraforming Mars remains a sufficiently far-future project as to make such ethical concerns of secondary importance to us now. It will be Martian colonists who make the crucial decisions over whether to bring their home planet to life.

Thus, after two centuries, this scenario also produces a Martian environment similar to a Precambrian Earth, but one warmer and wetter that that achievable by the standard paradigm. The Martian environment would still only be suitable for anaerobic life, but would be able to host such life in much greater abundance. For full terraforming, there remains the question of the creation of a breathable atmosphere. The best way to achieve this seems to be to manage the ecological succession of the biosphere to maintain it in a state of *pulse stabilized sub-climax*. In such a state, production outstrips consumption but a steady state is attained by cropping of excess biomass. If this excess is prevented from re-oxidizing, by storage or burial, then atmospheric oxygen will accumulate. At first, bacterial biomass would have to be sequestered to produce the first few percent of O_2. However, once sufficient is present to permit root respiration and to reduce the surface flux of UV radiation, oxygen production might be taken over by Martian forests. The succession timescale of a forest is ~100 years and its biomass per unit area is large, so cropping could be done at a more leisurely pace. Studies of the cropping of terrestrial forests to provide biomass energy suggest that, if properly managed, it could be done in a sustainable way.

The speed at which O_2 builds up in the Martian atmosphere also depends on the efficiency of net primary production (NPP). On Earth, the efficiency of conversion of the planet's entire insolation into biomass is ~0.05%. Applying this efficiency to a terraformed Mars, with its insolation increased by 30%, and assuming a 100% storage of biomass, gives a timescale of ~21,000 years to form ~160 mb of O_2. This might be reduced in a number of ways. The efficiency value for Earth's entire biosphere is low compared to the theoretical maximum for leaf-incident radiation (~5). This is because of such factors as atmospheric absorption and reflection; the lack of a full canopy; the diurnal cycle; cold seasons where temperatures fall below the photosynthetic threshold; and wide areas where production is limited by scarcity of water or nutrients. On Mars, all of these factors that reduce NPP from its theoretical maximum would also occur. However, when constructing a biosphere specifically as a planetary engineering tool, to work in concert with other techniques, it may be possible to optimize it for production. Possibilities include exploiting Mars' greater land/sea ratio; using space reflectors to provide night time illumination and to lengthen growing seasons; and supplementing natural nutrient flows with technology. If by such methods the process can be speeded up by an order of magnitude, then the target oxygen partial pressure might be reached in ~2000 years. At this point, the biosphere would be steered into a state of overall ecological climax, where production and consumption are balanced and the net effect of the biota on the environment is zero. Terraforming would be complete—the only planetary engineering remaining would be that required to maintain the planet's habitability. This issue has been discussed in detail elsewhere [11,50] and appears to be well within the capability of a planetary civilization.

4.3 Outstanding Problems and Prospects

Two problems with terraforming Mars, common to both paradigms, are most often cited. The first is the question of nitrogen. Atmospheric nitrogen is going to have to be available in some minimum quantity to support biological nitrogen fixation and in much larger amounts to act as a buffer gas in a breathable atmosphere (see Table 8). The partial pressure of N_2 in the present Martian atmosphere is a tiny ~0.12 mb and would have to be substantially increased. This small amount is unlikely to represent the total past nitrogen inventory of Mars; by scaling the amount of volatiles per unit mass with that of the Earth, one might expect the planet to have outgassed ~300 mb N_2 [10]. The fate of Mars' nitrogen is not known, and so most terraforming scenarios side-step this issue for want of knowledge that is not yet available. Two extremes are possible—one is that nitrogen has ended up mineralized in the crust as nitrates and the other is that it has been lost to space. If the former situation is true, then nitrate deposits might be mined for their volatiles using methods similar to those proposed for mining carbonates [11,56,59]; if Mars has no more than its present inventory of nitrogen, then importation of the gas from elsewhere in the solar system becomes mandatory and terraforming becomes a much more distant future prospect. However, the need for supplementation of nitrogen from beyond Mars does not necessarily end the dream of a fully habitable planet. The rate of nitrogen importation only needs to keep up with the gradual accumulation of oxygen which might take over $\sim 10^3$–10^4 years to reach breathable levels. Given this amount of time, it is reasonable to suppose that a future civilization with access to the resources of the outer solar system would be able to provide Mars with any missing volatiles.

The other commonly cited problem with the concept of terraforming are the long periods of time that will be required for the completion of planetary engineering projects. The contention goes that a return on initial investments will only be achieved in the remote future and centuries of millennia of patient work may be required before the benefits of terraforming become fully apparent to the local population—thus there will be no motivation to start planetary engineering in the first place. Whilst this argument has considerable merit, it is a viewpoint that arises naturally from the paradigm of terrestrial culture, able until recently to take its life-support system for granted. On Mars, this luxury will not be present and will have to be created. It is often a feature of newly established societies that they are unafraid of upholding certain principles and of entertaining a long-range vision. Perhaps if Martian terraforming can be sub-divided into discrete ~20 year steps, each of which produce some noticeable improvement in the environment, then motivation can be sustained over a longer timespan than suggested by our current experience of macro-engineering projects. History however does show at least one socially motivating force that does transcend the generations, that of religion. These belief systems that promise such things as "eternal life" and a "heaven on Earth" at some indeterminate time in the future are not totally dissimilar to the utopian goals implicit in terraforming. Religions have demonstrated the capability of surviving for thousands of years, harnessing immense amounts of energy and producing ideologies particularly resistant to change. Can religion be shorn of its many negative and destructive aspects and be adapted to the lengthy task of taking terrestrial life out into the universe? It is an open question... and one of the reasons why terraforming is presently easier to study from the scientific, rather than the social stand-point.

Terraforming Mars, therefore, represents a distant, but probably not impossible, dream. However, the prospect of an earthlike Mars holds out such immense advantages for the long term habitability of the planet that the issue will not go away, so long as human interest remains. Whilst approaches to the problem vary, all researchers would agree that current scenarios are no more than possibilities, poorly constrained by imprecise knowledge of Mars and the general phenomenon of planetary habitability. They are mere illustrations of what *might* be done, assuming sustained space settlement becomes a reality, *not* blueprints for what should happen. The present day uses of terraforming as a planetological thought experiment with applications in environmental studies and education amply justify the modest research and negligible funding it currently receives [13].

5. CONCLUSIONS

The dilemma facing the nations of Earth is the maintenance of an existing technological civilization, approaching limits to its growth within a global system. The ultimate goal of *The Case for Mars* is to create a new technological civilization that is also planetary in character and which will be constrained by similar, or more severe, limitations.

The analogy between the future of civilization on both planets is quite striking. Both will only survive if it becomes practical to generate large and sustainable amounts of net energy from non-fossil fuel resources. Both need to maintain, and use wisely, their supplies of raw materials and to develop methods of mineral extraction from sources of low quality. Both must take the issue of their life-support system seriously—in the case of Earth, management of the stressed biosphere is becoming essential; and on Mars, the most desirable aim of colonization will be the creation of an unconstrained biosphere and its subsequent maintenance.

Although there appear to be no fundamental scientific and technological barriers in the way of a sustained settlement of Mars (and space in general), the difficulties and unknowns inherent in the problem should not be underestimated. The future could take humanity either way—constrained to the surface of the Earth forever, or, as the slogan has it, "From Mars to the Stars!" In any event, the thought experiment as to what is involved in the enterprise, from initial visits, to the long-term habitation of Mars, is of use in understanding our present condition. *The Case for Mars*, therefore, goes beyond being a challenge to human curiosity, creativity and ingenuity—it also has a direct relevance to "Case for Earth".

ACKNOWLEDGEMENTS

The author would like to thank Drs. Steve Gillett, Chris McKay, Sandy Smith and Richard Taylor for helpful discussions and Michels/Bollinger Architecture for supplying me with information about the Hyperion Project.

REFERENCES

1. C. R. Stoker, J. M. Moore, R. L. Grossman, And P. J. Boston, "Scientific Program for a Mars Base." in C. P. McKay (Ed.) *The Case for Mars II*, AAS *Science and Technological Series*, Vol. 62, pp. 255-282 (1985).
2. M. H. Carr, "Scientific Objectives of Human Exploration of Mars," in C. R. Stoker (Ed.), *The Case for Mars III*, AAS *Science and Technology Series*, Vol. 74, pp. 267-275 (1989).
3. J. B. Pollack, J. F. Kasting, S. M. Richardson and K. Poliakoff, "The Case for a Wet, Warm Climate on Early Mars," *Icarus*, 71, 203-224, (1987).
4. T. A. Heppenheimer, *Colonies in Space*, Warner Books Inc., New York (1977).
5. A. R. Oberg, "The Grass Roots of the Mars Conference," in P. J. Boston (Ed.), *The Case for Mars*, AAS *Science and Technology Series*, Vol. 57, pp. ix-xii (1984).

6. National Commission on Space, *Pioneering the Space Frontier*, Bantam Books, New York, (1986).

7. M. Nelson, T. L. Burgess, A. Alling, N. Alvarez-Romo, W. F. Dempster, R. L. Walford and J. P. Allen, "Using a Closed Ecological System to Study the Earth's Biosphere," *Bioscience*, 43 (4), 225-236 (1993).

8. J. Allen, *Biosphere 2—The Human Experiment*, Penguin Books, Harmondsworth, UK (1991).

9. "Designing a City for Mars," *The Futurist*, 27 (2), 29-32, (1993).

10. C. P. McKay, O. B. Toon, and J. F. Kasting, "Making Mars Habitable," *Nature*, 352, 489-496 (1991).

11. M. J. Fogg, "A Synergic Approach to Terraforming Mars," *JBIS*, 45, 315-329 (1992).

12. R. H. Haynes and C. P. McKay, "The Implantation of Life on Mars; Feasibility and Motivation," *Adv. Space Res.*, 12, (4)133-(4)140 (1992).

13. M. J. Fogg, "Terraforming: A Review for Environmentalists," *The Environmentalist*, 13, 7-17(1993).

14. B. C. Clark and D. C. van Hart, "The Salts of Mars" *Icarus*, 45, 370-378 (1981).

15. M. H. Carr, "Mars: A Water-Rich Planet?" *Icarus*, 68, 187-216 (1986).

16. B. M. Jakosky and P. R. Christensen, "Are the Viking Lander Sites Representative of the Surface of Mars?" *Icarus*, 66, 125-133 (1986).

17. B. C. Clark, "Chemistry of the Martian Surface: Resources for the Manned Exploration of Mars, "in P. J. Boston (Ed.) *The Case for Mars*, AAS *Science and Technology Series*, Vol. 57, pp. 197-208 (1984).

18. B. C. Clark, "Survival and Prosperity Using Regolith Resources on Mars," *JBIS*, 42, 161-166 (1989).

19. T. R. Meyer and C. P. McKay, "The Atmosphere of Mars—Resources for the Exploration and Settlement of Mars," in P. J. Boston (Ed.), *The Case for Mars*, AAS *Science and Technology Series*, Vol. 57, pp.209-232 (1984).

20. A. Banin, "Mars Soil—A Sterile Regolith or a Medium for Plant Growth?" in C. R. Stoker (Ed.), *The Case for Mars III*, AAS *Science and Technology Series*, Vol. 74, pp. 559-571, (1989).

21. R. C. Boyd, P. T. Thompson and B. C. Clark, "Duricrete and Composites Construction on Mars," in C. R. Stoker (Ed.) *The Case for Mars III*, AAS *Science and Technology Series*, Vol. 74, pp. 539-550, (1989).

22. T. Volk and J. D. Rummel, "The Case for Cellulose Production on Mars, "in C. R. Stoker (Ed.) *The Case for Mars III*, AAS *Science and Technology Series*, Vol.75, pp. 87-94, (1989).

23. J. R. French, "Rocket Propellents from Martian Resources," *JBIS*, 42, 167-170, (1989). Also published as AAS 97-375 in R. M. Zubrin (Ed.) *From Imagination to Reality: Mars Exploration Studies of the Journal of the British Interplanetary Society* (Part II: Base Building, Colonization and Terraformation), AAS *Science and Technology Series*, Vol. 92, pp. 43-51, (1997).

24. A. G. Smith, "Settlers and Metals—Industrial Supplies in a Barren Planetary System," *JBIS*, 209-217, (1982).

25. B. M. Cordell, "A Preliminary Assessment of Martian Natural Resource Potential," in C. P. McKay (Ed.) *The Case for Mars II*, AAS *Science and Technology Series*, Vol. 62, pp.627-640 (1985).

26. P. J. Mouginis-Mark, "Volcano/Ground Ice Interactions in Elysium Planitia, Mars," *Icarus*, 64, 265-284, (1985).

27. S. W. Squyres, D. E. Wilhelms, and A. C. Mosman, "Large-Scale Volcano-Ground Ice Interactions on Mars," *Icarus*, 70, 385-408, (1987).

28. S. S. Nedell, S. W. Squyres and D. W. Andersen, "Origin and Evolution of the Layered Deposits in the Valles Marineris, Mars," *Icarus*, 70, 409-441 (1987).

29. J. M. Goldspiel and S. W. Squyres, "Ancient Aqueous Sedimentation on Mars," *Icarus*, 89, 392-410, (1991).

30. A. G. Smith, personal communication, (1993).

31. T. R. Meyer and C. P. McKay, "The Resources of Mars for Human Settlement," *JBIS*, 42, 147-160 (1989). Also published as AAS 97-373 in R. M. Zubrin (Ed.) *From Imagination to Reality: Mars Exploration Studies of the Journal of the British Interplanetary Society* (Part II: Base Building, Colonization and Terraformation), AAS *Science and Technology Series*, Vol. 92, pp. 3-29, (1997).

32. H. W. Haslach, Jr., "Wind Energy: A Resource for a Human Mission to Mars," *JBIS*, 42, 171-178 (1989). Also published as AAS 97-376 in R. M. Zubrin (Ed.) *From Imagination to Reality: Mars Exploration Studies of the Journal of the British Interplanetary Society* (Part II: Base Building, Colonization and Terraformation), AAS *Science and Technology Series*, Vol. 92, pp. 53-70, (1997).

33. J. A. Angelo, Jr. and D. Buden, "Power Requirements for the Conquest of Mars," in C. P. McKay (Ed.) *The Case for Mars II*, AAS *Science and Technology Series*, Vol. 62, pp. 497-516 (1985).

34. Y. Ishikawa, T. Ohkita and Y. Amemiya, "Mars Habitation 2057; Concept Design of a Mars Settlement in the Year 2057."*JBIS*, 43, 505-512 (1990). Also published as AAS 97-378 in R. M. Zubrin (Ed.) *From Imagination to Reality: Mars Exploration Studies of the Journal of the British Interplanetary Society* (Part II: Base Building, Colonization and Terraformation), AAS *Science and Technology Series*, Vol. 92, pp. 83-97, (1997).

35. R. J. de Young, E. J. Conway, W. E. Meador and D. H. Humes, "Laser Power Transmission Concepts for Martian Applications," in C. R. Stoker (Ed.) *The Case for Mars III*, AAS *Science and Technology Series*, Vol. 74, pp. 697-708 (1989).

36. H. T. Odum and E. C. Odum, *Energy Basis for Man and Nature*, McGraw Hill Book Co. New York, (1981).

37. T. B. Johansson, H. Kelly, A. K. N. Reddy and R. Williams (Eds.) *Renewable Energy—Sources for Fuels and Electricity*, Island Press, Wash. D.C. (1993).

38. J. Mouginis-Mark, "Recent Water Release in the Tharsis Region of Mars," *Icarus*, 84, 362-373 (1990).

39. B. K. Luchitta, "Young Volcanic Deposits in the Valles Marineris, Mars?" *Icarus*, 86, 476-509 (1990).

40. J. B. Plescia, "Recent Flood Lavas in the Elysium Region of Mars," *Icarus*, 88, 465-490, (1990).

41. R. A. Heckman, "Deposition of Thermal Energy by Nuclear Explosives" in *Proceedings of the Third Plowshare Symposium*, TID-7695, pp. 289-304, US Atomic Energy Commission (1964).

42. J. B. Burnham and J. B. Stewart, "The Economics of Plowshare Geothermal Power," in *Engineering With Nuclear Explosives*, CONF.-700101 (Vol. 2) pp.1376-1383, American Nuclear Society & US Atomic Energy Commission (1970).

43. F. T. Powell, "Life-Support System Considerations and Characteristics for a Manned Mars Mission," in C. Stoker, (Ed.), *The Case for Mars III*, AAS *Science and Technology Series*, Vol. 74, pp. 135-156 (1989).

44. P. J. Boston, "Low Pressure Greenhouses and Plants for a Manned Research Station on Mars," *JBIS* 34, 189-192 (1981).

45. R. D. MacElroy, H. P. Klein and M. M. Averner, "Evolution of CELSS for Lunar Bases," in W. W. Mendell (Ed.), *Lunar Bases and Space Activities in the 21st Century*, pp. 623-633, Lunar and Planetary Institute, Houston, (1985).

46. D. T. Smernoff and R. D. MacElroy, "Use of Martian Resources in a Controlled Ecological Life-Support System (CELSS)," *JBIS*, 42, 179-184 (1989). Also published as AAS 97-377 in R. M. Zubrin (Ed.) *From Imagination to Reality: Mars Exploration Studies of the Journal of the British Interplanetary Society* (Part II: Base Building, Colonization and Terraformation), AAS *Science and Technology Series*, Vol. 92, pp. 71-82, (1997).

47. R. L. S. Taylor, "Paraterraforming: The Worldhouse Concept," *JBIS*, 45, 341-352 (1992).

48. "Lean Times for Glasshouse People," *New Scientist*, p.5, 5th Dec. (1992).

49. J. Ramage, *Energy—A Guidebook*, Oxford University Press (1983).

50. M. J. Fogg, "Dynamics of a Terraformed Martian Biosphere," *JBIS*, in press (1993).

51. I. G. Simmons, *Changing the Face of the Earth*, Basil Blackwell Ltd., Oxford (1990).

52. A. Goudie, *The Human Impact on the Natural Environment*, 3rd Ed., Basil Blackweell Ltd., Oxford (1990).

53. J. A. Burns and M. Harwit, "Towards a More Habitable Mars – or – The Coming Martian Spring," *Icarus*, 19, 126-130 (1973).

54. C. Sagan, "Planetary Engineering on Mars" *Icarus*, 20, 513-514 (1973).

55. M. M. Averner and R. D. MacElroy, *On the Habitability of Mars: An Approach to Planetary Ecosynthesis*, NASA SP-414 (1976).

56. M. J. Fogg, "The Creation of an Artificial Dense Martian Atmosphere: A Major Obstacle to the Terraforming of Mars," *JBIS*, 42, 577-582 (1989).

57. R. H. Haynes, "Ecce Ecopoiesis: Playing God on Mars," in D. MacNiven (Ed.) *Moral Expertise*, pp. 161-183, Routledge, London and New York (1990).

58. C. P. McKay, "Does Mars Have Rights? An Approach to the Environmental Ethics of Planetary Engineering," in D. MacNiven (Ed.) *Moral Expertise*, pp.184-197, Routledge, London and New York, (1990).

59. P. Birch, "Terraforming Mars Quickly," *JBIS*, 45, 331-340 (1990).

60. C. Sagan, O. B. Toon and P. J. Gierasch, "Climatic Change on Mars," *Science*, 181, 1045-1049 (1973).

61. F. P. Fanale, J. R. Savail, W. B. Banerdt and S. R. Saunders, "Mars: The Regolith-Atmosphere-Cap System and Climate Change," *Icarus*, 59, 381-407 (1982).

62. A. P. Zent, F. P. Fanale and S. E. Postawko, "Carbon Dioxide: Absorption on Palagonite and Partitioning in the Martian Regolith," *Icarus*, 71, 241-259 (1987).

63. V. R. Baker, R. G. Strom, V. C. Gulick, J. S. Kargel, G. Komatsu, and V. S. Kale, "Ancient Oceans, Ice Sheets, and the Hydrological Cycle on Mars," *Nature*, 352, 589-594 (1991).

64. R. M. Zubrin, "The Outer Solar System and the Human Future," *Ad Astra*, 5 (2), 19-23 (1993).

65. F. Winterberg, *The Physical Principles of Thermonuclear Explosive Devices*, Fusion Energy Foundation, New York (1981).

66. J. B. Pollack and Y. L. Yung, "Origin and Evolution of Planetary Atmospheres," *Ann. Rev. Earth Planet, Sci.*, 8, 425-487 (1980).

67. H. J. Melosh and A. M. Vickery, "Impact Erosion of the Primordial Atmosphere of Mars," *Nature*, 338, 487-489 (1989).

AAS 93-847

PARATERRAFORMING: CONSTRUCTION, ENERGY, ENVIRONMENT AND HAZARD CONTROL STRATEGIES FOR A QUASI-GLOBAL MARTIAN WORLDHOUSE

Richard L. S. Taylor*

The 'Worldhouse' concept of paraterraforming provides a means of creating and maintaining habitable environments on small planets, such as Mars, where indigenous volatiles exist but their inventory is, or may be, insufficient either for the production of a 1 bar gravity-bounded atmosphere of near-terrestrial composition, or for its longer- term geochemical cycle stability. Although, when completed, the deliberately restricted ecospheric environment (DREE) within the quasi-global worldhouse will function close to the point of self-stability, it will be necessary to ensure that the biosystems within the worldhouse mimic closely the biological/geochemical cycles of the Earth, and that the total energy inputs and outputs are closely matched with each other. During the construction phase, energy requirements and the maintenance of metastable biospheres within sub-global worldhouse structural units pose significant problems for determining the best overall campaign strategies. Additionally, consideration must be given to the prevention and circumvention of any potential hazards that might affect the short- and long-term integrity of the worldhouse structure. An attempt is made to present an optimized scenario, the outcome of which should be a DREE system close to a fully Earth-like state of dynamic equilibrium.

1. IS THE HUMAN SETTLEMENT OF MARS A REALISTIC OPTION?

1.1 The Earth Paradigm

The biosphere of the Earth is the paradigm for a 'complete' life-support system. Although conditions within the system undergo periodic and near-cyclic fluctuations, as well as longer term systematic changes, at any given instant the entire bio-geophysical system can be regarded as being stable and self-regulating. The most important aspect of the terrestrial ecosystem is that this essential background of life-support is provided free of charge. There is no overhead that must be paid to maintain the conditions within the system and which must be met by an input from resources exterior to it.

The requirements of life-support systems for inhabited planetary bases, and by extension for permanent human settlement and colonization of Mars, can thus best be assessed in comparison with the terrestrial paradigm. Considered in these terms Mars settlements of the type envisaged by Y. Ishikawa *et al.*,[1] or the Biosphere II experiment Allen and Nelson,[2] must be judged as being incapable of approaching anything like the

* Centre for Extra Mural Studies, Birkbeck College, University of London, U.K.

required degree of self-regulation and stability for indefinitely long periods of habitability. They will demand and need continuous high level intervention and control to hold their internal environments within limits acceptable for even a small range of living biota.

The closest approach we can envisage to the terrestrial life support paradigm is a completely terraformed planet. Ideally this should possess a stable self-regulating ecosphere in which the prevailing conditions are all within the range of human physiological tolerance over a significant percentage of the planetary surface. This in turn suggests that the rotational and orbital dynamics and characteristics may need also to be broadly similar to those of the Earth for which the majority of terrestrial organisms have been fully adapted by evolution. The importance of such gross parameters as the diurnal and seasonal cycles play in the maintenance of the self-stable and self-regulating biosystem of the Earth is often overlooked.

These geophysical/solar cycles are so obvious that it is easy to ignore the fact that all aspects and phases of the biological activity, growth-and-decay, and so on, are able to go on at the same time in the different zones of a single environmental system that results from the physical nature of the Earth and its rotational and orbital dynamics. Because of these factors the life-support system of the Earth always contains countervailing cycles which in global terms amount to a complete feedback system. Such a regime is not possible in the contained biosphere of a small space colony but would come into existence automatically on a terraformed Mars.

1.2 Terraforming Mars - Disincentives

Assuming Mars is capable of being terraformed to match closely the composition and surface of pressure of the terrestrial atmosphere some of the unchangeable gross physical and dynamic characteristics of the planet will prevent a zero-cost life support system being attained. Even with the maximum percentage of CO_2 that can be tolerated by humans and higher animals, the 'greenhouse' temperature increment will be insufficient to increase the global temperature by the required amount. Technological intervention, on a continuous, or periodic basis will therefore by necessary. Thus the great advantage that the terraformed planet can be left to 'run itself' giving inhabitants an 'automatic background of air conditioning, climate regulation and the production of food,' Lovelock,[3] is not likely to be true for Mars. Terraforming Mars will be restricted to bringing about environmental changes within the constraints of the planet's existing inventories and a measure of continuing technological intervention will be required to maintain the desired equilibrium. Left to itself the resulting Martian ecosystem is more likely to be suitable for micro-organisms that for higher forms of life.

Even if an almost Earth-like outcome could be assured the purely biological terraforming process persuasively advocated by McKay *et al.*[4] is faced by probable time-scales for terraforming Mars completely which are likely to exceed the unbroken life-span of any previously known political system or civilization. Even technologically accelerated and synergistic terraforming scenarios, such as proposed by Fogg[5,6] face a politically significant time barrier. Those wishing to terraform Mars have to convince a funding agency that a project of potentially millennial duration, where the full economic, technical and manpower resource requirements must be committed and expended before habitability is achieved is worth undertaking. The extended timescale necessary to complete the transformation process means that the start of pay-back on the total investment will be very long delayed, Figure 1. A cost/benefit analysis cast in conventional economic terms would make terraforming appear to be a very unattractive proposition by those asked to fund the project. This is because they will be unable to perceive any tangible benefits for themselves. Hence no economic incentive will exist for them and they may be very reluctant to enter upon, and then sustain, an enterprise whose hoped for benefits would begin to arise only many generations into the future, if at all.

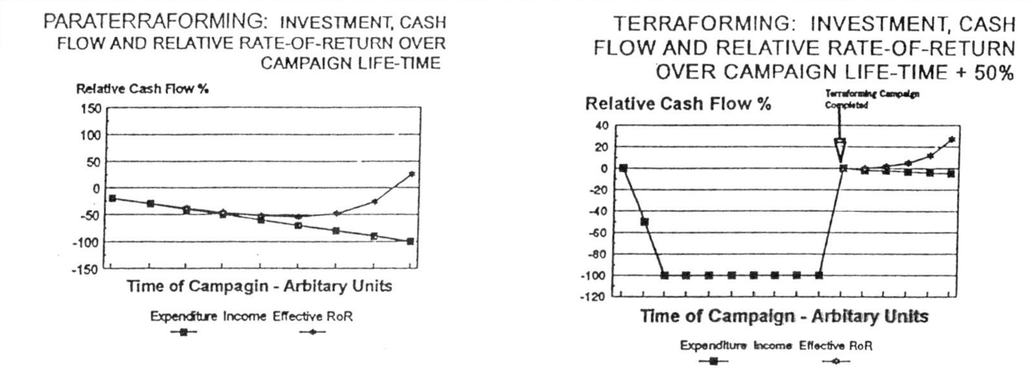

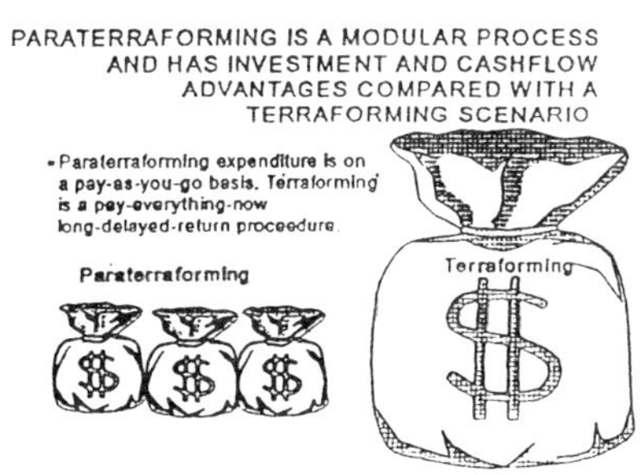

Fig.1. Habitability on a terraformed Mars is only achieved when the transformation process is completed and all the financial, technical and manpower rescources have been deployed. Paraterraforming is a modular process and some degree of habitability is achieved at an early stage of worldhouse construction. Because of this modularity investment is on a stage-by-stage basis. Additionally the partly completed worldhouse is able to provide local income and in some measure the process can be partially self-financing.

Aside from such considerations it is by no means certain as yet, that sufficient quantities of any of the volatiles necessary for the production of an atmosphere of terrestrial composition and 1 bar surface pressure exist currently on Mars. This would make it impossible to terraform the planet to a state where it is habitable for humans and advanced forms of life.

1.3 Human Habitability as the Sole Justification for Transforming Mars

Does this mean that the permanent human settlement of Mars is not a realistic option? In this paper the contrary view is expressed. I believe that there is no justification whatsoever for attempting to emplace a biosystem upon Mars unless the overriding purpose is to make it habitable for human beings and the complete suite of terrestrial biota essential for the support of human life. The justification and the end point of any undertaking to bring Mars alive must be the establishment of a fully Earth-like humanly habitable environment, producing a sentient planet. The questions as to what is an acceptable outcome to terraforming and the ethical considerations of the type, should we colonize Mars if. . . . are reviewed in Tables 1 and 2.

WHAT IS AN ACCEPTABLE OUTCOME TO TERRAFORMING?

1. ***A planet-wide microbial ecosystem*** - **'green gunge' - thin atmosphere and cold climate. Achieved by 'gentle' biological methods. A neo-creational attitude advocated by those content with conducting a giant experiment of bringing Mars to life but not necessarily making the planet habitable for humans.**

2. ***A humanly habitable ecosystem*** - at near terrestrial stability and climatic conditions, but achieved by a combination of biological, chemical, physical and engineering processes - the aided synergistic approach. Time scale is long, 10's to 100's of thousands of years, but much shorter than for case (1.) above.

3. *A sentient planet* - the sole justification for establishing an ecosystem on Mars or any other planet is seen as the establishment of a fully Earth-like humanly habitable environment as rapidly as possible and at the lowest possible rate of expenditure and the lowest possible economic cost. The time scale is of paramount importance as this will be decisive in raising the political will to turn terraforming from a possibility into a reality

Table 1. There are three possible scenarios for bringing Mars alive and each has its own time-scale and outcome. The purely biological method (1) would be ideal if a humanly habitable outcome could be guarantied and if the time from start to completion was not so long. The technologically assisted synergistic route (2) has many attractions but is also a lengthy process which cannot completely guarantee human habitability. The last outcome (3) is the result of a technologically intense approach. This encompasses the worldhouse concept of paraterraforming as well as other methods of planetary engineering.

The determination that Mars must be made humanly habitable implies a willingness to make use of whatever technological solutions are available to provide an environment in which the conditions necessary for the establishment and perpetuation of a planetary civilization can exist and survive beyond the confines of the Earth. The worldhouse concept of paraterraforming provides a method of creating such off-Earth habitable environments on volatiles deficient planets, as may be the case for Mars, and as a modular process it may make it possible to circumvent at least some of the political and economic difficulties confronting any campaign for the permanent settlement of Mars.

ETHICAL CONSIDERATIONS

- Should we colonize Mars if . . . . . ?

1. A planet-wide microbial ecosystem is discovered on the planet
2. A handful of microbial species are found with planet-wide distribution
3. One or two species of micro organisms are discovered in one or two separate and/or widely spaced locations.

From our knowledge of the basic requirements of life and the sterilizing nature of the surface chemistry of Mars none of these possibilities is in reality even remotely probable. If, however, we make the assumption that one of these possibilities is a realistic probability only (1) can be considered as constituting a fundamental. objection to human interference in the martian environment or occupancy of the planet.

Possibilities (2) & (3) do not constitute an ethical barrier to the human use of Mars. Both cases would call for the corralling or captive maintenance of the organisms and in the case of (3) the possible preservation of specific locations and their micro-climates where they exist.

Cases (2) & (3) can be considered as examples of a failed or stalled biological systems and allow a move over microbe' approach to terraforming Mars.

Table 2. In logical terms there are no ethical barriers to the conversion and colonization of planets devoid of life. Of the three possibilities listed here only (1) need raise ethical considerations. If through human agency terrestrial life can be made to flourish on worlds with stalled or failed biological systems there are no valid reasons why colonization should not take place as it accords with natural evolutionary principles.

2. MARTIAN INVENTORIES: PARATERRAFORMING AND THE WORLDHOUSE SOLUTION.

Paraterraforming and the "worldhouse" concept have been described elsewhere by Taylor.[7] Whereas terraforming offers the ideal solution to the establishment of a self-regulating and self-sustaining planetary life-support system on Mars it can only produce an environment suitable for human life if the volatiles inventory is sufficient for the purpose. Paraterraforming, Table 3, differs from terraforming in that the intention is to achieve conditions suitable for human habitability even though the transformation of the planet to a fully Earth-like gravity-bounded state may either be impossible because of materials limitations or is not achievable on a realistic time-scale. Instead the aim of paraterraforming is to construct an almost immediately inhabitable Deliberately Restricted Ecospheric Environment, or DREE, system.

PARATERRAFORMING: THE WORLDHOUSE CONCEPT
ADVANTAGES:
* Much less gas needed.
* Modular approach, no need to terraform all Mars simultaneously.
* Near-natural Earth-like environment attainable.
* Little new technology required.
* No-off planet rescources required.
* No need for relocation of settlers.
* Roof can assist greenhouse effect.
* Cheaper than conventional terraforming.
* Could form an intermediate stage prior to the ability to import volatiles for full terraforming.
DISADVANTAGES:
* Continuous [possibly high-level] maintenance required.
* Vunerable to fire, explosion, geological disturbance, meteor impact.
* Low mass/volume of the enclosed ecosphere requires rapid bio-geophysical recycling.

Table 3: As a technologically dependent environmental system paraterraforming must be judged to be a less than ideal solution to the problems of making Mars habitable the Worldhouse concept may be the only way of emplacing a human civilization upon Mars if the planet is now as deficient in volatiles as some planetary scientists estimate.

A DREE system is essentially a contained biosphere in which masses, mass-flows, and volumes are all limited by choice. The containing structure takes the form of a quasi-global sealed roof and is termed a 'worldhouse.' This allows the creation of habitable environments on planetary bodies that are volatiles poor and on which it would not be possible to create a gravity-bounded 1 bar atmosphere. This is particularly important in the case of Mars where on the basis of the least optimistic available data it appears to be possible that insufficient volatiles, in particular nitrogen,[8] Figure 2, may be present to allow for the production of an atmosphere of the desired surface pressure and composition that would permit the natural survival of humans and advanced forms of terrestrial life on the fully terraformed planet.

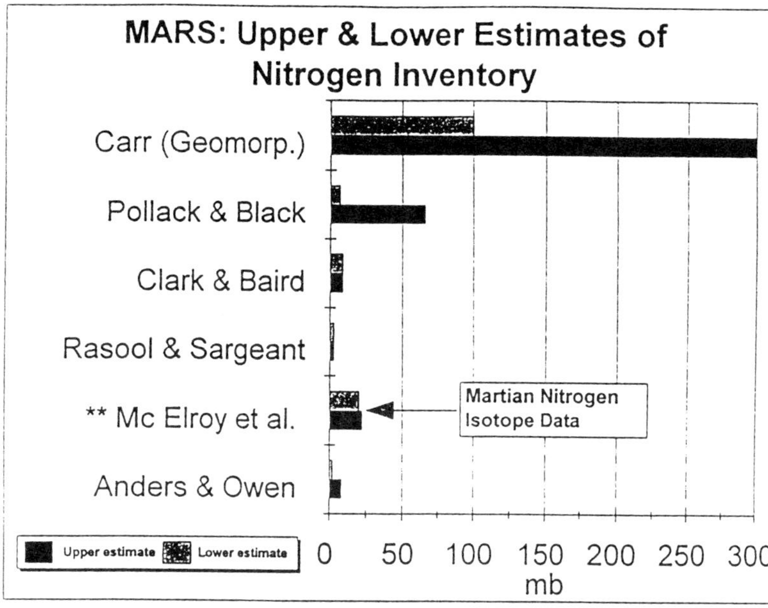

Fig.2. Only the geomorphological estimate of Carr appears to offer the possibility that sufficient nitrogen is available on Mars to form a terraforned atmosphere of terrestrial composition and surface pressure. If any of the more pessimistic estimates are correct even a worldhouse roof would have to be less than 3 km above the surface of the planet.

Figure 3 and Table 4 show the atmospheric masses necessary to produce an atmosphere of ~1 bar surface pressure on a terraformed Mars, or a contained pressure atmosphere within a paraterraformed Mars worldhouse of different internal heights of 1, 2 and 3 km. These are compared with the mass of the Earth's atmosphere. The ability of paraterraforming to make use of a very limited volatiles mass inventory is apparent. This ability to constrain the required mass of the atmospheric gases means that far smaller volatile reservoirs become capable of providing enough material to make the planet habitable. A 100% CO_2 filling to 1 bar of a 3 km high Martian worldhouse would require the release of ~194 mb which is within the anticipated quantity adsorbed in the Martian regolith.[9] to obtain the required 20% oxygen for a 3 km worldhouse atmosphere only a little more than 30 mb of CO_2 would be required from a Martian reservoir. The major problem lies in the location of a N_2 resource, but here too the required quantity is only the equivalent of a ~90 mb reservoir.

However, what is not immediately apparent is the fact a 1 bar surface pressure terraformed Martian atmosphere of terrestrial composition cannot produce the temperature increment required to make Mars truly habitable for humans. Even a pure CO_2 atmosphere at 1 bar surface pressure, quite unsuitable for anything but a limited range of microbial biota, would leave Mars a relatively cold place. McKay[4] considers that the additional temperature increment may be attainable by releasing quantities of CFC's or other yet to be developed super-greenhouse gases into the terraformed Martian atmosphere.

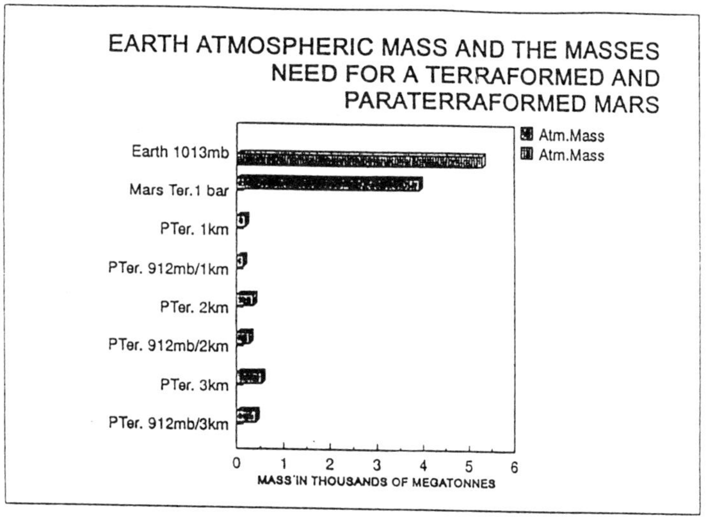

Fig.3. Because of the low surface gravity of Mars the mass of gas needed to form a fully terraformed atmosphere of terrestrial composition and 1 bar surface pressure is approximately 75% of the mass of the Earth's atmosphere. The inventory on the present Mars may be insufficient for the purpose. The ability of paraterrforming to make use of a very limited volatiles mass inventory is apparent.

ATMOSPHERIC MASSES NEEDED FOR MARTIAN TERRAFORMING AND PARRATERRAFORMING

	Atm.mass/10^{18}kg	% of terraformed Atm.
Terraformed (1 bar)	3.92	100
Paraterraformed (1 bar air)		
Worldhouse height: 1km	0.18	4.6
2km	0.35	8.9
3km	0.52	13.3
Paraterraformed (912 mb air, 84% coverage)		
Worldhouse height: 1km	0.14	3.6
2km	0.27	6.9
3km	0.4	10.2

Table 4. The ability of paraterrforming to utilize a very limited volatiles inventory to produce an ecosystem capable of supporting human and terrestrial biota can be seen from the above Table. The down-side of this ability is the need for the rates of recycling to be increased aproximately as the reciprocal of the mass.

The problem is that the quantities required are four to six orders of magnitude greater than the quantities of CFC's we currently produce on Earth. Combined with the shorter life-time, CFC's are expected to have in the upper Martian atmosphere this method of producing the additional temperature increment would appear to require an on-going manufacturing process of considerable magnitude and cost. One way around this problem might be to use perfluoro compounds, Ravishankara *et al.*,[10] which have an order of magnitude greater life-time in the Earth's atmosphere than CFC compounds. Another possibility proposed by McKay is that it might eventually be possible to genetically engineer a bacterium to produce the desired CFC greenhouse gases as a metabolic by-product of its living processes. At the moment this must be regarded as highly speculative but the possible use of micro-organisms as a means of carrying out at least some of the processes necessary for terraforming should not be overlooked. Friedmann *et al.*[11] have recently discussed the release of bound CO_2 from carbonate rock by the action of Matteia Sp. cyanobacteria, a microbiological process that might be applicable for the first stage of terraforming Mars.

An alternative to the use of large quantities of relatively short-lived greenhouse gases is to increase the amount of sunlight falling on the planet. This can be done by the use of an orbiting annular mirror or a solletta as described by Birch.[12] Paraterraforming presents a different temperature problem. Quite a substantial temperature increment can be produced by a single glazed roof and the use of double or triple anti-reflection and specific anti-transmission coated glazing materials may provide a greater temperature increment than required. Table 5 indicates how the worldhouse temperature increment is produced.

MARS:Temperature

The Required Temperature Increment of 55 to 60K is Produced by:

1. Control of transmission of radiation by the roof. Gain ~50K.

2. Control of surface albedo by biological planting stratagies - eg. Coniferous forests (albedo 0.12)

3. Addition of trace greenhouse gases (CFC's) to the contained atmosphere and/or in the roof voids.

Table 5. A worldhouse allows the rate of the input solar energy and its re-emission to be controlled a relatively direct and simple manner. Both positive and negative temperature increments can be produced raising or lowering the temperature of the internal environment. On Mars only the positive greenhouse effect is required.

2.1 Contained Biospheres and the Worldhouse DREE System Compared.

To date all the contained biospheres that exist (Mir, Biosphere II, etc.) have existed, or are planned, are single-state, or mono-polar, environmental systems. All these systems are of very small volume, and total mass, and the associated mass/energy flows are small in absolute terms. A single-state or mono-polar environmental system can be defined as one that does not contain two completely opposite physical, climatic, or biological sets of conditions simultaneously. Although in Biosphere II different biomes have been created and maintained adjacent to each other the environment is still mono-polar in that it is day or night, and the same season of the year throughout the entire structure. Opposing parts of these natural terrestrial feedback cycled do not occur within Biosphere II.

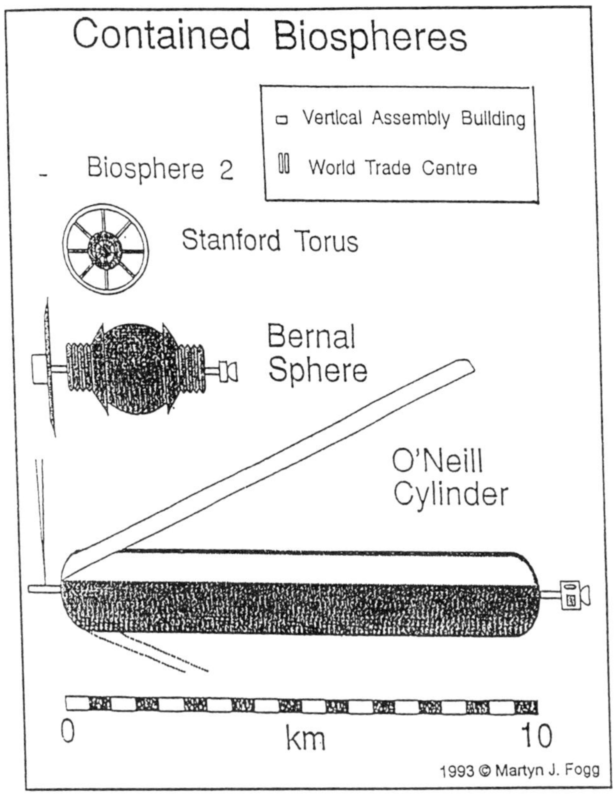

Fig.4. Biosphere II and three proposed contained biosphere space colonies are shown compared with the Apollo Missions Vertical Assembly Building and the World Trade Center, the world's tallest fully inhabited building. The small size and single-state nature of these environmental systems can be compared with the Mars worldhouse shown in Figure 5.

Small size and the single-state condition holds true to date for all planned permanent self-contained planetary bases as well as for space colonies like the O'Neill Cylinder, the largest of the single-state contained biospheres yet proposed. In terms of mass, volume and thermodynamics these enclosures are little more than enlarged spacecraft and the difficulties of trying to make such low inertia systems stable and self-regulating are so great that a strategy of active intervention has to be adopted. There is little evidence for natural damping within such small contained biospheres. Corrective intervention has been necessary with Biosphere II for example. The comparative sizes of contained biospheres ranging from Biosphere II to the O'Neill Cylinder are shown in diagrammatic form in Figure 4 (courtesy of M. J. Fogg).

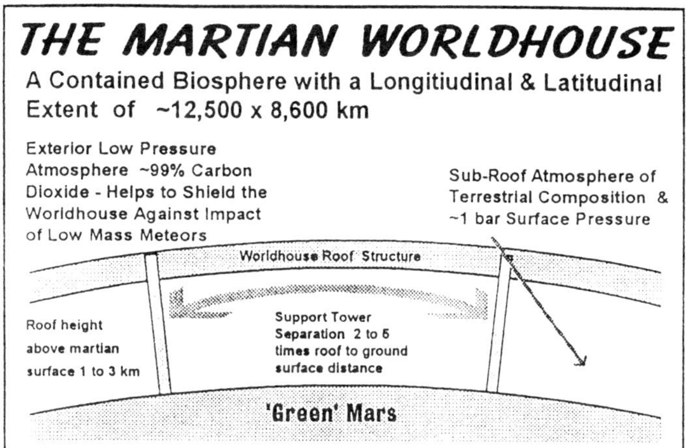

Fig.5. Although a martian worldhouse is a contained biosphere its sheer size and geometrical configuration cause contrary sets of physical and biological conditions to exist simultaneously within it. Because of this the deliberately restricted ecospheric environment (DREE) system of a paraterraformed world cannot be compared with any other type of currently existing or proposed contained biosphere. For example Biosphere 2 and space colonies like the O'Neill Cylinder contain single state biosystems with artificially created and maintained micro-biomes. The DREE system on a paraterraformed Mars is more closely comparable with the frree environment of the Earth in terms of the behaviour of its biosystem and will differ little from conditions of a fully terraformed Mars.

A completed Mars worldhouse, Figure 5, on the other hand, encompasses virtually the entire surface of the planet having an approximate longitudinal and latitudinal extent of 12,500 x 8,600 km and a mean internal height of 1 to 3 km. A mature Mars worldhouse will contain a very large biospheric mass (the enclosed biospheric mass/volume will lie somewhere between 12 and 24% of the terrestrial value). The quasi-global nature of the worldhouse ensures that the system is multi-polar and that at any one time half the enclosed biosphere will be in light and half in darkness, and, except at times of equinox, there will be marked seasonal difference between the northern and southern hemispheres, Figure 6. Thus like the Earth's biosphere the Mars worldhouse environment will support divergent micro-biomes in different zones and locations on the planet's surface, Figure 7.

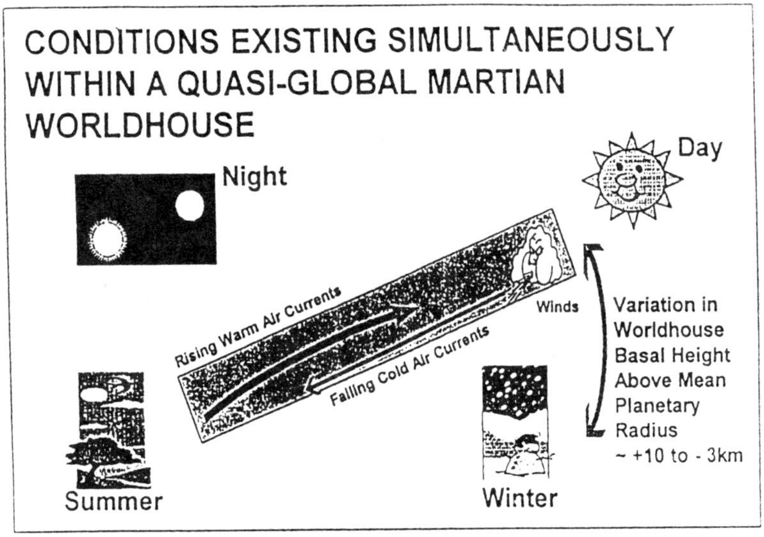

Fig.6. The quasi-global nature of the Mars Worldhouse ensures that the enclosed ecosystem is naturally a multi-polar and not a single state environment. Day and night and the cycle of the seasons are experienced within the worldhouse just as they are on Earth allowing a complex network of micro-biomes to come into existence and flourish.

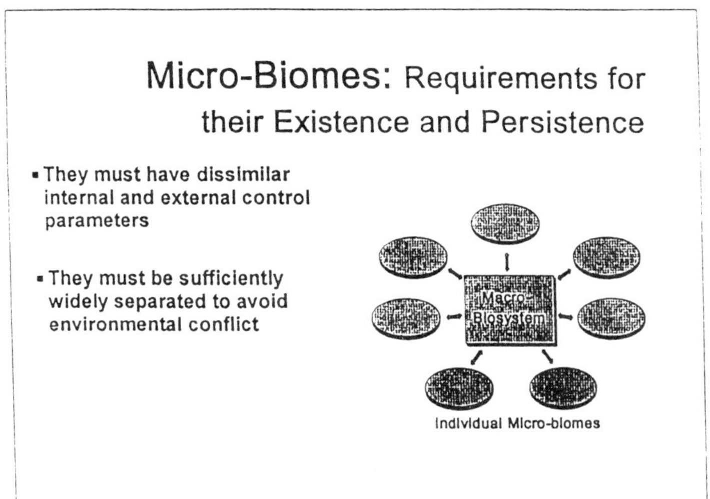

Fig.7. Whereas small contained bispheres can only maintain micro-biomes artificially through a process of continuous intervention and control they can exist naturally on a terraformed Mars or in the quasi-global Mars Worldhouse.

Although the atmosphere contained within a worldhouse is nominally at a uniform pressure the irregular and fluctuating distribution of the radiant energy input and the variation of the basal altitude of the worldhouse above the mean planetary radius, will give rise to the establishment and persistence of local differences of barometric pressure. The non-uniform surface friction encountered by the resulting atmospheric flows can be expected to give rise to quite marked weather systems characterized by diurnal and seasonal wind flow patterns giving a range of climatic variation, thermal and mass flows that would not be manifest, or realizable, within small single-state biospheres.. Moreover, in the completed worldhouse it will be possible to extract power from these atmospheric mass flows. Indeed, it will be both desirable and necessary to do so in order to keep the interior stresses on the worldhouse structure within acceptable levels. Solar driven energy flows within a Mars worldhouse are summarized schematically in Figure 8.

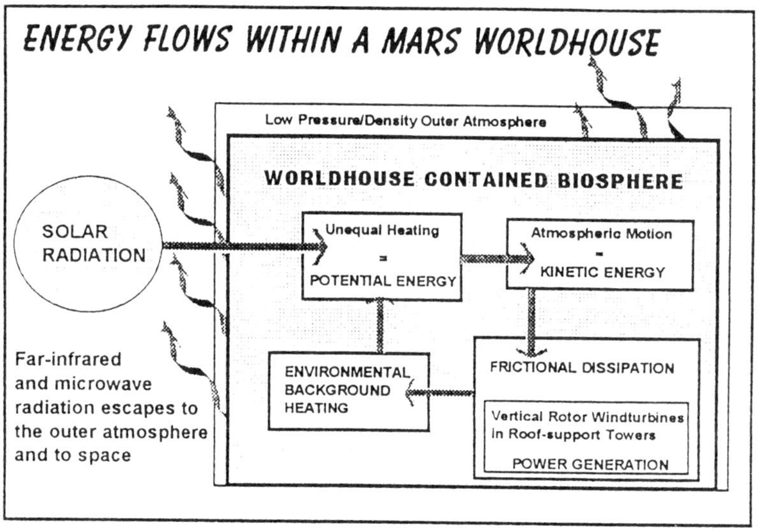

Fig.8. The solar input and resultant energy flows within a Mars worldhouse are depicted schematically in the above diagram. The use of CTSPG (Central Tower Solar Power Generators) on exposed highland regions exterior to the worldhouse environment will provide an additional energy input which is not shown in this diagram.

3. THE COMPLETED WORLDHOUSE STRUCTURE

A paraterraformed Mars will comprise an exterior atmosphere, (approximating to the original 6 mb of CO_2 plus any surplus quantities of the gas not immediately required as an input to the ecosystem or for industrial purposes) exposed highlands with altitudes > 7 km uncovered polar latitudes and certain canyon floor regions amounting to a little less than ~20% of the total surface area of the planet. The remaining ~80% of the Martian surface will be under the worldhouse roof set at an preferred average height above the surface of ~1 to 3 km. The completed worldhouse structure will consist of inhabited tower roof support units. Together with uninhabited support towers, some of which will incorporate vertical axis wind-turbines, they will be distributed across the Martian surface in a basic hexagonal cell pattern.

The roof, most probably a cable-braced and supported spaceframe structure, will be a double or triple glazed structure at a nominally constant height above the surface of the planet. The roof level will not take account of small undulations in the terrain which do not exceed 20% of the roof to ground clearance height and it will not actually follow the general gradient of the Martian surface but will be held normal to the planetary radius with periodic step-wise increments in the roof height level. This avoids a build-up of unresolvable gravity induced directional loads in the structure. The outer limits of the worldhouse will terminate in a cable-braced or cable-stayed wall which will allow air-lock entry and egress to those regions of the planet that are exterior to the worldhouse which form part of the operation of the total planetary energy/industrial system.

The contained atmosphere of terrestrial composition will be at a nominal pressure of ~912 mb equal to the terrestrial atmospheric pressure at the average height above sea-level of inhabited land. Enclosed regions where the elevation is below the mean planetary radius can be expected to form small seas if sufficient water is available for this purpose.

3.1 The Contained Biosphere

Because of the quasi-global nature of the worldhouse the distribution of solar energy into the ecosystem will be Earth-like in all respects. There are, however, three principal differences in the Martian situation that may affect the functional efficiency and equilibrium of the biosphere. They are the lower average intensity of illumination, the greater asymmetry in the duration of the seasons and the associated greater range of variation in seasonal temperatures, and the average temperature. As in the case of a terraformed Mars the function of the worldhouse biosphere will be brought as close to a condition of self-regulation and stability by biological means. However, on Mars neither a worldhouse or terraformed environment can operate in the manner of the bio-geophysical system of the Earth. So far as is known there is no equivalent of many, or perhaps any, of the geochemical cycles which play a vital part in ensuring the long-term stability of the terrestrial biosystem. There is for example no closed CO_2 + rock cycle and probably no major volcanic recycling of other volatiles. Because of these differences in geologic processes a degree of geochemical intervention on a small but continuing scale will be required to prevent serious longer term deviation from the desired path of environmental equilibrium. A comprehensive discussion of some of these issues can be found in a paper by Fogg.[13]

4. PLANNING AND STRATEGIC CONSIDERATIONS

The concept of paraterraforming is predicated upon the idea that it is possible to construct an effectively gas-tight roof 1 to 3 km above the surface of the chosen planet. The resulting worldhouse enclosure has to meet some stringent and rather unusual structural demands and loads: see Table 6. The idea that high-rise buildings can provide a means of constructing and supporting a near-planet-wide roof, a Worldhouse, falls well within current civil and structural engineering capabilities as well as materials technology. It is an available option which can be developed to construct off-Earth habitable environments without the need for any fundamentally new science although it will require some areas of present technology to be extended to new limits.

WORLDHOUSE STRUCTURAL REQUIREMENTS
A Worldhouse structure must:
(i) be capable of sustaining a near 100% gas-tight roof both with or without a ~1 bar pressure atmosphere beneath it. The mass of the structure to be kept to the minimum consistent with the aim of achieving near zero-fail-rate structural integrity;
(ii) have a roof structure which offers the minimum obstruction to incoming radiation and which is as near as possible to being 100% transparent in all the required regions of the solar spectrum. Panels must be cleanable and replaceable without allowing major atmospheric leakage;
(iii) have a roof structure permitting effectively unrestricted passage across the surface of the planet and must also allow for the possible use of aerial transportation systems within the worldhouse;
(iv) permit some of the roof support towers to be used to provide secure and self-safe accommodation modules. These will constitute martian 'tower-cities' and thus help to reduce the amount of structural building work and materials consumption;
(v) allow the use of specific support towers to provide access points to the exterior environment and vehicular access to space;
(vi) permit the use of areas of the planetary surface not within the worldhouse to serve as materials reserves and output sinks contributing to the overall operation and control of the metastable habitat system. These external areas will be used also to perform mineral processing and any toxic or pollutant producing manufacturing operations;
(vii) be configured in such a way that the enclosed environment is not totally vulnerable to impact, explosion, or fire damage.

Table 6. Some of the structural requirements of a worldhouse are quite different from those required in a conventional terrestrial building. When filled with a 1 bar pressure atmosphere the roof has to be held down rather than supported as is the case during construction.

At the present time the tallest terrestrial buildings are less than 0.6 km high,[14] Table 7, but from as far back as 1965 construction materials and building, structural and civil engineering technology have been sufficiently powerful to permit the erection of buildings ~3.5 km high here on Earth. Thus from the technical viewpoint the erection of a whole series of 3.5 km high roof support towers on Mars is not a difficult technological problem but one of logistics. It is a matter of devising the most appropriate and cost effective strategies to produce, in the shortest possible time, the desired result of a quasi-global habitat capable of supporting human life under 'normal' terrestrial environmental conditions.

NOTABLE EXISTING HIGH-RISE BUILDINGS COMPARED WITH THREE PROPOSED TALLER TERRESTRIAL STRUCTURES AND THE MARS WORLDHOUSE ROOF SUPPORT TOWERS.		
Building and Location	**Year**	**Height m.**
Existing Buildings:		
Chrysler Building. New York, USA.	1930	318
Empire State Building, New York, USA.	1931	381
John Hancock Center, Chicago, USA. (*World's tallest exterior frame structure*)	1968	344
World Trade Center, New York, USA.	1973	417
Sears Tower, Chicago, USA. (*World's tallest fully inhabited building*)	1974	443
CN Tower, Toronto, Canada. (*World's tallest building*)	1975	553
Proposed New Buildings:	***Planned for:***	
Merry Hill Tower, Dudley, UK.	1995/96	610
Millennium Tower, Tokyo, Japan.	2001	800
Aeropolis 2001, a 'Vertical City' proposed by Ohbayashi, Japan.	2001	2000
Mars Worldhouse Roof Support Tower	2099 ?	3500

Table 7. Existing terrestrial high-rise buildings are only about 15% of the height that can be built with existing materials and technology. A 3.5 km Mars Worldhouse support-tower is about equivalent to a 1.5 km high terrestrial building in terms of load and stress levels.

4.1 Construction, Materials and Energy Strategies

Apart from the initial input of supplies, equipment and technological knowledge from Earth the entire paraterraforming campaign will be conducted on the basis of using indigenous Martian resources. Of these resources ensuring the availability of sufficient energy will be the most difficult task, Table 8. Otherwise only essential high-tech items or products that are needed only in quantities that are too small to justify manufacture on Mars will be resourced from Earth.

ENERGY: **The Major Difficulty is the Absence of Fossil Fuel Resources on MARS.** **All Available Martian Energy is a Real-Time Resource.** **Constructional Phase will require Nuclear, SPS, and ground based CTSPG Systems.** **When in operation the Worldhouse Energy requirements will come frcm Nuclear, Biological and Meteorological Resources.**

Table 8. Energy for terraforming or paraterraforming Mars provides would-be planetary engineers with significant difficulties. All available martian energy is a real time resource. Small scale nuclear power systems, solar power satellites, and central tower solar power generator systems will provide the energy required during construction. Exploitation of the meteorological mass flows, winds, within

In proposing two basic outline strategies it is assumed that complete geologic, mineral and seismic surveys will have been made and that the condition of the Martian subsoil, particularly in relation to the extent a depth of any permafrost layer, will have been accurately determined. Display 1 shows in outline two possible types of construction, materials and energy strategy - Type I Slow-track Modular, and Type II Fast-track Global. In the case of the Type I strategy the initial financial input from the Earth is moderate, and the subsequent partial financing of the manpower requirement will be less than 100%. The process of paraterraforming the entire planet will be done at a rate and as determined by the permanent settlers—the Martians—themselves. The Type II strategy will involve a far larger initial investment and the commitment of far larger manpower resources. The continuing costs throughout the campaign will also be greater. However, the advantages of the fast track approach are that a quasi-global extent can be planned and achieved before 100% of the planned surface coverage is achieved. This has the major advantage of allowing the closure of the bio-climatic environmental cycles at an early stage of the establishment of the biosphere, something that cannot be achieved early on in the Type I non-global modular approach.

4.2 Establishment of the Biosphere and Ecosystem Development

Display II sets possible options for Type I and II strategies. Efficient and maximum possible photosynthesis to capture incoming solar radiation is vital to the establishment of a long lasting near-self-stable ecosystem on a terraformed or paraterraformed Mars. Intensity of radiation is the key factor in obtaining satisfactory plant photosynthesis, and hence growth performance. However, Mars receives a mean flux only about half that of the Earth. The problem of light intensity is not completely intractable and there is evidence that the terrestrial plant gene pool contains sufficient variation for us to develop high growth lower light intensity varieties. The greater length of the Martian seasons, and the marked asymmetry between seasons of the northern and southern hemispheres may be more difficult to solve for there is nothing already existing in the terrestrial situation which could have selected for an adaptability for seasons of such asymmetry or length. The problem is likely to be more serious for long lived species, like trees, which have an essential part to play in the balance of a terraformed or paraterraformed environment.

Temperature is also an important growth factor. For most terrestrial plants over a limited range of temperature an increment of 10°C doubles the rate of CO_2 uptake. On Earth many plants benefit from higher CO_2 concentration than the 300 ppm atmosphere but above certain levels (2000 ppm) further increase in CO_2 concentration will in some species produce no further benefit and in some cases adverse effects. On Mars even under the same temperature conditions the ability of plants to take up CO_2 will be lower than on Earth. This is a result of the reduced irradiance and is an example of the principle of limiting factors as conceived by Blackman and quoted by Mastalerz,[15] Figures 9 and 10. Many of the terraforming proposals have failed to appreciate that for higher plants the acceptable levels of CO_2 are far lower than they have supposed.

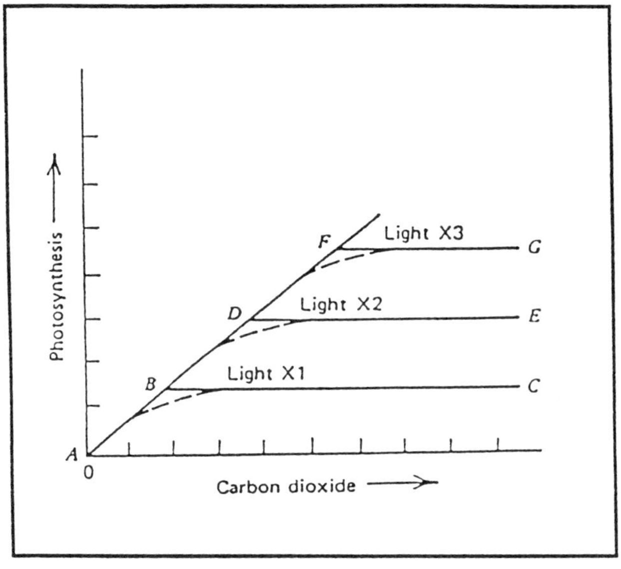

Fig.9. The principal of a limiting factor as conceived by Blackman (After Mastalerz) 1977

A similar misapprehension appears to exist with regard to the minimum level of atmospheric oxygen and the partial pressure of oxygen in the Martian soil that is necessary to permit the growth of higher plants and trees. On Earth the root systems of many trees and higher plants only function satisfactorily if the partial pressure of oxygen in the soil is above a minimum of ~27mb. A majority of trees and higher plants only achieve satisfactory root growth and extraction of vital minerals from the soil through reverse osmosis, if the partial pressure of O_2 is ~50 mb. Both the atmospheric CO_2 and soil O_2 parameters are to a large extent dependent of the exact nature of the soil microbiota and the level of competition and/or interdependence between the organisms in the soil and the higher plants growing in it. Diaz, Grime *et al.*[16]

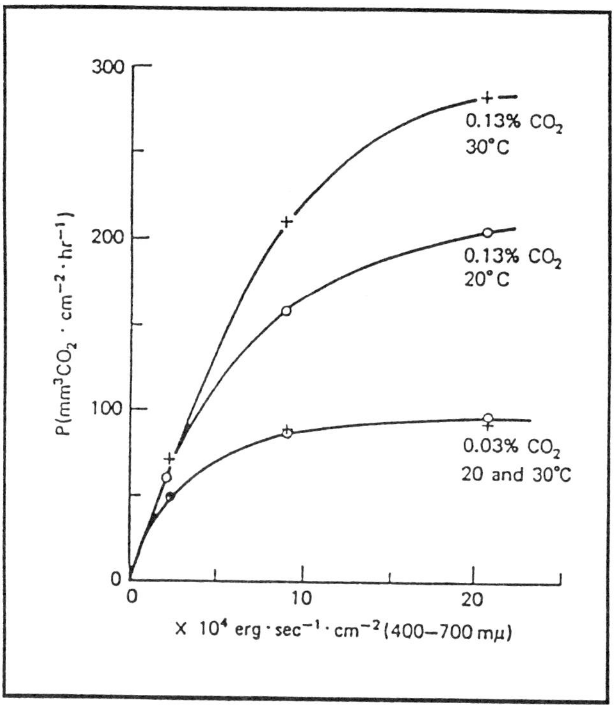

Fig.10. Photosynthesis of a cucumber leaf in relation to radiant flux density and temperature at limiting (0.03%) and a saturating (0.13%) CO_2 concentration. Note the close correspondence with Blackman's proposed principal of a limiting factor. (Gaastra 1959, from Mastalerz 1977)

The need for rapid development of soils on Mars is a further important aspect of the establishment of a worldhouse biosphere. Natural soils can be slow to develop under terrestrial conditions but some of the techniques and procedures that have been developed in the study of restoration ecology will find application in the reclamation of Mars for life. There are three types of higher plants which will play a key role in the establishment of a near-stable largely self-regulating ecosystem. They are forest trees, high photosynthetic yield plants, and nitrogen fixing plants. The species selected from within each of these type will be chosen for any additional characteristics that make them better adapted to live under the conditions found on Mars, and for the ability to thrive in very poor soils.

4.2.1 Forest Trees

The Type II strategic approach will call for the early establishment of high latitude forest cover in the northern and southern hemispheres. These forests will eventually be made up of a mixture of non-deciduous coniferous trees and deciduous hardwoods. They will play an important part in binding carbon and in raising the average temperature of the high latitude zones of the planet. The initial species to be selected however, will need to be able to maintain a high photosynthetic yield under low light conditions, be drought tolerant and able to grow rapidly on very poor soils. They should be able to grow to give a high stand mass, 150-250 t ha^{-1}, when mature and be relatively long-lived, life-span exceeding ~100 Earth years. Two species of Betula, B. pendula and B. pubescens, appear to be almost ideally suited hardwoods and Chosenia arbutifolia a well-adapted conifer.

The particular advantage of B. pendula is its ability to maintain its level of photosynthesis, as measured by CO_2 uptake, under reduced levels of sunlight. According to Haritonovitsh[17] on Earth a reduction of the level of illuminance from 100% to 30% only reduces CO_2 uptake by 1/3 of the intake in full sunlight. This species is also fast growing and can reach a height of 20 to 30 m in about 1/5 of its life-span of ~135 years. B. pubescens, equal in size to B. pendula, has in some respects an even stronger claim to be chosen, it not only tolerates very poor acidic soils but is naturally a pioneering arboreal colonizing species. The conifer Chosenia arbutifolia is tolerant of very low light levels, and permafrost conditions. It is a large tree reaching 35 to 37 m in 30 to 40 years. On Earth it colonizes regions closer to the North Pole than any other tree where it grows to a maximum height of 8 to 10 m. It remains unknown how these otherwise 'Mars-ready' species would cope with the long asymmetrical Martian seasons.

4.2.2 High photosynthetic yield plants

These plants, for example varieties of sugar cane, have a vital part to play in the establishment functioning of the bio-energy and industrial materials manufacturing industries it will be necessary to establish on Mars. Cane can form the input to an ethanol based petroleum-like fuel, chemical and polymer industry, Figure 11, helping to some degree to overcome the absence of fossil fuels resources on the planet.

4.2.3. Nitrogen fixing and food plants

Nitrogen mineralization in terrestrial boreal forests, where nutrient availability is generally low, is often poor particularly where the soils become frequently water-logged and where permafrost conditions exist at relatively shallow depths. In the interior of Alaska Van Cleve *et al.*[18] and Flanagan and Van Cleve[19] found nitrogen mineralization ranges from 89 kg ha^{-1} yr^{-1} in cold wet spruce forests to 24-58 kg ha^{-1} yr^{-1} in Betula papyrifera (birch), forests on warmer permafrost soils. On Mars the absolute quantity as well as the biological availability of nitrogen in all forms may be the limiting factor as to whether, or to what level, we can bring Mars alive. If the apparent low abundance of N_2 is real, then the need for nitrogen as a major atmospheric constituent could severely limit, or even prevent, its use in the production of high technology construction materials that would withdraw it permanently from participating in the biological environmental cycle within the worldhouse. This could prevent the large scale use of aramid fibres for worldhouse support cables and other high strength low density fibre materials might have to be selected.

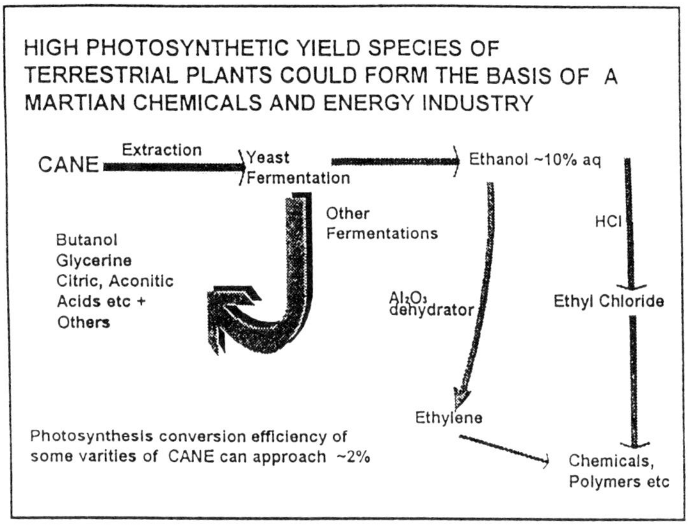

Fig.11. High photosynthetic yield species of terrestrial plants like cane may form the basis of an indegenous martian energy and chemicals industry. A combination of biological and chemical processes as show above could be used to produce a wide range of different materials.

Assuming that sufficient nitrogen is available to form ~80% of the worldhouse atmosphere and that through the use of pioneering leguminous species, varieties of lupin may be well adapted to Martian worldhouse conditions, sufficient mineralized nitrogen can be made available in the soils to allow the growth of a wide range of higher plants including cereals and other agricultural and horticultural products. Given the low mass of the worldhouse atmosphere and the low rate at which the nitrogen biochemical/physical cycle operates it will almost certainly be necessary to develop methods of assisting the nitrogen recycling process. Figure 12 indicates the much more rapid recycling necessary in an environment where the atmospheric mass is limited.

4.3 Ecosystem Stability

The ecosystem of the Earth is made up of a wide range of competing feedback systems and the sheer size and complexity of the many processes that are going on defeats any attempt to quantify or understand the system in its entirety. However, the inertial effects resulting from this size and complexity confers an appearance of stability which is in reality a dynamic equilibrium made to run within relatively narrow boundaries by a myriad of competing and conflicting feedback loops. Odum[20] in discussing ecosystem stability and the Gaia hypothesis of Lovelock has used the term superecosystem to describe the behavior of the total complex of the different types of ecosystem which are present on the Earth. Citing the part biotic community plays in the nitrogen cycle he concludes that the biotic community plays a major role in biospheric homeostasis. On a more local level he suggests that different forms of stability can be quite clearly identified. Rapid recovery from disturbance, where species not usually present in the original undisturbed environment invade and colonize the disturbed ecosystem can be seen to be enhanced by the presence and availability of many different species. This is a case of *resilience stability*. High species diversity is believed by many ecologists to enhance also the ability of an ecosystem to remain stable, that is to resist disturbance, this is consequently termed *resistance stability*. But species diversity and the size of the ecosystem are probably not the only factors which operate to produce stability. Just as a man riding a bicycle finds it easier to retain his balance the faster he is moving there is some evidence that the faster an ecosystem is operating the better it is able to recover from at least minor perturbations. This implies that the more productive an ecosystem is the more stable it will be. Moore *et al.*[21] have examined data from studies of actual soil food webs and suggest that there is evidence that most ecological production world-wide is from ecosystems that are sufficiently productive to recover from perturbations.

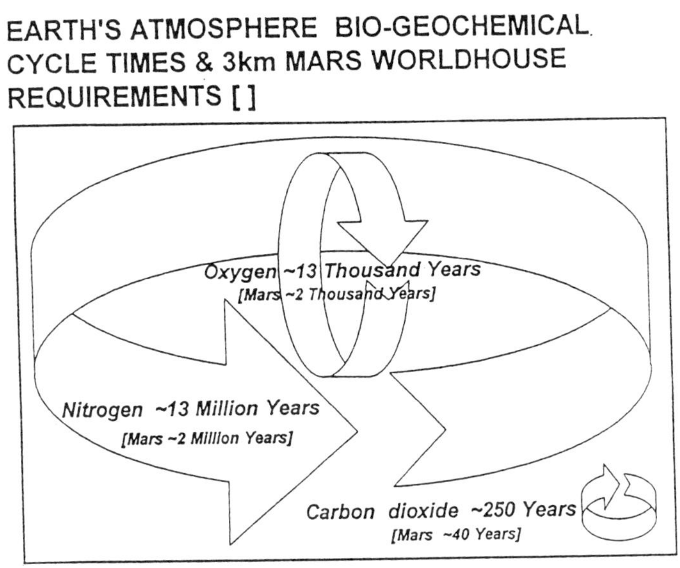

Fig.12. The approximate recycle times of the constituents of the Earth's atmosphere and those necessary within a 3km high Mars worldhouse at ~80% cover.

On the basis of current knowledge it would seem that to ensure the long-term stability of a Martian worldhouse ecosystem it will be necessary to ensure that it contains the greatest possible diversity of biota and is as productive as limiting boundary conditions will allow.

5. HAZARD CONTROL STRATEGIES—SOURCES OF RISK

The possibility of serious structural failure of any vital component parts of the Martian worldhouse, and of the entire structure must be reduced to negligible levels. Most foreseeable disasters involve the formation of quite large holes in the roof structure but these are in principle, technologically simple to deal with. Temporary repair can be made well within the flow time for significant atmospheric loss to the exterior environment.

The principal major hazards we have to prevent, or contain to manageable proportions are listed in Table 9 and shown graphically in Figure 13, and assessed in Display III. Studies of putative solutions or methods of dealing with the first three of these hazards listed in Table 9 already exist. The last hazard is a problem which can be largely avoided by the correct choice of construction materials at the outset of design and construction. The most effective strategy for anticipating, preventing and overcoming hazards to the worldhouse is to make the structure 'intelligent.' Automatic and continuous monitoring of the structure by a set of independently instructed computers (programmed by different teams of programmers) will keep check on all failure-critical parts of the worldhouse. Any disagreement between members of this 'computer parliament' (CP) will be resolved heuristically and may involve human judgment if the CP cannot arrive at a decision. The one aspect of human behavior that could threaten the integrity of the worldhouse is our propensity to violence in the form of war and revolution. However, as the inhabitants will be aware that they all depend much more directly on the worldhouse ecosphere than is the case for the citizens of Earth they may be more inclined to be cooperative rather than aggressively competitive. People who live in a worldhouse may learn not to throw stones!

MAJOR HAZARDS TO BE TAKEN INTO ACCOUNT
WHEN PLANNING A WORLDHOUSE

(i) **Fire,**

(ii) **Interior meteorological stresses**

(iii) **Geological disturbance,**

(iv) **Cosmic Impact,**

(v) **Explosion**

(vi) **Exterior heating and erosion**

(vii) **Interior collision damage**

(viii) **War,**

(ix) **Dereliction.**

Table 9. Nine of the most significant hazards are identified above.

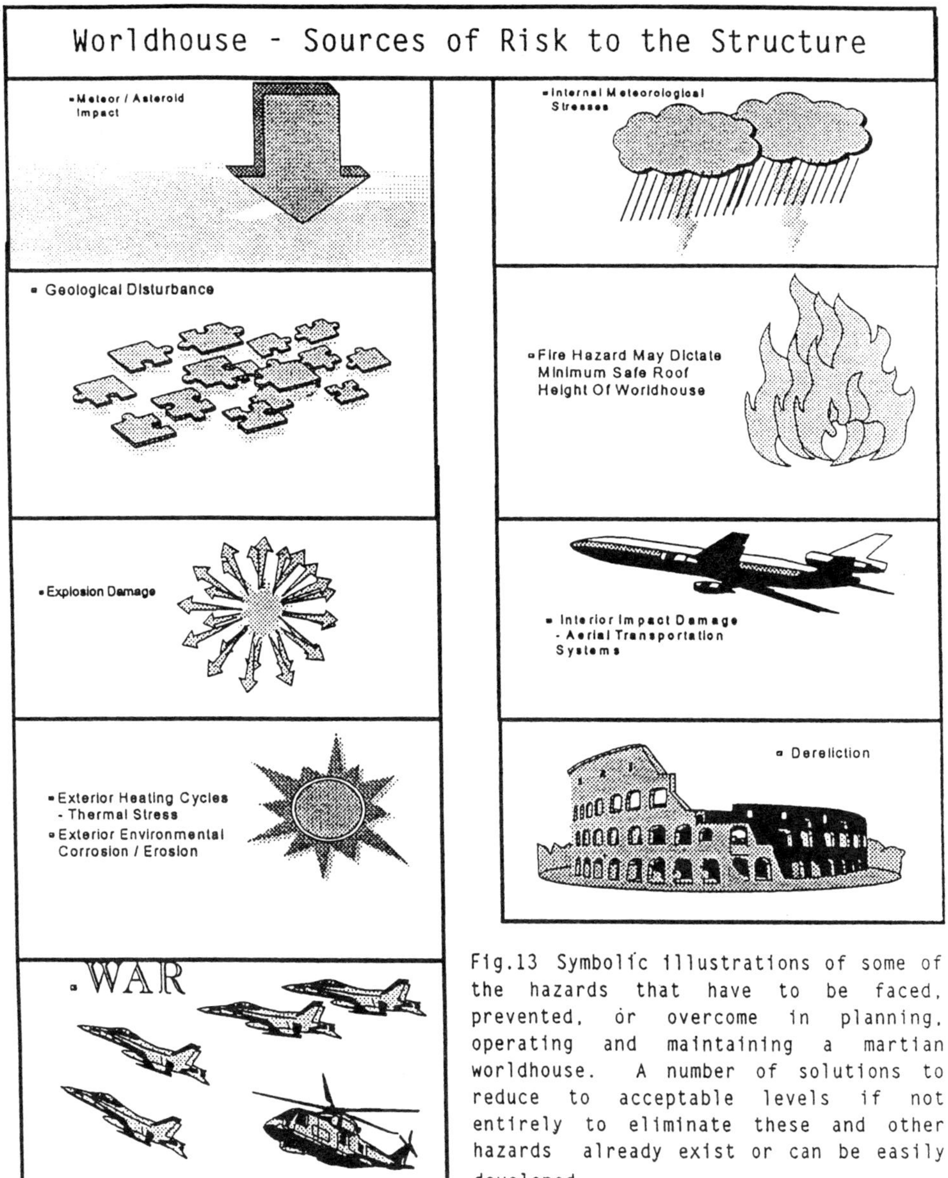

Fig.13 Symbolic illustrations of some of the hazards that have to be faced, prevented, or overcome in planning, operating and maintaining a martian worldhouse. A number of solutions to reduce to acceptable levels if not entirely to eliminate these and other hazards already exist or can be easily developed.

Perhaps the hazard that attracts most attention and speculation is that of the risk of a large cosmic impact. Surface precautions against such an eventuality will not be enough: Mars will have to be protected by some form of space based meteor strike prevention system. This has been studied in respect of preventing asteroid impact on the Earth.[22,23,24] Possibly of more relevance to the protection of the enormous capital investment the Martian worldhouse would represent would be thc development and application of somc of the techniques of 'star-wars' technology. Although terrestrially a failed military defense concept similar and less costly orbital techniques could be implemented to protect a paraterraformed Mars from major impact damage.

Obviously, it will be impossible to entirely guarantee the worldhouse environment against a decompression catastrophe. However, the Mars infrastructure, the support towers, their foundations and other surface structures within the worldhouse will all be designed to function as refuges in the event of an environmental disaster. Since no individual will be more than 1 to 3 km from a refuge, there will be plenty of time to reach safety unless no one is situated in the immediate danger area.

6. CONCLUSION

The worldhouse concept of paraterraforming is a less than ideal solution for providing Mars with a fully habitable environment suitable for permanent human settlement. But given insufficient volatiles inventory to produce on Mars a terraformed atmosphere of terrestrial composition and surface pressure a worldhouse solution enables us to utilize the planet for human purposes. I contend that there is no justification for transforming any planetary body from its existing state unless it is to form and establish a new dominion of human life and civilization.

7. ACKNOWLEDGEMENTS

I would like to thank Martyn Fogg of the Probability Research Group and Paul Birch of the British Interplanetary Society for helpful discussion and criticism, and Ian Burns of the Croyden Library Service for invaluable help with the literature research.

DISPLAY 1

Paraterraforming Mars:

Construction and Energy Strategies

There is no unique strategy that can be adopted for constructing a worldhouse on Mars. However there are two basic approaches to the task, these can be classified as Slow and Fast Track scenarios.

The Slow Track scenarios limit expenditure and use lower energy solutions to carry out the engineering, physical, chemical processes involved. The main disadvantage of these low rate of expenditure routes is that randomly scattered worldhouse cell units, erected near mineral / volatile rescources, must remain space-colony-like contained biospheres. That is they contain an ecosystem in a single-state condition. The stabilization premium to keep the ecosystem in balance will be very high and will so remain until the individual cell units are joined in a configuration that is contiguous and quasi-global.

Fast Track scenarios are conceived as being a high intensity campaign with the intention of reaching a global web of strips of worldhouse cells as rapidly as possible. This permits much faster progress towards the eventual establishment of a fully Earth-like biosystem which will maintain its own near-equilibrium with the minimal technological intervention to close geochemical cycles.

TYPE I

SLOW TRACK

RANDOM MODULE CONSTRUCTION

Low Initial Investment

Semi-permanent Living & Assembly Buildings

Ground Based Energy Generation Nuclear & Solar-Cells

Limited Mineral Extraction and Materials Manufacture

Physical, Chemical & Biological Volatiles Extraction

Individual Worldhouse Cells 'Scattered' in the Equatorial Region

Extension of Worldhouse on an 'as required' basis

Non-Global Cell Systems carry high Ecospheric Stabilization Premium

Complete Cover in ~750 - 1000 years

TYPE II

FAST TRACK

NON-RANDOM MODULE CONSTRUCTION

Large Initial Investment

Fugitive Accommodation & Assembly Buildings

Ground & Space Based Energy Systems

Biological Chemicals & Hydrocarbon Fuel Industry

Large Scale Mineral & Materials Manufacture.

High Rate Volatiles Extraction

Initial Worldhouse cells configuration of Circum-Polar and Circum-Equatorial Rings connected by 8 Longitudinal Corridors thus allowing the early establishment of a global atmospheric circulation system.

Extension by 'Filling-in' the Gaps

Fully global Macro / Micro-biome System Minimum Ecospheric Stabilization Premium

Worldhouse Completed in ~200 - 400 years

DISPLAY 11

Paraterraforming Mars: -

Biological Strategies

The initial steps to establish the earliest forms of ecosystems within a Type I random modular growth worldhouse structure and in the incomplete global-network in the early stages of a Type II construction strategy are similar. But as the two different strategies progress it is the Type II approach that rapidly becomes more favourable for the establishment of a self-stability seeking biosphere.

The global network worldhouse structure permits diurnal and seasonal climatic variations to modulate the internal environmental conditions, although with some restriction, well before the fully global structure is complete. This allows individual micro-biomes to arise naturally and will allow ecospheric development to proceed in a ways similar to the techniques of restoration ecology used on the Earth. The extension of the plant ecosystems will be able to advance freely in ways similar to natural succession processes on the Earth.

In this way individual plant species will be able to pioneer regions of the worldhouse environment and the biosphere within it will develop its own natural momentum. The final outcome may differ from our expectations but will also be more fully Mars adapted.

TYPE I

SLOW TRACK

INDIVIDUAL MODULE ECOSYSTEMS EVENTUALLY LINKED TO FORM A GLOBAL BIOSYSTEM

Distinct Ecospheric Environment in each cluster of Worldhouse Cells which will consequently exhibit Agro-Horticultural behaviour rather than that of a free-running biosphere

Climatic behaviour within each Cell-Cluster without forced circulation will tend towards that of a hot or cold arid desert environment

Development of the individual & separate Cluster-Ecosystems will have to be managed to ensure that when linked they will show trajectory stability so that the entire system will move towards the desired end point

The number and range of species within each Cell-Cluster will be limited by the size & extent of the enclosed biosphere. Natural succession and colonization processes will await global linkage when a single planet-wide biosphere will be established

TYPE II

FAST TRACK

RAPID ESTABLISHMENT OF A GLOBAL NETWORK OF LINKED WORLDHOUSE CELLS WITH A SINGLE BIOSPHERE

Fully global Macro / Micro-biome System brought into existence as early as possible.

Techniques similar to those of Restoration Ecology used to promote a 'martian natural succession'

Initial Selection of Plant Species will be directed at etablishing biomass sinks and long term biomass recycling systems.

Aforestation of high latitudes with mixed hardwoods and non-deciduous coniferous species will reduce mid-to-low latitude temperature gradients

Species suitable for Low & Equatorial Latitudes will include food species & plants capable of forming the basis of energy and chemicals manufacture.

Stable Biosphere Achieved ~ 200 years

DISPLAY III Paraterraforming Mars: - Hazard Control Strategies

Natural Hazards - Dealing With

COSMIC IMPACTS

Orbital Asteroidal Detection Instrumentation + Orbiting Interception, Deflection, & Fragmentation Devices - 'Star Wars Technology'
Inflatable Sealing Systems within roof spaceframe structure to deal with small meteorite punctures.

GEOLOGICAL DISTURBANCE

The completed Worldhouse is a tension structure and not prone to collapse under seismic conditions. Components in compression can be mass inertia-stabilized.

METEOROLGICAL STRESSES

The gravitational and thermally induced atmospheric mass flows can have their energy reduced by fitting uninhabited Mars Roof-support Towers with a series of verticle axis wind turbines. Power generation ~1 to 2 Tw will be easily achieveable.

EXTERIOR THERMAL STRESSES, EROSION & CORROSION

The problems can be minimized if not entirely eliminated by correct materials design when planning the construction of the Worldhouse. Complete 'intelligent self-monitoring' will be built into the structure.

Natural or Man-made Hazards - Dealing With

FIRE

The possibility of fire partially dictates the height of the Worldhouse. Liquid CO_2 extinguisher systems and continuous monitoring of fire-risk regions will enable any outbreaks to be brought quickly under control.

Man-made Hazards - Dealing With

DERELICTION

Only likely to arise if the martian civilization collapses and the population loose interest in their own survival. Cure regular maintenance.

EXPLOSION

This problem can be minimized if not entirely eliminated by ensuring that large concentrations of potentially explosive materials are not accumulated in one place.

INTERNAL IMPACT

Only the roof spaceframe structure would be significantly vulnerable to an aerial impact, and then only if it was normal to the roof inner surface. Aerial collision with a roof-support tower is a more probable event but would have no serious structural consequences even though the aircraft would be destroyed.

WAR

Only lunacy could lead to the entire population committing collective suicide. Indeed, awareness of the vulnerability of their Worldhouse civilization may result in the Martians being a good deal more cooperative and helpful towards each other.

REFERENCES

1. Y. Ishikawa, T. Ohkita and Y. Amemiya, "Mars Habitation 2057: Concept Design of a Mars Settlement in the Year 2057," *JBIS*, 43, 505-512, 1990. Also published as AAS 97-378 in R. M. Zubrin (Ed.) *From Imagination to Reality: Mars Exploration Studies of the Journal of the British Interplanetary Society* (Part II: Base Building, Colonization and Terraformation), AAS *Science and Technology Series*, Vol. 92, pp. 83-97, (1997).
2. J. Allen and M. Nelson, *Space Biospheres*, Synergetic Press, London (1986).
3. J. E. Lovelock, "The ecopoiesis of Daisy World," *JBIS* 42, 583-586, (1989).
4. C. P. McKay, O. B. Toon and J. F. Kasting, "Making Mars Habitable" *Nature*, 352, 489-496, (1991).
5. M. J. Fogg, "The creation of an artificial dense Martian atmosphere: A major obstacle to the terraforming of Mars," *JBIS*, 42, 577-582, (1989).
6. M. J. Fogg, "A synergic approach to terraforming Mars," *JBIS*, 45, 315-329, (1992).
7. R. L. S. Taylor, "Paraterraforming: The Worldhouse concept," *JBIS* 45, 341-352, (1992).
8. T. Owen, K. Biemann, D. R. Rushneck, J. E. Biller, D. W.Howarth and A. L. Lafleur, "The composition of the atmosphere at the surface of Mars," *J. Geophs. Res.* 82, 4635-4639, (1977).
9. M. H. Carr, *The Surface of Mars*, Yale Planetary Exploration Series, New Haven (1981).
10. A. R. Ravishankara, S. Solomon, A. A. Turnipseed, R. F. Warren, "Atmospheric lifetimes of long-lived Halogenated Species," *Science*, 259, 194-199, (1993).
11. E. I. Friedmann, M. Hua, and R. Ocampo-Friedmann, "Terraforming Mars: Dissolution of carbonate rocks by cyanobacteria," *JBIS* 46, 291-292, (1993).
12. P. Birch, "Terraforming Mars Quickly" *JBIS* 45, 331-340, (1992).
13. M. J. Fogg, "Dynamics of a terraformed Martian biosphere," *JBIS* 46, 293-304, (1993).
14. W. S. Ellis and N. Ben, "Skyscrapers: Above the crowd," *Nat. Geographic*, 175, 143-173, (1989).
15. J. W. Mastalerz, *In the Greenhouse Environment*, John Wiley & Sons, New York, (1977).
16. S. Diaz, J. P. Grime, J. Harris & E. McPherson, "Evidence of a feedback mechanism limiting plant response to carbon dioxide," *Nature*, 346, 616-617, (1993).
17. F. M. Haritonovitsh, "Biology and Ecology of Tree Species," *Lismaya promyshlenost*, Moscow, (in Russian), (1968).
18. K. Van Cleve, R. Barney, & R. Schlenter, "Evidence of temperature control of production and nutrient cycling in two interior Alaskan black spruce ecosystems," *Canadian Journal of Forest Research*, 11, 258-273, (1981).
19. P. W. Flanagan & K. Van Cleve, "Nutrient cycling in relation to decomposition and organic matter quality in taiga ecosystems," *Canadian Journal of Forest Research*, 13, 795-817 ((1983).
20. E. P. Odum, *Ecology and our Endangered Life-Support Systems*, Sinauer Associates, Mass. (1989).
21. J. C. Moore, P. C. de Rutter, H. W. Hunt, "Influence of productivity on the stability of real and model ecosystems," *Science* 261, 906-908, (1993).
22. L. A. Kleiman, (Ed.) M.I.T. Report No.13 - *Project Icarus*, The MIT Press, Cambridge, Mass. (1968).
23. T. J. Ahrens & A. W. Harris, "Deflection and fragmentation of near-Earth asteroids," *Nature*, 360, 429-433, (1992).
24. D. Steel, "Project Spaceguard: The fiction and the fact," *Spaceflight*, 35, 218-223, (1993).

TECHNOLOGICAL REQUIREMENTS FOR TERRAFORMING MARS*

Robert M. Zubrin and Christopher P. McKay

The planet Mars, while cold and arid today, once possessed a warm and wet climate, as evidenced by extensive fluvial features observable on its surface. It is believed that the warm climate of primitive Mars was created by a strong greenhouse effect caused by a thick CO_2 atmosphere. Mars lost its warm climate when most of the available volatile CO_2 was fixed into the form of carbonate rock due to the action of cycling water. It is believed, however, that sufficient CO_2 to form a 300 to 600 mb atmosphere may still exist in volatile form, either adsorbed into the regolith or frozen out at the south pole. This CO_2 may be released by planetary warming, and as the CO_2 atmosphere thickens, positive feedback is produced which can accelerate the warming trend. Thus it is conceivable, that by taking advantage of the positive feedback inherent in Mars' atmosphere/regolith CO_2 system, that engineering efforts can produce drastic changes in climate and pressure on a planetary scale.

In this paper, we propose a mathematical model of the Martian CO_2 system, and use it to produce analysis which clarifies the potential of positive feedback to accelerate planetary engineering efforts. It is shown that by taking advantage of the feedback, the requirements for planetary engineering can be reduced by about 2 orders of magnitude relative to previous estimates. We examine the potential of various schemes for producing the initial warming to drive the process, including the stationing of orbiting mirrors, the importation of natural volatiles with high greenhouse capacity from the outer solar system, and the production of artificial halocarbon greenhouse gases on the Martian surface through in-situ industry.

If the orbital mirror scheme is adopted, mirrors with dimensions on the order of 100 km radius are required to vaporize the CO_2 in the south polar cap. If manufactured of solar sail like material, such mirrors would have a mass on the order of 200,000 tonnes. If manufactured in space out of asteroidal or Martian moon material, about 120 MWe-years of energy would be needed to produce the required aluminum. This amount of power can be provided by near-term multi-megawatt nuclear power units, such as the 5 MWe modules now under consideration for NEP spacecraft.

Orbital transfer of very massive bodies from the outer solar system can be accomplished using nuclear thermal rocket engines using the asteroid's volatile material as propellant. Using major planets for gravity assists, the rocket ΔV required to move an outer solar system asteroid onto a collision trajectory with

* This article appears in full as paper number AAS 97-390, in *From Imagination to Reality: Mars Exploration Studies of the Journal of the British Interplanetary Society* (Part II: Base Building, Colonization and Terraformation), R. M. Zubrin, Ed., AAS *Science and Technology Series*, Vol. 92, pp. 309-326, 1997. The article was first published in the *Journal of the British Interplanetary Society*, Vol. 50, No. 3, pp. 83-92, March 1997.

Mars can be as little as 300 m/s. If the asteroid is made of NH_3, specific impulses of about 400 s can be attained, and as little as 10% of the asteroid will be required for propellant. Four 5000 MWt NTR engines would require a 10 year burn time to push a 10 billion tonne asteroid through a ΔV of 300 m/s. About 4 such objects would be sufficient to greenhouse Mars.

Greenhousing Mars via the manufacture of halocarbon gases on the planet's surface may well be the most practical option. Total surface power requirements to drive planetary warming using this method are calculated and found to be on the order of 1000 MWe, and the required time scale for climate and atmosphere modification is on the order of 50 years.

It is concluded that a drastic modification of Martian conditions can be achieved using 21st century technology. The Mars so produced will closely resemble the conditions existing on the primitive Mars. Humans operating on the surface of such a Mars would require breathing gear, but pressure suits would be unnecessary. With outside atmospheric pressures raised, it will be possible to create large dwelling areas by means of very large inflatable structures. Average temperatures could be above the freezing point of water for significant regions during portions of the year, enabling the growth of plant life in the open. The spread of plants could produce enough oxygen to make Mars habitable for animals in several millennia. More rapid oxygenation would require engineering efforts supported by multi-terrawatt power sources. It is speculated that the desire to speed the terraforming of Mars will be a driver for developing such technologies, which in turn will define a leap in human power over nature as dramatic as that which accompanied the creation of post-Renaissance industrial civilization.

CHAPTER 6:
HUMAN MISSIONS IV:
LONG-TERM HABITATION - MARS BASE CONCEPTS

AAS 93-883B

MARS 2008 SURFACE HABITATION STUDY - PART 5 MISSION DESIGN EVALUATION PROCESS AND CRITERIA*

Marc M. Cohen†

This report, Mars 2008 Surface Habitation Study, MISSION DESIGN EVALUATION PROCESS AND CRITERIA documents the results of the 1992 "Summer Study," August through November 1992, sponsored by the Center for Mars Exploration at the National Aeronautics and Space Administration's Ames Research Center in coordination with the Office of New Initiatives at NASA-Ames, undertaken in addition to their regular jobs.

This effort to define a Mission Design Evaluation approach for Human mission to Mars began at the Case for Mars IV Conference at the University of Colorado in the summer of 1990. Larry Lemke led a small working group in exploring various evaluation approaches and metrics. This fifth position paper is the inheritance from that effort, revised in light of the Mars 2008 effort, into which this report incorporates it.

These people contributed also to the MISSION DESIGN EVALUATION PROCESS AND CRITERIA:
Geoff Briggs, Center for Mars Exploration
Marc Cohen, Advanced Space Technology Office
Larry Lemke, Space Projects Division
Brent Sherwood, Boeing, Huntsville.

The Mars 2008: Surface Habitation Study position papers cover these topics:
1. Objectives and Overview
   Appendix A, Surface Missions Study Outline
2. Mission Design Logic
3. Safety Philosophy
4. Habitation Strategy and Requirements
   Appendix B, Data Requirements from Precursor Missions
   Appendix C, Pre-Human Arrival Preparation Missions
5. **Mission Design Evaluation Process and Criteria.**

The ultimate goal of human space exploration is to discover if life exists on other worlds, to understand the genesis and evolution of the universe and to learn to live on other planets. Mars offers the closest opportunity to pursue these goals realistically. The capabilities to define, design, develop, build, test, contract out, manufacture and operate new technologies are the means to

* Center for Mars Exploration Position Paper, NASA-Ames Research Center, Final Review Draft, May 21, 1993.

† Prepared by Marc M. Cohen, Architect, Advanced Space Technology Office.
Approved: Geoffrey Briggs, Director, Center for Mars Exploration.
Concurrence: Anthony R. Gross, Chief, Advanced Space Technology Office.
Concurrence: Roger Arno, Manager of Advanced Projects, Space Projects Division.
NASA-Ames Research Center, Moffett Field, California 94035-1000.

achieve this set of goals. The purpose of this set of criteria is to evaluate mission design and exploration technology proposals to ensure that *the means support the goals* and do not obstruct them.

ACRONYMS

CELSS: Controlled Ecological Life Support System, based on plants for atmosphere revitalization, food production, and water and waste processing.

CO_2: Carbon Dioxide.

DC3: Distributed Command, Control and Communications.

EVA: Extravehicular Activity—activities that crew members conduct in space or on a planetary surface in a pressurized suit after exiting the pressurized IVA Habitat.

FLO: First Lunar Outpost.

FMO: First Mars Outpost.

H_2O: Water.

ISRU: In Situ Resource Utilization extracts O_2, N_2, CO_2, and H_2O from the Mars atmosphere and anything useful and processable from the Mars soil or regolith.

IVA: Intravehicular Activity—activities that crew members conduct inside a pressurized space vehicle or planetary surface habitat.

MEV: Mars Excursion Vehicle—Mars lander from "mother ship."

MSM: Mars Surface Mission.

MT: Metric Ton(s).

N_2: Nitrogen, buffer gas constituent of earth-normal atmosphere.

O_2: Oxygen.

TBD: To Be Determined.

TEIV: Trans Earth Injection Vehicle.

TMIV: Trans Mars Injection Vehicle.

INTRODUCTION

This evaluation will help us to determine not only the best designs but also to verify the requirements they must meet and the capabilities they must provide.

Perhaps the most important goal of these Mars Mission Evaluation Criteria is to move the program goals from an ill-defined to a well-structured problem state. Without such clarity, the program may wander aimlessly, in intermittent forays. Instead, this effort seeks a clear vision of both the exploration path and the exploration process. Path and process are distinct but equal imperatives. Path includes Technical Credibility and

Design Selection Logic while Process includes Design Management and all its ramifications. Process is vital because ultimately, people get what they do, not what they intend.

This approach is to identify domains of evaluation. While these domains ultimately inter, mission planners can evaluate them both separately and in parallel. These criteria represent an attempt to guide that evaluation. See Figure 1 for an illustration of the design evaluation process that this position paper proposes.

Marc M. Cohen

Mission Design Evaluation Process

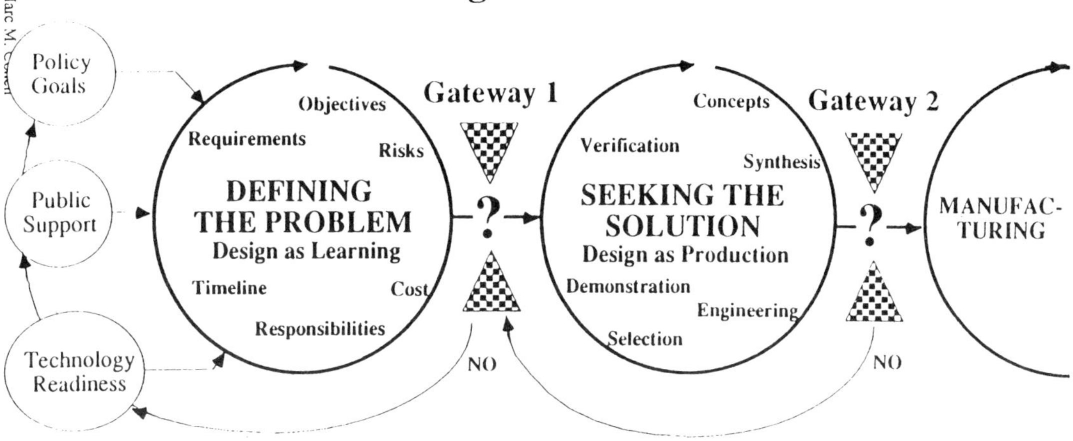

Evaluation Gateway 1
1. Is the problem sufficiently well-defined to allow solution-seeking?

2. Are the requirements clear and are they the right ones?

3. Is this problem definition well-documented?

Evaluation Gateway 2
1. Is the solution sufficiently well-defined to begin manufacturing?

2. How well does the design solution meet the requirements?

3. Is this design solution well-documented?

May 21, 1993

Figure 1 Mission Design Evaluation Process.

Design Theory

Design is a cyclical process in which the designer or the design organization iterates a sequence of conception, representation, and evaluation until arriving at a satisfactory solution. Evaluation occurs at several levels and at multiple points along the way. However, this approach to the design evaluation process concerns two major gateways that control the *design problem definition* and the *design solution.* Design problem definition involves design as learning about the problem, while design solution-seeking involves design as production of a project or product. Gary Klein offers a valuable insight into the nature of ill-defined problems:

> Ill-defined problems require two simultaneous processes: 1) Goal clarification and 2) option development. Designers cannot always wait for goals to become perfectly well-specified before starting to work. They must expect to learn more about their

> goals during the design process itself. This helps to make the design process challenging and maddening. The designer is trying to give the sponsor/user a product, and at the same time, is trying to learn more about what the user really wants and to help the user through this learning cycle.[1]

While Klein's observation about two processes seems right on the mark for ill-defined problems, they appear to reinforce Herbert Simon's conclusion that *there is no sharp boundary between well-defined and ill-defined problems*.[2] It is difficult or impossible to know whether a problem state is ill-defined or well-defined without engaging in design as learning—to develop design ideas as a way of testing approaches to the problem.

It is essential to distinguish between the evaluation of one's work that a designer does instinctively and the impartial evaluation of the design problem definition or the design solution. During the cognitive dynamics of the design process it is extremely difficult to prescribe evaluation techniques for the designer to use because a talented designer uses a highly subjective approach to his or her work. In this sense, the design process is unknowable. Because designing is unknowable, design managers and teachers attempt to rationalize design and make it transparent.

J. Christopher Jones presents this dichotomy that he calls *Designers as Glass Boxes* and *Designers as Black Boxes*. Jones poses a goal to achieve from this fundamental dialectic for systematic methods: to make design process explicit by strengthening the resources and methods available to the designer operating in rationalist *Glass Box* mode so that he will not revert to the intuitive, internalized *Black Box*.[3]

Yet, the black box plays an essential role in the creative process. Within the black box, the designer comprehends subjectively the gestalt in which his contribution must fit. According to Don Schön, designers create a whole "world" in which they construct their own protocols in the form of hypothetical rules and typologies for design solutions.[4] This form of heuristic reasoning in design is a necessary step in the solution seeking process, especially in the common situation "...in which it is unknown beforehand whether a particular sequence of steps will yield a solution..."[5] Thus, designing is intrinsically a process of discovery and learning, but no design text can prescribe how an individual learns through design, which is the crucial feature of defining a design problem.

In contrast to black box design as learning, Herbert Simon observed the glass box phenomenon in computer science:

> ...we as designers, or as designers of design processes, have had to be explicit as never before about what is involved in creating a design and what takes place while the creation is going on.[6]

This emphasis upon explication of design—or design as explication—affects profoundly how the designer works. It becomes especially acute when the designer must produce rationalistic constructs to explain and justify his work.

System engineering is the quintessential glass box method. System engineering provides an excellent set of tools to select and integrate design solutions. System engineers tend to assume that all problems come well-defined, and that any design effort is susceptible to the characteristic weighted comparisons, trade studies and iterations of reference configurations. These techniques prove successful for well-structured problems but not for ill-defined ones. The ill-defined problem typically displays unclear objectives, conflicting requirements, unreduced complexity or all of these difficulties and more. The misapplication of system engineering to ill-defined problems interferes with design as learning, and may produce perfectly consistent but incorrect results.

The systematic method's literature concentrates on elaborating the glass box model through bubble diagrams, functional flow block diagrams, PERT charts, GANTT charts, flow charts, and many other constructs. By describing the context and limits of design, design managers attempt to make it manageable. Ironically, because glass box methods are so persuasive and powerful, they often substitute for the actual design product. In a very large design project that imposes elaborate rituals to pass atomized design elements through many layers of review, it becomes easier for people to focus upon the bureaucratic process and not to think about the product. Witness Space Station *Freedom.*

Therefore, it is further important to distinguish *control of process* from *control of product*. Karl Popper, the philosopher of science, argues that "the problems of produced structure are as a rule more important than the problems of production."[7] Thus, the mission design evaluation gateways differ from the traditional project reviews in the form of preliminary design reviews, critical design reviews, final design reviews, etc. The need to manage people and control their interactions motivates these kinds of project review as much—or sometimes more—than the need to find the best design solution. The design evaluation gateways control the progress of the design product alone, without reference to who takes responsibility for which pieces of the puzzle.

These evaluation gateways offer an opportunity to step back from the whole design flow and give the project a sanity check. The experience of Space Station *Freedom* underscores the importance of such a sanity check, especially the necessity for a clear, well-defined problem definition before starting to seek design solutions.

Mission Design Evaluation Process

In this evaluation process, the design must satisfy three top level criteria at each gateway. These criteria take the form of questions that demand mostly yes or no as answers. The gateways require three unequivocal yeses before opening to allow the mission design to pass on to the next cycle. This gateway approach tends to be incompatible with "fast-track" methods in which the project begins building some pieces while other pieces are not yet even sketches.

This design evaluation process differs substantially from the traditional systematic sequence of pre-Phase A "Needs, Attributes and Architectural Options" studies,[8] Phase A Concept Development, Phase B Definition Studies, and Phase C/D Design/Build. The difference between these evaluation paradigms is that in the present bureaucratized design process, designers and design managers iterate endlessly between requirements and solutions without ever passing through a problem definition. It is all too easy to become caught up in the avalanche of details and to lose all sight of the project as a whole enterprise.

Gateway 1 controls whether the mission design is ready to pass from problem definition to solution seeking. The questions at Gateway 1 ask:

1. Is the problem sufficiently well defined to allow solution seeking to begin?
2. Are the requirements clear and are they the right requirements?
3. Is this problem definition well documented?

Question 1 demands a judgement of the clarity, completeness, consistency, and credibility of the design problem definition. Question 2 reviews whether the requirements are still correct which may shed a very different light than the early requirements review. Question 2 may prove the most difficult because it compels the evaluators to question the fundamental reasons for the mission. Question 3 is important to show that the solution-seeking designers will be able to understand the problem definition, especially the origin and relationship of the requirements. At Gateway 1, it is vital to build a consensus among all the stakeholders upon the general form and specific content of the design problem definition. Three yeses make this consensus possible (although not automatic). Without this consensus to confirm the problem definition, dissatisfied participants or stakeholders will often attack and undermine any design solution.

If a candidate problem definition fails to generate a yes to the three questions, it is not ready for solution seeking. Should the problem definition fail, two choices emerge. If, despite the failure, the problem definition is relatively within range of three yeses, design team may continue refining the "product" within the problem definition cycle. However, should a candidate problem definition receive a resounding no—and the designers feel they made their best effort—they have no choice but to revisit the basis of the whole endeavor; the policy goals, the technology, and the objectives.

Gateway 2 controls whether the mission design is ready to pass from design solution to manufacturing. The questions at Gateway 2 ask:

1. Is the solution sufficiently well-defined to begin manufacturing?
2. How well does the design solution meet the requirements?
3. Is this design solution well-documented?

Question 1 requires a careful assessment of the totality and detail of the mission design solution. It takes the watchword clarity, completeness, consistency and credibility several steps further by including technology verification, especially integrated technology test beds that demonstrate the functionality of the proposed systems and subsystems. Question 2 stems from a simpler and more stark inquiry: Does the design solve the problem? Is it possible for a design to meet all the requirements in a fragmented way and still not solve the problem? However, most designs that solve the problem will meet the various requirements with differing degrees of success. Thus, this question accepts the most "gray scale" answer of the set. Question 3 ensures that the design solution is ready for manufacturing, that the fabrication engineers and workers can comprehend what they need to do to produce the hardware and software. It also helps to verify that the solution solves the initial problem and will meet the project's goals and objectives.

If a design solution fails to pass Gateway 2, it may continue the design cycle. However, if it fails drastically, it should return to the first cycle to reexamine whether the problem definition—as stated—is solvable.

CROSS-CUTTING ISSUES

Within the overarching design evaluation approach, a number of Mars mission-specific evaluation criteria apply to the two gateways. In addition, a number of other Mars mission-specific issues cut across the design process itself. These cross-cutting issues include: technical credibility, operational practicality, participatory process (consensus) and program management.

The Mass to Orbit Fallacy

The great—and perhaps unique—difficulty in developing a design problem definition for a space mission is that so much depends upon launch capability and the subsequent diminutions in a payload capacity as it approaches its destination. The system engineer asks: "What could you do with 100 kg on the surface of Mars?" This mode of thinking places the delivery system ahead of the mission objectives. This paradox is the space age equivalent of "putting the cart before the horse." This approach is so engrained in the NASA and Space culture that any discussion of exploring Mars with most engineers quickly degenerates into a comparison of reconstituted or resurrected heavy launch vehicle systems. This way of thinking is unlikely to produce either the human Mars exploration mission OR the HLLV system anytime soon.

Instead, it is vital to take a new approach that looks at each major part of the problem separately and determines what is the best way to conduct each one, without prejudging how the technology for one major portion should influence the technology for another portion. The key question in this instance is:

To what extent is the design problem definition independent of any specific solution, idea or technology application?

Does the candidate design (either problem definition or design solution) offer a significant reduction in complexity over other approaches?

Exploration Modes

Each Mars exploration mode poses a potentially valuable contribution: manned or machine; orbital, atmospheric or surface; autonomous robot or telepresence; manned sprint or permanent base. Each specific exploration mode raises the key questions:

1. What is the best mode for a general or specific reconnaissance to pursue scientific exploration or to prepare for a permanent base?

2. Does a manned mission offer any advantages over unmanned Mars site examination, selection, survey or preparation?
3. Does a sprint mission offer any advantages over a permanent base approach as a first manned Mars mission?
4. Does the Mars surface habitat share common hardware with the space station, the lunar surface habitat, the TMIV, or the TEIV?
5. Is this commonality between entities a benefit or a penalty?
6. What is the appropriate safety philosophy and strategy for a Mars surface mission?

TECHNICAL CREDIBILITY

The essence of technical credibility is that a proposed Mars Mission design should be capable of doing what its proponents claim it will do. This criterion differs from judging proposals on whether they appear to be the smartest or best way to go to Mars. If a mission design proposal is fundamentally credible, it is important to evaluate it fairly on the broader aspects of technical credibility: completeness, temporal design and cost-effectiveness.

Completeness

To achieve completeness, a Mars Mission Architecture must address both participatory process and system engineering. These planning criteria include:

1. The objectives to accomplish on Mars, particularly science.
2. The most efficient and productive way to accomplish them.
3. The best way to deliver this capability to Mars.

Completeness means an explicit approach to consistency and continuity in the design and conduct of Mars exploration. Continuity implies that the exploration continues steadily over a sufficiently long period to accomplish all the scientific goals. Consistency implies providing an equally rigorous and reliable treatment to each aspect of exploration but without burdening individual investigations with unnecessary components. Finally, the directness of the approach will be crucial to implementing complete and timely Mars missions. Technical credibility means internal consistency from end to end for each mission and program.

Consistency

Is the mission design consistent in handling the issues—or does it make necessary that the exceptions outnumber the rules?

How many NASA engineers does it take to screw in a light bulb?
Answer: Two—one screws in the light bulb and one signs the waiver.

This little joke illustrates a profound problem in the way the bureaucracy can constrain a program so that predictable and necessary changes become "discrepancies" or "contingencies" or "exceptions" that require "waivers and sign-offs." These shortsighted constraints go a great way toward keeping thousands of civil servants and contractors in secure jobs processing paper, but they can make it difficult or impossible for the working technical people—the designers, engineers, manufactures and astronauts—to do their jobs. Safety should be the highest priority throughout the program, but often the ritual observances for trivial matters can obstruct the more significant factors that affect safety.

New Technology Development

Does the mission design treat technology development as a wasted cost or as an investment opportunity?

Should technology development occur in a focused technology program? Should this effort happen before selecting a "point design" or after selecting one? Since one can rarely determine impartially beforehand which alternative will serve best over the long run, this approach encourages the parallel development and testing of multiple options. No one can know or specify all requirements before undertaking these design and development activities. Rather, this technology development serves as inquiry by design research that will help to discover new requirements and technical possibilities.

The Mars surface mission critical technologies are: propulsion, aerobraking, regenerative life support, EVA, robotics, human factors, and crew health maintenance.[9] In situ resource generation, storage & utilization are emerging also as essential elements of the Mars Surface Mission. The vital step is comparing promising new technologies to off-the-shelf items for cost and improved performance, maintainability, producibility, reliability, quality and safety. The burden of proof should rest with the developers of new technology, although the mission design managers should adopt an openness to being persuaded that a new technology may be equal or better than an existing flight technology.

System Integration

System Integration, particularly the integration of multiple technologies, raises two sets of issues: Design Logic and Temporal Design. The mission design must resolve both sets of issues together.

Design Logic: The design logic articulates the assumptions about the MSM design approach and the interface constraints between the transportation and surface systems. It also applies at lower and more detailed levels to the whole chain of design decision-making and evaluation. The Design Logic must satisfy these six questions:

1. Does the technology and system integration approach accurately respond to the design problem definition and decomposition?
2. Does the system integration occur at "natural" intersections of the architectural elements and technologies or is it forced?

3. Do the partitions or separations of technologies occur at "natural break points"? Do these break points give the program "clean interfaces"?

4. Are there any artificial or forced integrations that reflect the external bias rather than the design problem definition, the mission design solution and the technology selection (e.g., pork barrel politics or parochial turf claims)?

5. Are the mission elements modular, interchangeable and replaceable or are they so highly integrated as to negate a real availability/maintainability strategy?

6. What approach to availability, maintainability, redundancy, reliability, repairability and safety best suits the MSM objectives and requirements? How does this strategy correspond to the system integration approach?

Temporal Aspect of System Integration: When and how should system integration occur in the design, development, test & engineering process? The traditional approach, based primarily upon the strictures of federal procurement law and a cultural tradition, places the system integration process late in the Phase C/D Design-Build phase of a hardware contract. Any changes or corrections in integration so late in the manufacturing cycle are extremely expensive and difficult. These questions concern the fundamental approach to system integration and its relation to technology development.

1. While Phase C/D is unavoidable for flight hardware integration, is it the best time to begin system integration?

2. Can the system integration occur early—"pre-Phase A"—to establish the functionality desired for the mission?

3. When can the Mission to Mars start?

4. Is NASA's traditional 8 point Technology Readiness scale an appropriate gauge for a mission-oriented technology development program?

Temporal design also concerns the long-term conduct of the whole Mars exploration program. A successful Mars Mission architecture should reflect a sustained commitment to exploring Mars. The MSM design will approach the temporal aspect of exploration, not as a series of arbitrary target dates, but as an assessment of appropriate and deliberate speed. This assessment involves meaningful milestones with a vigorous pursuit of precursor exploration and technology development missions. For each exploration activity, will a time delay allow the development of a new enabling, safer or more cost-effective technology within a reasonable time? To what extent should technology development pace Mars exploration? To what extent are missions repeatable—not just single flight opportunities—but for a plan for efficient, reliable and timely follow-on expeditions?

Performance Measures

All these considerations under Technical Credibility boil down to establishing performance measures for every significant aspect of the design and the technology. Performance measures require clear indices to evaluate both if the design succeeds or fails and how well it succeeds or how badly it fails. Then the program manager must decide whether to continue pursuing that approach or to try an alternative. This decision often requires courage, first to admit making a mistake and second to write off the costs of an unsuccessful design or unsatisfactory technology investment. The ability and willingness to make these evaluation decisions imply that the mission design program must study several alternative designs and technologies for many of the critical performance measures. This approach requires the willingness to devote more people and resources up front to ensure choosing the best approach instead of paying greater penalties downstream in the design process to correct poor decisions.

An example of an architectural design performance measure comes from the American Institute of *Architect's Handbook of Professional Practice*. The Guideline for the Architectural Areas and Volumes or Buildings provides specific rules for measuring different attributes of tenant-usable areas versus infrastructural or common areas and volumes. The evaluation of each habitat design concept should adopt a similar consistent approach to measuring spatial efficiency.

OPERATIONAL PRACTICALITY

Operational practicality involves several issues: cost-effectiveness, crew health and safety, human productivity, and life cycle costs.

Cost-Effectiveness

What is the appropriate criterion for cost-effectiveness? Every Mars mission proposal and technology is analyzable for cost, pay-off, and risk. The analytical criteria shall be fair and impartial in both as written and as applied. The benchmarks for assessing cost and benefit include: mass in Earth orbit as a cost, mass deliverable to Mars as a benefit, and delivery time as a variable. This evaluation must compare each specific exploration step for the risk of failure against the risk (to the next phase of exploration) of not taking it. The key cost question is whether the size of the infrastructural investment for humans to go to Mars implies that a base-building approach from the first landing will be most cost-effective.

Does the Mission design "trip over a pound to save a penny?" This question is particularly important for ideas of system integration and commonality. On Space Station *Freedom*, it was an article of faith that commonality and standardization would yield substantial "cost avoidances." However, standardization most often meant eliminating a capability that did not fit within a certain framework of what should be common or whose hardware should be common and whose should not.

Protect Human Health and Safety

How well does the mission design provide protection, support and multiple solutions to ensure the crew's health and safety? What criterion does the mission design use to baseline these protections, and are they the appropriate performance measures? The ***Mission Design Logic*** identifies these crew health milestones as interface constraints, but they also arise as design evaluation criteria insofar as they constitute performance criteria for the crew habitats on the TMIV, the Mars Surface and on the return TEIV. Thus, does the mission design provide clear assessment measures or criteria for crew health at each of the four major way points:

1. What is the baseline condition of the crew upon launch from Earth? (+0 days)
2. What is the crew's condition upon arriving on Mars? (+120 to +200 days)
3. What is the crew's condition before lift-off from Mars? (+720 to +800 days)
4. What is the condition of the crew upon return to Earth? (+920 to +1100 days)

Human Productivity

In assessing cost-benefit considerations, especially for science, the conventional input-output profit model of productivity is neither appropriate nor relevant. Rather, the evaluation must look to new values in productivity including: long term-returns, making opportunities for creativity, discovery and serendipity, the growth of capabilities and skills, planned reliability and quality, and enhancing the lives of the explorers and the people on Earth. Human performance will determine the success of manned missions of 1,000 days or longer more critically than any other element. The definition of cost effectiveness must incorporate human factors, habitability, and sustained performance as the highest priority. Habitability, the qualities of work life, crew autonomy, democracy, and teamwork are all components of productivity in the high technology environment. For the architectural design of Mars spacecraft and habitats, this priority means that the human performance quality of the living and working environment takes precedence over short-sighted economies such as maximizing equipment packaging or shortening utility runs.

Sustainable Operational Capability

The purpose of this technology development is to ensure thoroughness in the study of Mars through a substantial operating capability. The benchmarks of this capability will include self-sufficiency, mobility and the benefit of a variety of human skills. This mix of capabilities should attain a harmony in working together to accomplish the exploration objectives.

Is this mission a "flag & footsteps" enterprise that sows the seeds of its own demise even as it achieves its highest goal, or does it present persuasive long term benefits and staying power?

Growth Path

An important aspect of both credibility and design decision-making is the mission growth path. is it an evolutionary capability; that allows the project technology to grow and mature? Or is it a stable or closed-end approach in which accommodating a new technology means a whole new mission start?

Pre-judging this question is difficult at best. However, at least design decisions should examine which future options they foreclose and which they keep open. However, evolution is not an absolute value; it is important to recognize that after a particular lifetime, the Mars surface habitat and associated hardware may become obsolete or too expensive to maintain. Then it will be time to build the next generation, rather than limp along with the old status quo.

Evolutionary capability also means a progression in thinking about using machine missions to prepare for human missions. Increasing international cooperation may be another measure of evolutionary capability and reduction of the support staff on Earth may be another such measure. These two measures may conflict; resolving such contradictions is an essential aspect of planning for evolutionary capability.

Completion and Stopping Point

When is the build up of a Mars habitat or Mars base complete? How is the completion of construction different from the completion of exploration—at a particular site or region—or as part of a particular program?

Life Cycle Costs

Life-cycle costs, especially for logistics, operations and resupply, are equally important as initial costs for early explorations and even more important for the permanent Mars Base. This analysis will compare the economics of short term projects against other short term projects and the economics of a permanent base against other permanent base ideas.

1. Does the Mission design reduce Mission resupply cost through ISRU? How effective is this reduction in cost and mass?
2. Does the Mission design use Space or In-Situ Resources to enhance the safety and viability of the mission as a whole?
3. How do the life cycle costs for a series of short expeditions compare to fewer but much longer missions?
4. How do the comparative mass budgets compare for renewable versus non-renewable consumables compare, especially for life support and energy systems?
5. How long is a realistic "life" on which to base life-cycle costs?
6. Is the baseline life-cycle the same across the entire Mars mission?
7. **Is it possible to factor scientific return or accomplishment of mission goals into the cost-benefit equation?**

PARTICIPATORY PROCESS

A design problem definition or a design solution can live through the design process only so long as all the people involved agree to it. This design lifetime becomes the province of participatory process. The ideal outcome is a consensus upon the products of the two design evaluation gateways that will sustain the project all the way through to completion. Participatory process helps ensure that the stake-holders "buy into" both the problem definition and the design solution.

Dimensions of Problem Solving

This overarching criterion distinguishes between the two dimensions of problem solving: technical difficulty and participatory complexity. Technical difficulty is the domain of system engineering, and social or participatory complexity is the domain of participatory and democratic process. The relationship between these domains is that the participatory design research process will reduce ill-defined Mars exploration complexity to well-structured goals upon which system engineering can operate. A rigorous system engineering method will then focus upon specific analysis of technical requirements and capabilities to produce the mission design.

Participatory Process is also key to persuading all investigators who wish to explore Mars have an opportunity to contribute their ideas and resources to the Mars exploration program. It can also ensure—indeed, it may be the only way to ensure—that an appropriate and innovative problem definition can lead to a creative Mission Design solution. The absence or failure of participatory process that leads to a national or international consensus means that "business as usual" will smother any new human space program in its cradle under bureaucratic overload. Participatory process is support of a clear consensus upon a design solution may be the only way to build the momentum to go to Mars.

Participation in Mars exploration involves developing agreement upon both national and international goals. Building trust, consensus and openness among domestic and international partners is a key to optimizing the use of Earth resources. Participation in Mars exploration leads us to develop new educational, technical training and employment opportunities. The exploration program must serve to educate the public to appreciate the scientific objectives and benefits of Mars exploration and to provide the competent people to carry the work through future generations. Participation includes sharing technology development efforts with less developed economies on Earth to broaden the base of communities and nations that can play a role in Mars exploration that will be meaningful to them and that will help them develop economically.

Professionalism in Design

However, it is essential to maintain the highest professional standards of objectivity to avoid a rush to judgment about mission strategies or technologies. The main goal of design methodology for future space missions may be to avoid a premature division of the work into packages before defining the problem (on the model of the space station program). However, neither is it reasonable to wait until the program can identify every requirement for every mission situation. Rather, the best hope is to discover and better understand requirements through design synthesis. This design research strategy encourages the most innovative engineering and architectural design approaches including: concurrent engineering, designing—in quality, multidisciplinary teamwork, peer review, democratic-participatory (instead of hierarchical) decision-making, and research in design method and process. These innovations can serve as a competitive and cooperative strategy to gain the positive technical benefits of both competition and cooperation while avoiding their negative aspects.

International Cooperation

Peaceful cooperation and exploration beyond the Earth are symbiotic. They demand international trust and openness to cooperate successfully. Both national goals and international goals can strengthen all the partners and be mutually reinforcing. Technology development and technology transfer can help a number of potential participating nations to develop their own industrial base and train Mars exploration candidates and scientific and technical interns. In this context, the Mars program should help revitalize educational systems around the Earth and increase industrial efficiency. Ultimately, however, the design and execution of the Mars exploration objectives will determine how well the participants meet these goals.

How will various exploration modes require different technology development efforts? Conversely, how would a particular new technology enable different exploration modes than currently are possible? What do particular technology mixes imply for international cooperation? Will international participation be an asset or an added difficulty to designing or conducting Mars exploration? How might this participation vary over time? It is important for Mars mission architecture proponents to be explicit about these questions.

PROGRAM MANAGEMENT

What is the most appropriate approach to Mars Exploration Program management? Judging from the most recent human space project experiences in NASA, it would be easy to erect an immense bureaucracy that produces tons of reports and studies but does nothing to go to Mars. Consider that the Space Shuttle Program has three "levels," and has struggled with turf problems, including the *Challenger* accident. The Space Station Program consists of four complete levels of "management" spread over five locations. The prospect of multi-agency collaboration on a Mars expedition (NASA, DOD, DOE) raises the specter of a five-tiered bureaucracy. Substantial international participation raises the ante to a six-tiered bureaucracy. Without a radical change in the approach to program management, a Mars exploration enterprise will become a nightmare for everyone involved —except the paper and copier industries.

Mike Griffen and James French emphasize this point:

> It is too easy in the modern era to be seduced into creating overly complex and unnecessary paper systems. (The computer is particularly dangerous in this case with the multitude of software systems to "help the manager manage.") Once created, these paper systems seem to take on a life of their own and expand and propagate. Even with computerized concepts, huge amounts of time and money can be wasted in excessive documentation. The systems engineer should think through the documentation requirements for his activity and implement a plan to meet the requirements. Avoid unnecessary "bells and whistles" that sound great but do not contribute. They will exact their price later.[11]

Program Management for Mars exploration will not succeed as "Business as Usual." Instead, the Mars exploration program must seek out new approaches to design, management, and planning that will accommodate the new challenges of Mars exploration. These approaches most likely will include extensive international participation, vast complexity, sustained design and operations effort over several decades, and great technical difficulty. The key to approaching this challenge is to understand the nature of problem solving, particularly the relative advantages and disadvantages of participatory process and system engineering, and to frame them in the context of international cooperation.

All the great, successful aircraft and spacecraft programs share a single factor: a great design idea. Usually, there was a small group of designers or even a single designer who provided the intellectual and goal-seeking leadership that kept the program on track, and out of the many paralyzing detours of top-heavy management. Stated simply: once there is a consensus upon a design problem definition—and that consensus is completely independent of special interests—the designers should specify the requirements for program management.

Questions for Mission Design Management

1. Who are the Participants? Who are the Stakeholders?
2. What scientific and technical disciplines are necessary to provide analysis, design, engineering and management capabilities?
3. What governmental organizations should play a role—and what role should each play?
4. What industries can contribute to producing hardware, software, & systems integration?
5. How well does the management shape the design organization to fit the design problem and to develop its solution?
6. To what extent does the management map the design work packages onto existing organizations as welfare programs, without regard for the mission characteristics?
7. How should program management respond to all these factors?

CONCLUSION

The ultimate test of a space mission architecture is the degree to which it responds directly to the mission objectives as embodied in the design problem definition. Factors that may detract from success include: mapping the solution onto the performing organizations rather that addressing the design problem, pork-barrel distribution of tasks rather that picking the best people for the job, and subordinating critical design judgments to external considerations. The success of a Mars Mission architecture depends on how well it defines the problem, then solves it, while combating institutional inertia and political entropy.

REFERENCE NOTES

1. Klein, Gary A. (1987), "Analytical vs. Recognitional Approaches to Design Decision Making" in William Rouse and Kenneth Boff, editors, *System Design: Behavioral Perspectives on Designers, Tools and Organizations*, New York: North-Holland, Elsevier Publishing Co. p. 179.

2. Simon, Herbert (1973), "The Structure of Ill-Structured Problems," in Nigel Cross, editor (1984), pp. 145-166. *Developments in Design Methodology*, Chichester UK: John Wiley and Sons Ltd. [verify citation].

3. Jones, J. Christopher (1970), *Design Methods: Seeds of Human Futures*, New York: Wiley-Interscience.

4. Schön, Donald A. (July, 1988) "Designing: Rules, Types and Worlds," *Design Studies*, Vol. 9, No.3, London: Butterworth & Co. Publishers Ltd.

5. Rowe, Peter G. (1987) *Design Thinking*, Cambridge, MA: MIT Press, p. 75.

6. Simon, Herbert (1973) *The Science of the Artificial*, Cambridge, MA: MIT Press [check pg.].

7. Miller, D. (1985) editor, *Popper Selections*, Princeton, NJ: Princeton University Press, p.62.

8. This term derives from the *Space Station Needs, Attributes and Architectural Options Studies* by eight aerospace companies working in parallel but separately before the Space Station Phase A began in May of 1983.

9. Gross, A. R., Harper, L. D. Shafto, Michael G., Vernikos, J., Webbon, B. W., and Berry, W. E., (1992) "Human Support for Mars Exploration: Issues and Approaches," in Pritchard, E. B., *Mars; Past, Present, and Future*, Volume 145, *Progress in Astronautics and Aeronautics*, Washington, DC: American Institute of Aeronautics and Astronautics, pages 269-296.

10. "Guideline for the Architectural Area and Volume of Buildings, AIA D101," (1980 edition?) in *Architect's Handbook of Professional Practice*, Volume III, Washington, DC: American Institute of Architects. [check for most recent].

11. Griffen, Michael D. and French, James R., (1991) *Space Vehicle Design*, Washington DC: American Institute of Aeronautics and Astronautics, pages 8-9.

THE NOMAD EXPLORER ASSEMBLY ASSIST VEHICLE: AN ARCHITECTURE FOR RAPID GLOBAL EXTRATERRESTRIAL BASE INFRASTRUCTURE ESTABLISHMENT

Madhu Thangavelu*

Traditional concepts on lunar bases describe scenarios where components of the bases are landed on the lunar surface, one at a time, and then put together to form a complete stationary lunar habitat. Recently, some concepts have described the advantages of operating a mobile or "roving" lunar base. Such a base vastly improves the exploration range from a primary lunar base. Roving bases would also allow the crew to first deploy, test, operationally certify, and then regularly maintain, service and evolve long life-cycle facilities like observatories or other science payload platforms that are operated far apart from each other across the extraterrestrial surface. The Nomad Explorer is such a mobile lunar base. This paper describes the architectural program of the Nomad Explorer, its advantages over a stationary lunar base and some of the embedded system concepts which help the roving base to speedily establish a global extraterrestrial infrastructure. A number of modular autonomous logistics landers will carry deployable or erectable payloads, service and logistically resupply the Nomad Explorer at regular intercepts along the traverse. Starting with the deployment of science experiments and telecommunication networks, and the manned emplacement of a variety of remote outposts using a unique EVA Bell system that enhances manned EVA, the Nomad Explorer architecture suggests the capability for a rapid global development of the extraterrestrial body. The Moon and Mars are candidates for this strategy. The lunar case is emphasized in this paper.

NOMENCLATURE

AMCL - Autonomous Modular Common Lander
DIPS - Dynamic Isotope Power System
ETO - Earth to Orbit
EVA - Extra Vehicular Activity
ECLSS - Environmental Control and Life Support System
GNC - Guidance, Navigation and Control
LOI - Lunar Orbit Insertion
MOSAP - Mobile Surface Applications Traverse Vehicle

* Consultant, Space Systems Division, Rockwell International, Downey, California 90241. Research Associate, Institute of Aerospace Systems Architecture and Technology, School of Architecture and the School of Engineering, University of Southern California, Los Angeles, California 90089-1191.
AAA*VISIONEERING, 207 S. Hoover Street, Los Angeles, California 90004-6158. An earlier version of this concept was presented at the World Space Congress, Washington, DC Aug. 28 - Sept. 5, 1992.

POS - Point of Start
RFC - Regenerative Fuel Cell
RMS - Remote Manipulator System
TLI - Translunar Injection
VLTV - Very Long Traverse Vehicle

INTRODUCTION

Permanent Lunar base establishment will entail detailed terrain exploration, sampling and analyses. Current studies expect all the site analyses to be performed robotically. Though unmanned precursors and remote sensing satellites would provide valuable information on possible sites for lunar base location, in order to "live off the land," as President Bush directed, detailed manned exploration and formal site analysis of selected candidate sites will be required. Since site selection is a critical task, initial manned missions will have to "site-hop" before settling on the prime candidate site or sites. Global mobility then becomes imperative in the initial manned mission. Following this initial activity, long duration facilities like optical/ultraviolet interferometry observatories emplaced on the extraterrestrial surface will require regular on-site manned supervision for maintenance, refurbishment, and for steady evolutionary enhancement of capability. As these science and observatory platforms would be spread out over rather large stretches of extraterrestrial surfaces, again global mobility would greatly enhance such activity. And finally, these remote autonomous science outposts, observatories, technology testbeds and pilot manufacturing plants would have to be in constant communication with the Earth. As most of these facilities are stationary, extraterrestrial fiber optic networks linking them all together offer promise. Again, the mobile rover strategy would make it viable to lay fiber optic cables across the extraterrestrial surface, interlinking all of these outposts and providing simple, reliable yet very high quality telecommunication with the Earth.

Examining the program requirements which have already been developed by NASA [1], it might be possible to design a single manned mission assisted by an autonomous logistics lander system that would conduct all the required tasks as well as establish a global infrastructure of remote science outposts, observatories, technological testbeds for long duration missions and pilot indigenous materials manufacturing plants, and all of the necessary telecommunication networks to support such activity, all of which activities could be simultaneously manifested in a combined manned/unmanned mission architecture.

DEVELOPMENT OF THE NOMAD EXPLORER STRATEGY

NASA exploration studies indicate the need for manned rovers to assist in the development of extraterrestrial bases. Three classes of rovers have been identified and studied. Point design concepts have also been proposed in all three categories [1]. Studies include the short range rover for around-the-base activity with a range of 50-100 km, the long range vehicle with an operating range of 1000 km and finally, the very long range traverse vehicle (VLTV) capable of covering 3000-10,000 km on a single traverse. The mobile surface application traverse vehicle (MOSAP) is a vehicle that

NASA has proposed for this purpose [2]. Capable of all the exploratory functions and normal EVA that is carried out from a conventional stationary base, the VLTV, by virtue of its long range and enhanced manned crew systems capability, is in essence, a "roving lunar base"[3]. Unlike a stationary base, from which detailed manned terrain exploration is limited by the range of the rover, this mobile concept for a primary exploration-oriented base offers unlimited range for exploratory traverses (Figure 1). It is an extension of the program requirements of this very long range traverse vehicle that leads to the possibility of the Nomad Explorer strategy.

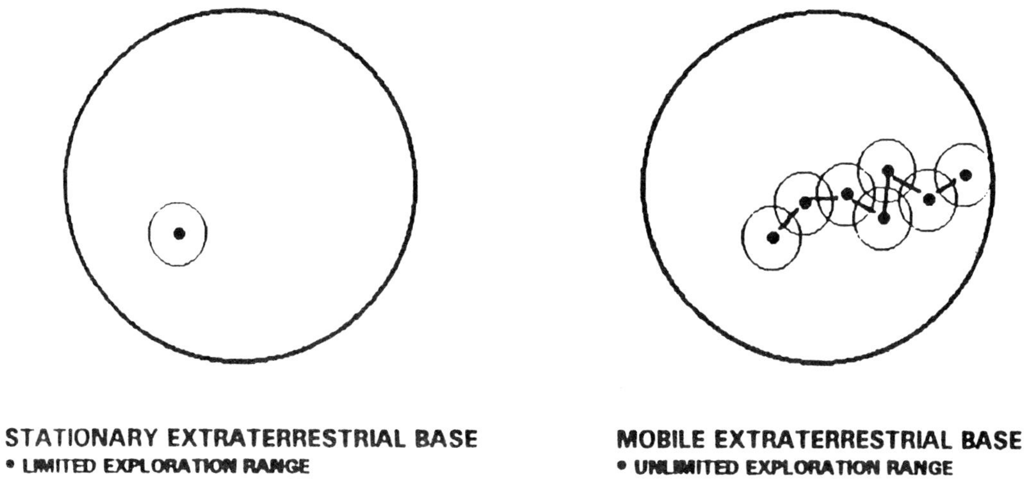

Figure 1

The Nomad Explorer Strategy is a synthesis of two major system architectures. They are:

1. The Nomad Explorer Vehicle.
   The Nomad Explorer Vehicle is derived from the MOSAP vehicle developed by NASA. Several other studies done by NASA contractors on long range rovers are also available. Very long range traverses are possible using this vehicle. The architecture of this vehicle is adaptable to both manned or autonomous unmanned operations. The configuration studied in this paper portrays a manned Nomad Explorer Vehicle. See Schematic in Figure 2.

2. The Autonomous Modular Common Lander System.
   The Autonomous Modular Common Lander (AMCL) is a common vehicle that can deliver both crew or cargo depending on the modular configuration employed. Logistics, consumables, and erectable/deployable science outpost components that are launched from Earth are delivered to the lunar surface using the AMCL system. The concept is derived from the lunar lander [3] and the Common Lander Study [4] which NASA has developed. The AMCL is outfitted with modular payload on the Earth, landed autonomously on the lunar/Mars surface, intercepted by the Nomad Explorer and the payload unloaded and deployed or transferred to the roving vehicle. The modular capability allows the lander to be sized for any mis-

sion, ranging from 5 MT to 25 MT depending on the requirements during the traverse of the Nomad Explorer vehicle (Figure 2).

SCHEMATIC OF NOMAD EXPLORER

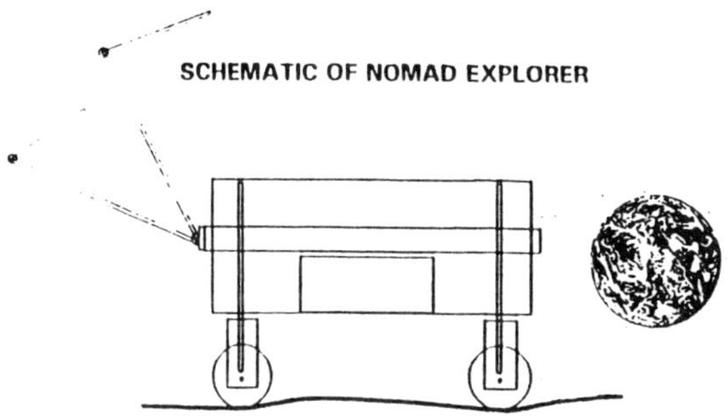

SALIENT FEATURES

1. VERY LONG RANGE TRAVERSE VEHICLE
2. CREW OF THREE, SIX MONTHS, 10,000 KM.
3. MANNED ASSEMBLY ASSIST VEHICLE.
4. BELL ATTACHMENT FOR ENHANCING MANNED EVA.
5. NUCLEAR POWER.
6. LOGISTICALLY RESUPPLIED BY MODULAR REPLACEABLE UNITS DELIVERED BY THE AUTONOMOUS LANDER.

MISSION

1. "SITE-HOP" TO LOCATE BEST SITE FOR PERMANENTLY MANNED LUNAR BASE.
2. DURING TRAVERSE, ESTABLISH SCIENCE OUTPOSTS ALONG THE WAY, DEPLOY EXPERIMENTS, AND ESTABLISH EXTRATERRESTRIAL GLOBAL TELECOMMUNICATION INFRASTRUCTURE USING FIBER OPTICS AND LASER TECHNOLOGY.

SCHEMATIC OF AUTONOMOUS MODULAR COMMON LANDER

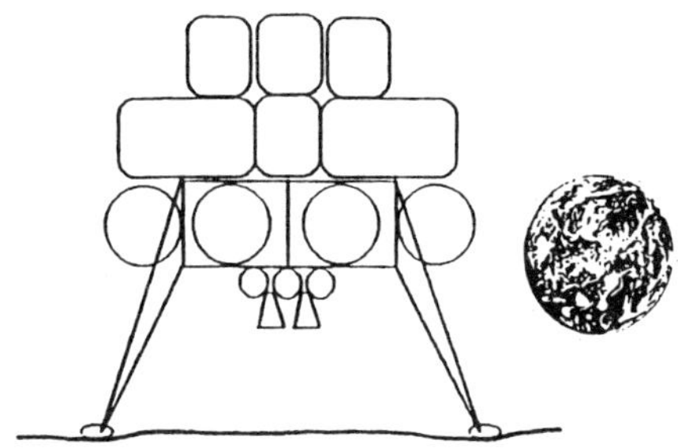

SALIENT FEATURES

1. AUTONOMOUS LANDING CAPABILITY
2. MODULAR STRUCTURAL/PROPULSION DESIGN
3. COMMON CARRIER FOR NOMAD EXPLORER CREW TRANSPORT, LOGISTICS, ERECTABLE OR DEPLOYABLE SCIENCE PAYLOADS
4. PAYLOAD CAPABILITY - 5,10,15,20,25 MT DEPENDING ON MODULAR CONFIGURATION

MISSION

1. PROVIDE CREW TRANSPORT, LOGISTICS SUPPORT FOR THE NOMAD EXPLORER
2. CARRY SCIENCE/OTHER PAYLOADS TO BE DEPLOYED/ERECTED BY CREW OF NOMAD EXPLORER.

Figure 2

The Nomad Explorer strategy is aimed at establishing a global scientific and telecommunication network, even during the initial "trail blazing" permanent base selection run. The Nomad Explorer, besides being a VLTV, is also conceived as a manned EVA and assembly assist vehicle.

Exploratory activity like soil sampling, soil mechanics, locating natural formations that enable habitation like lava tubes [5], is followed by the deployment of science outposts and habitation facilities during the course of the same mission. This vehicle, during the course of its very long traverse, will intercept autonomously landed payloads ("modular common lander" payloads?) along the traverse route, assemble the payloads or deploy them, check them out at the site, certify them for operations, doing all of these functions using the crew of the vehicle. During the later stages of the extraterrestrial base evolution, the strategy could be used for maintenance, repair activity and evolutionary enhancement of these remotely based highly sensitive scientific experiments like optical interferometry observatories [6]/pilot plants for manufacturing lunar indigenous materials and components [7]/telecommunications relay platforms, etc.

The Nomad Explorer Global Extraterrestrial Basing Strategy is as follows:

1. Precursory high resolution mapping of the entire lunar/Mars surface using polar orbiting satellites.

2. Analysis of terrain information, followed by determination of alternate likely candidate sites for lunar/Mars base location. Alternate rover traverse routes established with the aim of maximizing scientific returns [8] along the route while conducting detailed terrain surveys for locating alternate sites for a permanent base. A lunar polar traverse might be considered to explore trapped volatiles [9].

3. One or more Nomad Explorer VLTVs launched from ETO, coupled with lunar lander tankage in LEO, and landed on the lunar surface at the predetermined point of start of traverse.

4. Lunar lander with crew accompany Nomad Explorer to the point of start (POS) of traverse.

5. Crew transfer to Nomad Explorer and begin traverse. First telecommunication Earth link established at POS.

6. Modular autonomous landers deliver modular payload along traverse route. Alternate base sites surveyed. Crew intercept payload, retrieve consumables/logistics modules, prepare and carry out assembly and deployment of science experiments and telecommunication relay stations, using manned EVA enabling systems which are part of the Nomad Explorer vehicle architecture.

7. Nomad Explorer completes traverse. Establish global fiber-optic telecommunication network. Install scientific outposts. Conduct detailed geologic surveys and prospecting. Determine location for a permanent lunar settlement.

8. If more than one Nomad Explorer is landed, parallel activity would further speed up the base site selection and global infrastructure development. If required, the teams could assist each other by congregating at a particular site of interest, for exploration, for need of added manpower, or for establishing a permanent manned lunar base at a most suitable location.

9. At end of traverse, mothball Nomad Explorer into energy conserving "hibernation mode" or set up vehicle for remote traverse operations. Permanent base site established.

10. Crew transfer to Earth return vehicle that has been autonomously landed at end of traverse. Crew depart for Earth. Mission complete. Permanent base establishment activities commence.

11. Nomad Explorer crew return to vehicle and proceed on traverses at prescribed intervals to repair, maintain or evolve remote outpost elements.

The Nomad Explorer Strategy for lunar basing is depicted in Figure 3.

THE NOMAD EXPLORER MISSION PLAN

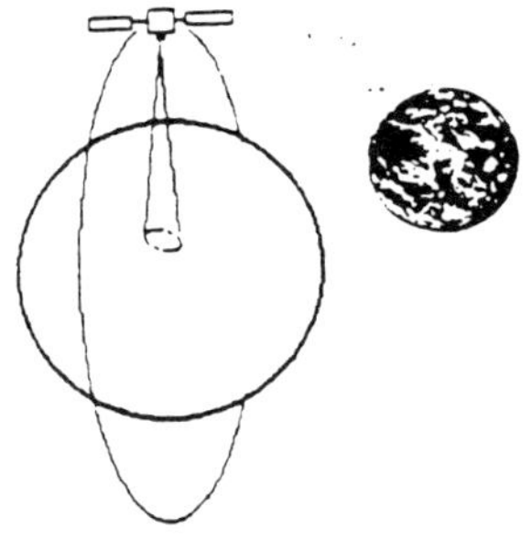

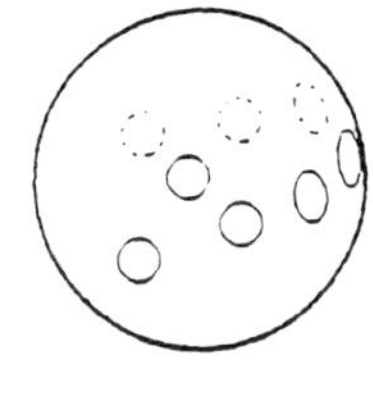

1. HIGH RESOLUTION IMAGING OF EXTRATERRESTRIAL SURFACE. DETAILED GLOBAL MAPS PREPARED.

2. SEVERAL CANDIDATE BASE SITES IDENTIFIED. BOTH ON LUNAR NEAR SIDE AND FAR SIDE.

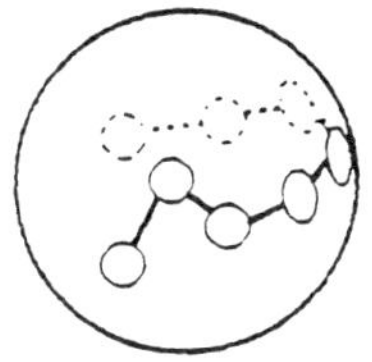

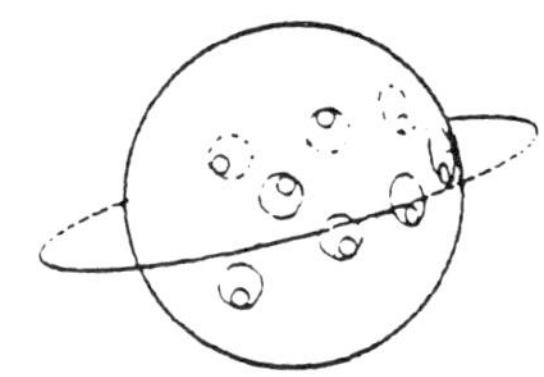

3. ALTERNATE NOMAD EXPLORER TRAVERSE ROUTES EXAMINED. FINAL TRAVERSE ROUTE IDENTIFIED.

4. AUTONOMOUS LANDER(AMCL) DELIVERS LOGISTICS TO CANDIDATE SITES.

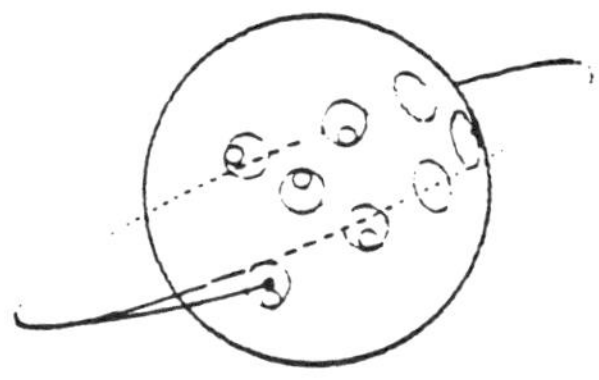

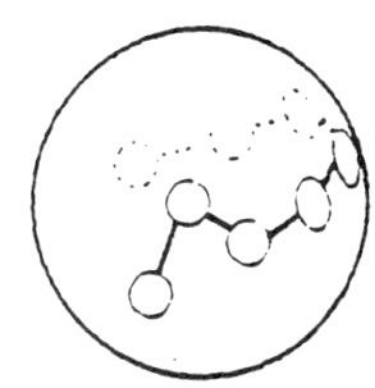

5. NOMAD EXPLORER LANDED AT POINT OF START OF TRAVERSE. ALL SYSTEMS CHECKED OUT FROM EARTH. CREW LANDS NEXT TO VEHICLE IN AMCL CONFIGURED FOR CREW TRANSPORT. TRANSFERS TO NOMAD EXPLORER.

6. DETAILED MANNED EXPLORATION OF NATURAL TERRAIN FORMATIONS(LAVA TUBES, RILLES ETC.) SOIL MECHANICS AND OTHER "HANDS-ON" EXPERIMENTS CONDUCTED. TELECOMMUNICATION LINK DEPLOYED AS MISSION PROCEEDS.

7. AS MISSION PROCCEDS, NOMAD EXPLORER REGULARLY RENDEZVOUS WITH AMCLs CARRYING SITE SPECIFIC PAYLOADS OF SCIENCE EXPERIMENTS AND RELATED HARDWARE. NOMAD EXPLORER INTERCEPTS AMCLs, ASSEMBLES AND DEPLOYS SCIENCE PACKAGES. LOGISTICS REPLENISHED. IF MISSION DURATION EXCEEDS SAFE STAY TIMES, SECOND FULL CREW REPLACEMENT ARRIVES DURING MIDDLE OF MISSION AND THE FIRST FULL CREW RETURN TO EARTH IN AMCL.

8. NOMAD EXPLORER COMPLETES TRAVERSE. VEHICLE SWITCHED TO "HIBERNATION MODE" TILL NEXT MISSION. CREW RETURN TO EARTH IN AMCL THAT IS READY AND WAITING AT END OF TRAVERSE.

Figure 3

THE NOMAD EXPLORER VEHICLE SYSTEMS ARCHITECTURE

The Nomad Explorer is a VLTV with an essentially unlimited operating range which depends entirely on the number of logistic resupply missions that are flown to it during the course of the mission. Assisted by the AMCL system, the Nomad Explorer carries only enough consumables, logistics and spares required by it between regular intercepts of the AMCL; much like the optimum pitstops and refueling operations carried out in automobile racing. A 10,000 km traverse could be used as an example to demonstrate the proposed capability of the Nomad Explorer. Duration of traverse could be six months to a year with the possibility of a complete crew changeout during the middle of the mission. Two regenerative fuel cell (RFC) power plants with a total peak output in the range of 50 kW are required for powering the drive train and all of the manned and unmanned systems. Advanced photovoltaic arrays would assist the rover systems during the lunar day cycle. Mission architecture dictated a crew of four for optimum performance and the manned systems and life support are provided for four crew. A simple exploded schematic in Figure 4 shows the basic systems of the manned Nomad Explorer.

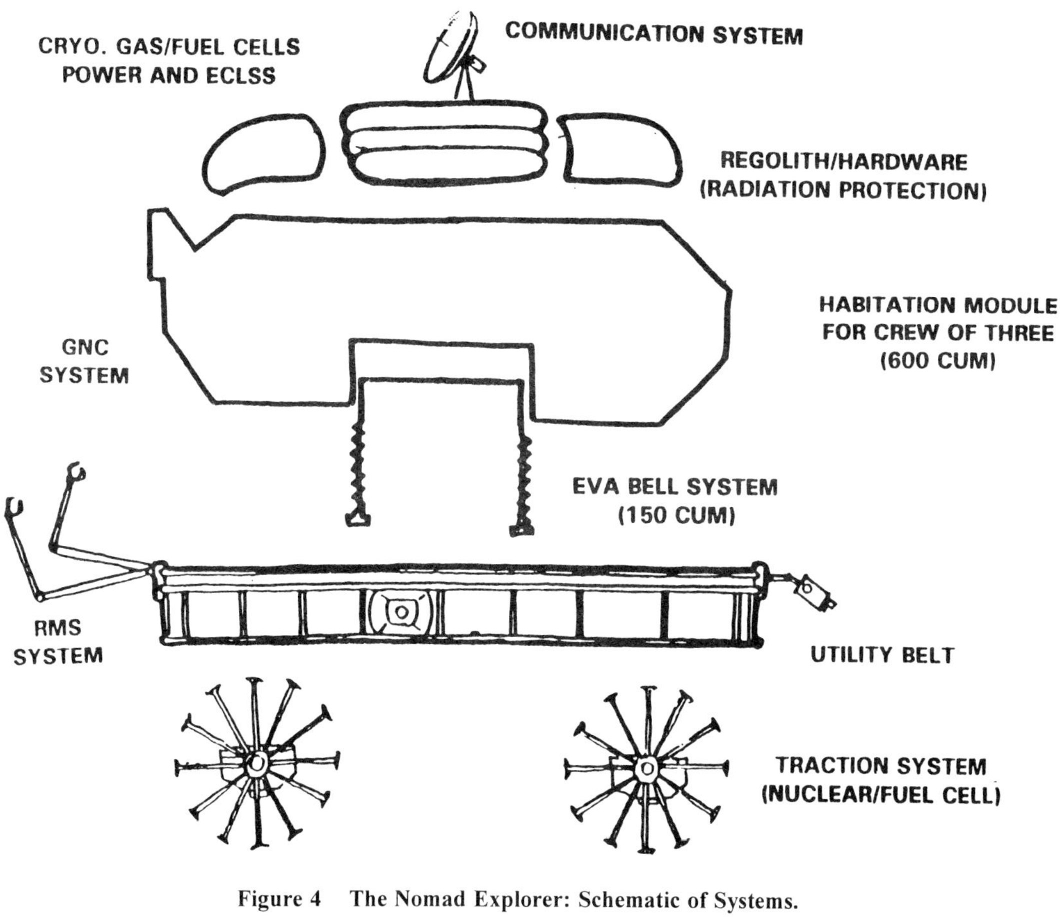

Figure 4 The Nomad Explorer: Schematic of Systems.

The main features of the Nomad Explorer are as follows:

1. Habitation
The pressurized volume of the vehicle is about 300 cum.
This volume contains long term accommodation facilities for three crew, work and conferencing areas, a command and control center, and ample storage space. Besides a galley, hygiene and waste management facilities, the long term accommodations include a health maintenance facility and a recreation space.

2. Environmental Control Life Support System (ECLSS)
The ECLSS will handle the needs of four crew. Cryogenic nitrogen, oxygen, hydrogen and water are available from RFC operations. Though complete closure of the ECLSS is not envisaged, it may be possible to operate with a 90% efficiency [1]. Space station Freedom will provide the basis for the Nomad Explorer ECLSS [10,11,12.13].

3. The EVA Bell
The EVA Bell for enhancing manned EVA is a prominent feature of the vehicle and is described later in detail. The invention allows the astronaut crew to perform EVA in a more comfortable manner by providing a shirtsleeve environment around the payload to be assembled and deployed during the earlier mission, or serviced, repaired or enhanced during later extraterrestrial base evolutionary activity (Figure 6).

4. The Utility Belt
An utility belt is strapped around the perimeter of the vehicle. This belt has modular racks which carry "plug on" modules for logistics, consumables and waste management that are replenished by replacement nodules arriving on autonomous landers which the vehicle intercepts for time to tome along the traverse. The belt also carries tools and accessories for EVA. Two small unpressurized rovers for "around the base" traverses are also part of the EVA accessories. These rovers have a range of about 100 km and are powered by fuel cells and advanced photovoltaics.

5. The Remote Manipulator System
Two remote manipulator systems that are capable of assisting manned EVA, loading and unloading cargo, and providing anchoring or scaffolding support during assembly/deployment operations, run along two tracks on the top and bottom through the entire length of the utility belt. High resolution cameras mounted on this track as well as in other strategic points provide video support during the traverse as well as during assembly operations.

6. Traction System
Traction could be provided using several options. The Nomad Explorer configuration in Figure 5 depicts independently powered and steerable large variable diameter wheels which arc unfurled and deployed after landing. Figure 7 schematic shows a telescopic traction system that is capable of adjusting the height of the Nomad Explorer chassis for enhancing traverse as well as assembly operations.

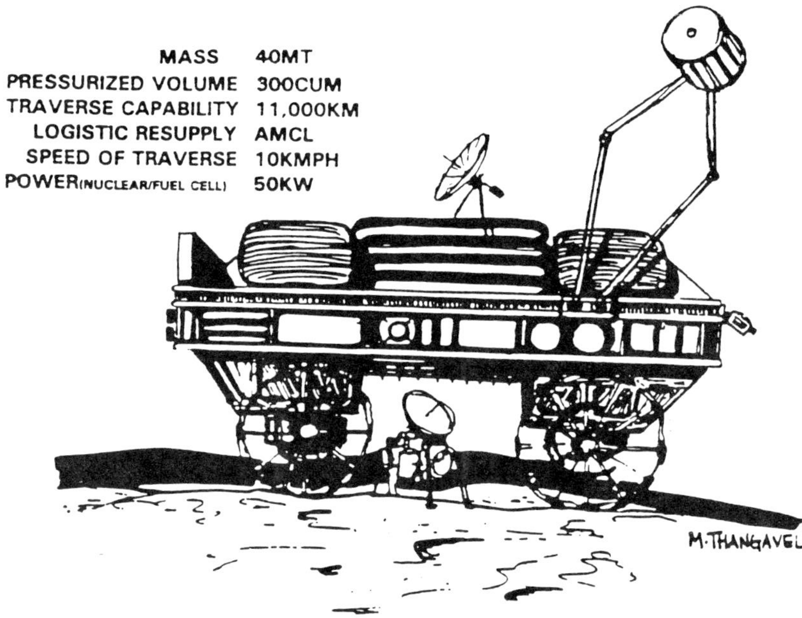

A Longitudinal Section

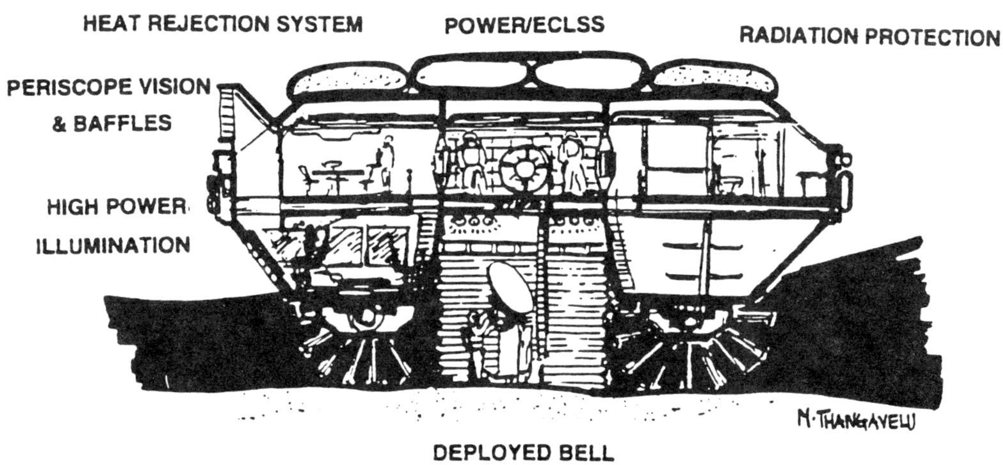

Figure 5 The Nomad Explorer: A Possible Configuration.

7. Radiation Protection
   Radiation protection is provided by skillful placement of system hardware on the vehicle so that they provide sufficient mass for protection. Tankage might be employed to enhance radiation protection [14]. Regolith bags could be packed and laid in area that require additional protection during solar particle events which might occur during the traverse.

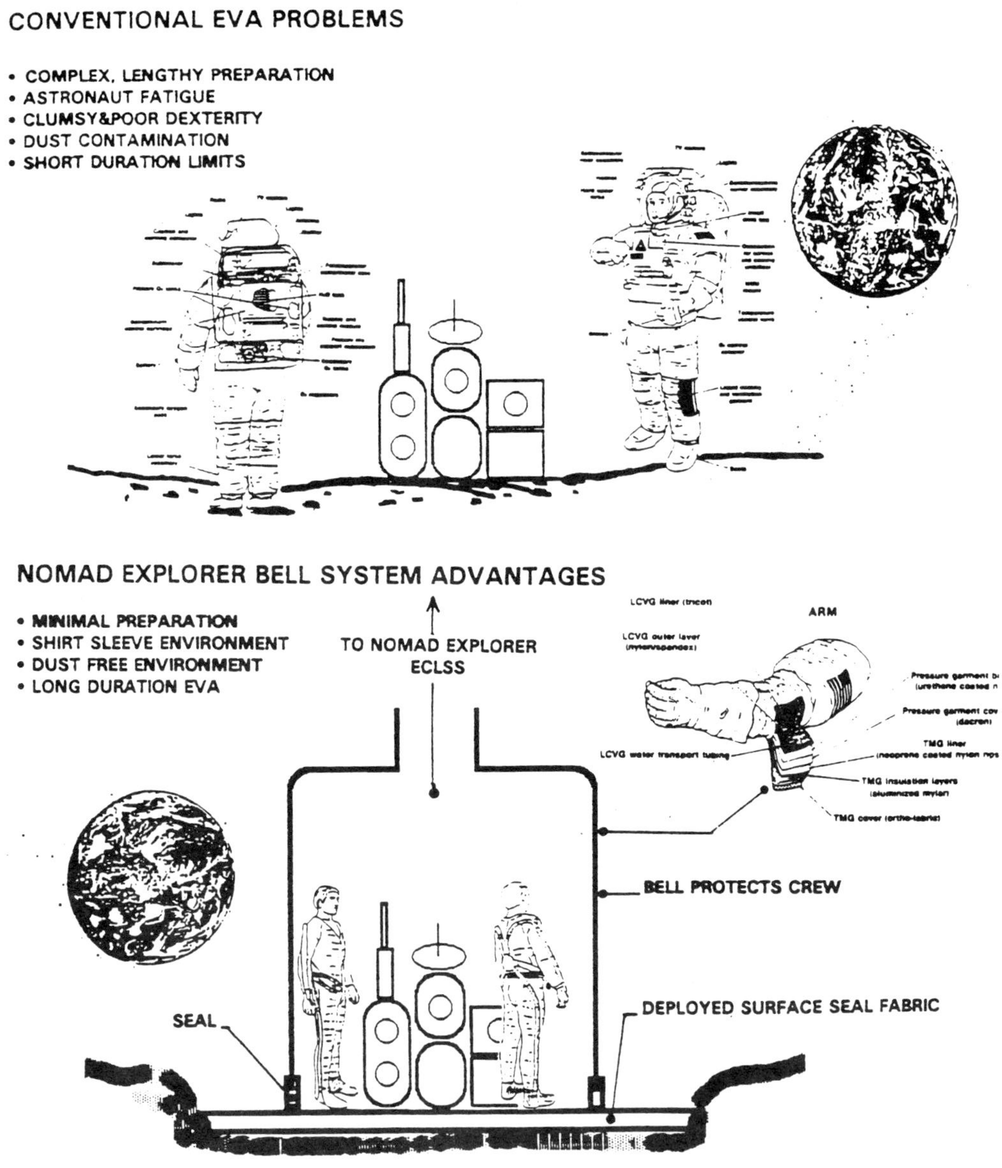

Figure 6

8. Guidance, Navigation and Control System

Guidance, navigation and control (GNC) of the vehicle is achieved through appropriate systems. Visual feedback could be direct or augmented by video support. Real-time telecommunication is possible through a 3 m Earth pointing antenna and a chain of surface-based relays or low orbiting satellites that could be employed for contact when line of sight communications is not possible. Electronically steered planar array antennas could be employed to maintain contact with low orbiting satellites or stationary extraterrestrial relay stations while the rover is in motion, even over rugged terrain. Fiber optics/free space laser communication and related technologies could enhance Nomad Explorer operations.

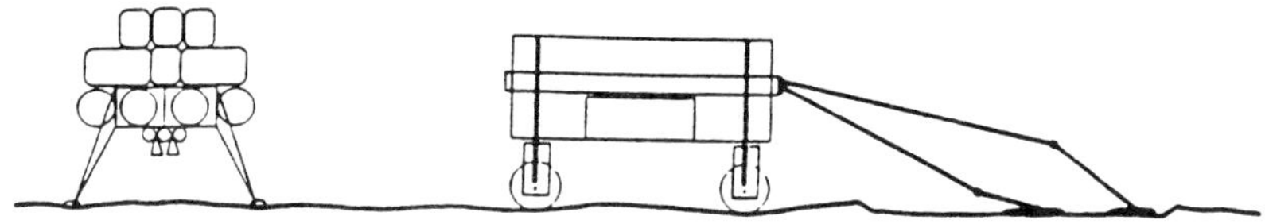

1. NOMAD EXPLORER INTERCEPTS AND PARKS CLOSE TO AUTONOMOUS MODULAR COMMON LANDER(AMCL) WHICH CARRIES LOGISTICS AND SITE SPECIFIC SCIENCE PAYLOADS. MODULAR LOGISTICS RETRIEVED. SUPPLIES REPLENISHED. DEPLETED LOGISTICS MODULES/WASTE DISCARDED.
2. NOMAD EXPLORER PREPARES TERRAIN FOR PAYLOAD ASSEMBLY/DEPLOYMENT.

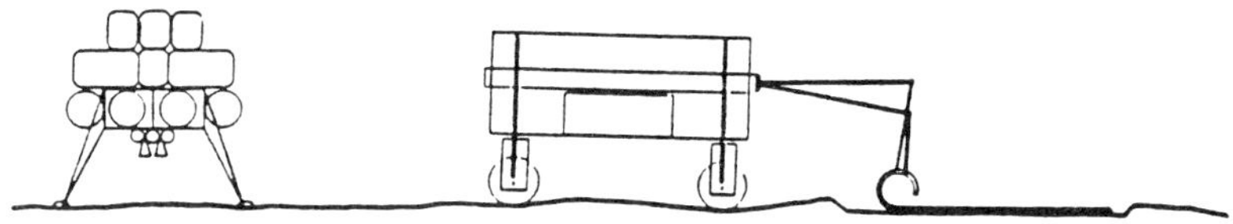

3. NOMAD EXPLORER LAYS SURFACE SEAL FABRIC OVER PREPARED TERRAIN. THIS FABRIC WILL HELP TO CONTAIN THE PRESSURE INSIDE THE DEPLOYED EVA BELL.

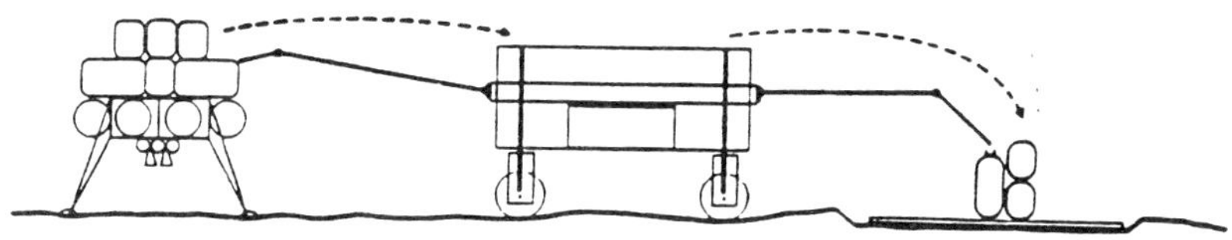

4. NOMAD EXPLORER UNLOADS PAYLOAD FROM AMCL OVER THE SURFACE SEAL FABRIC AT THE PAYLOAD ASSEMBLY SITE.

5. NOMAD EXPLORER IS ELEVATED, SLOWLY MOVES TO PREPARED SITE AND ALIGNS EVA BELL OVER THE PAYLOAD AND THE SURFACE SEAL FABRIC.

6. NOMAD EXPLORER DEPLOYS EVA BELL WHICH PROVIDES CREW A SHIRT-SLEEVE ENVIRONMENT FOR MANNED EVA. CREW ASSEMBLE, CHECK OUT, CERTIFY PAYLOAD INSIDE DUST FREE, THERMAL, MICROMETEORITIC, AND RADIATION PROTECTED AND PRESSURIZED EVA BELL.

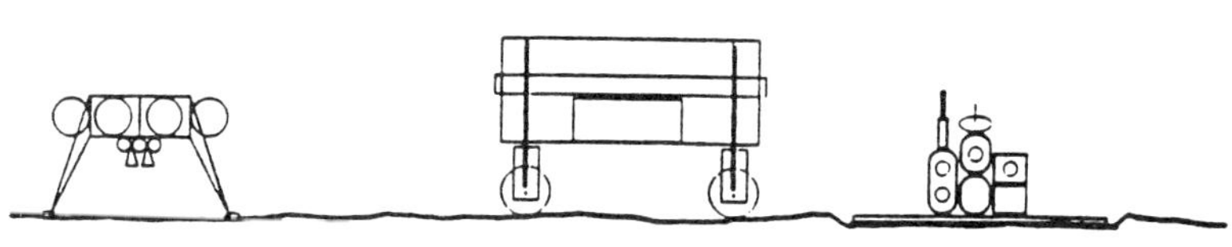

7. ASSEMBLY COMPLETE. EVA BELL RETRACTED. NOMAD EXPLORER MOVES AWAY FROM PAYLOAD ASSEMBLY. PAYLOAD IS OPERATIONAL. PROCEEDS TO NEXT SITE.

Figure 7 Nomad Explorer Payload Assembly Assist Sequence.

9. Power System
   Advanced Regenerative Fuel Cell (RFC) technology, advanced photovoltaics, as well as nuclear technology are suggested for the Nomad Explorer power system. The nuclear power option using the Dynamic Isotope Power System (DIPS) needs further study [1]. Shielding requirements need to be considered for nuclear power systems. A set of RFC batteries could power the Nomad Explorer between AMCL intercepts. At each intercept of the AMCL, these RFCs could be recharged. In addition, advanced photovoltaic arrays could be employed to provide support to the RFCs. In generating about 50 kW of power, (25 kW for drive train and 25 kW for the manned systems including the ECLSS), the heat rejection system would have to handle about 16 kW. High efficiency heat rejection systems are required. Dust contamination of radiator surfaces will require study and appropriate design. Recent developments in "power beaming" a technique whereby a microwave/laser beam from an external source is used to transmit power to the vehicle during traverse operations could substantially improve the performance of the Nomad Explorer by reducing the payload associated with power generation and storage equipment [15,16,17,18].

10. Mass and Payload Configuration
    All these systems are designed to fit within a HLLV (Energiya or revived Saturn V-B technology) payload shroud that is 30 m tall and 10 m in diameter. The payload mass at launch is about 35MT + propulsion, tankage and structure for TLI, LOI and lunar landing. This mass is above the NASA lunar lander capability of 25 MT. Though this Nomad Explorer study suggests that larger landers are required for the mission, it would be possible to scale down the vehicle and still preserve the mission architectural strategy for a smaller vehicle. The unmanned version of the Nomad Explorer is such a small vehicle. Substantially smaller (5-7 MT), the robotic Nomad Explorer would employ the same mission strategy. However, reliability and the capability to handle contingencies require demonstration. A possible manned Nomad Explorer configuration is depicted in Figure 5.

In the next section, the paper will discuss certain special features of the Nomad Explorer vehicle which were developed to enhance mission capabilities.

THE PROBLEM WITH CONVENTIONAL EXTRAVEHICULAR ACTIVITY

Conventional EVA is a most time consuming, hard and inefficient yet essential part of human activities in space or on the extraterrestrial surface. Complex and lengthy preparatory procedures are part of EVA. Present studies aimed at establishing extraterrestrial bases will require substantial EVA during buildup operations. Long hours of continuous EVA are expected during the early development phase of these projects. It is well known that present day designs for space suits are simply inadequate for these operations [19]. When inflated to the optimum operating pressure of about 8 psi, the suit becomes very stiff and quite difficult to flex at the required joints. The astronaut then has to work against this suit pressure stiffness as well as the forces which are required of the task to be performed. Loss of dexterity is the result and almost every component in a payload package to be assembled or deployed has to be designed to adapt to the limitations imposed by this loss of dexterity (Figure 6). In past EVA missions, astronauts and cosmonauts have complained about the difficulty of working in the EVA suit. Lunar dust is notorious for degrading astronaut as well as vehicle performance [20,21]. Compounded by the fact that future missions are expected to be more complex and substantially longer in duration, it is imperative that alternate methods for conducting EVA be studied. Hard suits, where the fabric is substituted for a metallic shell with articulation mechanisms at the essential joints have been studied but they have their limitations too. Compact modules with appropriate life support and remote manipulator systems (the so-called "man-in-a-can" concepts) have also been suggested as an alternate means for enhancing long duration EVA. Fully robotic systems have yet to prove their ability to handle contingencies, and until then, manned EVA will continue to play the leading role in assembly operations in space and on the extraterrestrial surface. It is also possible that hybrid concepts employing both robotic and manned systems may prove to be more effective, using a mutual support strategy, during extravehicular buildup activity.

RATIONALE FOR AN ALTERNATIVE MANNED EVA SYSTEM

The EVA Bell concept proposed in this paper is a concept for enhancing manned EVA operations on the extraterrestrial surface. The rationale for this concept are as follows:

1. Conventional EVA is an extremely inefficient way of utilizing astronaut capabilities. EVA time is expensive and must be used more efficiently.
2. Conventional EVA requires that components to be assembled/deployed be designed to respond to the limitations of the EVA suited astronauts. Such a strategy limits efficient design and operation and therefore should not be a design driver.
3. Extraterrestrial base development will surely involve much more complex and arduous EVA tasks that will heavily tax the physical and mental capabilities of the best astronauts. Though the conventional EVA suits may be ample for some of the envisaged activity, alternate concepts for EVA are required to handle different and diverse EVA scenarios, which are a natural implication of the plethora of necessary EVA tasks required that will eventually lead to a final establishment and operation of the base.
4. "Back to stay" missions entail long duration missions, and most importantly, long life cycle facilities. These facilities will require servicing repair activity as well as associated evolutionary enhancement and modification procedures. EVA must be simplified in order to help the crew repeat these functions with ease.
5. Dust contamination from the extraterrestrial surface has and will continue to pose a serious threat to successful assembly and maintenance operations [20,21]. EVA concepts are needed that will effectively combat this problem during assembly/repair operations.

6. It may not be possible to tackle all EVA scenarios using the same strategy (i.e. use of only the conventional EVA suit). Therefore, at the planning stage, it is prudent to have as many alternate concepts for manned EVA as possible, so that tasks may be designed for efficient execution.

7. Many payloads to be deployed and maintained remotely on an early lunar base could be designed to be quite small in their physical dimensions (e.g. remote data relay stations, small science experimental platforms, photovoltaic arrays, optical interferometry array components).

8. It should be possible to provide an EVA environment for the assembly/repair/maintenance crew which is less taxing and more comfortable. Concepts are required which would enable the astronaut crew to perform more precise and delicate tasks on the site without the strain imposed by EVA suit constraints.

9. A rapid and highly flexible global extraterrestrial telecommunication/scientific experimental station network infrastructure establishment may be possible if a manned and robotic hybrid architecture is adopted during the primary phase of buildup activities.

It is on the basis of these premises that the Nomad Explorer architecture for rapid lunar/Mars global infra-structure development and the EVA Bell concept for enhancing manned EVA is developed.

THE EVA BELL ARCHITECTURE

In its simplest manifest, the idea is to separate the astronaut from the suit and try to provide as close to shirtsleeve environment as possible during EVA. In order to accomplish this, we will provide a pressurized shack, referred to as the "Bell" (programmatically similar in many ways to the diving bell used underwater), at the place where the EVA is to be performed (Figure 6). Obviously, then, this Bell becomes an integral part of the Nomad Explorer!

Though only one configuration of the EVA Bell is addressed in this paper, several other ways exist in which to provide this protective enclosure around the payload and the astronauts during assembly activities. They are being studied at the institute.

The Bell works in the following manner:

1. The Nomad Explorer drives to the EVA/Assembly site. The payload to be deployed could have been landed at the site separately (common lander concept?) or carried in the vehicle.

2. The surface is roughly leveled by the RMS on the vehicle. A surface seal fabric is unrolled on the smoothed out terrain. (This fabric could be part of each common lander payload).

3. The payload/experiment to be assembled/deployed is then unloaded on top of the prepared surface seal fabric.

4. The Nomad Explorer then aligns itself with the prepared surface seal and gently lowers the Bell so that the complete payload is covered by it with space around the payload to spare. The volume inside the Bell is about 150 cum.

5. The Bell is lowered until it uniformly contacts the surface seal fabric all around the payload. The Bell is then secured to the surface seal fabric in such a way as to produce a nominal pressure seal between the inside of the Bell and the extraterrestrial surface. Studies are underway that examine several ways of establishing this pressure seal.

6. The Bell is then pressurized (8 psi nominal).

7. After assuring that the seal is operational and that the nominal leakage rates are not exceeded, an airlock into the Bell allows the astronaut crew to access it from inside the Nomad Explorer.

8. The assembly/deployment activity is performed by the crew wearing minimal EVA garments (an emergency pressure suit?). After test and checkout of the experiment/setup or system (e.g. optical IF, VLBI components, relay stations, remote monitoring equipment) the crew get back into the vehicle.

9. The Bell is depressurized, retracted and the Nomad Explorer is on its way to the next assembly assist/experiment setup/maintenance/repair site.
This sequence of operations is illustrated in Figure 7.

CHALLENGES POSED BY THE EVA BELL

1. How do we mitigate counter pressure? At 8 psi, a 4 x 6 m Bell footprint would produce a total uplift of nearly 300,000 lbs.!! However, we have a substantial surface contact perimeter of 20 m to device an anchoring mechanism. External as well as internal anchoring mechanisms for the EVA Bell are being explored.

2. How do make sure of the seal? We will require a 100% reliability on the seal mechanism if a shirtsleeve environment is the goal.

3. How large a Bell can we practically build and operate? It has to be compatible with payloads that we intend to assemble and deploy, of course!

4. The surface seal fabric will have to be left in place after the assembly activity.

5. The payload cannot contact the surface during the assembly operation.

ADVANTAGES OF THE EVA BELL SYSTEM

1. Shirtsleeve environment for assembly/deployment activity. No prebreathing or associated EVA preparations are required.

2. The fully enclosed EVA Bell system provides a completely dust contamination-free environment during assembly and checkout of sensitive experimental equipment.

3. Lacking post landing serviceability, conventional missions carrying sensitive scientific equipment need to be designed to accommodate the shock of lunar lander impact. This translates directly as additional mass or design for shock absorption. Alignment, and recalibration of equipment could pose a problem. The EVA Bell strategy provides a way to circumvent this problem.

4. If the 8 psi pressure is too much to handle, then the crew could still work inside of the Bell wearing a pressure suit just enough to combat the differential pressure between the Bell and the suit. For e.g., if the Bell can withstand up to 4 psi, then the crew needs roughly another 4 psi inside the suit. Such a decrease in suit pressure will make the suit less stiff during operation and enable more comfortable EVA. This method of operation is also very safe in the event of a Bell pressure seal failure. New, more maneuverable and comfortable pressure suits could be designed to use with the Bell.

5. All of the multilayer insulation used in the conventional EVA garment for radiation and micrometeoric protection are eliminated from the suit which then retains only the pressure garment.

ADVANTAGES OF THE NOMAD EXPLORER STRATEGY

1. The Nomad Explorer strategy completely eliminates the need for conventional buildup equipment and infrastructure. All the heavy machinery and associated equipment, roads, launch and landing pad facilities associated with previous studies are not required during the initial stages of development.

2. Nomad Explorer technology is mature and does not require heavy investment to realize. NASA has studied long range vehicles like MOSAP in enough detail to be able to build and test Nomad Explorer prototypes.

3. Prototypes based on existing vehicles used for special terrestrial purposes (like the MX Missile transporter and other advanced recreational vehicles) could also be modified and studied in order to design and build the Nomad Explorer economically.

4. The Autonomous Modular Common Lander (AMCL) is not a new concept. NASA has been working on several lunar landers including the 'common lander' and Artemis using existing RL-10B or equivalent engine technology. Modular clustering of engines and tankage need more study in order to realize the AMCL concept. Innovative ways of landing bulk payload that is not so sensitive to higher than average terminal impact velocities (5-10 m/s) need further study [22].

5. If the AMCL system is employed to land modular payload containing experiments, logistics and consumables at various locations along a predetermined traverse route, then the Nomad Explorer could assemble and deploy the payload at regular intercepts along the path. Such a strategy will provide unlimited range for the Nomad Explorer.

6. The autonomous common lander approach would bring to the Nomad Explorer architecture a powerful design flexibility. During the course of the traverse, if unforeseen events require different logistics, science or consumable payloads, the landers could be outfitted and flown to intercept the Nomad Explorer with the required mission specific hardware at short notice.

7. During the course of the Nomad Explorer global traverse, if a globally accessible fiber-optic cable could be laid by the vehicle along the traverse, then it may be possible to eliminate the need for the deployment and maintenance in orbit of a constellation of telecommunication satellites.

8. This flexible open-ended mission architecture can be tailored as the mission proceeds. The if-then philosophy is best suited for exploration oriented missions and can be used to alter the traverse to best suit the exploration and extraterrestrial base site selection as well as the science mission goals.

9. The EVA Bell on the Nomad Explorer will provide a less stressful environment for astronaut crews on long duration EVA missions. The crew will have more freedom to alter traverse routes on the basis of their own exploration results, making the mission more exciting and eventful for both astronaut crew as well as mission control.

10. This ultra-dynamic strategy will also hold the fascination of the public because of the continuously changing terrain vistas, the regular rendezvous with the AMCL system, the number of mission goals which are rapidly met, and the spontaneous nature of the tasks that may have to be performed by the crew during this mission.

11. In this way a speedy global infrastructure may be established on the lunar/Mars surface.

TECHNOLOGY FOR THE NOMAD EXPLORER STRATEGY

The Nomad Explorer vehicle and the Autonomous Modular Common Lander System employ state-of-the-art and mature technologies. NASA has been working on similar concepts and sufficient data exists within the US which could be used to build and test prototypes.

The Nomad Explorer vehicle requires a heavy lift launch vehicle (HLLV). HLLVs are required in order to minimize otherwise costly and risky on-orbit assembly and rendezvous procedures for which the infrastructure is not in place yet [23,24]. Reviving the Saturn V-B and incorporating new modifications to the twenty year old technology could result in a work horse launcher that NASA needs and that is essential for any permanent manned presence on the Moon or Mars. The Russian Energiya is a typical example of an HLLV that could provide support for the Nomad Explorer operations [25].

Communications technology has come a long way since Apollo. Fiber-optics and free space laser communication systems can provide dependable and very high bit rate communications for the Nomad Explorer strategy. Furthermore, if the Nomad Explorer is used to lay a fiber optic network as the traverse proceeds, eventually it might be possible to have a global extraterrestrial communication system that would eliminate the need for a constellation of satellites in unstable orbits (in the lunar case) and poor life times. Optical line-of-site free-space laser communications without atmospheric disturbances and associated attenuation is possible on the lunar surface. Laying such extraterrestrial fiber optic cable/free-space links requires serious study. Preliminary studies indicate that fiber optics is a feasible option for the moon [26].

THE NOMAD EXPLORER BUDGET

Project Apollo cost $100 billion in 1990 dollars [27,28]. Much of the technology base had to be built up from scratch. Though much of the hardware and infrastructure associated with the project is no longer with us, much wealth in hardware and experience from the project remains dormant within NASA and the space industry. In addition, NASA and the space industry have already done much study on long range rovers and landers. MOLAB and MOSAP are some of the long range rovers which have been studied in depth. Several designs have also been developed by NASA for lunar landers. Artemis, lunar lander and the Common Lander are some of the NASA studies underway. The Nomad Explorer strategy will rejuvenate NASA and the manned spaceflight hardware builders of the world by tapping into research that is already underway. Using Apollo as the gauge for the Nomad Explorer, it should be possible to return to the Moon using the Nomad Explorer and the Autonomous Modular Common Lander for about the same price tag.

CONCLUSION

The Nomad Explorer Strategy for Extraterrestrial Base Evolution is an alternate strategy for global extraterrestrial infrastructure establishment. Capable of all the functions normally conducted from a stationary base, a mobile base like the Nomad Explorer, when coupled with an autonomous payload/logistics lander system like the AMCL system, had unlimited exploration range as well as a highly tailorable mission plan. Maximum flexibility is the chief attribute of this "open architecture." The "if-then" capability allows the mission to be tailored as it proceeds while maintaining close contact with mission control. Autonomous modular common landers carry mission specific hardware as dictated by the crew. The EVA Bell, a new system concept for enhancing manned EVA in the Nomad Explorer vehicle will help to assist assembly and deployment of science experiments and pilot projects while the vehicle traverses from site to site, examining them in detail for establishing a permanent manned base. An extraterrestrial fiber-optic/free-space global telecommunication infrastructure may be laid during the course of the exploratory traverse which will eventually eliminate the need for expensive operation and maintenance of a constellation of telecommunication satellites which would otherwise be required as complex extraterrestrial projects evolve. Rapid extraterrestrial development that will hold the interest and excitement of the public as the mission proceeds from site to site is the consequence of using such a mobile strategy. The various systems constituting this architecture require further study and analysis.

ACKNOWLEDGEMENTS

This study was done as a task for the extraterrestrial surface transportation group at Rockwell International, Space Systems Division, Downey, California. Thanks are due to Ed Repic, Project Manager for the Space Exploration Initiative, Steve Kent who supervised my work, John Pulley, manager of surface transportation group, and all the others including Todd Wise and Ron Jones who helped to critique the concept. Thanks are due also to the participants of the Institute of Aerospace Systems Architecture and Technology at the University of Southern California who are embarked on a brave new voyage, exploring and charting new territory in the field of Aerospace System Architectures.

REFERENCES

1. a. NASA (1989). Exploration Studies Technical Report TM 4170, Volume III: Planetary Surface Systems, NASA Office of Exploration.
   b. NASA (1988) Exploration Studies Technical Report TM 4075, NASA Office of Exploration, FY 1988. Volumes 1 & 2.
   c. Eagle Engineering (1988). Lunar Reports. NASA contract NAS9-17878. Eagle Engineering Inc., Webster, Texas.
   d.3. Landis, G. A. (1989). Solar Power for the Lunar Night, NASA TM 102127. 9th Biennial SSI Princeton Conference on Space.
   e. Thangavelu, M. (1991). The Very Long Range Traverse Vehicle: Option for Primary Lunar Base Establishment. Space Systems Division, Rockwell International, Downey, CA.
   f. Landis, G. A. (1989). A Proposal for a Sun Following Moon Base. Journal of the British Interplanetary Society.
2. Alred, John.(1989) Lunar Outpost. Systems Definition Branch, Advanced Programs Office, NASA Johnson Space Center.
3. a. Cohen, Aaron *et al.*, (1989). Report of the 90 Day Study on Human Exploration of the Moon and Mars. NASA Johnson Space Center, Houston, Texas.
   b. National Research Council (1990). Human Exploration of Space. A Review of NASA's 90 Day Study and Alternatives. National Academy Press, Washington, D.C.
4. Common Lander Study (1991). NASA JSC. NASA Office of Exploration.
5. Horz, Fredrich (1985) Lava Tubes: Potential Shelters for Habitats. *Lunar Bases and Space Activities in the 21st Century.* Ed. Mendell, W. W., Lunar and Planetary Institute, Houston.
6. Burke, F. B. (1985). Astronomical Interferometry on the Moon, *Lunar Bases and Space Activities in the 21st Century.* Ed. Mendell, W. W., Lunar and Planetary Institute.
7. Haskin, L. A. (1985) Toward a Spartan Scenario for Use of Lunar Materials, *Lunar Bases and Space Activities in the 21st Century.* Ed. Mendell, W. W., Lunar and Planetary Institute.
8. Cintala, M. J., Spudis, P. D., Hawke, B. R. (1985). Advanced Geologic Exploration Supported by a Lunar Base: A Traverse Across the Imbrium-Procellarum Region of the Moon. *Lunar Bases and Space Activities in the 21st Century.* Ed. Mendell, W. W., Lunar and Planetary Institute.
9. Burke, J. D. (1985) Merits of A Lunar Polar Base Location. *Lunar Bases and Space Activities in the 21st Century.* Ed. Mendell, W. W., Lunar and Planetary Institute.
10. Humphries, R. (1991). Life Support and Internal Thermal Control System Design for Space Station Freedom. NASA MSFC, Huntsville, USA. 4th European Symposium on Space Environmental Control Systems. ESA SP-324.

11. Rummel, J. D., Averner, M. (1991) Regenerative Life Support: The Initial CELSS Reference Configuration, Life Sciences Division, NASA Hq. Regenerative Life Support Systems and Processes. Society of Automotive Engineers. SP-873, paper # 911420.

12. Schwartzkopf, S. H., Brown, M. F. (1991) Evolutionary Development of a Lunar CELSS. Regenerative Life Support Systems and Processes. Society of Automotive Engineers. SP-873, paper # 911422.

13. Hypes, W. D. and Hall, J. B., (1988). ECLS Systems for a Lunar Base - A Baseline and some Alternate Concepts. SAE 881058. 18th Intersociety Conf. on Environmental Systems, San Francisco, CA.

14. Griffin, B. & Appleby, M. (1991). The Pressurized Lunar Rover. Radiation Analysis using the Boeing Radiation Exposure Model (BREM). Advanced Civil Space Systems Division, Boeing Co.

15. Stone, C. (1991). ALPS: Advanced Lunar Power System. Space Systems Division, Rockwell International, Downey, CA.

16. E. P. Coomes, J. E. Dagle, J. A. Wise (1992) Design Study of a Laser Beam Powered Lunar Exploration Vehicle. Pacific Northwest Laboratory. IAF-92-0742.

17. Brown, W. C. (1969) "Experiments involving a Microwave Beam to Power and Position a Helicopter." *IEEE Transactions*, Aerospace Electronics Systems, Vol. AES-5.

18. Glaser, P. E. (1974) Feasibility Study of a Satellite Power Station, NASA CR-2357.

19. Kuznetz, L. H. (1990) Space Suits and Life Support Systems for the Exploration of Mars. NASA Ames Research Center.

20. Wise, Todd (1991). Dust Contamination Control. IDEEA Conference, Houston, TX.

21. Glaser, Peter (1991). Dust Contamination and Control. IDEEA Conference, Houston, TX.

22. Thangavelu, M. (1991). The Multicellular Airbag Landing System. A concept for landing extraterrestrial payloads. Space Systems Division, Rockwell International, Downey, CA.

23. Stafford, T. P. (1991). America at the Threshold: Report of the Synthesis Group on America's Space Exploration Initiative, Washington, DC.

24. Augustine, N. (1991). Report of the Commission for the Future of the Space Program.

25. Gobanov, B. I. (1989). Energia - A New Versatile Rocket Space Transportation System, IAF-89-202.

26. Lutes, G. F. (1991). Communicating on the Moon via Fiber Optics. Lunar Fiber Optic Link Study. NASA JPL. California Institute of Technology. NAS7-918.

27. Crawley, Ed (1988). Aerospace Systems Analysis and Costing. Lecture to the International Space University at MIT, Boston.

28. Sellers, W. O. and Keaton, P. W. (1985). The Budgetary Feasibility of a Lunar Base. *Lunar Bases and Space Activities in the 21st Century.* Ed. Mendell, W. W., Lunar and Planetary Institute, Houston.

AAS 93-886

CONSTRUCTION METHOD AND PROCEDURE FOR EARLY-PHASE MARS BASE

Yoji Ishikawa,* Owen Gwynne,† Yukinobu Yamamoto* and Hisashi Uyeda*

Construction of a Mars Base is one of the major tasks the travelers to Mars confront, however, it has never been the subject of intensive research. In this paper, we specify the types of construction works for major infrastructure of a Mars Base, and estimate the time necessary to accomplish these works. Most of the procedures are found to be completed in a reasonable time scale, however, covering the habitats with soil needs tremendous time using a conventional dozer-type machine. We suggest that most of the construction machines be transported before the first human landing, and that covering the module with soil needs some alternative method such as a snow blower (other than dozer-type construction machines).

INTRODUCTION

The purpose of a Mars Base will be to do science and to establish the fact that people can live and work on another planet for extended periods of time. A major step in human history will be the colonization of another planet besides Earth. The most likely candidate for this critical role is Mars. Mars has the unique aspect of having all the resources required to support long term human presence in abundance and in easily obtainable forms.

Mars will be someday another home for humans, by continuing our presence on the surface of this planet. To accomplish this long term goal, the establishment of human habitats on the Martian surface, sometime in the next century, will be a necessary step. It is not an easy step; however, we must start now to prepare for the time to come.

We face many technical challenges to take a step towards this goal, one of which is undoubtedly a construction of a Mars base. "A Construction" on Mars includes many works: ground work, rock breaking, preparing roads, excavation, loading/unloading, hauling, grading, covering modules with soil, and so on. However, there is hardly any study done in this area. We, as engineers involved both in space and in construction, focus on the construction method of a Mars base, especially at its early stage, here.

First, we show a long-term Mars base development scenario and clarify the significance of the construction of a Mars base in human history. Then, we present a surface infrastructure which consists of an early-phase Mars base. A mission sequence to estab-

* Obayashi Corporation, 2-8-3 Kanda Tsukasa-cho, Chiyoda-ku, Tokyo 101, Japan.

† SETI Institute, NASA Ames Research Center, Moffett Field, California 94035.

lish such an infrastructure follows next. After the assumed construction machines are shown, the construction time is estimated.

MARS BASE DEVELOPMENT SCENARIO

Mars development surely cannot be done in a day. Careful planning is definitely required before an actual base development is realized. Mars development never profits any single country in terms of its significance, and it must be pursued by some international entity. The true difficulty does not only lie on the development's technical aspects, but also on its political and financial situations. Since Mars development will easily take up to a half century, strong commitment and long-term financial support are a must.

The long term Mars development will proceed as follows:

Robotic Exploration
First Human Landing
Outpost Construction
Expansion of Base
Base of Complete self-sufficiency
Ordinary People's Immigration
Asteroid's Resource Utilization
Independence of Martian Civilization

It is not easy to predict when each phase will be realized; however, we believe the first human landing will occur in the first half of the next century, and in the latter half ordinary people may begin to immigrate Mars. The driving force for the immigration must be some kind of economical benefit, for example, asteroid resources. In the 22nd Century, a quite large population of people will live and stay on the surface with its own cultural spirits.

SURFACE INFRASTRUCTURE IN THE EARLY-PHASE MARS BASE

One of the authors has shown one possible form of a Mars Base, which houses 150 habitants in the middle of the next century (Ishikawa, Ohkita and Amemiya, 1990). We, in this paper, focus on the Mars base infrastructure in the earlier stage, right after the first human landing. Another one of the authors examined all of the necessary facilities for the Mars base (Gwynne *et al.*, 1990). Below is the infrastructure required just after the first landing in order to facilitate survival.

1. Rigid habitat module:
   cylindrical, covered with soil 6 m diameter, 16m length,
   15 tons weight per module

2. Inflatable habitat module:
   spherical, covered with bagged soil buried semi-underground
   16 m diameter, 50 tons (dismantled into less than 1 ton pieces)

3. Greenhouses:
   sausage-shaped, inflatable 5m diameter, 20m length

4. Nuclear power station:
   SP-100, 4 tons per piece, 100 kW output per piece

5. Solar panel arrays:
   folding, 0.7 tons per 1,000 m^2

MISSION SEQUENCE

The base infrastructure described in the previous section will be erected in the following mission sequence:

The 1st mission
- 4 crew
- approx. 1 year time frame for stay
- 30 tons transported onto the surface
- first human landing
- scientific exploration around the landing site

The 2nd mission
- 4 crew
- approx. 1 year time frame for stay
- 30 tons transported onto the surface
- expansion of the base
- scientific exploration using rover
- deployment of solar arrays

The 3rd mission
- 8 crew
- approx. 1 year time frame for stay
- 60 tons transported onto the surface
- scientific exploration at the longer distance
- more automated system for exploration
- deployment of solar arrays

The 4th mission
- 12 crew (8 continue to stay since previous mission)
- approx. 1 year time frame for stay
- 60 tons transported onto the surface
- scientific exploration
- erection of inflatable habitat module (#1)

The 5th mission
- 16 crew (8 continue to stay since previous mission)
- approx. 1 year time frame for stay
- 60 tons transported onto the surface
- scientific exploration
- erection of inflatable habitat module (#2)

Each mission will transpire approximately every 2 years based on a launching window, although it is not easy to predict when these will occur. I propose here the target year for the first human landing is in the 2020's.

CONSTRUCTION METHOD AND PROCEDURE

Assumption

The first Mars base is likely to be constructed at the site of lower latitude, because access is easier and the climate is milder. The site also must be relatively flat within 1 km to make construction possible. The site may have a small bump, however, it should not have a gradient to avoid any unnecessary leveling work.

The Martian soil was analyzed by the Viking landers and we estimated the Martian soil characteristics (Gwynne *et al.*, 1991).

The soil on the surface seems to be too soft to do construction work compared to the terrestrial case, however, it does not make the work completely impossible. Pressing the soil or freezing it with water may change the soft characteristics of the surface soil. The important soil characteristics for construction are those under the surface, which are not known—even with the Viking results. Completely different construction techniques are required, especially for constructing support of the building, depending on the softness/hardness characteristics of the subsurface soil. For example, the inflatable habitat module requires support for its erection. If the Martian subsurface soil is hard, the anchoring will not be difficult. In the case of soft subsurface soil, however, long anchors must be thrust into the ground until reaching a hard rock bed (Figure 1). Information on the subsurface condition, therefore, is necessary beforehand to spot the base site. In order to make accurate estimates, information on the subsurface condition of only up to 10 m is necessary.

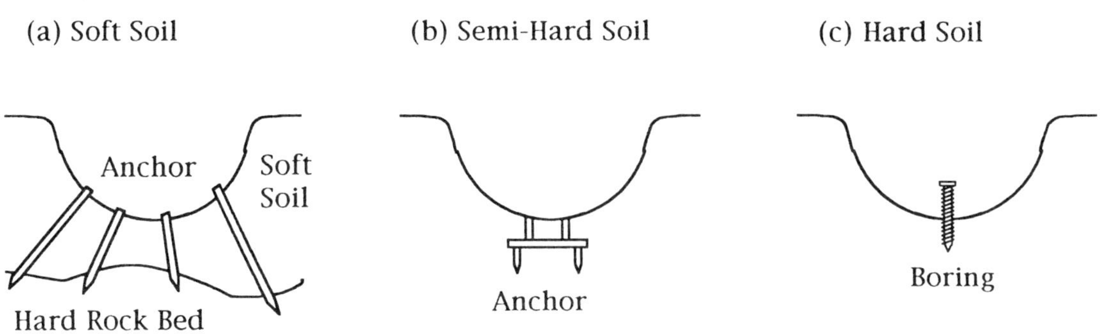

Figure 1 How to make support for the Inflatable Habitat Mudule. In the case of the soft soil model, support is almost impossible without a subsurface hard rock bed.

Another unknown factor for designing the construction machines and processes is the working conditions of Mars inhabitants. Working while wearing space suits considerably limits work time and mobility. Hard suits are essential as space suits to avoid fatal disasters due to hard construction work. Besides, the inhabitants do not seem to be able to work eight hours a day outside with space suits because of the limitation of life support systems, a danger of radiation exposure, and mental stress. Automation, thus, is a key word for the design of the machine and its operation.

Base Configuration

The purpose of this paper is to show the general standard for the construction method and procedure. However, the ground work strongly depends on the configuration of the base, or the transportation distances. We thus assume the configuration of the base (Figures 2 and 3). This design is based on the following principles:

1. The facilities of the habitat area are connected to each other, and make a habitat cluster.
2. The installment of each facility must follow the mission sequence described earlier.
3. The launching pad is located at least 1 km east from the habitat area.
4. The nuclear power station is located at least 300 m from the habitat area.

The configuration shown in Figures 2 and 3 does not reflect any artistic view or unique idea, and is used only as a standard basis for the estimation which will be described later.

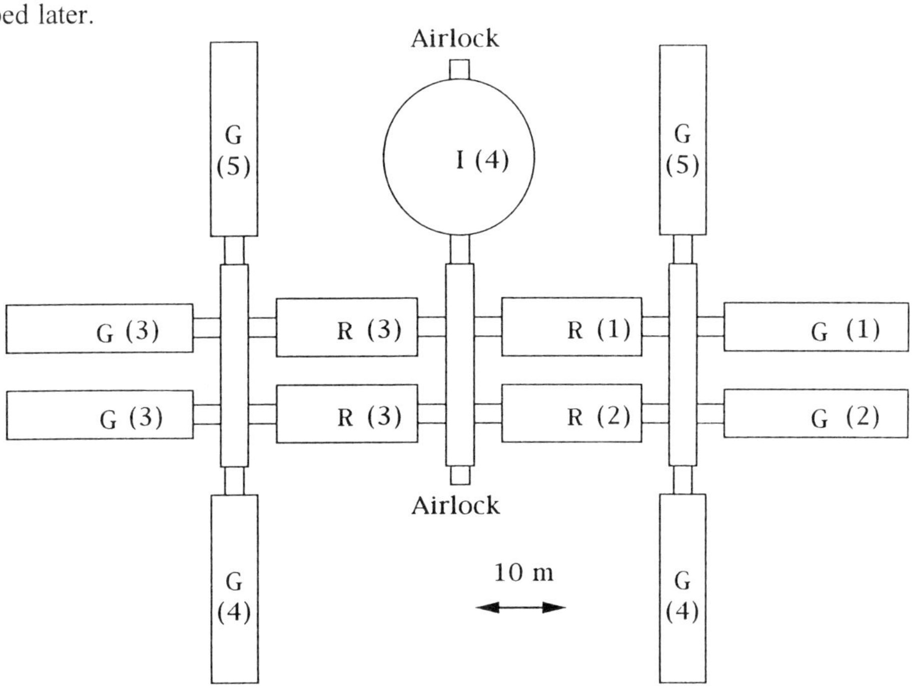

Figure 2 Configuration of Mars Base Habitat Area. Actually, both the rigid and inflatable habitat modules are covered with soil and bagged soil, respectively. R=Rigid Habitat Module. I=Inflatable Habitat Module. G=Greenhouse. The numbers in parentheses mean the mission number when the facilities are transported and installed.

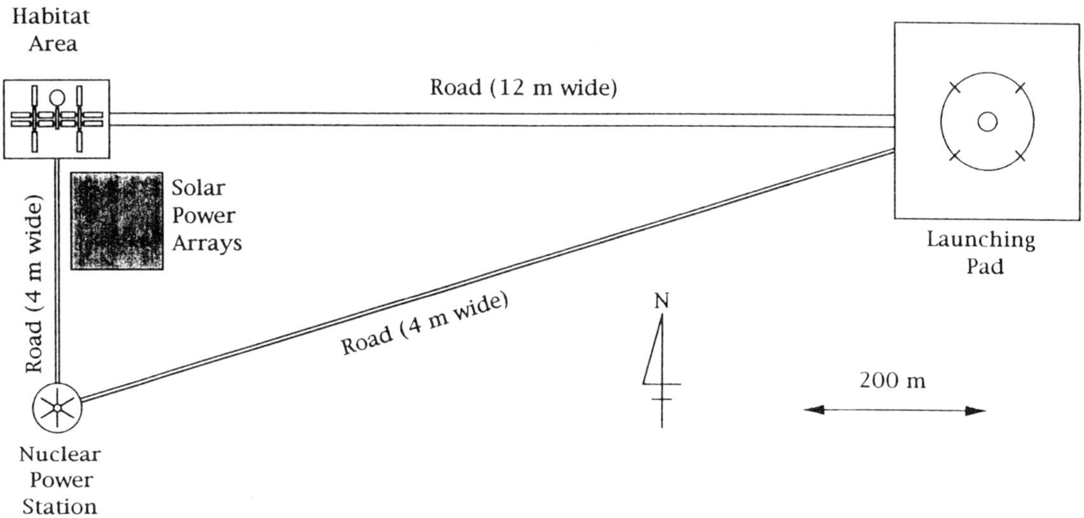

Figure 3 Mars Base Configuration.

Construction Machines

In order to estimate the construction time and work, construction machines have to be specified. There are many kinds of construction machines, needless to say, for use on Earth (for example, Church, 1981; Martin *et al.*, 1982), however, the machine to be used for Mars must be designed considering the following conditions:

low gravity (3/8 G)
low atmospheric pressure
soil characteristics
protection from sand storms
limited power source on the surface
easy operation (automated operation preferable)
low cost of transportation from Earth (low weight).

Among these conditions, gravity gives essential design criteria. Under lower gravity, the traction will be smaller compared to Earth's case. The weight the machine can lift will be theoretically the same as the weight of the machine itself, however, in effect, half will be lifted at most. In order to lift and carry the base infrastructure a heavy construction machine is required, which will contradict the other criteria requiring low weight for transportation. Although the design of the Martian construction machines will be the research area where intensive study is required, the basis for the conceptual design is the same as the machines used on Earth.

Here we do not intend to design Mars construction machines. Instead, we assume here the conventional construction machines which are at work at many construction sites on Earth, because these are more familiar to us and can be analyzed easily. We list three types of construction machines in Table 1 which we assume for work time estimate as well as the additional tools.

Table 1
ASSUMED CONSTRUCTION MACHINES

Machine I Excavator(Mini Backhoe Type)	Crawler, Weight 3 tons Bucket 0.08 m3, Power 23 kW 4.5 m(L) x 1.5 m(W) x 2.5 m(H)
Machine II Wheel Loader(Backhoe Specification)	4-wheels, Weight 5.5 tons Bucket 0.8 m3, Power 40 kW 5.5 m(L) x 2.0 m(W) x 3.0 m(H)
Machine III Excavator(Backhoe Type)	Crawler, Weight 10 tons Bucket 0.4 tons, Power 55 kW 7.0 m(L) x 2.5 m(W) x 2.0 m(H)
additional tools	winch, pulley, sled, boring machine, compressor, cables, batteries, lights

Construction Work and Time

The estimate is carried out by extrapolating our experiences on Earth to the Martian conditions. The soil mechanics is one of the strongest factors which influences the time estimate. Our estimates follow both the soil characteristics from the Viking results and our previous analysis on the Martian soil mechanics (Gwynne *et al.*, 1992).

Common foundation: The first task the Mars builders take on is to prepare roads. Roads need to be built between the habitat area, the launching pad, and the nuclear power station. The total area to be prepared is 18,000 m^2. The surface is scraped to 0.5 m depth and leveled. The excavator (Type 1) will do this work in 60 hours total.

Table 2
CONSTRUCTION WORK AND TIME FOR COMMON FACILITIES

work	machine	time(hour)
Prepare roads	I	60

Rigid habitat module: To prepare the emplacement location, 200 m^2 must be scraped, flattened, and pressed. The distance between the launching pad and the habitat area is 1,000 m, and such a heavy object (15 tons) cannot be carried without a heavy construction machine (at least 30 tons). The habitat module is not considered a fragile item, it should be moved on sled by winch and pulley system (and hopefully with cylindrical runners). It takes 10 hours to move one module for the distance of 1,000 m. The installment and the alignment are done in 2 hours by winch system per module.

Covering module with soil for radiation protection is a time consuming process. To completely shield the module, the soil thickness above the module must be more than 3 m. The soil of 2,500 m^3 has to moved to cover the module from a simple geometrical calculation. In addition, a slope with a slight gradient (15 degrees) is required for the excavator to climb up. This volume is 420 m^3. If the Type I excavator (covering ability of 7 m^3/hour) is used, it takes 450 hours. Even if a larger excavator (weight 20 tons, covering ability of 76 m^3, not listed in Table 1) is used, 40 hours are required. We have to conclude that covering the module needs an alternative method other than using excavators.

Table 3

CONSTRUCTION WORK AND TIME FOR RIGID HABITAT MODULE

work	machine	time(hour)
prepare the location	I	10
move the module	winch	10/each
install the module	winch	2/each
cover with soil	I	450/each

Inflatable habitat module: The area to be prepared is 450 m^2, however, the inflatable model is buried semi-underground. The volume to be dug is estimated as 2,100 m^3, which includes a grading approach to the site. This is done by the machine type II in 220 hours. In order to move the module (total weight 50 tons) it must be disassembled beforehand. A larger machine type III weighs 10 tons, and is able to carry the parts of 5 tons at one time. It requires 50 hours to move such parts (5 tons) from the launching pad to the habitat area. Installing support (anchors) strongly depends on the hardness of the soil as described before. If the soil is not so soft, the anchoring will be done in 50 hours. To raise a center post, 20 hours are required using a winch system. Installing the interior and exterior strongly depends on the module design. Based on our experience on Earth, it may take on Mars a full week of work by 4 people. Inflating the module is done within 14 hours by compressor. Covering the module with bagged soil for radiation protection takes 530 hours by using the wheel loader (type II). This figure is again unrealistic, and some alternative method should be sought.

Table 4

CONSTRUCTION WORK AND TIME FOR INFLATABLE HABITAT MODULE

work	machine	time(hour)
prepare the location	II	220
move the parts	III	50/5 tons
install the supports		50
raise a center post		20
install interior/exterior		4 persons x 1 week
inflate	compressor	14
cover with bagged soil	II	530

Greenhouse: The area of 150 m^2 should be prepared to accommodate one greenhouse module. It takes 10 hours/each. Moving is done by Type II wheel loader in 5 hours/each, and the module is inflated by compressor in 14 hours each.

Table 5

CONSTRUCTION WORK AND TIME FOR GREENHOUSE

work	machine	time(hour)
prepare the location		10/each
move equipment to site	II	5/each
install and inflate	compressor	14/each

Nuclear Power Station: The area of 1,000 m^2 will be prepared for a cluster of SP-100. The required time is 14 hours. Each SP-100 weights 4 tons, and should be handled very carefully. A winch system is not appropriate for this work, and the rover with crane attachment with at least 10 tons weight is required. Machine III can carry one SP-100 for 5 hours from the launching pad to the nuclear power station. Installing equipment requires 7 hours/each. To protect the habitat area from nuclear contamination, the soil wall of 3 m high must be constructed. It is done by the wheel loader (type II) in 50 hours.

Table 6

CONSTRUCTION WORK AND TIME FOR NUCLEAR POWER STATION

work	machine	time(hour)
prepare the location		13
move equipment to site	III	5/each
install equipment		7/each
construct soil shield wall	II	50

Solar power arrays: Preparing the location takes 30 hours. The excavator (type I) carry each sheet for 3 hours, and installing is done in 7 hours per sheet.

Table 7

CONSTRUCTION WORK AND TIME FOR SOLAR POWER ARRAYS

work	machine	time(hour)
prepare the location		30
move the arrays	I	3/each
install the arrays		7/each

Additionally, the passageway (see Figure 2) between habitat modules and greenhouses must be installed and covered with soil in the same way as the rigid habitat modules. Covering with soil is also a time consuming process, requiring moving half the amount of soil as the rigid habitat module. Another method must be sought here, too.

MISSION CHARACTERISTICS

Based on the present study, the characteristics of each mission are summarized in Table 8. The weight listed in this Table is the total of the base infrastructure, and does not include the weight of the construction machines.

It seems to be unavoidable to transport most of the construction machines with or before Mission 1. The reason for this is that we have to rely on nuclear power as an energy source even from the first mission. SP-100, the only practical power source in use for space, weighs 4 tons and requires transportation by heavy machine (10 tons) on the Martian surface. The recommendation here is that the necessary construction machines should be sent in advance as another cargo mission to be ready for construction upon the arrival of the first Mars travelers.

Table 8
MISSION CHARACTERISTICS AND CONSTRUCTION MACHINES
Please note the weights do not include the construction machine weights.

	Mission 1	Mission 2	Mission 3	Mission 4	Mission 5
Rigid Habitat	1 unit	1 unit	2 units		
Inflatable Habitat				shell	support equipment
Greenhouse	1 unit	1 unit	2 units	2 units	2 units
Nuclear Power	2 units	1 unit	1 unit	1 unit	1 unit
Solar Power		1 unit	2 units		
Crew on Surface	4	4	8	12	16
Weight (tons)	32.5	28.5	48	45	35
Habitat area (m^2)	452 (4)[1,2]	904 (9)	1,808 (18)	1,808 (18)	1,808 + Infl. (30)
Cultivation area (m^2)	500 (5)[3]	1,000 (10)	2,000 (20)	3,000 (30)	4,000 (40)
Power (kW)	200(20)[4]	400(40)	600(60)	700(70)	800(80)
Machines	Type I (Type III) Winch Pulley Sled Boring machines	Backhoe Repair tools Batteries Cables	Type II Repair tools Cables Lights	Tools	Tools

[1]The figures in parentheses mean the number of person which can be supported.
[2]100 m^2 of habitat area is assumed to be necessary per person.
[3]100 m^2 of cultivation area per person.
[4]10 kW per person.

CONCLUSION

1. Most of the construction works are done in a reasonable time scale.
2. Covering the rigid and inflatable habitat modules with soil requires an unusually long time by using excavator. Alternative methods or apparatus like snow blowers should be sought.
3. A series of five missions is designed to establish the habitat for 16 people.
4. Most of construction machines should be sent to the surface as another cargo mission before a human presence starts.

REFERENCES

Y. Ishikawa, T. Ohkita, and Y. Amemiya, "Mars Habitation 2057: Concept Design of a Mars Settlement in the Year 2057," *J. British Interplanetary Soc.* Vol. 43, pp. 505-512, 1990. Also published as AAS 97-378 in R. M. Zubrin (Ed.) *From Imagination to Reality: Mars Exploration Studies of the Journal of the British Interplanetary Society* (Part II: Base Building, Colonization and Terraformation), AAS *Science and Technology Series*, Vol. 92, pp. 83-97, (1997).

O. Gwynne, C. McKay, L. Lemke, C. Stoker, Y. Clearwater, M. Sims, R. Ramos, "Major Components of a Human Base on Mars," The Case for Mars IV, Boulder, Colorado, June 4-8, 1990. An abstract of this paper was published as AAS 90-181 in T. R. Meyer (Ed.) *The Case for Mars IV: The International Exploration of Mars* (Mission Strategy and Architecture), AAS *Science and Technology Series*, Vol. 89, p. 337, (1997).

H. Church, *Excavation Handbook*, McGraw-Hill, 1981.

J. W. Martin, T. J. Martin, T. P. Bennett, and K. M. Martin, *Surface Mining Equipment*, Martin Consultants, Inc., 1982.

O. Gwynne, Y. Ishikawa, Y. Yamamoto, H. Uyeda, and T. Bongiovi, "Construction and Development of a Human Base on Mars," Engineering, Construction, and Operations in Space III, Denver, Colorado, 31 May- 4 June, 1992.

AAS 93-887

DESIGN AND CONSTRUCTION FOR A PERMANENT PRESENCE ON MARS

Ethan W. Clifton*

All early human exploration of Mars will rely upon aircraft technology similar to Apollo for living environments on the surface. The high tolerances of these dual-use crafts are necessary for the transit to Mars and return trip protection on the surface. They do not, however, provide two essential elements for the long-term habitation of science stations. The first is radiation protection, which must overlay the aeroshell, and the second is endurance, which requires repair and replacement of working parts such as doors. Successful long term occupation of a site has been dependent upon the systematic use of local materials in the construction of semi-permanent encampments, and ultimately useful urban centers for communication and resource management.

Concrete construction may be used on Mars for permanent enclosures. Portland cement material has been identified [1] in soil samples, and may be recovered in sufficient quantities to provide building materials. Precursor missions will establish the location and accessibility of these and other useful mineral deposits. The equipment to recover such minerals exists in well tested and fully operational forms.

Draglines [2], batch plants, concrete pumps, lift-slab and tilt-up technology are well understood and will be expanded to include flexible formwork in the production of safe, repeatable and permanent buildings for the expanding scientific community at Mars. A mature base will begin to test locally available erz, sulfur, and Portland cements as candidates for use in this adaptive concrete construction.

The incremental addition of flexible formwork to the lexicon of concrete construction technology will reduce imports and lead to more elaborate structures over time. Validating this technology in small-scale habitats is essential to perfecting the construction technology necessary for large buildings intended to integrate habitation, research, and biotic environments. Permanent buildings will represent the third phase in the evolution of Mars from a scientific outpost to a mature research station.

INTRODUCTION

A second-generation building system for permanent Martian basing by an expanding scientific and engineering population is required within a few years of initial arrival. Surface level is proposed for the location of these buildings for two reasons; one, excavation of soils requires equipment, power, and time, and two, knowledge of subsoil

* Architect, 870 Market Street, Suite 1268, San Francisco, California 94102.

conditions requires equipment, power, and luck. Borings are notorious for missing hidden objects in substrate.

The orderly approach to permanent base design, with special emphasis on life safety, will require an infrastructure which uses local materials for pressure containment and radiation protection. Pressure vessels may be assembled from precast concrete segments integrating post-tension technology common to Earth-based construction [Figure 1]. The radiation protection envisaged commonly in the burial of habitations may be achieved at the surface through use of typically massive concrete construction.

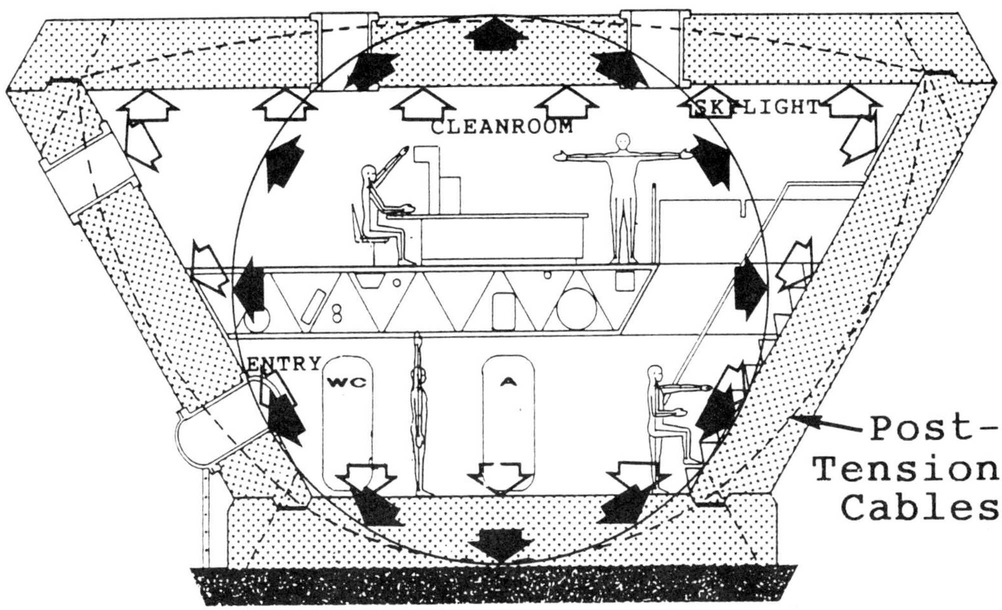

Figure 1 Internal Pressure Distribution in a Cylinder and Building Section.

Life-safety issues must be addressed on the Martian surface to assure that the limited number of personnel are protected. Loss of pressure, atmospheric degradation, and fire will pose real threats to survival. Two means of egress and other uniform building code requirement will be introduced [Figure 2] so that the available Martian technologies are not isolated from conventional practice and experience. No new assumptions about safety are required. Surface level buildings afford easy connection to one another through pressurized but light-weight tunnels which allow quick egress without suiting up.

TECHNOLOGY

A combination of extant construction and building systems will provide a permanent facility at Mars. Diesel engines operating on locally separated oxygen and carbon monoxide will be available for the high torques needed in construction activity.

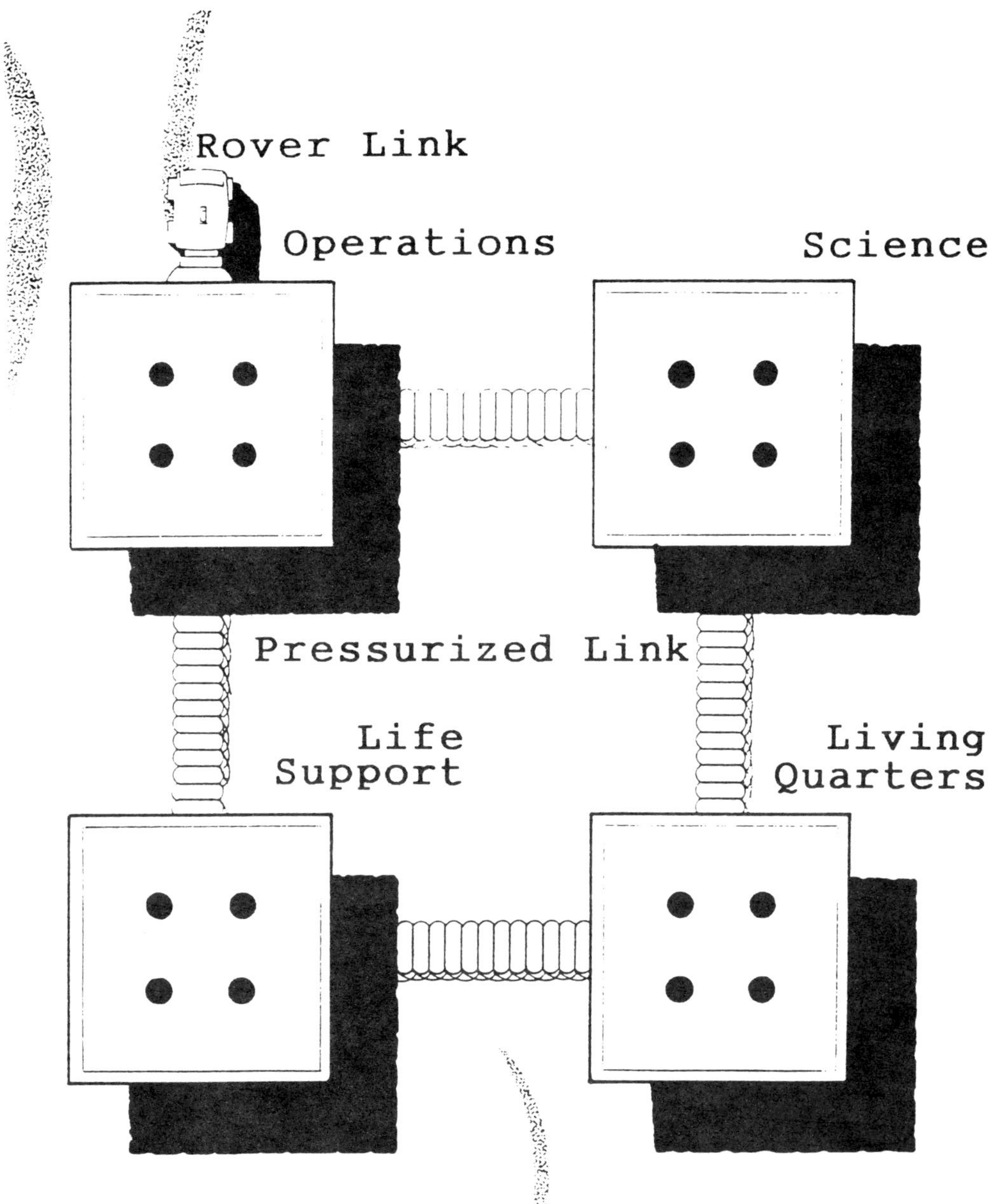

Figure 2 Buildings Connected by Two Links for Safe Egress in Emergencies.

Draglines are light-weight means of mineral and aggregate collection. The cables are recovered from tethered vehicles used in the human transfer stages to Mars. With a steady supply of salvaged cable, wear and tear on the system will not pose a threat to continuous operations.

In-situ concrete production may be achieved by adapting batch plants [3] to the Martian environment. Using local cement and aggregate mixed with potable water condensed from the atmosphere, structural concrete panels may be produced [Figure 3]. Pumping [4] is commonly used for placement of concrete. The reach of the hoses is especially useful where formwork for slabs is widely dispersed, and immediate rolling access is limited.

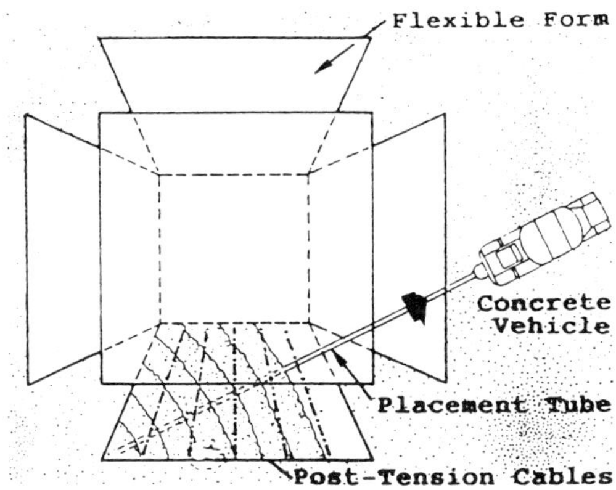

Figure 3 Form Layout and Placement of Concrete.

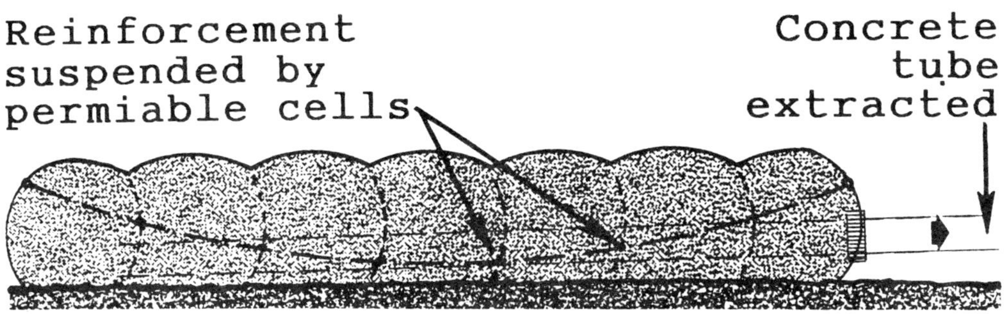

Figure 4 Inflatable Formwork With Tension Strands Suspended.

Lift slab construction for roofs in another common technology useful in lifting the massive radiation shielding for the roof of the building. Tilt-up walls are common in light industry. The identical technology is applicable to the erection of walls on mars. The same tether cables available for draglines may be utilized for assembling slabs [Figures 5-10].

Compressible sealants are imported initially, and ultimately produced from waste and local sulfur deposits. The accuracy of casting technology will greatly reduce the quantities of sealants required to close joints. The fabric lining [5] of the precast panels may be coated with diluted sealant material where attachments or wear and tear have broken its integrity.

Windows and hatches are imported pre-tested systems, as they are here. These complex elements may be used in the transfer vehicles, disassembled, and delivered to the surface for re-use in buildings. Uniform size and shape make integral casting of the portals feasible.

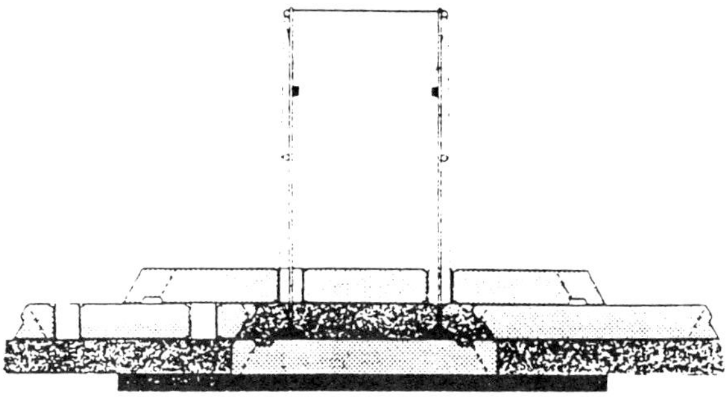

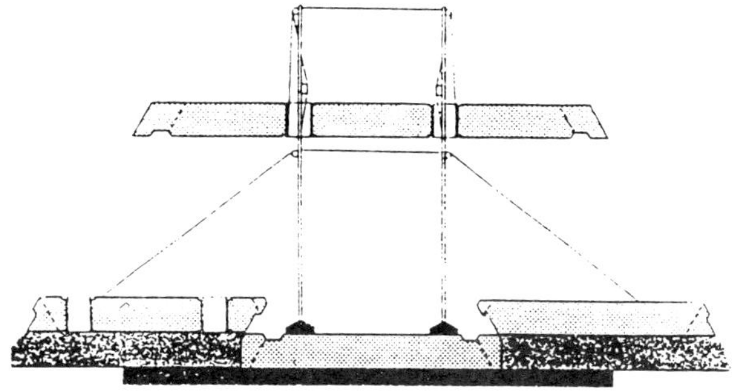

Figure 5 (left) Precast Stack Ready for Lifting.
Figure 6 (right) Roof Slab Lifting.

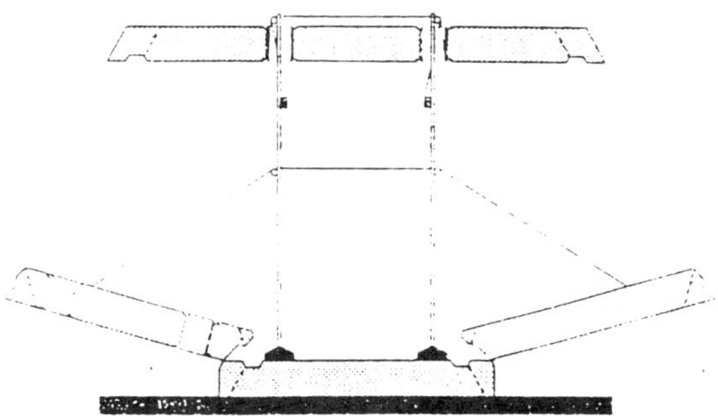

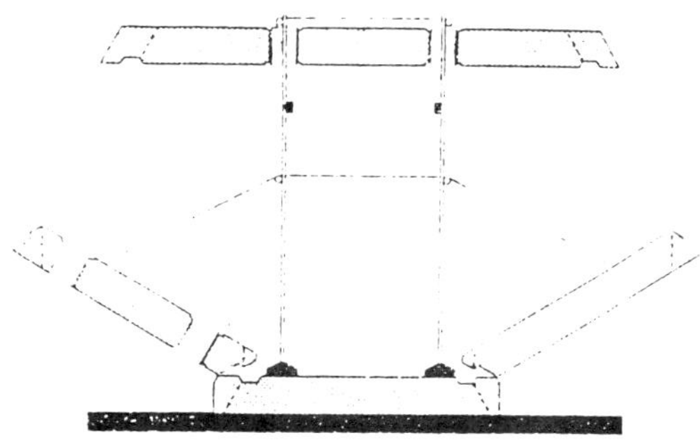

Figure 7 (left) Roof Slab at Maximum Height, With Walls Tilting.
Figure 8 (right) Walls Tilting.

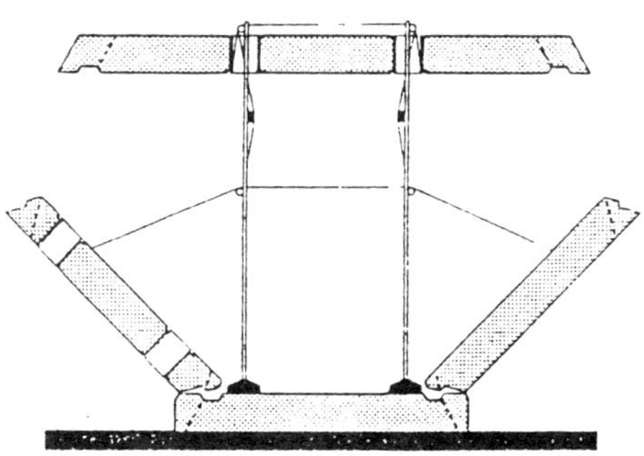

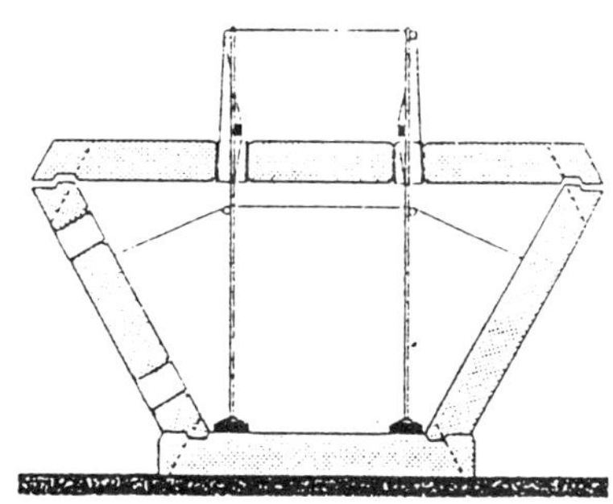

Figure 9 (left) Walls Tilting.
Figure 10 (right) Walls in Place With Roof Slab Lowering onto Keys.

The precast lift-tilt construction proposed is dependent upon the availability of formwork and accurately placed reinforcement. The introduction of a new building system is required by this scenario. This new system, *flexible formwork*, will achieve all of the essential criteria for precast concrete production on the Martian surface.

The containment of volatiles, water, or other hydrating medium which will evaporate in the low-pressure Martian atmosphere, is essential. True cement produces heat of hydration which would quickly boil-off water before the chemical bonding process was complete. The pressure containing fabric formwork will contain the fluid concrete mix and prevent outgassing. An insulation layer within the fabric will protect the hydration process from the temperature drop in the Martian diurnal cycle.

Accurate suspension of reinforcement is feasible through Earth-based fabrication of the formwork. The configuration of tension cords is critical to resisting bending in slabs. Further, the tension cords must be pulled after curing of the concrete to pre-stress the slab in compression. The half or better atmospheric pressure to be contained within the final building requires an effective tension matrix within each slab.

The formwork must be delivered to Mars reasonably intact and ready for use [Figure 10]. Layout, staking, and batting forms is not feasible on the surface. Light-weight forms with complex shapes are best prepared on Earth.

BUILDING DESIGN

Tilt-up and lift-slab construction is best utilized for rectilinear buildings [Figure 11]. For the purpose of pressure containment, the resulting volume should closely resemble a sphere, so that bending stresses in panels are not extreme. For this reason, the proposed buildings are compact, relying upon numbers for creation of volume. This multiplication enhances safety through isolation of failed units, and also allows specific use of each unit. Individual units may be roughed in for high-bay plant growth or process facilities.

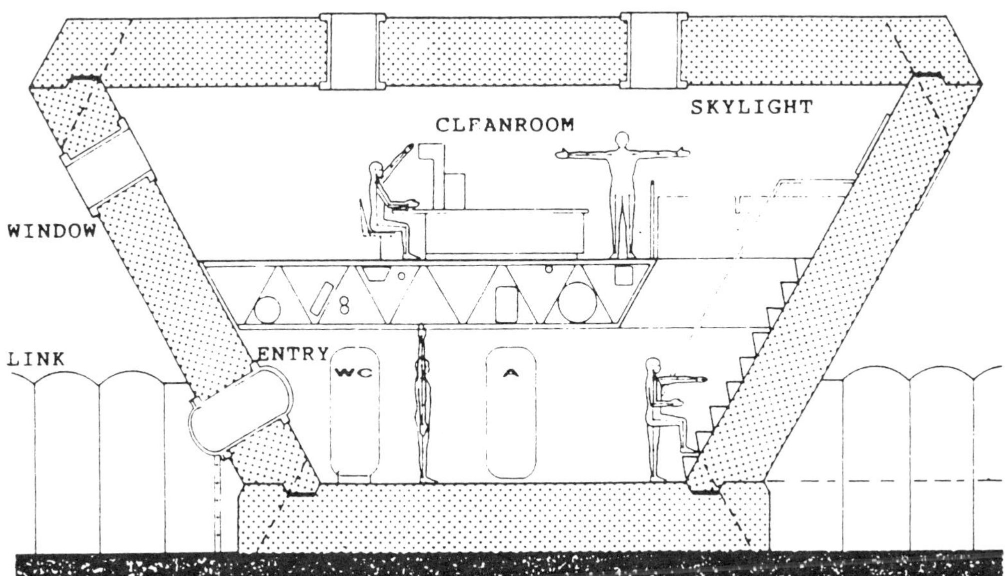

Figure 11 Operational Building.

TRANSPORT

Transportation cost is, and will remain, the principle cost of Martian exploration. There is no likelihood of economic return from the red planet, and on-going exploration will pose little tangible reward for non-technology users on Earth. Equipment and parts to be transported for second-generation infrastructure on Mars must be minimized. Therefore each component must be recycled where possible, or designed for long-term low-maintenance service.

Draglines are in place from previous tethered flights. The cable, pulley, and drive mechanisms will have been perfected in shielding work for the first phase camps.

Batch plant and pump technology will be modified for lighter weight where practicable. Endurance of concrete production equipment is directly related to the iron content of its working parts. Existing designs should be used wherever possible, with penalties in launch costs recovered by long-term replacement savings. The batch plant can be designed for Martian concrete mixes without re-engineering. Re-engined and sealed, existing equipment will perform well in the Martian environment. Admixtures such as plasticizers and gas entrainment products to aid in concrete fluidity and curing will have to be perfected on Earth, and delivered in bulk to Mars. This low-value cargo is well suited to electro-magnetic launch and solar sail transport.

Forms will be imported with tension members, penetrations, and complex edge conditions pre-molded into the fabric. Pretested and produced in greater quantities than required for immediate use, the light-weight forms will be delivered as parasails for common cargoes landed accurately in work areas.

Lifting gear will be adapted from dragline components for compatibility of winches, connectors, and cable. No new technology is required.

Floor trusses, ports, partitions, equipment, and furnishings will be imported and integrated for the specific functions of each building. As in other science stations, many items will be adapted from scavenged primary landing vehicles and equipment.

CONCLUSION

Concrete construction offers advantages to the long-term exploration of Mars and its potential as a terraformed environment following the discovery period. Principal among these is permanence.

Concrete structures are inert. Local concrete does not react with the caustics elements in Martian soils. Minimal maintenance is required, specifically in the areas of entry ports or inspection of sealants and tension members. The high figure and thick walls resist wind blown sand and do not erode significantly in time. The structure is fireproof. While the contents will always burn in the presence of oxygen, a burnt-out shell can be repaired with new hatches and ports. Repeatable units suit a number of applications, including habitation, laboratories, administration, growing areas, and factories.

The construction system is simple, with a minimum of imported material. The use of spoils from mineral processing for other base needs is especially applicable to cast concrete technology.

CHALLENGES

The human presence on Mars is best served by a statement of place, identifiable by the clear imprint of humanity. The presence at the surface, rather than in hollowed-out cave dwellings commonly proposed, asserts an intent not to survive, but to expand human presence.

The industrialization necessary for advancement to a secondary level of exploration and permanence is manifold. The concrete system proposed carries with it the expectation of a continuous supply line of building materials and high technology fittings, as well as a diverse collection of individuals to use them, and to expand our presence in the solar system.

REFERENCES

1. Clark, B. C., "Chemistry of the Martian Surface: Resources for the Manned Exploration of Mars," AAS 81-243, in *The Case for Mars*, P. J. Boston, ed., AAS *Science and Technology Series*, Vol. 57, San Diego, 1984, pp. 197-208.
2. Clifton, E. W., Candor Chasma Camp, Engineering Construction, and Operations in Space, ASCE, New York 1988.
3. Zimmerman Industries, Inc. 196 Wabash Road, Ephrata, PA, 17522.
4. Koski, J. A., Shotcrete Pumps and Related Equipment, Concrete Construction, Addison, 1993.
5. Boyd, R. C., Thompson, P. S., Clark, B. C., "Duricrete and Composites Construction on Mars," AAS 87-213, in *The Case for Mars III: Strategies for Exploration—General Interest and Overview*, C. Stoker, ed., AAS *Science and Technology Series*, Vol. 74, San Diego, 1989, pp. 539-550.

COVERED CRATER BIOME: A MARTIAN LIFE SUPPORT SYSTEM

Rollin P. Van Zandt

Providing acceptable living quarters for a relatively permanent Mars colony is a major problem demanding a minimum of imported weight and power requirements from the Earth. To implement this, already existing materials and resources on Mars should be used as much as possible.

This paper proposes a means by which impact craters, abundant on Mars, could be converted into living biomes by providing them with a flattened dome-like cover consisting of an underlying plastic or fiber glass seal and loaded down with about 30 feet of Martian soil to balance the atmospheric pressure trapped within. A slight over-pressure would be desirable, balanced by a system of hold-down cables to provide stability. The interior would be flooded with sunlight by a system of guided mirrors on top of the cover which would concentrate sunlight to a central tower, like presently existing solar collectors on earth. Instead of converting this into steam for power generation, this light would be directed down the tower and scattered uniformly throughout the interior by a suitable reflecting system under this light column at the center of the dome.

The initial construction would start with an air-tight film laid down on the crater floor and ceiling. The space between the floor and the ceiling is gradually inflated with Martian atmosphere and pressurized as Martian soil is placed, initially by flingers, over the cover until the desired atmospheric pressure is achieved. The mostly carbon dioxide atmosphere could be converted into a livable oxygen atmosphere by suitable biochemical means. The already present nitrogen could be enriched from the outside atmosphere.

This paper will consist of a study of the earth material weight (including construction equipment), energy requirements on the construction of the biome made from a 1 km diameter Martian crater and a thermal study of the amount of solar energy that can be trapped inside and subsequently utilized.

The results show a material moving machinery weight of 95,000 lbs. for one set machines (3 sets of machines would be preferred) and a cover and seal weight of at least 540,000 lbs. Total energy requirements for cutting and transporting Martian material surrounding the crater and moving it to form a 13.4 million ton cover is 3.3 million kw-hrs. The biome would be an excellent heat trap losing a maximum of 13.1% of the unclouded solar heat conducted into the crater.

INTRODUCTION

This paper is a preliminary analysis describing the steps involved in creating a comfortable, roomy biome on Mars as a prototype for a multitude of nearby similar ones to constitute a Martian colony. It is an attempt to approach an earth environment for the emotional and physical well being of the inhabitants making maximum use of existing Martian materials and crater topography. Figure 1 shows the final form of such a biome created as sustainable living quarters for about 150 humans providing food production and maintenance of a breathable atmosphere. Solar energy will be used for photosynthesis and temperature control. Crater diameters between 1/2 to 1 mile have been contemplated as the most feasible and the construction appears to be within the limits of present technology. The biological problems of growing food and other plant life necessary to maintaining the proper oxygen balance and recycling impurities is probably the greatest unknown. These are being attacked by the well-known Biosphere 2 experiment in Arizona and solutions should be well in hand by the time such a biome would be constructed.

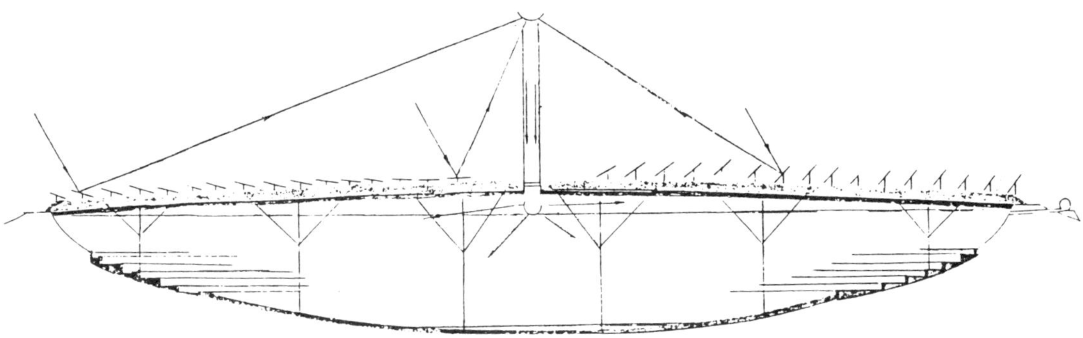

Figure 1 Completed biome ready for human occupancy.

The figure illustrates the essential ingredients: an earth-type atmosphere from compressed Martian air; water stored in a central lake acquired from the assumed plentiful permafrost on Mars; sunlight transported from the outside by a system of guided mirrors concentrated in a central tower and then dispersed to the inside; and a system of terraces to grow crops comparable to oriental farming practices long since developed on the earth. Hold down fiberglass cables from the ceiling would constrain the slight over-pressure for stability.

An important feature is the 30 ft cover blanket of Mars sand and rock and an assumed 12 ft thickness of soil and natural insulating materials on the crater floor which serves as a very effective heat trap to ensure livable temperatures inside. The cover will provide sufficient weight to balance an atmospheric pressure inside, of 8.5 psi or the equivalent of 15,000 ft altitude on Earth, arbitrarily low for economy reasons.

The steps leading to this final structure are embodied in this paper followed (1) by an "earth-moving" and material handling study to determine power requirements, (2) a rough thermal study of the percent heat loss through the cover and floor of the crater and (3) an estimate of weights of the machinery and fiber glass seal to build the cover and retain the atmosphere.

It should be emphasized that the material handling machinery once on Mars will have sufficient durability to be used in the construction of 10 or more biomes and can be thought of as an initial weight investment. Passageways connecting these biomes are also assumed but are not part of this study.

Other covered craters of similar construction could be dedicated to other biological purposes than human habitats such as specialized plant farms supplying the raw materials for plastics which will be needed in great abundance.

Other than to estimate the power requirements to construct such a biome, no study has been made of the precise nature of the power sources which must be imported from the earth to do this job. It is presumed to be some form of a nuclear-electric power plant similar to that used in certain present day naval vessels on earth. To save weight, it is further presumed that electricity will be generated from thermal energy by gas turbines rather than steam plants. Photo-ionic solar cells will be used for ongoing electric power needs once the biome is established.

STEPS IN BIOME CONSTRUCTION

Step 1. Preparation of Crater Floor

The crater floor must be evaluated for the type and grade of loose materials and how much is suitable for covering into fertile soil, and rocks and boulders for terrace walls. Figure 2 shows an assumed crater shape (copied after the Barringer crater in Arizona), for a crater diameter of 1 km and depth of 120 m. The underlying structure must be evaluated for permeability of biome atmosphere and suitably sealed where necessary. It is assumed that the total weight of sealant needed to make the floor air tight is only half that required for the cover. There is a very real possibility that water can be used as a sealing agent. As soon as a humidified atmosphere is developed while pressure is being built up under the cover, gas leaks will freeze out the water vapor and act as a seal.

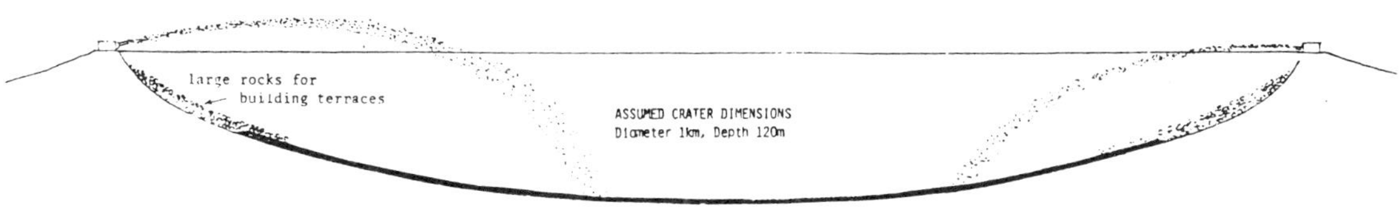

Figure 2 Step 1 - Sand lining for insulation and eventual soil.

Additional materials from the outside can be added and placed by flingers (as shown in the Figure) or tractor scrapers as needed but were not figured in with the energy requirements. Additional rocks and boulders if needed will be placed just below the rim to be used later for the construction of the agricultural terraces. Heat loss through the floor is based on an assumed average of a 10 ft layer of sand or equivalent.

Step 2. Placement of Support Netting and Fiberglass Seal Cover

As shown in Figure 3, a coarse netting of fiberglass is stretched over the crater rim strong enough to support a light-weight device capable of traversing the netting to lay strips of gas-tight film suitably cemented together, to form a sealing cover. Stretching can be done by scraper and winches available for later uses. The device to lay film has not been designed but it may involve only a stitching and cementing function along the strips after each one has been pulled over the cover from the rim. Rolling them across would require the netting to carry this additional weight.

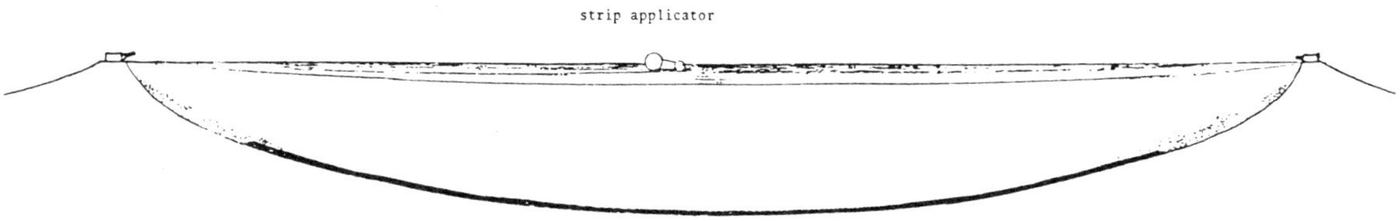

Figure 3 Step 2 - Support netting and fiber glass seal cover.

Step 3. Inflating the Cover and Placement of the First Layer of Blanket

A blower shown on the right of Figure 4 inflates the cover to a slightly convex outer surface using Martian atmosphere. Pumping continues to keep the cover in static balance as sand and fine gravel are next placed by flingers. Fiberglass hold-down cables are placed between the cover and floor to restrain a slight over-pressure needed to insure stability. Progressively coarser material can be applied after the initial protective under-layer. This first layer is built up to a 12 ft thickness assumed to be enough to support the weight of a scraper to spread the remaining thickness of required cover.

Stress measurements on the hold-down cables provide an additional function of precisely indicating how the material must be spread to maintain the static balance between the weighted cover and the constrained gas.

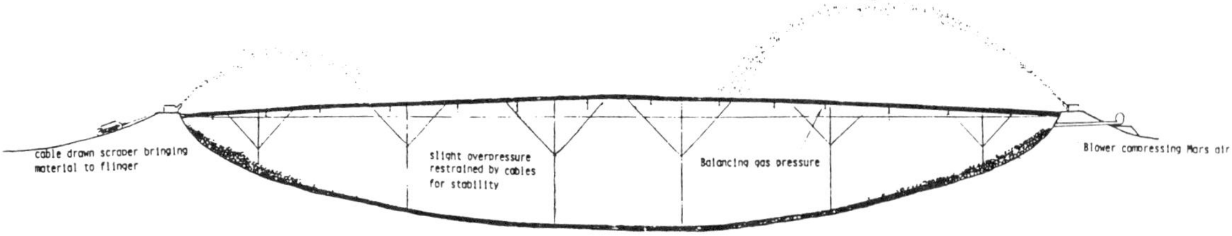

Figure 4 Step 3 - Inflating cover and flinging sand and fine gravel to a 12 ft. depth.

Step 4. Completing the Cover Using Scrapers

At this point scraper spreading will replace flinging to reduce energy requirements. As previously mentioned, final thickness will be 30ft to balance a livable atmospheric pressure assumed to be the same as the Earth at 15,000 ft altitude of 8.5 psi. Rocks and boulders as large as the skyscraper can handle are now acceptable. The initial 12 ft layer will have fiberglass netting top and bottom to spread and form a structure to carry vehicle loads as more material is added. Figure 5 illustrates this step with hold-down cables in place, a blower still supplying added pressure from the Mars atmosphere and material located around the crater being loaded by a cable drawn scraper.

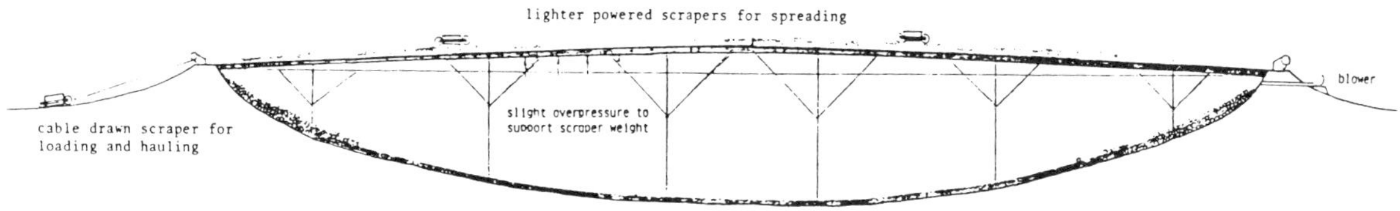

Figure 5 Completing cover with sand, gravel and rock being spreead by scrapers.

Step 5. Completed Cover with Light Collector Added

Figure 6 shows the completed cover with reflectors and a light tower installed (like solar power plants on Earth) to concentrate sunlight and direct it vertically downward. A transparent plug at the bottom retains biome pressure and filters out ultraviolet light. Light is reflected and dispersed at the bottom of the tower to spread it uniformly throughout the interior of the biome. A central lake of water will have suitable cyanobacteria, algae or other vegetation to reduce the CO_2 in the Martian atmosphere to breathable oxygen by photosynthesis. The naturally occurring nitrogen could be accumulated and further contracted to approach an earthlike condition and permit the human voice to function normally.

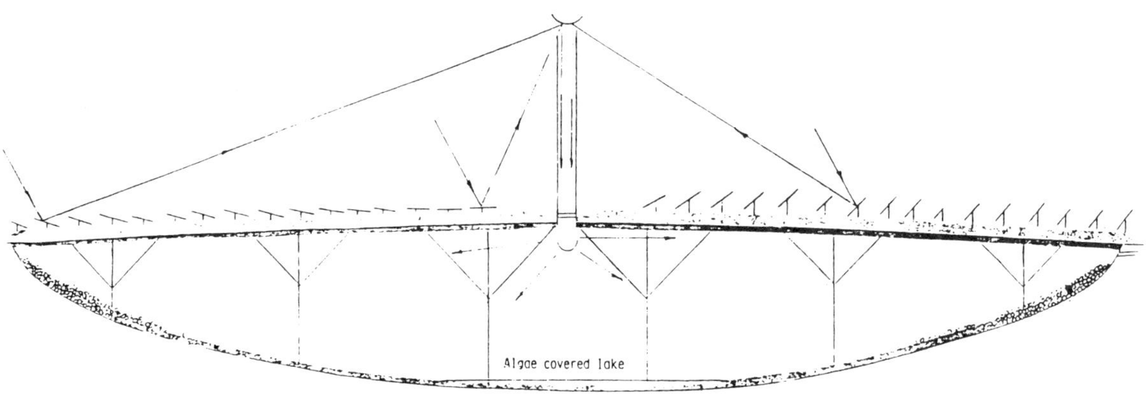

Figure 6 Completed crater with solar light collector and converting to living atmosphere through photosynthesis.

Step 6. Completed Biome with Terraces Built for Crops

Once the atmosphere is suitable for human occupancy, humans could construct circular terraces using the rocks already provided for laying up terrace walls. The lower gravity would be conducive to greater manual work and the material handling machinery could be correspondingly lighter. Once the terraces are built, suitable bacteria imported from the Earth could be introduced and further grown in the natural sand and clay to convert it into a fertile soil. Crops and orchards would provide the food and means of continued photosynthesis to result in a sustained ecology.

Terraced farming practices on Earth should be applicable here with water for irrigation flowing down from the highest terrace and percolating down to the central lake in a repetitive cycle. Humidity could be controlled as it increases from natural evaporation and respiration, by being condensed by the cold Martian atmosphere outside the biome. This would also offer a continuous supply of distilled water for human consumption.

Gas Leaks

Perhaps the most critical requirement is to prevent gas leaks through the cover. A freeze out of the water vapor will not protect the cover as it does the floor because it will upset the static balance of the cover and even cause separation of the part of the cover layer below the freeze zone.

A warning system such as sensitive pressure monitors must set off a leak alarm and identify its location. A sealant can then be applied by some means such as a flinger or manually by cable access. Preventative leak maintenance of the underside of the cover will be an essential requirement.

MATERIAL MOVING ENERGY

Criteria for Selection of Low Atmospheric Pressure

A low as possible air pressure was selected to minimize the energy requirements to build a cover and the amount of gas leakage to the outside. The equivalent 15,000 atmospheric pressure on Earth is assumed survivable. Whether this is adequate on a continuous basis and to allow sufficient expenditure of physical energy will require more evaluation. The low gravity on Mars may make it livable there where it would be inadequate on Earth. This would be a factor in the selection of the original pioneers to be sent to Mars and the make up of the imported gene banks. Additional oxygen could be provided in emergencies.

Magnitude of the Material Moving Job

The volume of the material to be placed on the cover 1 km (3279 ft) in diameter and 30 ft thick amount to 253 x 10^6 cubic feet or 9.37 x 10^6 cubic yards (usual contractors units) which is fairly sizable by Earth standards. In terms of weight, assuming a density of 1.7 for the material near the crater, this converts to 2.68 x 10^{10} lbs. (13.4 x 10^6 tons) on Earth and 1.02 x 10^{10} lbs. (5.1 x 10^6 tons) on Mars with 0.38 the Earth's gravity. The amount of additional material required for the crater floor is uncertain and the material already in place may suffice. Assuming the material surrounding the crater would be sufficiently unconsolidated to use for the biome cover, it could be cut, loaded and hauled by the highly developed scraper designs used on the Earth.

The low gravity on Mars would exclude self loading because of very poor traction. The scrapers could be provided with adequate cutting force by electrically powered winches anchored along the crater rim. See Figure 7. The system is represented by three equal work areas within pie shaped boundaries. The winches and screens and other rim equipment will be semi-portable to provide closest access to the scrapers as needed. The internal power requirements for the scrapers will be electrical and served by power cables from the winches.

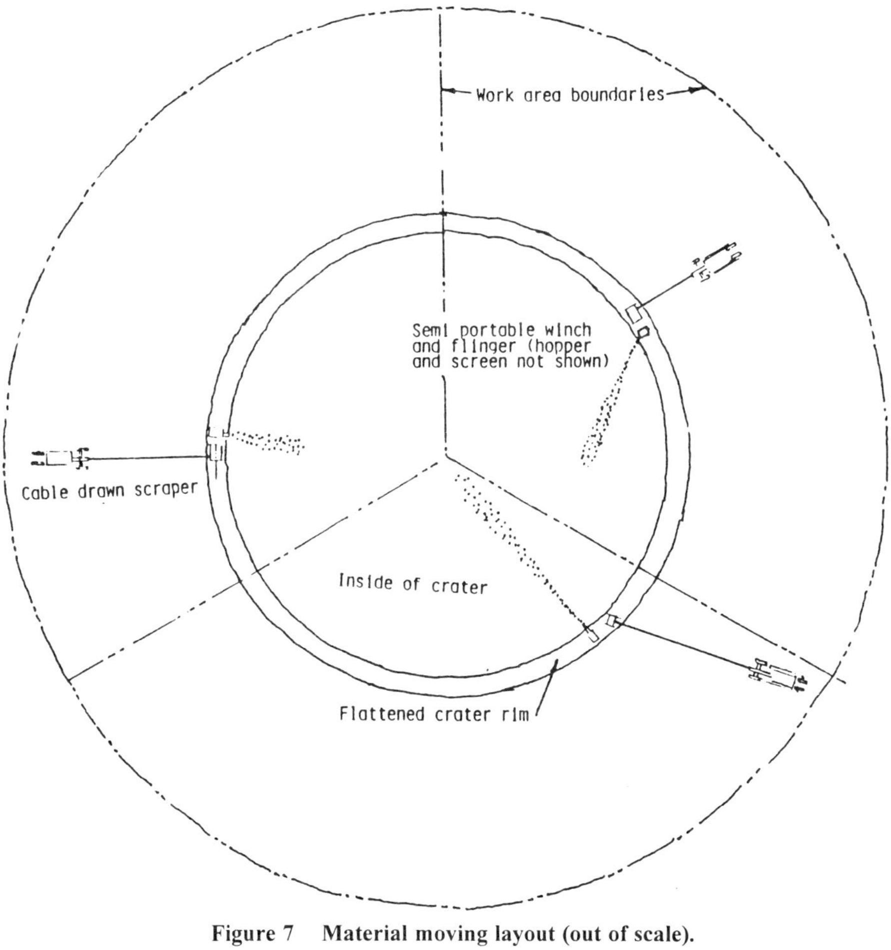

Figure 7 Material moving layout (out of scale).

Once the required cover thickness has been achieved by the flingers, the scrapers will then use their own motive power in travelling over the cover where traction requirements for spreading are minimal. Three spreads of equipment are suggested in order to get the material moved more quickly. The estimated time to completion is one Martian year.

Scraper Spreader Design

Figure 8 is a sketch of a proposed design based on well-known earth technology.[1] The scraper is converted to cable haulage rather than wheel traction, already mentioned as being inadequate in the low gravity field on Mars. The front wheel will be steered by the towing cable in cutting and hauling but equipped with a normal tractor steering system used in spreading and maneuvering. The rear wheels will be lightly powered for the latter requirements.

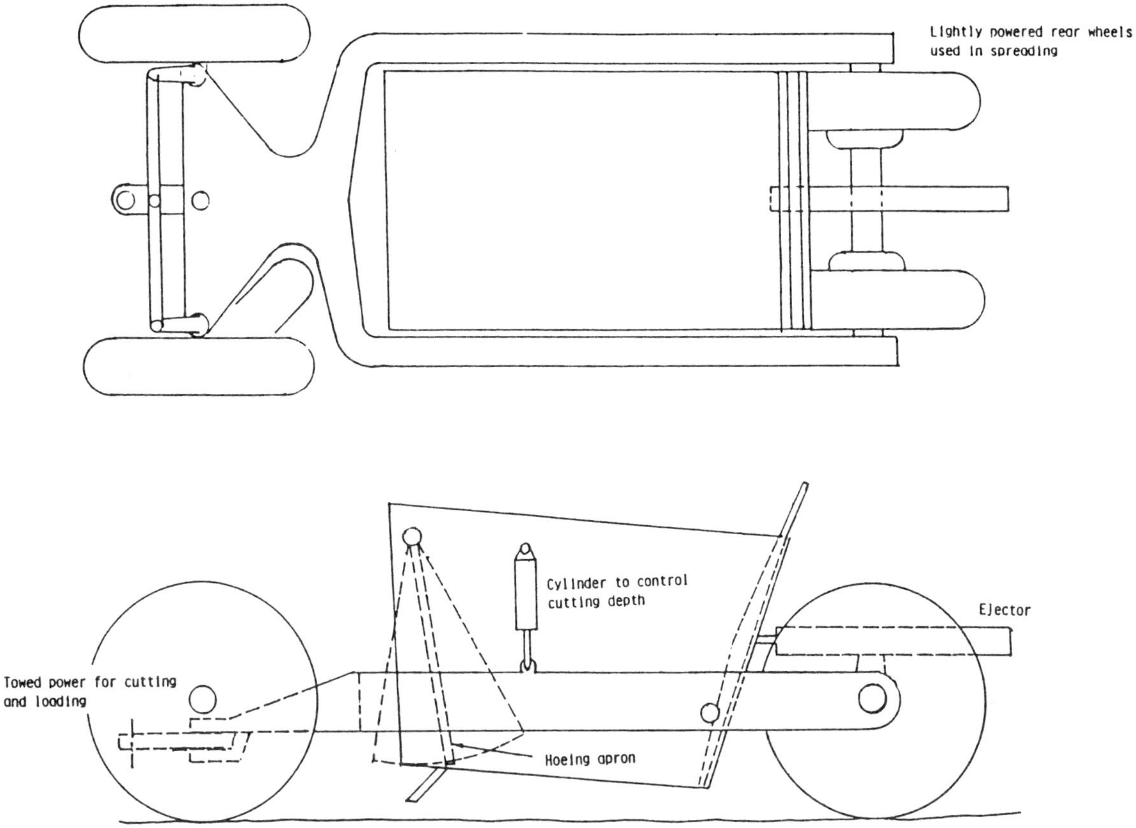

Figure 8 Scraper spreader design.

An important loading feature will be a "hoeing apron" at the front of the bowl to assist in drawing in material including boulders and depositing it in the bowl. Scrapers of the size contemplated can handle at least 1 ft boulders. An hydraulically actuated ejector will discharge and spread the material from the bowl which is the usual practice on earth machines. For minimum earth transport weight and energy requirements on Mars, these vehicles can be made mostly of aluminum or titanium instead of the usual steel except for the carbide tipped cutting edge.

The material moving study is based on a 50 heaped yard scraper capacity weighing an estimated 55,000 earth pounds compared to an 82,000 all steel vehicle.

Material Moving Energy

Referring again to Figure 7, the sketch shows a source of material extending to 1 km past the rim of the crater. An average depth of cut to acquire the desired volume comes to 3.75 ft which is presumed to be a reasonable depth accessible to the scrapers.

Cutting energy: Assume the scrapers have 10 ft wide blades requiring 4000 lb./ft cutting force at a depth of 1 ft which is an extrapolation of an actual instrumented traction test in tough silty clay.[2] The total cutting energy = 253 x 10^6 x 4000/33000 x 60 or 510,000 hp-hrs.

Loading energy: The hoeing apron will force the material into the bowl, pushing it along the 15 ft long floor. Based on a 0.5 soil to floor friction coefficient and using Mars material weight, it takes 0.227 hp-hrs per load or a total of 42,600 hp-hrs.

Hauling to the crater rim: Figure 7 shows how the material will be hauled in by cable. The winch (and flinger at the outset) will be moved around the rim as needed to reduce the amount of scraper haul distance. A slope to lower ground surrounding the rim is assumed to be 240 ft maximum at about 1900 ft from the rim where the material must be raised. A common rolling resistance of 60 lbs. per ton on the tires was included and by graphical means not shown it takes 1.467 x 10^6 hp-hrs to haul this material and adding the empty weight of the scraper, the total hauling energy is 2,028,000 hp-hrs.

Flinging material on to the cover: In order to save imported weight the fiber glass cover seal should be made as thin as possible and the first step in applying cover material will require considerable care. The method already mentioned was to fling the material from the crater rim as it is brought up from the scrapers. To start with, it should be screened to a fine enough size so as not to damage the seal but not so fine as to be blown away by the tenuous atmosphere—probably at a time when the winds are low. The flinger would allow accurate placement of the sand and gravel at all locations to keep the weight of the cover in accurate balance with the gas pressure inside.

Flinger ballistics: The flinger is conceived as an inclined rotating disk with vanes accelerating material applied near the center and escaping at the rim with appropriate guides to keep the stream from spreading. It would look something like a common automotive supercharger blower and be variable in speed to control the exit velocity and distance the material would travel. In this study the blower is tilted at a constant 45° angle assumed to require the least amount of energy. No correction was made for the drag of the thin atmosphere in the belief that even if corrections had to made for the finer materials they would not significantly increase the energy requirements.

The exit velocity is v with v_v and v_h being the vertical and horizontal components, 0.707v. The time of flight to rise and fall is 2t, and using the well known x = $1/2gt^2$, is t = the square root of 2x/g. The horizontal component of speed is constant during this time, v_h = gt = square root of 2gx. The horizontal distance, y, covered during the time of flight thus simplifies to y = 2 x square root of 2x/g. 2gx = 2 times the square root of $4x^2$ = 4x. The kinetic energy required to impart this exit velocity = $1/2mv^2$, noting that m is inertial mass and thus the same regardless of gravity. The procedure followed was to lay out the cover into 5 concentric rings or annuli, eyeball the horizontal distance traveled from the rim, and calculate the energy required to fling the material this distance. No correction was made for additional sliding upon impact.

As an example, the energy requirement to fling the full 30 ft thickness of material to the center ring at an average distance eyeballed to 1500 ft calls for a flight height, x, of 375 ft and an exit velocity of (sq rt) 2gx/.707 = 135.5 ft/sec (note that the lower gravity value of Mars is used here). The kinetic energy per earth pound = $135.5^2/64.4$ = 285.1 lb.-ft/lb. This converts to 285/33,000 x 60 = 0.00014 hp hrs/lb. The earth weight of the center ring is 1.052 x 10^9 lbs. and the energy requirement = 147,000 hp-hrs.

Following this procedure for all 5 rings gives a total of 2,634,000 hp-hrs which increases, assuming a flinger efficiency of 70%, to 3,763,000 hp-hrs. Flinging is not a very efficient way of moving material, suggesting the alternative method of flinging only enough material to support the scrapers and letting them finish the job. Assuming that flinging 12 ft of cover is adequate, the flinging requirement was reduced to 1,505,000 hp-hrs.

Final Cover Spreading Using Scrapers. There are three energy components in completing the cover: ejection of the material from the bowl; spreading the material as it is forced out of the bowl; and the rolling resistance of the scraper as it spreads and the tires compact it. It turns out that the ejection energy with this design of scraper is negligible leaving only the latter two.

The scraper is assumed to spread a 1 ft cover thickness (1 ft elevation of the cutting edge) and a frictional drag coefficient between the cutting edge and the ground of 0.5. This gives 50 x 27/10 = 135 ft spread length per load. Spreading energy per load (on Mars) calculates to 3.67 x 10^6 lb.-ft. No. of loads to produce 30-12 = 18 ft. thickness are 112,000. Total spreading energy = 3.67 x 10^6 x 112,000/33,000 x 60 = 208,000 hp-hrs.

The hauling and compacting energy while spreading followed the same graphical procedure as flinging for the determination of haul distances. A rolling resistance of 60 lb. per ton of Mars weight remaining on the wheel times the net distance for each of the 5 concentric zones totaled 84,000 hp-hrs.

Summary of Material Moving Energy

	Horsepower hours	kw hours
Cutting	511,000	381,000
Loading (hoeing)	43,000	32,000
Hauling to the rim	2,028,000	1,513,000
Flinging 12 ft thickness	1,556,000	1,151,000
Remaining cover spreading (using scrapers)	208,000	155,000
Cover hauling and compacting	85,000	63,000
Total	4,431,000	3,295,000

Note that these figures do not appear excessive when the total power is equivalent to 373 kw running for a continuous earth year. Also notice the substantial savings by using the scrapers rather than flingers to build most of the cover.

CRATER THERMAL BALANCE

Heat Loss Through the Cover

Using a figure for conductivity for dry sand of 0.188 btu per hr per square foot per ft of depth per degree F temperature difference gives a heat loss of only 1.065 btu/ft^2 hr for a 30 ft cover thickness and temperature differential of 170°F. The average outside Martian temperature was assumed at –100°F.

The solar heat input was based on a solar constant of 3.9 x 10^{33} ergs per second which converts to 9.49 x 10^{11} btu per sec. The heat flux at aphelion distance to Mars of 155 x 10^{6} miles works out to be 158 btu/ft^{2}hr. The biggest uncertainty is how much energy can get through the mirrored lighting system. If this is assumed 50%, the heat loss is only 4.2% of the incident energy.

Heat Loss Through the Floor

A greater loss is taken for the 10 ft floor using the same conductivity figure but an average underlying floor bottom temperature of –500°F gives 8.9%. Note that no correction for the slow heating of the underlying crater floor was made which should gradually reduce this thermal loss. On the other hand, no consideration was made for the reduced solar input during Martian dust storms but this would not appear to be of dangerous significance.

Total Heat Loss

The total heat loss for the floor and cover is 13.1% due to conduction. Thus, with its thick cover and moderately thick insulation for soil and vegetation on the terraces and floor, the covered crater biome serves as a good solar heat trap with enough excess to be used for other purposes.

MAJOR IMPORTED WEIGHTS OF MACHINERY AND SEALS

Earth weights will be used throughout since these are the ones involved in the required transport energies to Mars.

Scraper weight: Based on converting an advertized Caterpillar machine weight to the configuration shown in Figure 8 and the use of light weight metals wherever possible:
55,000 lbs

Flinger: The weight was an educated guess since no attempt was made to design the device:
15,000 lbs

Hopper and screen - also a guess:
25,000 lbs

Basic material handling machinery - per set
Total: 95,000 lbs

Cover netting: 0.10 in. Dia spun. Strands woven in 1 ft squares - assumed density of 2.7:
78,000 lbs

Cover seal: Also fiberglass 0.003 in. Thick:
360,000 lbs

If three sets of machinery, two nets (one below and one above the 12 ft lower cover layer) the total imported weight is:
801,000 lbs.

This is a rather formidable amount but it compares with 27 billion lbs. (10 billion lbs. Mars weight) of Mars materials used to produce a truly livable human habitat. Machinery weight could be reduced by 1/3 if only one set were imported, taking a longer construction time. There may be a further weight saving if the seal and netting were manufactured on Mars, if the necessary processing machinery weighed less.

CONCLUSIONS

1. The imported weight requirements place more demands on the system than power. The weight of the cover and floor sealing strips is high and call for careful design to minimize thickness. Exotic materials other than fiberglass may be required.
2. A scale model test on earth could simulate the delicate initial flinging operation as well as to determine the design of the ceiling cables for maximum stability. It may well turn out that once the cover is completed and static stability realized, the hold down cables can be eliminated. A suitable binder in the 30 ft of cover material could act as a "truss" and provide rigidity against flexing or wave action in the cover blanket. Small changes in the volume of trapped atmosphere could be accommodated by a slight "breathing" action of the whole cover.
3. Along with conclusion 2, the scale model test should include a structural study to determine more accurately the required thickness of the initial layer sufficient to support the weight of the scrapers.
4. The biome is a sufficiently effective heat trap to insure acceptable living temperatures at all times even during a dust storm. Normally there will be a generous supply of solar thermal energy to be available for other purposes.

REFERENCES

1. Derived from author's 35 years experience at Caterpillar, Inc. Research Department, where he was involved in design and testing of a wide range of earth moving equipment.
2. Testing done on a full scale vehicle at Caterpillar, Inc. Experimental Proving Grounds near Peoria, Illinois.

MARS 2008 SURFACE HABITATION STUDY - PART 3 SAFETY PHILOSOPHY*

Marc M. Cohen†

This report, Mars 2008 Surface Habitation Study, SAFETY PHILOSOPHY documents the results of the 1992 "Summer Study," August through November 1992, sponsored by the Center for Mars Exploration at the National Aeronautics and Space Administration's Ames Research Center in coordination with the Office of New Initiatives at NASA-Ames, undertaken in addition to their regular jobs.

NASA Ames Mars Architecture and Habitation Working Group:
Marc M. Cohen, Chair, Advanced Space Technology Office
Yvonne Clearwater, Aerospace Human Factors Research Division
Betsy Dunsky, Advanced Life Support Division
Paul Espinosa, Flight Experiments Branch
Vladimir Garin, Space Projects Division
Andy Gonzales, Advanced Life Support Division
Butler Hine, Information Science Division
Michael Sims, Information Science Division

Special Acknowledgement:
Robert F. Raasch
Lisa A. Rockoff

The Mars 2008: Surface Habitation Study position papers cover these topics:
1. Objectives and Overview
   Appendix A, Surface Missions Study Outline
2. Mission Design Logic
3. **Safety Philosophy‡**
4. Habitation Strategy and Requirements
   Appendix B, Data Requirements from Precursor Missions
   Appendix C, Pre-Human Arrival Preparation Missions
5. Mission Design Evaluation Process and Criteria.**

---

* Center for Mars Exploration Position Paper, NASA-Ames Research Center, Final Review Draft, May 17, 1993.

† Prepared by Marc M. Cohen, Architect, Advanced Space Technology Office.
Approved: Geoffrey Briggs, Director, Center for Mars Exploration.
Concurrence: Anthony R. Gross, Chief, Advanced Space Technology Office.
Concurrence: Roger D. Arno, Manager of Advanced Projects, Space Projects Division.
NASA-Ames Research Center, Moffett Field, California 94035-1000.

‡ The text for this paper appears in full in paper AAS 95-491, in *Strategies for Mars: A Guide to Human Exploration*, C. R. Stoker and Carter Emmart, Eds., AAS *Science and Technology Series*, Vol. 86, pp. 490-500, 1996.

** The text for this paper appears in full as paper AAS 93-883B, in this volume, pp. 401-417.

APPENDICES

APPENDIX A

ABSTRACTS OF THE CASE FOR MARS V CONFERENCE PRESENTATIONS

1. STRATEGIC PLANNING

AAS 93-801
STATUS REPORT ON STRATEGIC PLANNING AT NASA
Charles Pellerin, Office of the Administrator, NASA.

This talk will provide a status report on strategic planning at NASA. A strategic planning process which assures implementation of strategy will be outlined. An overview of perspectives on the strategic purposes of the Agency by Congressional staff, space pundits, NASA employees and the participants in "town meetings" will be summarized. Analysis of the needs of NASA's customers will then form the basis for a proposed re-alignment of the Agency around `Economic Benefits' as an optimal approach to meeting the broadest goals of the civil space program—working in, and colonizing space.

2. SOCIAL, POLITICAL AND SPACE POLICY ISSUES

AAS 93-802
THE LONG GOODBYE: A WRITER'S PERSPECTIVE ON THE PSYCHOLOGICAL IMPERATIVES OF SPACE EXPLORATION
Thomas Simmons, University of Iowa

Although space travel has often been sold to the public as a miraculous combination of science and technology, that combination has nearly been its doom; the more marvelous it's been, the farther it's drawn from the lives of ordinary people. What, then, is space about? Beginning with my own experience of growing up in the space age—including recollections of Apollo and Challenger—explore the possibility that, despite every argument in favor of the practical value of space exploration, space travel is really about faith as much as science. It's about becoming human. But what's involved in this becoming, and how do imagining space and going into space change human beings?

Although I approach this question with specific references to the Mars project—focusing on the work of Prof. Bruce Lusignan of Stanford University—I concentrate on wider philosophical issues: I draw on a variety of texts, from those of Loren Eiseley to those of children's book authors, to suggest psychological parameters for the future of space exploration. I argue that the essential characteristic of human experience is learning to say good-bye: to leave old selves, old visions of the world, and even the world itself, without discarding that which is central to the self.

AAS 93-803
THE THIRD ERA OF EXPLORATION
Dwayne A. Day, Space Policy Institute

A political environment conducive to large scale planetary missions such as a human flight to Mars will have to be substantially different from that which led to previous major space decisions such as Apollo, the Space Shuttle, and Space Station. The end of the Cold War has eliminated most of the former justification for ambitious space missions. The reasons why countries choose to explore space are being redefined and are still not clear, as the experience with the Space Exploration Initiative indicates. Nevertheless, it is still possible to broadly outline the political criteria which must be met in order for human exploration of the solar system to occur. These requirements are: a strong national leader dedicated to the goal and willing to expend political capital to initiate it; a Congress which either agrees with the President or is willing to support him due to his popularity; a national consensus that the goal is a worthy one and that large expenditures on space are justified; and an international consensus that such a mission is worth undertaking. Beyond these there are several bureaucratic/institutional requirements for success, including: a competent and enthusiastic NASA administrator; a flexible and dynamic international framework to carry out such a mission; a program plan which is reasonably low cost and has low peak funding requirements; and technology which has commercial relevance to further justify its funding.

A close examination of these criteria indicates that a large scale human mission to Mars is unlikely in the foreseeable future. Furthermore, the changes in the political environment necessary to achieve these criteria and lead to a commitment to such a mission are beyond our grasp at the moment. There are several steps which can be taken to decrease the political opposition to such a commitment in the future, but none which can assuredly *lead* to such a commitment. Unfortunately, from a political standpoint, the future of human planetary exploration is very grim indeed.

AAS 93-804
EARTH AND MARS IN HISTORICAL PERSPECTIVE
Thomas R. Meyer, Boulder Center for Science and Policy

Now that the cold war has ended, it is no longer nuclear confrontation but world population growth that is the predominant threat to mankind. The Earth's unprecedented population expansion, made possible in this century by the advance of modern technology and the availability of cheap energy, is placing exponential demands on the planet's finite reserves of resources and fossil fuels. Simultaneously, as a consequence of industrial development, overuse of resources, and pollution, the regenerative capability of the biosphere to supply our renewable resources is also being eroded at compound exponential rates.

Because the doubling times for many of these effects are mere decades, we are fast approaching the time when the demand for resources will begin to exhaust the Earth's endowments, and the deleterious effects to the environment will reach a scale that will challenge the natural processes that maintain the very stability and viability of the biosphere. For some parameters this may already be happening.

As population continues its exponential climb, we will move ever more rapidly toward a global crisis. We face the prospect of environmental catastrophes, economic and social instabilities, sky rocketing costs, a declining quality of life - particularly for the poor, and a high potential for global conflicts and war.

In a world where mankind's numbers have grown so large as to threaten the entire biosphere, there is no choice but to ascend to the moral responsibility of maintaining our planet in balance. All of mankind's activity on the planet, all of our creations, products, wastes, and uses must be conducted in a manner that assures human existence in perpetuity.

This is the kind of historical perspective upon which our future space policies must be built. It is only with a clear understanding that we must bring our own planet into balance that can properly assess the opportunities presented by Mars, the other planets, and space itself.

Although we praise President George Bush for having set a course to return humans to the moon and then to go on to visit Mars, in light of the impending environmental crisis it is perhaps for the best that this program is now being abandoned. It is good that NASA's Code X, the Office of Exploration has been disbanded. It is good that Congress choked the funding for the Space Exploration Initiative. It is good that under the Clinton Administration we have abolished the National Space Council.

Good, because all of these activities were the apparatus of a space policy rooted in the Cold War era that has now ended. In view of our need to deal with the environmental crisis that lies ahead, they are clearly the wrong objectives for the wrong reasons.

Thus in the death of the old, a new space program is being born. "The Phoenix arises," as Chris McKay said last winter when we planned this conference.

Our immediate priorities are first to understand the Earth and second, to learn how to fix it. For these tasks, Mars has a great deal to offer. Mars is the planet which will yield a rich harvest of scientific information about planetary evolution, climate change, and the origin of life. Critically, in this time of endangered global environment, Mars also provides a second viewpoint for understanding Earth and its precious biosphere.

The study of Mars also provides an assessment of Mars as a future habitat for humans and a second home for life in our solar system. To learn how to make Mars habitable also teaches us how to repair our own planet.

Mars is already known to have all of the resources necessary to sustain life. It has a surface area equal to the land surface area of the Earth and a climate that does not preclude human life. In the long term the exploration and settlement of Mars promises to bring mankind another great era of prosperity. Mars offers us a second chance, not simply to consume and waste, but ultimately to learn to live in harmony and balance with nature as a multi-planet species.

AAS 93-805
IS MARS A SOLUTION TO THE PROBLEM OF EXPONENTIAL GROWTH ON EARTH?
Albert A. Bartlett, Professor Emeritus, University of Colorado, Dept. of Physics

One of the simplest but most poorly understood concepts in mathematics is the law of exponential growth. The doubling time is easy to calculate, simply divide the percent growth rate into 70.

The Earth's population is growing exponentially at the rate of 1.7 percent per year which means our population is doubling about every 40 years. How many people can the planet hold?

From about 1880 to 1980, world oil consumption grew at the rate of 7 percent per year, which means the total amount consumed doubled every ten years for a century. America has built a fine standard of living on oil that is the envy of the world. Can this continue and how long will it last?

The depletion of our renewable resources is growing at exponential rates, in some cases faster than they can be resupplied. Environmental damage is occurring on the Earth at exponential rates.

If all these factors continue, what will be the consequences? What are our options? As the population grows, can we transport people to Mars? When we run out of resources on Earth can we turn to Mars as a new source? And how long will Mars last? Can we continue in our present ways or must we develop a new ethic of ecology and the human spirit in managing our planets?

AAS 93-806
MARKETING MARS: A PRELIMINARY SURVEY OF PUBLIC ATTITUDES
David T. Cadden, Quinnipiac College

Any United States commitment to a Mars landing program, either by itself or in conjunction with other nations, would require considerable and sustainable public support. Marshalling this support translates into a need to go beyond merely educating the public. It requires a deep understanding of the level of national willingness to pay the bill, support for international cooperation, and mission justification. Support would not be homogeneous and it would be extremely useful for program managers to appreciate the segmentation of the public's attitudes.

This research examines the attitudes of (481) college students from two colleges in Connecticut. It focuses on three areas - attitudes toward the budget requirements for a Mars landing program, willingness to work with various international partners, and various mission justification. The results indicate the manner in which the mission's costs are presented can have a significant impact on level of support. There was also significant variation in student support for the proposed international partners and their perception of different justifications for a manned mission to Mars.

This research is being extended to a national phone survey. The author believes that these results indicate the need for a benefit segmentation marketing approach to muster strong public support.

AAS 93-807
FOOTSTEPS TO MARS: AN INCREMENTAL APPROACH TO MARS EXPLORATION
Geoffrey A. Landis, Sverdrup Technology, NASA Lewis Research Center

A program for the exploration of Mars will fail unless it can pass three critical feasibility criteria: (1) technical feasibility, (2) political feasibility and (3) economic feasibility.

Most discussions of programs for Mars exploration have focussed on only the first of the three necessary criteria. Political and economic feasibility, although equally important, are usually ignored.

In terms of political and economic reality, a Mars exploration plan must learn from the lessons of the Apollo program and of the Shuttle program. The lesson of Apollo: If you accomplish your goal, your budget will get cut. The lesson of Shuttle: Infrastructure is boring. If you do the same thing over and over, the public will focus on your failures and ignore your successes.

A middle course is suggested, where a series of well-defined intermediate goals can be accomplished, resulting in the incremental development of the technology required for full-scale Mars exploration.

AAS 93-808
IS THERE A SHORT TERM ECONOMIC AND SOCIAL JUSTIFICATION FOR MANNED SOLAR SYSTEM EXPLORATION?
Robert E. Becker, 4-Sigma, Inc.

SEI is now officially dead, totally unfunded, widely viewed as unfounded. While one may disagree on the last point, we can all agree the case has not been made extremely well: Human wanderlust, or economic benefits from mining asteroids thirty years down the road, pale in comparison with the world's quotidian economic and social problems.

Proponents of exploration sometimes have a tendency to declare "not my job" with respect to these problems. Yet, exploration is fundamentally a human endeavor, and should embrace possible solutions that may derive from it, even if this requires broadening the definition of that endeavor. Serious investigation of the short-term benefits of going full thrust with SEI might provide a means for transforming a perceived negative into a positive.

By this is meant something far more than a jobs program for the aerospace industry. Many of the benefits would derive from broadening the industrial base of space exploration during the development phase beyond the aerospace industry. This becomes possible because of the vast scope of SEI itself and the necessity to solve, for the first time on a large scale in space, many of the problems associated with life and society on earth: SEI is actually the space mission most closely related to the man in the street, not the least. In turn, SEI naturally encompasses those industries which developed over time to provide such solutions. And this opens the way for incorporation of "spin-on" technologies from the civilian economy, while naturally transferring aerospace workers much closer to the commercial sector.

This paper does not answer the question of the title. Rather, these points and others provide a rationale for initiating a coordinated effort by the exploration community to provide an answer convincing to government and the public. A framework for this study, a generalization of the structures of Concurrent Engineering, is proposed.

If the answer were affirmative, the resulting Mission From Planet Earth, what might be called "Mars Holistic," coupled with its mate, a similarly broadened Mission to Planet Earth, explicitly embracing terrestrial problems and their solutions, could be announced with trumpets blaring, not with a bureaucratic whimper.

AAS 93-809
THE NEED FOR POLICY RESILIENCY IN MARS EXPLORATION (OR WHY HAVEN'T PEOPLE LANDED ON MARS YET)
Matthew Fisk Marshall

Over the years, there have been a large number of proposed Mars missions. There have also been a wide variety of national reports and committees that have sought to give direction to NASA and the space program. While many of these missions would have been technically feasible, humans have yet to set foot on Mars. Furthermore, we have not gotten much closer to the realization of this goal. Clearly the traditional approaches to mission or programmatic design have yet to prove successful in their application to a Mars exploration strategy. Therefore to successfully implement a Mars exploration program, new approaches and strategies to programmatic design will be required.

This paper addresses the issue of policy resiliency as a key ingredient to the success of Mars exploration architectures. Policy resiliency builds-in flexibility to adapt to changes in national and world priorities and goals towards space. Heretofore no space program has been truly policy resilient. Some programs, such as the Apollo Moon landings, have been implemented during periods of policy stability with favorable results. Other programs, such as Space Station *Freedom*, have been conceived in eras of policy instability with potentially adverse effects on the program's performance or success. This paper will discuss strategies for the development of implementation options that will eventually lead to a successful Mars exploration program.

3. COMMERCIAL OPPORTUNITIES

AAS 93-810
A COMMERICIAL BASIS FOR MARS DEVELOPMENT
Mike H. Ryan, University of North Texas

Patterns of trade and industrial development mark the passage of time as readily as the years. Commercial space operations have been slower to evolve than was once expected. Transportation issues remain the single greatest obstacle to commercial operations in space, on the moon, and ultimately on Mars. The lack of a "Business Rated" transportation system is a chasm over which no bridge has yet been built. The process of industrial transformation is an ongoing, although somewhat haphazard, process. There are nations boldly striding into the twenty first century while others struggle desperately out of sixteenth century economies. This disparity is likely to become more pronounced over the next decade. Space operations, primarily the province of the wealthy, industrialized countries, are being viewed increasingly as expensive luxuries which can no longer be afforded. Yet, these operations are the very ones with the potential to address the economic disparity among developed and developing nations. Unfortunately, the perceived gap between space related expenditures and their economic payoff has increased. Budget cutters, the world over, are using short term return arguments to curtail and/or eliminate space activities with little regard for their potential long term benefits. Government resources available for Mars ventures (and other space efforts) will diminish unless a significant return (generally financial) can be obtained. Consequently, traditional approaches for funding Mars efforts may not be practical for the foreseeable future. Private sector initiatives established in concert with international public sector programs may become the preferred mechanism to foster Mars activities. Given the time frames involved, pure commercial ventures using a stair step approach may be the most effective and efficient means to ensure Mars exploration and development early in the next century.

4. EDUCATION

AAS 93-811
CASE FOR MARS V — EDUCATION OUTREACH AND WORKING GROUP
Coordinator: Barbara Sprungman, Space Data Resources and Information

Long-range goals are drivers — they have a propulsion of their own. The goal of a future mission to Mars is motivational when students learn that voyages to Mars will happen within their lifetime. Designing missions to venture out into our Solar System captures students' interest in subjects that often seem irrelevant when studied in isolation.

The Case For Mars (CFM) Conference offers a unique opportunity for educators to meet with leading space scientists, engineers and policy experts. It also provides an opportunity for educators, as well as other Mars enthusiasts, to confer and purpose recommendations related to the important role of space science studies in both national and international education reform. This is an hot topic today with the U.S. President's new technology policy and his concern for educational improvement — and new funding for education-related projects is available.

Anyone interested in contributing to the CFM V education recommendations should attend the Education Session from 9 to 11 AM on Saturday (check schedule for location), or leave a message for the Education Coordinator Barbara Sprungman in the CFM Press Room. Other informal meetings may be held at earlier times during the conference — check announcement board. Recommendations will be published in space publications, as well as being submitted to key U.S. decision makers such as the NASA Administrator, Secretary of Education, White House Science Advisor and the President.

PAST CFM EDUCATION RECOMMENDATIONS: At CFM in 1987, education track participants recommended that scientists and engineers bring their expertise into the schools. The last six years have seen a remarkable movement of these experts into the schools through individual efforts as well as major educational outreach programs supported by government, industry and academia.

In 1990, the CFM IV Education Working Group produced a series of recommendations focused on the importance of education in the Space Exploration Initiative. These were published in a Space News Commentary and *Ad Astra* magazine.

These recommendations were also presented to the White House National Space Council and the Synthesis Group in 1990 when Case for Mars organizers gave briefings of the findings and general recommendations of the CFM IV scientists, engineers, and space policy experts. Many of the educational recommendations were incorporated into the final report "America at the Threshold - Report of the Synthesis Group on America's Space Exploration Initiative" (1991).

MARSVILLE: Educators who are involved in the classroom project called MARSVILLE - The Cosmic Village will also be participating in the conference. MARSVILLE is a project of the Challenger Center for Space Science Education. Thousands of students worldwide have participated in MARSVILLE since early 1992.

Working in teams, students research, design, and build models of life support systems crucial to successful human settlement of Mars that include air and water supplies, food production and delivery, temperature control, waste management, communications, transportation, and recreation.

The Mars mission simulation brings students from several schools together in an event to construct Marsville. The Martian habitats are constructed from clear, heavy-gauge plastic and inflated with large fans. The Boulder area MARSVILLE project sponsored by Susan Solari of the Colorado Space Grant Consortium held their event here at the University of Colorado on April 30, 1993.

POSSIBLE MARSVILLE LINK-UP DURING THE CONFERENCE: On Friday May 28th, two other MARSVILLE events will take place in California and Massachusetts. We are attempting to establish a link-up between students at these two events and CFM participants for about an hour probably sometime between 9 AM and 3 PM. If you are interested in participating or viewing this event, watch for announcements, and leave a message for Barbara Sprungman in the CFM Press Room.

Information will also be available on the Mars-related education projects of The Planetary Society, a major CFM Conference sponsor.

Comments on problems and solutions to incorporate into CFM V education recommendations are welcomed from everyone. Education Coordinator Barbara Sprungman can also be reached during the conference by leaving a message at Space Data Resources & Information at (202) 546-0363.

AAS 93-812
THE EDUCATIONAL IMPETUS FOR MARS EXPLORATION: USING THE PLANETARIUM AS A TOOL

Shawn Laatsch, Bishop Planetarium

In order for manned exploration of the planet Mars to take place, proper education of the public is needed. During the 1960's great pains were taken to educate the public about space and space exploration which led to a "mental readiness" to go to the Moon and take on other space exploration projects. The result of this "mental readiness" was the Apollo missions. The current spirit of our society does not support manned exploration of Mars, therefore, proper education of the public is needed to develop ourselves into a society motivated for the task.

This paper will suggest strategies for educating the public through the use of planetariums, schools, and the media. Specifically it will discuss how planetariums can create the "mental readiness" required for a Mars mission by educating the public about the benefits of such exploration, and by showing people the need for such a project.

AAS 93-813
EDUCATION STRATEGIES FOR MANAGEMENT OF A GLOBAL BIOSPHERE
Dana Richard, Earth Counsel, Boulder, CO

This talk will describe Earth Counsel (EC) and the educational aspects of its program including:

1. Evolution of Earth Counsel
2. Functions of Earth Counsel
3. Earth Counsel University
    - Modeled on ISU,
    - Branch campus on each continent,
    - Teaches the functions of EC's three principal organs,
    - Its interaction with EC publications,
    - Multidisciplinary,
    - Four spheres of study (lithosphere, hydrosphere, atmosphere, magnetosphere),
    - To evolve education in furtherance of planetary consciousness.

5. PRECURSOR MISSIONS

AAS 93-814
THE MARS ENVIRONMENTAL SURVEY (MESUR) NETWORK AND PATHFINDER MISSIONS
Richard A. Cook and John B. McNamee, Jet Propulsion Laboratory

The objective of the Mars Environmental Survey (MESUR) Network mission is to establish a global network of small science stations on the surface of Mars to operate concurrently over a minimum of one Martian year. MESUR Network is viewed as an evolutionary and affordable step in the scientific characterization of the Martian environment following Viking and Mars Observer and preceding sample return and human exploration missions. The full network is envisioned to consist of 10-20 landers providing pole-to-pole coverage of the planet. The broad science objectives of the MESUR Network mission are to characterize the Martian environment in terms of atmospheric structure, internal structure, global atmospheric circulation, surface chemistry, and surface morphology. The strawman science payload for the Network mission includes an atmospheric structure package (pressure, temperature, and acceleration measurements during descent), cameras for descent and surface imaging, 3-axis seismometer, meteorology package (surface pressure, temperature and wind velocity), Alpha/Proton/X-Ray Spectrometer, Thermal Analyzer/Evolved Gas Analyzer, radio science experiments, and others. The MESUR Network project start is targeted for FY 1996 with the first launch anticipated in 1999. A precursor to the MESUR Network mission, designated MESUR Pathfinder, is targeted for an FY 1994 project start. The objective of the Pathfinder mission is to conduct the engineering demonstrations required prior to the full commitment of funds to develop and proceed with the MESUR Network mission. The primary engineering test performed by Pathfinder will be of an entry, descent, and landing approach which employs an aeroshell, parachute, air bags, and a lander petal system; but no propulsion. This passive entry, descent, and landing system is required to decelerate the vehicle for high entry velocity, achieve a semi-hard landing of the Martian surface, and establish an upright configuration for the surface operational phase of the mission.

AAS 93-815
MARSNET: A EUROPEAN CONTRIBUTION TO MARS EXPLORATION
A. F. Chicarro, Space Science Department, ESA/ESTEC

Following an ESA preliminary study on the possible areas of European participation in the future international exploration of Mars (1), and an ESA call for ideas of new missions, MARSNET, a network of small surface stations, was selected for further in-depth scientific and technical Assessment (2) and Phase-A (3) studies as a potential European contribution to an international network of surface stations on Mars.

Network Concept: The MARSNET mission consists of a network of 3 semi-hard landers to be placed on Mars surface, each station being located about 3000 km apart from each other, thus defining a global seismological network designed for an operational lifetime of one Martian year (687 days). Each semi-hard lander would be carried on-board its own 2 m in diameter aeroshell and deployed at 15 km altitude for landing at about 25 m/s. Each lander would be approximately 1.2 m in diameter and 30 cm in height (excluding booms), and 67 kg in weight, including a scientific payload of about 9 kg. The distribution of these three elements would take place from low Mars orbit or arrival hyperbola depending on the selected mission scenario. The small stations would be targeted for impact and landing in scientifically interesting sites in the Tharsis region of Mars, which is the most likely area to still show tectonic activity that would allow the seismometers to determine the internal structure of the planet.

Scientific Objectives: In addition to the study of Mars internal structure, the other main scientific objectives would focus on science from the surface, and in particular the mineral and chemical composition of rocks and soils, the magnetic properties of minerals, and surface meteorology at sites of varied latitudes and altitudes. Also, atmospheric pressure and temperature profiles would be obtained during entry and descent, and surface and descent imagery would allow to interpret the geological setting of each landing site. The scientific payload would also include an instrument deployment device to allow sampling of rocks and soil away from the station by the chemical analysis instruments on-board. Also, the amount of solar-UV radiation reaching the surface would be monitored to infer the present exobiological conditions on the planet. The MARSNET mission would therefore contribute to provide a global perspective of Mars, and in particular would allow to answer outstanding questions concerning the internal structure and present activity of the planet; the elemental composition and most common minerals of volcanic and sedimentary rocks; the atmospheric pressure, temperature and dynamics, as well as other meteorological parameters; the present surface and subsurface water-ice distribution and the conditions allowing or not allowing organic compounds to exist on Mars surface.

Future Mars Exploration: The MARSNET mission could be part of a precursor mission to the *Mars Rover Sample Return* and *Manned Exploration* missions. This initiative for possible European development would complement the US *Mars Observer* and Russia's *Mars 94* missions. International cooperation stands as a sensible approach for Europe, in order to accomplish its planetary exploration goals. Therefore, MARSNET could translate into a major and independent contribution to a larger international Mars exploration program. Potential partners have already shown significant interest in the joint development with ESA of a *Global Mars Network* mission.

References: (1) *Mission to Mars: Report of the Mars Exploration Study Team*, ESA SP-1117, European Space Agency, 140 pp., 199. (2) *MARSNET: Report on the Assessment Phase Study*, ESA SCI(91)6, European Space Agency, 103 pp. 1991. (3) *MARSNET Report on the Phase-A Study*, ESA SCI(93) European Space Agency 119 pp., 1993.

AAS 93-816

MESUR-BASED MARS SAMPLE RETURN STUDY

Philip C. Knocke, Jet Propulsion Laboratory;
R. Terry Gamber and Benton Clark, Martin Marietta Astronautics Group

In 1996, the MESUR Pathfinder mission will test a new U.S. spacecraft for delivering small payloads to the Martian surface. A study is currently in progress which examines the feasibility of using a Pathfinder derivative, as well as the techniques and infrastructure expected to be developed for MESUR Network, to support a small, low-cost sample return mission launching as early as 2003. The cost goal for this mission is $1B, total life cycle cost.

The baseline scenario starts with the launch of an Atlas IIAS launch vehicle containing a Mars Ascent Vehicle (MAV) and an Earth Return Vehicle (ERV). The MAV is packaged inside a derivative of the Pathfinder lander. Like Pathfinder, the lander enters directly from hyperbolic approach, deploys a parachute, and lands on airbags. At the same time, the ERV is propulsively inserted into Mars orbit. The lander stays on the surface for up to one month, collecting samples by means of a modified MESUR Network microrover or other device, and then ascends into a low Mars orbit. The ERV rendezvous with the MAV, aseptically transfers the sample, and sends the sample capsule on its way to a direct Earth entry and collection via air snatch.

Preliminary mission design and flight system concepts are described, along with sampling goals and technology development needs for this scenario. Flight system designs developed by the Martin Marietta Corporation in support of this study include a Mars ascent vehicle (MAV) weighing approximately 100 kg.

AAS 93-817
MESUR PATHFINDER MICROROVER FLIGHT EXPERIMENT - A STATUS REPORT
Donna Shirley, Jet Propulsion Laboratory

The MESUR pathfinder microrover is to be carried by the Mars environmental survey (MESUR) pathfinder mission to the surface of Mars, where it will perform technology, science, and MESUR mission engineering experiments MESUR pathfinder is funded by NASA OSSA (office of space science and applications) and is cost-capped at $150m in FY92$ for phase C/D. The microrover flight experiment (MFEX) is funded by NASA OACT (office of advanced concepts and technology) and is cost capped at $25m in real year $ for phase A through operations. MESUR pathfinder will launch in November 1996, with a single spacecraft which will fly directly to Mars, enter the atmosphere with a Viking-type heat shield, and land with the aid of parachutes and airbags. During flight (launch through landing) the rover will be provided structural support and thermal stability by the spacecraft, plus any required communications of rover heath telemetry. After impact, the lander will open and its camera will take a panoramic image of its surroundings and begin transmitting it directly to the earth at a few 10us to a few hundreds of bits per second. The initial portion of the lander image to be transmitted will allow the site for rover deployment and initial operations to be selected. The rover will be deployed from its attach point on the lander structure and will thereafter be independent except for using the lander data and communications functions for command and telemetry. The rover is constrained to a few kilograms and a small volume by the MESUR spacecraft delivery capability, which poses considerable challenges for mobility, computation, control, power and thermal control. The rover is a NASA OACT flight experiment for autonomous mobile vehicle technologies, whose prime function is to determine microrover performance in the poorly understood Martian terrain such that future rovers can be designed to be effective in navigating and moving over the surface of Mars. The primary means of collecting this technology information is by instrumenting the rover mechanisms to determine wheel-soil interactions, detect hazards, determine navigational errors, etc. Because the rover is part of the MESUR pathfinder payload it will also gather science data by deploying and alpha-proton-x-ray spectrometer against one or more rocks. Finally, because MESUR pathfinder is an engineering test of a transport and landing system for an envisioned network of 16 meteorology/geoscience stations, the rover will image the lander to allow its condition to be assessed. The current rover design is based on a demonstration (rocky 4s) conducted in June 1992 of a 6-wheeled vehicle using for control a combination of operator path and task designation, and on-board autonomy for task execution and hazard avoidance. Operator designation of targets for rover operations is currently based on using stereo images from the lander. The MFEX is being conducted as a JPL in-house project, with involvement from members of the mobile robot technology community to develop, and analyze the results of, the technology experiments. The MFEX is being conducted as an element of the MESUR pathfinder project and reports jointly to the JPL office of flight projects-FPO (for OSSA) and the JPL office of technology and applications programs-tap (for OACT).

AAS 93-818
THE MARS AERIAL PLATFORM MISSION: UTILIZING SUPER-PRESSURE BALLOONS TO CARRY OUT A GLOBAL RECONNAISSANCE OF THE RED PLANET
Robert Zubrin, Ben Clark, Steve Price and Terry Gamber, Martin Marietta Astronautics; Jim Cantrell, Space Dynamics Lab

This paper presents the design, technology, and potential science return of the Mars Aerial Platform (MAP) mission. MAP is a low cost, Discovery class mission whose purpose is to help revolutionize our knowledge about Mars by generating tens of thousands of very high resolution (20 cm/pixel) pictures of the Martian surface, mapping global circulation of the Martian atmosphere, and examining the surface and subsurface with ground penetrating radar, infrared spectroscopy, neutron spectroscopy, and other remote sensing techniques. The data is acquired by instruments which are carried by balloons flying at a nominal altitude of about 7 km over the Martian surface. Because new balloon and microspacecraft technology is now available, the balloon probes could be quite long-lived, lasting hundreds of days, producing an immense science harvest in the process.

The MAP mission will be carried out as follows: A low cost launch vehicle such as a Delta 7925 (1000 kg TMI) is used to propel a small spacecraft carrying 8 entry capsules onto trans-Mars injection. Approaching Mars, a spinner is used 10 days prior to arrival to release the capsules outward so that they enter Mars' atmosphere at widely dispersed locations. Each capsule then enters the atmosphere, and deploys a parachute which slows it down to the point where a balloon can be inflated. The inflation is accomplished during descent, so that no landing system is required. After the balloon is inflated, the parachute, capsule, and inflation equipment is jettisoned, and used to land a meteorology package consisting of pressure and temperature sensors, battery and transmitter on the surface. Each of the balloon probes will then commence their float around Mars at an altitude of about 7 km above the mean surface level. The probes, which are lofted by superpressurized balloons will need never land, carry a gondola of 8 kg, which includes 2 cameras, 1 kg of atmospheric science and additional instruments, data recording and transmitting equipment, a rechargeable battery and solar array.

When it is daylight two pictures are taken simultaneously every 15 minutes. One is taken with a high resolution CCD black and white camera with a nominal resolution of 20 cm per pixel; with a 1024 x 1024 pixel field, this gives a field of view of 204 m on a side. The other is taken with a color camera at a moderate resolution of 10 m per pixel; this gives a field of view of 10.2 km on a side. The two cameras are aligned so that the high resolution image is located at the center of the medium resolution picture, whose features in turn can be used to locate the region studied on a map of the planet. The data is stored and then is periodically uplinked to an orbiting satellite, which could either be Mars Observer or a MESUR comsat, and them is transmitted back to Earth.

It is anticipated that the science return of this mission will be large, with 32,000 high resolution pictures returned to Earth for each 100 days of operation of an eight balloon fleet, plus an equal number of pictures at resolutions superior to the best Viking images. The high resolution pictures will have a factor of 7 greater resolution than the best pictures returned from Mars Observer. Additional science return will result from use of other instruments, such as ground penetrating radar. Tracking the balloons will yield knowledge of Mars' global atmospheric circulation. The cost of the mission is anticipated to be low, falling below the $150 million (without launch) ceiling of the Discovery-class guidelines.

AAS 93-819
AEOLUS - AN UNMANNED HUMAN PRECURSOR TO MARS USING HIGH L/D ATMOSPHERIC ENTRY SYSTEMS
Marcus S. Murbach, NASA Ames Research Center
Robotic precursor missions to Mars using a lifting hypersonic entry vehicle is investigated. The basic vehicle, derived from a mature hypersonic flight system called SWERVE, offers several advantages over more conventional low L/D or ballistic entry systems. The proposed vehicle has greatly improved lateral and cross range capability (e.g., it is capable of reaching the polar regions during less than optimal mission opportunities), is not limited to surface target areas of low elevation, and is less susceptible to problems caused by Martian dust storms. Further, the integrated vehicle has attractive deployment features and allows for a much improved evolutionary path to larger vehicles with greater science capability. Analysis of the vehicle is aided by the development of a Mars Hypersonic Flight Simulator, from which flight trajectories are obtained. Atmospheric entry performance of the baseline vehicle is improved by a deceleration skirt and transpiration cooling system which significantly reduce TPS (Thermal Protection System) and flight battery mass. The use of the vehicle is also attractive in that the maturity of the flight systems make it cost-competitive with the development of a conventional low L/D entry system. Finally, the potential application of similar vehicles to other planetary missions is discussed.

AAS 93-820
A MARS PULSE SCIENCE BASE ARCHITECTURE EVOLUTION EMPLOYING THE NOMAD EXPLORER STRATEGY
Madhu Thangavelu, Institute of Aerospace Systems Architecture and Technology, University of Southern California

A most exhaustive scientific mission to study the planet as a whole, "measure its pulse," collect and telemeter large amounts of data to Earth in near real time is the ambitious aim of this Mars Science Base architecture. The "pulse" thus obtained will be used to build and investigate a real time interactive Mars model that will help to understand and predict clear evolutionary trends and the origin and destiny of other planets including the Earth.

Exploration and establishment of a variety of unmanned/man tended science outposts and stations on a global scale all across Mars from pole to pole is the physical objective of this architecture. Earth based telerobotic and telepresent systems are deployed in a global "seeding run" that is primarily conducted from a manned long range traverse vehicle called the Nomad Explorer.

The essential and primary components of this architecture are the Nomad Explorer Assembly Assist Vehicle, the Autonomous Modular Common Lander (AMCL), the various modular Science platform payloads and the extraterrestrial Fiber optic telecommunication system network. The Nomad Explorer vehicle is a long range manned traverse vehicle that is designed to assist the crew to deploy various scientific payloads that are landed at sites of interest by the unmanned modular common lander. The AMCLs also carry vehicle logistics as part of the payload, in order to resupply the Nomad Explorer frequently. The Nomad Explorer regularly intercepts these landers along its global traverse, unloading and erecting science outposts as well as refurbishing consumables for the crew and vehicle systems at each intercept of the autonomous lander.

During the course of the traverse, the Nomad Explorer will also continually lay a sturdy and very high quality fiber optic cable communication network linking the various sites so that at the end of the mission all the science platform stations are interlinked enabling better command, control, communication and intelligence networking between the outposts. At the end of the Nomad Explorer traverse the fiber network would make it possible to access all the experimental platforms from any one of the deployed stations or from the mission control on Earth. About thirty six primary stations are planned globally.

Parallel traverses by a series of Nomad Explorers would accelerate the development of a global Mars Science Base and the establishment of the global telecommunication network would provide almost round the clock, very high bit rate, high quality Earth-Mars link availability using the wide bad capability of laser communication technology. The paper outlines the Nomad Explorer Strategy for Mars Science Base establishment.

6. PLANETARY PROTECTION

AAS 93-821
MARS SAMPLE RETURN AND BIOHAZARDS: A SOURCE OF PUBLIC CONCERN AND CONTROVERSY
Margaret S. Race, College of Natural Resources, University of California at Berkeley

Societal concerns about environment, health, and safety are likely to complicate planning for future Mars missions, especially those involving sample returns. Unless planners and engineers seriously consider "biohazard" and "planetary protection" concerns during the earliest mission planning stages, it's possible that public opposition, cost increases and missed launch windows will interfere with mission success. Using lessons learned from genetic engineering and past environmental controversies, it's possible to understand and anticipate how social and non-scientific factors could adversely impact future missions to Mars. In the face of important social trends of the past two decades and the public's growing aversion to perceived risks and biohazards, NASA should adopt a strategy that actively plans both the generation and subsequent management of planetary protection information to ensure that key audiences obtain needed information at appropriate pre-launch times.

AAS 93-822
RELEVANCE AND IMPACT OF PLANETARY ENVIRONMENT PROTECTION ON FUTURE MARS MISSIONS
Francois Spiero, ESA/ESTEC

Care and protection of the Earth environment have recently become major concerns. Indeed, human activities have often expanded at the expense of the preservation of natural balances, sometimes leading to irreversible damages. Mars is, as of today, an almost virgin planet. Little - if any - pollution has yet taken place there. But an increasing number of automatic and, later, manned missions to Mars are planned for the next decades. Eventually, human settlements will be built. Safe and comfortable settlements will be viable in the long term if they can develop in a clean environment. Such an environment may be achieved if, for all future Mars missions, adequate planetary protection measures are taken. Therefore, it would be most useful to soon define a long-term Mars environment policy.

The measures to protect the environment of Mars may depend on the type of missions. For most automatic spacecraft, relaxed Viking-like planetary protection requirements could be applied. But later, for manned missions, a more stringent policy may be needed in order to avoid the contamination of Mars by human sub-organisms. Finally, a comprehensive set of procedures to protect both the environment of Mars and the "settlers" might have to be enforced at the era of inhabited bases. In particular, such regulations would focus on how human activities on the planet should be performed in order to minimize the negative impact on the Martian environment.

7. PATHWAYS TO HUMAN EXPLORATION – MISSION STRATEGIES

AAS 93-823
MARS DIRECT: HUMANS TO THE RED PLANET WITHIN A DECADE
Robert M. Zubrin, Martin Marietta Astronautics

This paper describes a plan, known as "Mars Direct" (1,2,3), which could enable the initiation of human exploration of Mars within a decade. While requiring only present day or near term technology, Mars Direct is not a brute-force plan for a single high-cost, low capability mission. Rather it is a plan for a simple, robust mission architecture that puts a highly efficient space transportation and a sustainable and powerful Mars surface exploration capability into action starting with the very first mission.

In the Mars Direct plan two launches of a heavy lift booster optimized for Earth escape are required to support each 4 person mission. The first booster launch delivers an unfueled and unmanned Earth Return Vehicle (ERV) to the Martian surface, where it fills itself with methane/oxygen bipropellant manufactured primarily out of indigenous resources. After propellant production is completed, a second launch delivers the crew to the prepared site, where they conduct extensive regional exploration for 1.5 years and then return directly to Earth in the ERV. No on-orbit assembly or orbital rendezvous is required in any phase of the mission, and the same set of booster, crew hab, and ERV used to support Mars missions can also be used to support a lunar base.

This paper discusses both the Martian and Lunar forms of implementation of the Mars Direct architecture. Candidate vehicle designs are presented, and the means of performing the required in-situ propellant production is explained. The in-situ propellant process is also shown to present very high leverage for a Mars Rover Sample Return mission flown as a scaled down precursor version of the manned Mars Direct. Methods of coping with the radiation and zero gravity problems presented by a manned Mars mission are discussed. Prime objectives for Mars surface exploration are outlined, and the need for substantial surface mobility is made clear. Combustion powered vehicles utilizing the in-situ produced methane/oxygen are proposed as a means for meeting the surface mobility requirement. While the Mars Direct plan can be implemented utilizing only chemical propulsion, it is found that substantial improvement in mission capability can be achieved through the introduction of nuclear thermal rocket propulsion for trans-Mars injection, and that the architecture is highly amenable to the introduction of such technology whenever it becomes available. It is concluded that the Mars Direct architecture offers an attractive means of rapidly initiating the human exploration of the Moon and Mars, thereby opening the doors of the solar system to humanity.

References: (1) R. Zubrin and D. Baker, "Humans to Mars in 1999," *Aerospace America*, August 1990. (2) D. Baker and R. Zubrin, "Mars Direct: Combining Near-Term Technologies to Achieve a Two-Launch Manned Mars Mission," *JBIS*, Nov. 1990. Also published as AAS 97-366, in *From Imagination to Reality*, R. M. Zubrin, ed., AAS *Science and Technology Series*, Vol. 91, 1997, pp. 225-239. (3) R. Zubrin, D. Baker, and O. Gwynne, "Mars Direct: A Simple, Robust, and Cost Effective Architecture for the Space Exploration Initiative," AIAA 91-0326, 29th Aerospace Sciences Meeting, Reno, NV, Jan 7-10, 1991. Also published as AAS 90-168, in Proceedings of *The Case for Mars IV*, Boulder, CO, June 1990. ed. T. R. Meyer. American Astronautical Society, *Science and Tech. Series*, Vol. 89, pp. 275-314, 1997.

AAS 93-824
MARS EXPLORATION AND SETTLEMENT
Bruce Lusignan, Stanford University

The International Mars team at Stanford has developed an efficient approach to Mars exploration using existing technologies of Russia, U.S., Japan and Europe. The surface facilities would be transported to Mars in 2007. A crew of six would be sent in 2009. The crew will be on the surface for fourteen months before returning to Earth. The transportation systems are described in a companion session (Merrihew-Stoiko).

A second crew would be transported in 2009; permanent occupation would begin in 2011. The base site chosen, with input from an international panel is Candor Chasma II, in the bottom of the canyon Valles Marinaris. Exploration by surface vehicles to sites 75 km from the base would take two to three weeks each. A trip to a dry lake bed to explore for traces of early life, one to sample material exposed from different geologic eras, and one to the base of the 2 km high canyon walls would be done in the 14 months.

Two habitats, landed fully operational, provide the first accommodations. The crew assembles prefab units during their 14 month stay to increase living and work space and test equipment for the permanent base. In the second and third visit, facilities are increased to establish a permanent settlement. "Greenhouse" area is increased to allow food to be produced locally, chemical plants are run to create local sources of transportation fuel and life support oxygen from Mars' atmosphere and masonry plants are developed to make building materials from local quarries.

These facilities make use of currently available technology wherever possible to ensure a cost efficient program. General purpose cargo vehicles have a high development cost ($5.7 billion) but relatively low per copy cost ($150 Million) and launch cost ($300 Million). At an annual world cost of $3.5 Billion to 4 Billion per year, the first base can be expanded and new sites developed at a steady rate. Detailed design and costing of the systems will be given.

AAS 93-825
MARS EXPLORATION STRATEGIES: A REFERENCE PROGRAM AND COMPARISONS OF ALTERNATE ARCHITECTURES
Michael B. Duke and David B. Weaver, NASA Johnson Space Center

A strategy for the human exploration of Mars has been developed by the NASA Exploration Programs Office's Mars Study Team over the course of the past year. The Mars reference program incorporates many of the most ingenious and enabling approaches of the previous strategies (e.g. Zubrin, Lusiqnan, and the Synthesis Group. The reference program will serve as the basis for further refining issues and for comparing alternate mission concepts, particularly those that employ alternate space transportation strategies and international cooperation.

The proposed Mars exploration activities have two principal focuses for conducting planetary-scale scientific exploration and evaluating the potential of Mars to support extended human presence. By directly challenging the traditional obstacles to the near-term human exploration of Mars, the Mars Study Team has developed a strategy that eliminates the need for complex assembly operations in low-Earth orbit, significantly reduces the crew's exposure to the interplanetary space environment for any Earth-Mars opportunity, and decreases the total launch requirements from Earth. The reference program employs a 600-day stay at the Mars surface for a crew of six, with maximum

Earth-Mars and Mars-Earth transit times of 180 days. By focusing program resources at the surface of Mars, the strategy enables a robust set of activities and associated surface capabilities, complemented by the exploitation of indigenous resources for life support, safe-haven stores, surface mobility, and ascent vehicle propellant.

Space transportation to Mars exploits the use of nuclear thermal propulsion for both human and cargo flights. The reference program employs a split mission strategy where the cargo is delivered to the surface of Mars the opportunity prior to the human flight. Each Earth-to-orbit launch vehicle sends a payload (either cargo or human mission) direct to Mars. Mars ascent is accomplished with an oxygen/methane propulsion system that uses locally-produced propellants. Similarly, the Earth- return stage uses oxygen/methane propulsion, permitting the eventual use of Mars-produced propellants, but initially relying on Earth-derived propellants.

AAS 93-826
THE IAA MARS COSMIC STUDY
George E. Mueller, International Academy of Astronautics

"Provide ship or sails adjusted to the heavenly breezes, and there will be some unafraid even of that empty void (of space). So, for those who will come shortly to attempt this journey, let us establish the astronomy: Galileo you of Jupiter, I of the Moon." - open letter to Galileo Galilei from Johannes Kepler, 19 April, 1910.

The International Academy of Astronautics is embarked on a series of "Cosmic Studies," designed to fully explore the many aspects of the implementation, impact, and rewards of cooperative international programs for the exploration and ultimate possible habitation of the Solar System and beyond. In these Cosmic Studies the IAA is attempting to provide International leaders and decision makers of the space faring countries of the world an unbiased scientific and technological basis for their commitment to multinational space programs. These Cosmic Studies are designed to provide a knowledgeable technical perspective and assessment of proposed programs as a basis for governmental decisions. This Mars Cosmic Study examines the possibilities and implications of establishing an international robotic and human exploration program leading to the establishment of an International Mars Base.

The International Academy of Astronautics and the International Exploration of Mars Subcommittee thanks Aerospatiale, Boeing and other companies, agencies, and universities who have supported the participation of its experts in preparation of this Mars Cosmic Study.

AAS 93-827
BEING THERE: THE MISSION ON THE MARTIAN SURFACE FROM THE 1992 NASA MARS EXPLORATION MISSION STUDY
L. Lemke, Center for Mars Exploration, NASA Ames Research Center

This paper reports on results of the first Mars human exploration mission and system study conducted by the Office of Exploration under the tenure of Associate Administrator Michael Griffin. Specifically, results described in this paper were produced in the time period from approximately August 1, 1992 to October 1, 1992. Mission studies for Mars exploration naturally divide into two major components: the transportation system between Earth and Mars (getting there) and the mission on the Martian surface (being there). This paper addresses the latter activities. Some of the groundrules and constraints for the mission study were: (1) multiple visits on consecutive launch opportunities beginning in approximately 2008, (2) surface crew size of approximately 6 for each visit, and (3) long surface stay time (approx. 600 days) for each visit.

The highest level mission objectives were taken to be a combination of conducting comprehensive Mars-intrinsic scientific investigations ("science") and experimenting with Mars as another abode for humans ("habitation"). The objectives of science and habitation were given equal weight in the mission design. The team derived functional definitions of the qualities the surface mission would have to possess in order to meet these objectives. In the case of science, the two most important criteria are: (1) routine access by humans to multiple sites on the Martian surface on a spatial scale commensurate with the geophysical features of the planet itself (1000's of km) and (2) crew time and general-purpose science lab facilities available at the base adequate to support a wide range of anticipated science investigations.

The habitation objectives of the mission are satisfied if, (1) a range of Earth biota (including, but not limited to humans) could be shown to survive, (2) local planetary materials could be utilized in a productive manner, (3) the base could move in the direction of greater self-reliance over time, including the production of significant surpluses.

It became apparent that a design strategy to accommodate these functions would have to include a highly capable surface transportation system and the designed-in ability to process and store indigenous materials, both of which point to the need for a relatively energy-rich environment. Similarly, satisfying the other functions points to the need for a relatively large, efficient, and maintainable crew habitat. The overall picture that emerges is of a single base built around a "cache" of consumables (air, water, food, fuel), the necessary subsystems to replenish the cache, and a highly capable surface mobility system consisting of at least two pressurized rover vehicles capable of sustaining a crew of about 4 at a distance of hundreds of km from the base for a period of 10 days or more. The initial habitat, cache, and a surface power system is emplaced and powered up by remote control on the launch opportunity before humans arrive. Before humans are launched, the habitat can be verified to be minimally functional, with enough consumables in the cache to sustain the crew for the full duration of their surface stay. Subsequent missions will deliver more infrastructure along with crew, including a duplicate main habitat, and inflatable/erectable outbuildings. It is expected that the crew will divide its useful time between Mars surface exploration (both robotically and with humans) and operation and development of the base.

8. INTERNATIONAL COOPERATION

AAS 93-828
SUMMARY OF THE WORKSHOP ON INTERNATIONAL COORDINATION OF THE EXPLORATION OF MARS
Roger D. Bourke, Jet Propulsion Laboratory

The Interagency Consultative Group for Space Science (IACG) is comprised of NASA, the European Space Agency, Intercosmos Council (Russian Academy of Sciences), and the Institute of Space and Astronautical Science, Japan (ISAS). This Group has convened a workshop on the international coordination of the exploration of Mars to be held 10-11 May 1993 in Wiesbaden, Germany, two weeks prior to The Case for Mars V. The IACG workshop is intended to be a forum for space-faring nations to share information on current and future robotic missions to Mars and to generate and discuss ideas and opportunities for cooperation among the international community. The expectation is that the workshop will result in a higher level of cooperation across international boundaries.

At the time this abstract is going to press, neither the agenda nor the results of the IACG workshop have been established. However, the results of the workshop are likely to be of considerable interest to attendees at The Case for Mars V. This paper will summarize the IACG workshop.

AAS 93-829
LET'S MAKE MARS EXPLORATION TRULY INTERNATIONAL
Louis D. Friedman, The Planetary Society
John M. Logsdon, Space Policy Institute

Almost two decades after the U.S. Viking mission landed on Mars and explored the planet for four productive years — limited, however, to two sites — plans are afoot around the world to resume the *in situ* exploration of the Martian surface. Scientists and space agencies in Russia, Europe, Japan, and the United States are working on a suite of separate missions to Mars that together would represent a major advance in knowledge.

While the various space-faring nations carry out their current plans, coordinating them through both formal and informal arrangements, it may be time to consider a truly international mission which could be accomplished at less cost to each participating country. An integrated approach to a single exploratory objective — such as Mars — could provide experience to advance international cooperation in future solar system exploration — a development that we believe is needed.

We propose that an international team be organized to study an internationally conceived, developed, and implemented "Earth Mission to Mars" project. Such a project could and should incorporate elements of currently planned national missions — **Mars 94, Planet A**, and **MESUR-Pathfinder** — and later — **MESUR, MARSNET**, and **Mars 96**. First, consider the current situation that has led us to make this proposal:

The U.S. **Mars Observer** mission is on its way to the red planet. After it arrives in August 1993, it will explore the planet for two years characterizing the atmosphere and surface, and deducing a history of the Martian climate. Its readings will be taken from Mars orbit, 350-400 kilometers above the surface. In October 1994, the **Mars 94** mission will be launched as a collaborative venture of Russia, France, Finland, and, after extended negotiations, the United States. It will include the first spacecraft to return to the Martian surface since **Viking**. In addition to the large orbiter making remote sensing measurements, this mission will include two small surface stations and two penetrators to make measurements below the surface.

The international **Mars 96** mission is an important following step in Martian exploration. With its balloon and rover it will provide the first mobile exploration of Mars. The mission has extraordinary backing, both in Russia as part of the high priority **Mars 94/96** project, and among the European partners involved in different aspects of the mission. Notable is the large commitment of the French space agency responsible for the balloon and its instrumented guide-rope.

The United States is, finally, preparing to continue exploration of the Martian surface after the spectacular **Viking** missions of 1976. The American science community has developed a plan for a series of small landers to comprise a network, known as the Mars Environmental SURvey, **MESUR**. A technology "Pathfinder" mission for **MESUR** is proposed for a 1996 launch. The **MESUR-Pathfinder** will demonstrate and advance affordable technology for entry and landing systems, and for a new class of exploring vehicles known as microrovers — several kg vehicles with limited mobility but high sensor sophistication, using state-of-the-art micro electronics. Also in 1996, the Japanese Institute for Space and Astronautical Sciences is planning a Mars mission called **Planet-A** — an aeronomy orbiter in eccentric Martian orbit. The European Space Agency is considering **MARSNET**, a set of Mars landers, to be flown in conjunction with the **MESUR** mission during the years 1998 to 2005.

Considering all that is being planned for Mars, and the financial constraints pressuring all space agencies, it is reasonable to ask if the combined missions could be carried out with a higher degree of international collaboration, so that more exploration might be accomplished for less. The principles behind such cooperation are already in place. European countries, notably France, Germany and Finland are already major contributors to **Mars 94/96**. For the first time the US has become part of a Russian planetary mission with a new **Mars 94** lander agreement to fly an experiment on a small station. An international meeting to address cooperative Mars exploration specifically has been organized for May 1993, in Wiesbaden, Germany. New ideas for collaborative exploration are likely to emerge from that meeting.

But not all is positive. The high priority assigned to **Mars 94** and the short-term demand for resources to ensure that the mission goes forward could leave **Mars 96** on shaky ground. The balloon and rover are ambitious developments, requiring more resources than are now available. **MESUR-Pathfinder** is being attacked by some in the US planetary science community as lacking enough science value. (This is reminiscent of the killing of a US Halley comet flyby, and belies the entire technology pathfinder aspect of the mission.)

These possible weaknesses might disappear, however, if viewed from an international perspective. The combination of **Mars 96** and **MESUR-Pathfinder** provides tremendous science. They also embody tremendous technology. Russian entry and landing technology is impressive and proven — and is applicable to NASA's network plans. New interest in a Mars balloon has developed in the US — the French/Russian program can lead to a new series of explorers to complement **MESUR** landers. The combined robotics of the US microrover and Russian minirover have already spawned a number of suggestions for international team building — which will lead to more capable science and exploration.

Among the ideas now under study are the use of the microrover or some of its elements to serve as a mobile hand on the Russian **Mars 96** rover; combined development of some aspects of the navigation and telepresence system; experiments in a "virtual environment," which may be used to control the rover in the real environment; international development of a robotic arm; and international testing of telepresence and autonomy. Best of all, the money planned to be allocated (even in conservative scenarios) among the US, Russia, Europe (both in ESA and in the national agencies) and Japan, is sufficient to permit both reasonable and vital Martian exploration in 1996 and beyond.

Given the above, we propose that immediate study begin for *an international "Earth Mission to Mars" project, principally combining elements of* **Mars 96** and **MESUR-Pathfinder**. This study would not necessarily be a threat to existing national plans; it is far more likely that national plans will go forward than would a new, precedent-setting, international approach to a joint project. But the study should still be undertaken because: (1) It will help define cost-effective and scientifically beneficial international opportunities, whether used in a single project or in cooperation with many projects; (2) Should the national plans falter because of budget and other economic pressures, the study might lead to alternate suggestions for how to achieve high-priority scientific objectives; (3) The study of a single international project in planetary exploration is an important system engineering and management undertaking, which has not previously been done. A Mars robotic surface mission is ideal for such a study, because of the solid base of planning in **Mars 96** and **Pathfinder**, the existing elements of international cooperation, and the relatively small (or at least manageable) size of the projects; (4) This study would have high value in terms of thinking about the eventual international organization for the human exploration of Mars.

To perform such a study properly requires a small, internationally organized team, drawing on the resources of the larger existing national groups now working on future Mars missions. The international team should either work together in the same location for some significant period of time, or be linked together by a specifically dedicated and more advanced (than now employed) set of technology links, e.g. fax, TV, telephone, e-mail, computers. Even if collocated, some advancement of communications links with the national groups will be required. Rather than a study requirement, however, this should be looked upon as a study benefit — one that will facilitate future mission interaction.

The organization of a study team should not be difficult. The groups working on future Mars missions are reasonably well-known to each other, and have sufficient resiliency to detail people for such a study. Both the management and the structure of such a group needs discussion. The customers for the study, the national space agencies, will have to assign representatives to set up this structure, and to act as a steering group for its conduct. To maximize the value of results, all national agencies who wish to contribute should be encouraged and allowed to do so. There are also a number of existing international groups who might contribute to the study and its organizations, e.g. the U.N. Outer Space Affairs Division, the International Astronautical Federation, COSPAR, the International Academy of Astronautics, The Planetary Society, the Association of Space Explorers. However, at this stage, the national space agencies are the principal sources of resources, and they should organize the study.

Reform and revolution are going on throughout the space programs of the world. Driven by economics, growing complexity of missions, burgeoning objectives and constituencies, greatly reduced budgets, competing national priorities, and, most of all, by a post cold-war new world order, the need to do more for less is paramount. An international **Earth Mission to Mars** can fulfill that need and help create the means to carry out even more ambitious exploration in the next millennium.

AAS 93-830

THIRTY YEARS TOGETHER: A CHRONOLOGY OF U.S.-SOVIET SPACE COOPERATION

David S. F. Portree, Hernandez Engineering, Inc.

The chronology for the poster session is based on NASA Contractor Report 185707 (February 1993), "Thirty Years Together: A Chronology of U.S.- Soviet Space Cooperation." It provides a background briefing on the 30 years of cooperation between the United States and the Soviet Union (and its successor, the Commonwealth of Independent States, of which the Russian Federation is the leading space power). The chronology tracks successful cooperative space projects and failed attempts at space cooperation. Included are the Dryden-Blagonravov talks (1962-63); the UN Space Treaties (1961-1972); The Apollo-Soyuz Test Project (1970-1975); the abortive Shuttle-Salyut discussions (1973-1978); widespread calls for joint manned and unmanned exploration of Mars (1962-1992); conjectural plans to use Energia and other Russian hardware in ambitious future joint missions (including Mars missions); and contemporary plans involving the U.S. Shuttle, Russian Mir, and Soyuz-TM. The chronology also includes events not directly related to space cooperation to provide context.

9. SPACE POWER TECHNOLOGY

AAS 93-831
POWER SYSTEMS FOR THE FIRST MARS OUTPOST
Paul S. Gill, Rocketdyne Division, Rockwell International

NASA's current vision of the first human presence on Mars, labeled "First Mars Outpost" (FMO), calls for power requirements ranging from a few hundred watts for mobile and remote applications to hundreds of kilowatts for in-situ resource processing. Power systems to satisfy these requirements include non-nuclear systems such as the Photovoltaic (PV) array and the Regenerative Fuel Cell (RFC), and nuclear systems such as the Dynamic Isotope Power System (DIPS) and the SP-100 reactor with the Closed Brayton Cycle (CBC) dynamic conversion system. Based on the surface elements and their power needs identified for a given mission scenario, a number of different power systems can be selected. These power systems can then be used to establish a number of different power system architectures which can be evaluated and ranked based on criteria deemed important for the prescribed mission.

As the FMO mission objectives evolve through technical examination and political maneuvering, the FMO constituent elements may change and consequently, so may the power system architectures that meet the power needs of these elements. Therefore, the evaluation of appropriate power system technologies during initial assessment of the FMO, both in terms of their applicability to the mission and their impact on the mission, is essential. A flexible approach allowing for quick, top level evaluation of candidate power system architectures will be beneficial to the overall decision process in formulating the FMO.

The traditional method of evaluating and, often, selecting power systems based on initial system mass alone ignores a number of other important attributes which must be part of the decision making process. For example, system life cycle cost (LCC), life, reliability, safety, and availability also greatly impact the mission. System mass, which translates into transportation cost, may not be the dominant part of the overall system cost. In some instances, a power system technology having high initial mass may have low technical risk and, therefore, be desirable over a lighter, advanced technology. Therefore, a method which employs all of the important criteria which enhance the mission may be more meaningful than the mass comparison alone.

A simple approach which evaluates power system candidates for the FMO and demonstrates the impacts of different evaluation criteria on these candidates is presented. The current FMO power requirements established by NASA have been used to produce a number of different power system architectures which satisfy the power requirements. These architectures are then evaluated and ranked with respect to the established evaluation criteria and compared with each other to determine their relative merits. Sensitivity of the overall rankings of these power system architectures to variations in a number of these evaluation criteria is demonstrated to signify how the overall decision may be impacted by aspects which may be ignored otherwise.

AAS 93-832
POWER SYSTEMS FOR MARS EXPLORATION
Robert L. Cataldo, NASA Lewis Research Center

An intercenter NASA team is currently investigating options that will provide a logical, scientifically meaningful, cost effective approach to human exploration of Mars. This includes studying various transportation options and strategies for surface science and habitation. The goal of any human mission is to reduce risks, but it is particularly critical with human exploration missions beyond cislunar space. One method of reducing risk is know before you go. In one option being studied, a 26 month cache of life support resupply materials, i.e., oxygen, water and buffer gasses, is generated prior to the crews departure. Methane and oxygen are options for ascent vehicle propellants. Methane would be made from imported hydrogen and atmospheric gas using the Sabatier process. The oxygen would also be extracted from the atmosphere. The methane and oxygen would be generated, liquefied and stored prior to crew departure. The concept of functional redundancy, i.e., multiple uses for the same device, and backup systems are also used to reduce risks. This same philosophy is used in the development of the power system strategy for Mars.

The maximum power required by the Mars Base is 160 kWe. Options will be presented comparing the characteristics and features of nuclear versus solar power systems, including requirements for deployment and maintenance. Nuclear power systems have an advantage of low mass, low volume, and low deployed area as contrasted to solar power systems. Although the night time energy storage requirements are small resulting in low battery mass, solar arrays are very large due to Mars distance from the Sun and the solar irradiance attenuation caused by atmospheric conditions. Alternatives for powering surface mobility systems, ranging from long range pressurized rovers to piloted and robotic utility rovers, will also be compared.

A strategy will be presented that satisfies the Bases power requirements with minimum mass, volume, cost and complexity, while maintaining a high degree of reliability and safety. Power system commonality between surface elements and transportation vehicles will be discussed as well as their adaptability to lunar missions.

AAS 93-833
THE VIABILITY OF PHOTOVOLTAICS ON THE MARTIAN SURFACE
James R. Gaier and Marla E. Perez-Davis, NASA Lewis Research Center

Photovoltaic power sources are an attractive alternative for supplying the energy needs of Martian surface missions. They are relatively simple, lightweight, and have a long heritage both in deep space and on the terrestrial surface. Concerns about their viability on the Martian surface, however, have been raised because of the periodic dust storms which dominate Martian weather. Photovoltaic power could be sharply curtailed either by occlusion of the photovoltaic surface by accumulated dust, or by abrasion of the surface by wind driven dust. A series of studies on the accumulation, clearing and abrasion of photovoltaic surfaces have been undertaken utilizing the Martian Surface Wind Tunnel (MARSWIT) at the NASA Ames Research Center. The findings of these studies, coupled with surface data from the Viking Landers and terrestrial experience with photovoltaics lead to the conclusion that photovoltaics are viable on the Martian surface. These findings also provide guidance for optimizing the configuration of the arrays to minimize dust accumulation and abrasion.

AAS 93-834
DUST ACCUMULATION ON SOLAR ARRAYS ON MARS: THEORY AND INSTRUMENTATION
Geoffrey A. Landis, Sverdrup Technology, NASA Lewis Research Center

Understanding the operation of solar arrays on the Martian surface is important for the design of power systems for future missions to Mars. One critical issue is dust depositing onto the surfaces of solar arrays, resulting in degradation of performance. There are two mechanisms for sand and dust to be deposited on a solar array. Particles may accumulate on the array by the process of saltation. At the Mars atmospheric pressure of about 10 mbar, saltation requires a minimum wind of about 15 m/s.

Second, dust suspended in the atmosphere will settle onto exposed surfaces. The expected degradation in performance, using the best available information on Mars conditions, was calculated for 30-day and 2-year missions. Since the amount of dust in the atmosphere varies, the degradation of array performance could be between 1% and 38% over a 30 day mission, and between 20% and 88% over a 2-year mission.

Wind removal of dust from the array surfaces was considered. At the wind velocities observed at the Viking lander sites, it is unlikely that significant amounts of dust will be cleared from array surfaces. Accurate measurements of dust settling rates and the properties of atmospheric dust are also of scientific interest. Dust properties affect the atmospheric solar absorption and thus the heat balance of Mars, as well as serving as nucleation sites for water and carbon dioxide frost. Dust settling results in significant albedo changes over the planetary surface. Also, the size and shape of dust particles is required in order to characterize the degradation of mechanical parts of future Mars probes (rover bearings, mechanical arm grapplers, etc.) due to abrasion.

Dust settling on a solar array can be measured simply by measuring the output of the array over time as long as the incident solar flux is well characterized. It would be useful, however, to have an independent method for measuring the amount of dust accumulated on the MESUR array in order to calibrate the losses due to dust. Three instruments are proposed to measure the dust accumulation and then to quantify the properties of the accumulated dust. The three measurements provide complementary information on the dust accumulation: (1) Coverglass spectral transmission experiment; (2) Crystal monitor of accumulated dust loading; and (3) CCD microscope of individual dust particle size count and obscuration.

The experiments are designed to have low footprint, low mass, low power usage, and low telemetry requirement for returning data. The result will characterize the overall obscuration factor, spectral transmission, size, shape, individual particle obscuration, and the mass and number density of deposited dust.

AAS 93-835

EFFECTS OF DUST ON RADIATOR SURFACES ON THE MARTIAN SURFACE

James R. Gaier, Marla E. Perez-Davis, and Sharon K. Rutledge, NASA Lewis Research Center

Power systems that will be put in place on the Martian surface will generate waste heat which will be rejected through radiators. The most efficient radiators have convoluted surfaces, many with surface texture on the order of a few micrometers. Concerns have been raised about the effects of the ever present dust on the surface. The dust could either clog the radiators, degrading their efficiency, or wind driven dust could abrade them, degrading or even destroying their surface texture. A series of studies on the accumulation, clearing, and abrasion of radiator surfaces have been undertaken utilizing the Martian Surface Wind Tunnel (MARSWIT) at the NASA Ames Research Center. The findings of these studies, coupled with surface data from the Viking Landers lead to the conclusion that the effects of dust can be controlled such that serious degradation of the radiators need not occur. These findings also provide guidance for optimizing the configuration of the radiators to minimize dust accumulation and abrasion.

AAS 93-836

SOLAR ARRAY CORRECTIONS FOR MARS SURFACE

Dale R. Burger, Jet Propulsion Laboratory

Geoffrey A. Landis, NASA Lewis Research Center

OBJECTIVE: Modeling all of the atmospheric factors that influence solar array operation on the surface of Mars.

SCOPE: Atmospheric factors only will be considered since dust deposition data is scarce and experimental efforts have not yet provided enough inputs on deposition effects.

INTRODUCTION: The Mars Environmental Survey (MESUR) Pathfinder mission is considering the use of inexpensive solar arrays. Power generation by a solar array on Mars is affected by a number of environmental conditions. Corrections for these conditions are necessary to provide an accurate estimate of available power. A computer model has been constructed to make these corrections. Required inputs are solar array location (latitude) and time of year (areocentric longitude).

COMPUTATIONAL METHODOLOGY: The first quantities that are determined are optical depth and surface albedo. Values for these can be manually entered from the keyboard or calculated from either of two models derived from Viking lander data. Next, the sun declination angle and number of daylight hours are calculated. The number of daylight hours is used to set the limits of integration for total array power generation in one Martian day. When the zenith angle is found, the normalized net flux function value can be determined from a polynomial function of optical depth, albedo and zenith angle. A value for global insolation at the Mars surface is then calculated from the net flux function and the amount of light available above the atmosphere. The direct beam insolation is computed using Beer's law and the diffuse insolation is then the difference between the global and direct beam values.

On any but the most clear days, Martian atmospheric dust causes a spectral shift in surface light toward the red end (around 800 nm). This shift is significant, so a lengthy series of efficiency correction calculations must be made. First, the transmission function is determined for various optical depths and zenith angles. This calculation makes use of the delta-Eddington approximation for radiative light transfer. The transmission function is then used to achieve a modified solar spectrum using the AM0 spectrum as a start. The modified spectrum is rescaled to avoid intensity effects and then used with spectral response data to determine the total current. The ratio of total red-shifted current to AM0 current is used to make the efficiency correction.

A Fresnel reflectivity correction is made for direct beam light only. A baseline Pathfinder configuration is used to make shadow loss calculations. Total insolation and an estimated value for atmospheric cooling is used to calculate the array temperature. Once the array temperature is known then Voc and Isc temperature corrections are made. A fifth order polynomial fit was used to obtain coefficients for 2 ohm-cm and 10 ohm-cm silicon, and GaAs/Ge cells. These coefficients are then used to calculate correction factors for the array temperature and intensity conditions. The usual array losses from packing factor, radiation, UV, and assembly mismatch are also taken. Effects of tilting the array or use of a faceted array can also be calculated. Since Martian hours were used for power calculation they must be converted to Earth hours to keep units consistent.

RESULTS: Use of polynomial functions allowed consideration of extreme conditions such as polar sites. The red-shift efficiency correction can add about 30% to achievable power during dusty conditions. Calculation of array temperature was found to be vital each hour since the dawn to noon temperatures can range from –111°C to +2°C. Correction of Voc and Isc for both temperature and intensity over a wide range of conditions was found to be necessary for accurate total power prediction.

10. ROVERS AND SURFACE TRANSPORTATION SYSTEMS

AAS 93-837
GENETIC REINFORCEMENT LEARNING IN A ROBOTIC SYSTEM FOR SPACE EXPLORATION

Robert E. Smith and Lia Noemi Tusanotte Cisnero, University of Alabama

The objective of this presentation is to suggest that a general reinforcement learning controller achieved through the application of genetic algorithms could be an interesting control tool for Robots for Space Exploration. Space is a dynamic and unstructured environment with perpetually novel data stream. Space robotic activities could require the continuous execution of control strategies in real time. These control actions must be directed at multiple goals, possibly with conflicting priorities. Given these complexities, it is difficult to specify a sufficient fixed-structure controller. Classical control methods require accurate plant models. In effect, these models are used to predict plant behavior so that appropriate control decisions can be made. However, as indicated above, application domains for robots in space exploration make *a priori* plant model specification difficult. Space robots are constrained to interact with an unstructured environment through noisy and drifting sensors and effectors. A learning method could allow adaptive specification of a control strategies, without an explicit plant model specified beforehand. Such methods could allow for more effective, robust robotic applications in space systems. Reinforcement learning techniques allow an automatic system to refine control strategies towards a long term goal based on an immediate reward or cost signal. These techniques do not require that desired *actions* be provided to the controller, only that the controller can evaluate the effectiveness of actions that it provides. A reinforcement learning controller acquires a control strategy by exploring, possibly refining an internal plant model, and adaptively updating its control strategy. The generic nature of reinforcement learning could allow a robot to learn in unknown environments, to adapt its learning to the particulars of a task, and choose actions based on the currently an adapted evaluation of its perceived state. Most of the reinforcement learning techniques found in the literature are based on a lookup table mapping of sensory inputs and possible control actions to evaluations that allow for the determination of a control strategy. In practical systems, such a table-based approach is likely to be computationally infeasible. The approach proposed in this presentation is a learning method using genetic algorithms to vary the structure of a computationally realizable reinforcement learning controller. Genetic Algorithms (GA's) are robust, globally oriented search procedures whose mechanics are based on those of natural genetic evolution. Their most distinctive feature is that they do not require any explicit information about the relationship between input parameters and their effect on performance. Instead, they implicitly evaluate these relationships through a population-based approach involving simple selection, recombination, and mutation operators. GAs are unrestricted from requirements of continuity, modality, etc. that are common to other search techniques. The robustness of GAs has been demonstrated in a large number of practical applications. The broadly applicable character of GAs, and their ability to operate on performance-only feedback, make them uniquely suited to restructuring reinforcement learning controllers. This presentation will focus on how GAs can be integrated with reinforcement learning techniques, and how this combination can be used to supplement robotics for space exploration. The resulting systems hold promise for adaptive robots that can play a more substantial role in ongoing tasks in the perpetually novel environment of space.

AAS 93-838
RUSSIA'S MARS ROVER
Daphne Burleson, University of Colorado

In 1996, as part of a proposed Mars project, Russia plans to send their 200-pound, six-wheel, nuclear powered rover to Mars. Russia's rover is driven by a type of nuclear-powered battery called a radioisotope thermoelectric generator that allows the computer-controlled vehicle to operate in darkness or without solar energy. Russian scientists have been working on the $20 million prototype for over five years.

A smaller microrover similar to the Rocky 4 robot currently under development at the Jet Propulsion Laboratory (JPL) in Pasadena, California may be strapped onto the back or top of the larger Russian rover. The rovers could then either explore the Martian surface together or take separate paths. Rocky 4 is the most recent in a series of microrovers built by JPL. It weighs 15 pounds and has six wheels that can operate independently with an on-board computer control system and utilizes solar cells to charge its battery pack. Its current cost is approximately $2 million and a flight ready version would cost nearly $25 million. I propose to discuss in detail this joint Russian-U.S. mission as well as the nuclear power propulsion to be used by the Russian rover.

AAS 93-839
MINIMUM ENERGY PATHS ON A PLANETARY SURFACE
George W. Morgenthaler and Kevin K. Gifford, University of Colorado

For an extended surface exploration of a remote planetary surface by a vehicle on of the primary factors to minimize is the energy the vehicle expends while traversing the surface from the starting point to the desired destination. A method of solving this type of problem is to make use of minimum energy paths on a surface for which Euler's equation of the Calculus of Variations is used.

On a remote planetary surface there will be various soil types, i.e. hard rock, sandy soils, pebble fields, etc., and the difference in the available traction of these fields may be modeled via a drag function. The energy expended by a planetary surface vehicle, as it traverses on a path from a known starting point to some ending point can then be minimized if the topography and the surface conditions in the region to be traversed are known or can be estimated. The minimum energy path is also referred to as an energy geodesic.

The method of solving this problem is to build a repertoire of algorithms which find the minimal energy path for different scenarios. The different scenarios include planer surface areas, sloped surface areas, surface areas with multiple regions of differing drag functions, surface regions where the drag function is constant, and surface areas where the drag function is linearly increasing or is dependent upon some other factors. Once a suitable and well tested repertoire of algorithms to calculated energy geodesics for differing scenarios is developed the overall problem of minimizing the energy expended on a planetary surface can be solved with the aid of computers. The surface topography of the region of interest will be known as will the surface soil conditions. The problem then becomes an exercise in searching the region surrounding the starting and ending points of the planetary traverse and calculating a minimal energy path using the different algorithms for the various surface topography and soil conditions encountered.

The methodology of determining the drag functions for the varying surface regions will be discussed. Additionally, the methodology of the computer solving algorithms including the use of A* Pathfinder search techniques will be analyzed.

The completion of this research will provide NASA with a valuable tool to aid in the modeling of the required energy necessary for vehicle traversing of a remote planetary surface.

AAS 93-840
A "JUMPER" FOR THE MARS SURFACE EXPLORATION
G. Savu, Institute of Fluid Mechanics and Flight Dynamics, Bucharest, Romania

In the present paper, an autonomous vehicle capable to make jumps over the natural obstacles existing on the Martian rugged soil is described. In contrast with the recent proposals using rovers or winged machines to explore the surface of Mars, the present Martian robot utilizes the rarified Martian atmosphere to obtain gaseous jets for lifting and propulsion. The gas from the surrounding atmosphere is compressed by a pump driven by an electric motor, in a storage reservoir. The energy for the electric motor is ensured by a photovoltaic array. The compressed gas of reservoir is expanded through a Laval nozzle, which ensures sufficient thrust for lifting and longitudinal propulsion necessary for a given trajectory of the jump. Due to rugged Martian soil, a vertical takeoff/landing trajectory is proper, even if this kind of trajectory cannot ensure minimum power consumption.

Enforcing the jumper trajectory, the orientation angle of the nozzle and the valve opening is obtained by integration of the motion equations, taking into account the continuous decreasing storage pressure in the tank. A computer code was elaborated to simulate the gas reservoir supplying during the stand-by period and its discharging along the jump trajectory. Numerical experiments were performed for various payloads and reservoir's capacities, taking into account the actual performances of the solar cells [1] and the Martian solar constant. The mission of a jumper having about 20 kg of scientific equipment, like that proposed for MESUR (Mars Environmental Survey) program [2], is analyzed.

References: [1] Bennett, G. L., "Aerospace Power," *Aerospace America*, p. 37, 1992. [2] Boyer, W. "Surveying Mars on a Tight Budget," *Aerospace News*, 4, pp. 16-17, 1993.

11. EXOBIOLOGY

AAS 93-841
LIFE ON MARS: DID VIKING CLOSE THE BOOK?
Benton C. Clark, Planetary Sciences Laboratory, Martin Marietta Astronautics

The environment on Mars is clearly hostile to organisms.

So is hot, acidic water. Yet the Yellowstone hot springs are teeming with microbial life. Even prior to the Viking landings, enough was known about the environment of Mars that the likelihood of life forms was met with skepticism. Discoveries by Viking only made matters worse. Temperatures and humidity levels are such that organisms might persevere in a cryptobiotic state, but not as actively growing entities. Much worse is the fact that strong oxidants which destroy organic compounds apparently exist in the atmosphere and/or soil on Mars.

Is the quest over? No. the metabolic diversity of microorganisms on Earth is manifestly astounding especially in view of the never-ending competition for substrates. Where there is an environmental niche, an organism or microecology of organisms will be found.

The Viking life detection experiments were elegant. But they were not all inclusive. At least three types of life forms could theoretically exist on Mars and not have been detected by Viking. These are (1) endolithic organisms, which live inside rocks (discovered on Earth only after Viking flew to Mars), (2) sulfate-reducing microbes, and (3) hydrothermal vent dwellers. Viking was not equipped to test for any of these. Neither rocks nor duricrust materials were analyzed by the Viking biology experiment. Unusual metabolisms were not sought out. And the Viking landers were far from any volcanic features where vents or fumeroles might still provide a refuge for adapted life forms in analogy with the bizarre oceanic hydrothermal vent communities.

Will life detection be a high priority for human exploration on Mars? Certainly. Will humans corrupt such experiments by their intrinsic bioburden? Not if standard laboratory precautions are taken and sufficient analytical equipment is provided to the explorers. Could such experiments be conducted better by robots? No, even if the funds for robots were as high as for human exploration (which has never before occurred in the history of anyone's space program). Will the Search for Life ever end? Not until humans have gone to Mars and completed the exploration!

AAS 93-842

EXOPALEONTOLOGY AND THE SEARCH FOR A FOSSIL RECORD ON MARS

Jack D. Farmer and David J. Des Marais, NASA-Ames Research Center

There is substantial geological and climatological evidence that Mars was more like the Earth early in its history, with a warm, dense atmosphere and abundant surface water. Under such warm, wet conditions, life may have developed on Mars, only to become extinct in surface environments as the planet lost its atmosphere and cooled to the present state. If life developed on Mars, it is likely to have left behind a fossil record. A record of Martian life could potentially include not only the actual remains of organisms (body fossils), but also a behavioral (trace fossil) and/or molecular record. In developing a strategy to explore for a fossil record on Mars, we are guided by the nature of the early fossil record on the Earth. The most common fossils in the Precambrian are stromatolites - laminated sedimentary structures that are the trace fossils of microbial mats. They form when microbial communities, buried by detrital sediments or precipitated minerals, migrate upward in response to chemical gradients or light. In the process, microbes entrap sedimentary particles, forming a thin layer of sediment. Repeated cycles lead to sediment accretion, producing biolaminated sedimentary structures of varying complexity. Such structures comprise the major record of life on the early Earth and are obvious targets in the exploration for fossils on Mars.

Identifying environments where mineralization could have occurred in the presence of organisms is key in defining a strategy to search for body and/or molecular fossils on Mars. Long-term preservation of terrestrial microorganisms typically requires very early post-mortem mineralization of cell walls and/or extracellular materials. The best results are obtained when organic materials are replaced by fine-grained, stable phases (e.g. silica, phosphate, or carbonate) where mineralization is more or less contemporaneous with biological activity. The oldest body fossils on Earth were preserved in this way, being found as permineralized cells in siliceous sediments (cherts) associated with ancient volcanic terrains in Australia and South Africa. Excellently preserved fossil microbiotas are also known from Proterozoic dolomites and silicified dolomites of Greenland. These microbiotas are often found within stromatolites.

Target deposits to sample for body and trace fossils on Mars include subaerial thermal spring sinters, ambient temperature, sub-lacustrine spring carbonates, and mineralized paleosol horizons (e.g. calcretes, silcretes, duricrusts) where stromatolites tend to be especially common. Such deposits are often simple in composition, consisting of a fairly narrow range of pure end-member, relatively stable mineral phases, including silica, carbonate, and phosphate. Other potential host minerals include evaporite minerals which being relatively unstable, have comparatively short crustal "half-lives." Consequently, evaporites are typically lost from the record fairly easily, if not replaced during early diagenesis by more stable minerals, such as barite or silica. Ice offers another potential host mineral for fossils, although the relative instability of ground ice on Mars indicates that it may be less important as a long term repository of biological information, than the other minerals mentioned.

In exploring Mars for molecular fossils, priority is given to sedimentary environments that have the highest potential for maintaining a closed chemical system after deposition. On Earth, fine-grained lacustrine sediments (e.g. shales, mudstones) tend to exhibit comparatively high total organic carbon contents. In addition, certain classes of clay minerals which comprise such sediments, selectively bind organic molecules as interlayer cations, thereby retaining a biochemical signature of their primary depositional environment. Compaction of fine-grained sediments, accompanied by early mineralization (cementation), reduces permeability, and promotes the long-term preservation of organic materials. By terrestrial analogy, likely targets on Mars are the basin-central, deep water facies of terminal lake systems.

Initially, the identification of Martian surface mineralogy will depend on the interpretation of high resolution spectral data acquired from orbit. Minerals commonly associated with the above environments have characteristic spectral signatures within the near and mid-infrared that could be detected with high resolution orbital remote sensing. More detailed information may eventually be obtained using instruments delivered to the Martian surface. Rover-based infrared spectroscopy, in conjunction with X-ray diffraction and/or fluorescence, could be extremely valuable for providing *in situ* compositional information, as well as opportunities for real time decisions regarding the selection of samples for return to Earth.

AAS 93-843

THE TEMPORAL AND SPATIAL DISTRIBUTION OF MARTIAN PALEOLAKES

James W. Rice, Jr., Arizona State University

Photogeologic investigations of the Viking imagery data set reveal that fluvial processes have operated throughout the geologic history of Mars. The valley networks of the southern cratered highlands primarily formed during the Noachian System, the oldest division of Martian geologic history, and the Hesperian System, the middle stratigraphic system on Mars. Many possible sites of ponded water occur in the highlands where valley networks converge toward low areas or debouch into craters that cover the region (Goldspiel and Squyres, 1991; De Hon, 1991; Rice and De Hon, in press). These sites would have been the location of the oldest lakes on the planet.

During Hesperian time, the outflow, and fretted channels and to a lesser extent the valley networks flowed and contributed to the formation of large paleolake systems (Scott *et al.*, 1991). These great lakes had calculated maximum depths of 1 km and volumes of 9×10^5 km^3 (Scott *et al.*, in press). These lakes were located primarily in the northern lowlands of Mars.

Some of these great lakes continued to exist into Amazonian time, the youngest division of Martian geologic history. Other investigators (Parker *et al.*, 1989; Baker *et al.*, 1990) have also found evidence of enormous bodies of water (postulated oceans) in the northern lowlands. If oceans did exist on Mars then this system of great lakes was probably both the precursors and last vestiges of the former oceans as it dried up (Rice, 1992). The existence of these bodies of water would have had profound implications for the origin of life on Mars.

References:

Baker, V.R., R.G. Strom, S.K. Croft, V.C. Gulick, J.S. Kargel, and G. Komatsu, "Ancient ocean-land-atmosphere interactions on Mars: Global model and geological evidence," *Lunar and Planetary Science Conf.* 21, 40-41, 1990.

De Hon, R.A., "Classification of Martian lacustrine basins," *Lunar and Planetary Science Conf.* 22, 293-294, 1991.

Goldspiel, J.M., and S.W. Squyres, "Ancient aqueous sedimentation on Mars," *Icarus*, 89, 392-410, 1991.

Parker, T.J., R.S. Saunders, and D.M. Schneeberger, "Transitional morphology in west Deuteronilus Mensae, Mars: Implications for modification of the lowland/upland boundary," *Icarus*, 82, 111-145, 1989.

Rice, J.W., Jr., "Antarctic lakes (above and beneath the ice sheet): Analogues for Mars, MSATT Workshop on the polar regions of Mars: geology, glaciology, and climate history," LPI Technical Report Number 92-08, Part 1, 23-24, 1992.

Rice, J.W., Jr. and R.A. De Hon, Geologic map of Darvel quadrangle, Maja Vallis region, Mars (MTM 20052), U.S.G.S. Misc. Inves. Series, 1:500,000 scale, in press.

Scott, D.H., J.W. Rice, Jr., and J.M. Dohm, "Martian paleolakes and waterways: Exobiological implications," *Origins of Life and Evolution of the Biosphere*, 21 no. 3, 189-198, 1991.

Scott, D.H., J.M. Dohm, and J.W. Rice, Jr., "Paleolakes and channels: A new look at Mars' fluvial history," submitted to JGR.

AAS 93-844

LANDING SITES ON MARS FOR EXOBIOLOGY

Ragnhild Landheim, Departments of Botany and Geology, Arizona State University
Ronald Greeley, Department of Geology, Arizona State University
Jack D. Farmer, David J. Des Marais, NASA-Ames Research Center
Harold P. Klein, Department of Biology, Santa Clara University

One of the fundamental questions to be answered in the exploration of Mars is whether the planet was hospitable to life. The Viking biology experiments were inconclusive in their results (1) and thus provide an incentive to continue the search for evidence of a past or present biosphere on Mars. Future missions currently under planning may provide such an opportunity. This study was aimed at selecting sites on Mars that could be targeted for exobiological purposes by future missions.

The two basic assumptions of Mars exploration Exobiology are that liquid water is required for living systems to exist, and that water was present long enough that life could evolve. This study was aimed at identifying landforms created by water-driven surficial processes and/or their sedimentary deposits. The primary objectives were to detect geological settings where water could have been present, assess the relative duration of water occurrence, and prioritize sites for high resolution imaging and future surface exploration (2). The geological environments considered to be of highest priority for exobiological studies are fluvio-lacustrine, spring, and periglacial. Of these environments, fluvio-lacustrine sites are considered to be excellent targets for meeting the goals of Exobiology, because (1) water-laid sediments may contain fossils and/or organic compounds, and (2) large lacustrine basins that are unaffected by volcanism are good landing sites from an engineering standpoint. Potential sites for thermal spring activity were identified as "point source" channels with amphitheater-shaped headlands that occur near volcanic centers, or other potential heat sources. Periglacial environments are of great interest to Exobiology, because ground-ice may have served to inhibit diffusion of oxidants in the soil (3), thus favoring preservation of organisms and organic compounds.

The study evolved from the Mars Analog Site Study (MASS) (4), which was developed to improve the process of landing site selection on Mars, using the southwest United States as a Mars analog. The initial site selection was based on an exobiological assessment of sites presented in the Mars Landing Site Catalog (5). The subsequent phase searched all Viking Mars Charts (MC; scale 1:2M) for potential Exobiology significant sites. High resolution data were applied to the extent available and include the Mars Transverse Mercator (MTM) maps (scale 1:500,000) and Viking Orbiter (VO) images (resolution >~40 m/pixel). The final phase consisted of retrieval of Mosaicked Digital Image Models (MDIM's: resolution 231 m/pixel) of the respective sites. Geologic features used to identify each site type were assigned scores based on three subjective weighting factors, including visibility of feature on the image, uniqueness of the feature-process relationship, and importance of the feature in relationship to goals of Exobiology.

The study demonstrates the diversity of geological settings that are of interest in the selection of future Exobiology sites and provides a preliminary site list of 25 potential sites for exobiological studies on Mars (Table 1). At present, two of the sites proposed by this study are under consideration as potential landing sites for the MESUR-Pathfinder mission (6). The study also confirms the need for high resolution imaging, visible and infrared remote sensing data during Mars Observer and Mars '94/'96 missions.

References:

(1) Klein, H.P., "General Constraints on the Viking Biology Investigations," *Origins of Life* 7, 273-279, 1976.

(2) Farmer, J.D., D.J. Des Marais, R. Landheim, R. Greeley, and H.P. Klein, "Site selection for Mars Exobiology," *Proceedings of the World Space Congress*. 29th COSPAR Meeting, 1992 (in press).

(3) Squyres, S.W., "Urey Prize Lecture: Water on Mars," *Icarus* 79, 279-288, 1989.

(4) Greeley, R., R. Kuzmin, F. Costard, F.S. Anderson, M.A. Geringer, R. Landheim, and M.L. Wenrich, "Mars Analog Site Study (MASS)," *Lunar Planet. Sci. Conf. XXIV*, 567-568, 1993.

(5) Greelye, R., Mars Landing Site Catalog, NASA Reference Publication 1238, 1990. (6) "MESUR Science Definition Team," *Proceedings of Sixth Meeting*, Oct 1992.

Table 1
POTENTIAL MARS EXOBIOLOGY SITES FOR FUTURE EXPLORATION

*MLSC site no.	Location of interest area (lat., long)	Target (lat., long)	Potential deposits	Name of site region of priority	Relative level references	Suggested
142	8deg-9deg S, 158.5deg-160deg W	8.5deg S, 159.5 deg W	Waterlain sed.	Mangala Valles	High	MTM-10157
143	33deg-35deg N, 340deg-344deg W	33.5deg N, 342.5deg W	Waterlain sed.	Ismenius Lacus SW	High	MC-55W
1	35deg-38deg S, 227deg-231deg W	37deg S, 230deg W	Waterlain sed.	Eridania NW	High	MC-29NW
2	d1deg-23deg S, 9deg-14deg W	22deg S, 11deg W	Waterlain sed.	Margaritifer Sinus SE	High	MC-18SE
8	22deg-23deg S, 229deg-231deg W	22.8deg S, 230.8deg W	Waterlain sed.	Mare Tyrrhenum SE	High	MC-22SE
10	24.5deg-26.5deg S, 264deg-266.5deg W	24.8deg S, 265.8deg W	Waterlain sed.	Mare Tyrrhenum SW	High	MC-22SW
137	5.6deg-6.4deg S, 149.1deg-149.9deg W	6.3deg S, 149.5deg W	Waterlain sed.	Mangala Valles	High	MTM-05147
79	13.5deg-15.5deg S, 187deg-190deg W	15.5deg S, 188.5deg W	Waterlain sed.	Aeolis SE	High	MC-23NE, SE
138	13.5deg-15.5deg S, 183.5deg-185.5deg W	15.5deg S, 184. 5deg W	Waterlain sed.	Aeolis NE (Gusev)	High	MC-23NE, SE
140	6.5deg-8deg S, 302deg-305.5deg W	7.3deg S, 305deg W	Waterlain sed.	Iapygia NW	High	MC-21NW
144	35deg-40deg N, 130deg-143deg W	39.5deg N, 135.5deg W	Waterlain sed.	Diacria SE	Moderate/High	MC-2SE, SC
145	11deg-15deg S, 173.5deg-175.5deg W	14.5deg S, 175deg W	Waterlain sed.	Memnonia NW	Moderate	MC-18NW
5	10deg-11.5deg S, 277deg-280deg W	11deg S, 279.5deg W	Waterlain sed.	Iapygia NE	Moderate	MC-21NE
7	41.5deg-43.5deg S, 208deg-212deg W	43.2deg S, 208.1deg W	Waterlain sed.	Eridania NC	Moderate	MC-29NC
22	5.9deg-6.2deg S, 73.7deg-74deg W	6.05deg S, 73.75deg W	Waterlain sed.	Candor Mensa	Moderate	MTM-05072
141	0deg-2deg S, 73.5deg-778deg W	1.5deg S, 76.5deg W	Waterlain sed.	Hebes Chasma	Moderate	MC-18NW
147	5deg-10deg S, 333deg-337deg W	8deg S, 335deg W	Evaporite dep.?	Sirius S. (White Rock)	High	MC-20NE
148	20deg-25deg N, 5deg-10deg W	21.3deg N, 8deg W	Evaporite dep.?	Oxia Palus NE	Moderate	MC-11NE
139	22deg-23.5deg N, 34.5deg-35.5deg W	22.1deg N, 36.7deg W	Evaporite dep.?	Oxia Palus NW	Moderate	MC-11NW
32	33deg-35deg S, 264deg-268deg W	33.2deg S, 266.4deg W	Hydroth. dep.?	Dao Vallis/Hadriaca P.	High	MC-28NE
77	2deg-4.5deg N, 15.2deg-17deg W	2deg N, 16deg W	Hydroth. dep.?	Ares Vallis	Moderate	MC-11SE
149	43deg 47deg N, 315deg-337deg W	44.3deg N, 333deg W	Ground-ice?	Ismenius Lacus SC	High	MC-5SC
4	55deg-58deg S, 190deg-199deg W	57deg S, 197deg W	Ground-ice?	Eridania SE	Moderate	MC-29SE
150	31deg-42deg N, 249deg-260deg W	36deg N, 259deg W	Ground-ice?	Casius SE	Moderate/Low	MC-6SE

*Mars Landing Site Catalog

12. TERRAFORMING

AAS 93-845
THE LONG-TERM HABITATION OF MARS

Martyn J. Fogg, Probability Research Group

One follow-on objective of the human exploration of Mars might be the permanent settlement of the planet. This entails examining issues considered by previous CFM conferences over much longer timescales.

In this paper a systems approach is adopted to consider the acquisition of energy and material resources on Mars and their organization within life-support systems. Since such resources will have to be initially manufactured on Earth, an early settlement will represent a particularly expensive example of fossil fuel subsidized infrastructure. Thus, growth of a Martian civilization cannot be realistically expected unless near complete independence is attained. A crucial advance early in any settlement scheme will be the bootstrapping of all the industrial capacity required for survival and growth from a minimum inventory of terrestrial goods.

Similarly, whilst it seems possible to fabricate closed loop ecological life-support systems in which colonists might live, such energy-intensive and constrained habitats appear a far less attractive option for a large population and a long-term future. The need to perpetually maintain the integrity of habitat structure, mechanically recycle water and wastes and monitor and control the balance of gases cycling through the biota compares poorly with the benefits of the autonomous and fully bioregenerative life-support system that is the Earth's biosphere. Since Mars may be the only planet in the Solar System where the creation of an unconstrained planetary biosphere may be a realistic proposition, it seems likely that Martian colonists will be attracted to the possibility of terraforming their adopted world. Recent scientific progress on the subject of terraforming Mars is briefly reviewed.

It is suggested that the permanent settlement of Mars represents a thought experiment, largely unclouded by non-scientific details, of great relevance to the present day Earth. Like future Martian colonists, the people of Earth are faced with the need to find an alternative to fossil fuels and an ultimate requirement to maintain and manage a global life-support system.

AAS 93-846

TECHNOLOGICAL REQUIREMENTS FOR TERRAFORMING MARS

Robert M. Zubrin, Martin Marietta Astronautics
Christopher P. McKay, NASA Ames Research Center

The planet Mars, while cold and arid today, once possessed a warm and wet climate, as evidenced by extensive fluvial features observable on its surface. It is believed that the warm climate of the primitive Mars was created by a strong greenhouse effect caused by a thick CO_2 atmosphere. Mars lost its warm climate when most of the available volatile CO_2 was fixed into the form of carbonate rock due to the action of cycling water. It is believed, however, that sufficient CO_2 to form a 300 to 600 mb atmosphere may still exist in volatile form, either adsorbed into the regolith or frozen out at the south pole. This CO_2 may be released by planetary warming, and as the CO_2 atmosphere thickens, positive feedback is produced which can accelerate the warming trend. Thus it is conceivable, that by taking advantage of the positive feedback inherent in Mars' atmosphere/regolith CO_2 system, that engineering efforts can produce drastic changes in climate and pressure on a planetary scale.

In this paper we propose a mathematical model of the Martian CO_2 system, and use it to produce analysis which clarifies the potential of positive feedback to accelerate planetary engineering efforts. It is shown that by taking advantage of the feedback, the requirements for planetary engineering can be reduced by about 2 orders of magnitude relative to previous estimates. We examine the potential of various schemes for producing the initial warming to drive the process, including the stationing of orbiting mirrors, the importation of natural volatiles with high greenhouse capacity from the outer solar system, and the production of artificial halocarbon greenhouse gases on the Martian surface through in-situ industry.

If the orbital mirror scheme is adopted, mirrors with dimension on the order or 100 km radius are required to vaporize the CO_2 in the south polar cap. If manufactured of solar sail like material, such mirrors would have a mass on the order of 200,000 tonnes. If manufactured in space out of asteroidal or Martian moon material, about 120 Mwe-years of energy would be needed to produce the required aluminum. This amount of power can be provided by near-term multi-megawatt nuclear power units, such as the 5 MWe modules now under consideration for NEP spacecraft.

Orbital transfer of very massive bodies from the outer solar system can be accomplished using nuclear thermal rocket engines using the asteroid's volatile material as propellant. Using major planets for gravity assists, the rocket DV required to move an outer solar system asteroid onto a collision trajectory with Mars can be as little as 300 m/s. If the asteroid is made of NH_3, specific impulses of about 400 s can be attained, and as little as 10% of the asteroid will be required for propellant. A 5000 MWt NTR engine would require a 4 year burn time to push a 1 billion tonne asteroid through a DV of 300 m/s. About 40 such objects would be sufficient to greenhouse Mars.

Greenhousing Mars via the manufacture of halocarbon gases on the planet's surface may well be the most practical option. Total surface power requirements to drive planetary warming using this method are calculated and found to be on the order of 1000 MWe, and the required times scale for climate and atmosphere modification is on the order of 50 years.

It is concluded that a drastic modification of Martian conditions can be achieved using 21st century technology. The Mars so produced will closely resemble the conditions existing on the primitive Mars. Humans operating on the surface of such a Mars would require breathing gear, but pressure suits would be unnecessary. With outside atmospheric pressures raised, it will be possible to create large dwelling areas by means of very large inflatable structures. Average temperatures could be above the freezing point of water for significant regions during portions of the year, enabling the growth of plant life in the open.

The spread of plants could produce enough oxygen to make Mars habitable for animals in several millennia. More rapid oxygenation would require engineering efforts supported by multi-terrawatt power sources. It is speculated that the desire to speed the terraforming of Mars will be a driver for developing such technologies, which in turn will define a leap in human power over nature as dramatic as that which accompanied the creation of post-Renaissance industrial civilization.

AAS 93-847
PARATERRAFORMING: CONSTRUCTIONAL, ENERGY, ENVIRONMENTAL AND HAZARD CONTROL STRATEGIES FOR A QUASI-GLOBAL MARTIAN WORLDHOUSE
Richard L.S. Taylor, Probability Research Group, Centre for Extra Mural Studies
The "Worldhouse" concept of paraterraforming provides a means of creating and maintaining habitable environments on small planets, such as Mars, where indigenous volatiles exist but their inventory is, or may be, insufficient either for the production of a 1 bar gravity-bounded atmosphere of near-terrestrial composition, or for its longer-term geochemical cycle stability. Although, when completed, the deliberately restricted ecospheric environment (DREE) within the quasi-global worldhouse will function close to the point of self-stability, it will be necessary to ensure that the biosystems within the worldhouse mimic closely the biological/geochemical cycles of the Earth, and that the total energy inputs and outputs are closely matched with each other. During the construction phase, energy requirements and the maintenance of metastable biospheres within sub-global worldhouse structural units pose significant problems for determining the best overall campaign strategies. Additionally, consideration must be given to the prevention and circumvention of any potential hazards that might affect the short- and long-term integrity of the worldhouse structure. An attempt is made to present an optimized scenario, the outcome of which should be a DREE system close to a fully Earth-like state of dynamic equilibrium.

13. LIFE SCIENCES CHALLENGES FOR MANNED MARS MISSIONS

AAS 93-848
OVERVIEW OF LIFE SCIENCE CHALLENGES FOR HUMAN EXPLORATION
James E. Lynne, University of Tennessee
This paper will present an overview of several of the critical biological issues for manned voyages to Mars. Topics to be examined include: (1) radiation exposure and shielding requirements with an evaluation of the effects of the interplanetary trajectory, transit time, and propulsion system; (2) microgravity exposure effects including bone demineralization, muscular atrophy, immunological changes, cardiovascular deconditioning, and potential countermeasures; (3) a discussion of the possible need for artificial gravity and its implications for spacecraft complexity; (4) the applicability of the current American and Soviet database to Mars mission planning; (5) an analysis of the minimal requirements for further biological data prior to a manned voyage, and (6) the impact of both the overall mission duration and interplanetary transit times on the need for more detailed biological data.

AAS 93-849
RADIATION CONCERNS ON LONG TERM DEEP SPACE MISSIONS
Karl A. Pfitzer
Uncertainties in predicting the radiation dose for deep space missions will have a significant impact on mission design. During long-term deep space mission, astronauts are exposed to radiation from galactic cosmic rays and from solar flare proton events. The galactic cosmic rays represent a steady state feature that varies over the eleven year solar cycle. The galactic cosmic rays contain a very energetic and difficult to shield component. Ion chambers flown on balloons to high altitude encounter a peak in the radiation dose called the Pfotzer maximum when the atmospheric shield is approximately 30 g/cm^2 thick. This demonstrates that extra shielding may make the dose worse. Solar flares represent a very unpredictable but very intense danger. The average flare produces fluences that are relatively small and can be easily shielded. However, a single large flare may produce a radiation dose greater than the integrated dose from all other flares over 10 to 20 years. Radiation from solar flares is more easily shielded than the galactic radiation. However, protection against a great flare and the ability to predict the occurrence of a great flare may strongly drive system design. Protection against flares may also strongly impact surface exploration.

Only limited data on radiation effects for humans exists for long-term deep space missions. The Apollo missions to the moon were of limited length and these missions did not encounter any solar flares. The shuttle missions are very short term and the astronauts are not only protected by the extensive shielding of the shuttle but are also protected by the earth's magnetic field. The Soviet space missions are at much higher inclinations and thus are exposed to a more representative galactic flux. Soviet astronauts were in orbit during the three very large flares during 1989 and received small but non-trivial exposures. During a portion of the orbit when the Soviet space station is over the polar region, Soviet astronauts do receive a representative deep space dose. However, integrated exposure time in the polar regions for the Soviet space station is still small compared to the expected exposure time for a long-term mission.

Data on the effects of radiation on humans is primarily based on ground based data. Most radiation damage to humans has been caused by gamma rays. In deep space virtually all of the dose is due to the more strongly ionizing protons and heavy ions. Thus applicability of the existing data to deep space missions is uncertain. Much work is now in progress to attempt to understand the effects of heavily ionizing radiation on biological cells. Additionally, work at linear accelerators is being used to study the effectiveness of shielding. Furthermore, all shuttle missions now fly Tissue Equivalent Proportional Counters (TEPC) to help study the incoming radiation.

Present uncertainties in the radiation fluences, the effectiveness of shielding, and in the response of biological systems to highly ionizing radiation produce large uncertainties in mission planning and system design. Design uncertainties cost money. It is imperative that studies be implemented early in the design phase to reduce this uncertainty.

AAS 93-850
IS BONE LOSS IN SPACE A PROBLEM?
Chris Cann, University of California San Francisco

A significant concern in any consideration of long-term exposure to microgravity has been musculoskeletal deterioration, with possible bone loss leading to the clinical consequence of fracture being an unresolved problem. However, all direct space flight data regarding bone loss indicate that the problem may be ignored for 6-12 months as long as some prophylactic measures, notably exercise, are included in the mission profile. The exercise used currently is about 2 hours per day for each person, and while this may be a significant problem for orbiting laboratories trying to do microgravity experiments it may be a psychologically as well as physiologically beneficial use of time on an interplanetary mission. Current concepts within NASA's medical operations program aim for specific muscle exercise to be done in 10-15 minutes per day with minimal impact on the microgravity environment, but a different approach may be optimal on Mars flights.

For the skeletal system, the dynamics of bone processes (bone breakdown and formation) must also be considered. Skeletal adaptation to microgravity begins within the first 24-48 hours and can vary among individuals. The adaptive mechanism appears to be "normal" physiologically, that is, the primary effect is an increase in breakdown of bone secondary to mechanical unloading, with the body trying to rid itself of the excess baggage of a no-longer-needed skeleton. If this is the only response, we may be able to prevent it using bone-loading exercises, electrical stimulation for muscle pulls on bone, or drugs. However, there is some evidence that microgravity may also induce effects on bone and its surrounding tissues independent of simple loading; if this occurs prophylactic measures will be much more complex, and may affect mission planning if the current "no risk" policy for human spaceflight is maintained.

AAS 93-851
ARTIFICIAL GRAVITY AS A COUNTEMEASURE TO THE EFFECTS OF SPACEFLIGHT ON LIVING SYSTEMS
Carl Rudiger, Jr. and Phillip A. Wagner, Lockheed Missiles and Space Company, Inc.

The elimination of gravity during spaceflight creates a number of known physiological effects. Flight times to Mars will last six months or longer, with roundtrip times of two years or more including time spent in 3/8 gravity on the surface. Mere survival in these conditions will not be sufficient - effective performance, safety, and long-term health must also be considered. Substituting centrifugal force for the lack of gravity may be the most effective countermeasure, but at what gravity levels, radii, and durations (continuous vs. intermittent) are not known.

Researching the effectiveness and effects of artificial gravity as a countermeasure requires facilities in space. The first major facility for this research is the centrifugal facility under development by NASA Ames Research Center for Space Station *Freedom*. This paper will describe that facility and its capabilities. We will also explore a facility designed for use on Spacelab and look at future larger human-rated centrifuges or rotating spacecraft.

AAS 93-852
SPACE STATION FREEDOM:
A KEY STEP FOR A MARS MISSION LIFE SUPPORT SYSTEM
Richard P. Reysa, Boeing Defense and Space Group

Mars transit and base missions present significant challenges for a Life Support System (LSS). A Mars mission LSS must be regenerative to reduce or eliminate expendables and be low risk, demonstrating high reliability and long-term performance. A regenerative physicochemical LSS for Space Station *Freedom* (SSF) is a key step toward meeting these requirements. In the development of the SSF regenerative LSS program, the challenges have proven to be both technical and budgetary. The SSF regenerative LSS has been down-scaled from totally closed water and air recovery systems to open loop life support systems for Manned Tended Capability (MTC). Water Recovery Management (WRM) system closure is planned for the Permanently Manned Capability (PMC), but the Air Revitalization (AR) system will not be closed unit Eight Man Crew Capability (EMCC). In the near term, not taking the LSS closure steps will impact the SSF logistics costs and, in the long term, will not generate the hardware and integrated system experience necessary for reducing Mars mission program risk.

The Mars mission budgetary realities for a technical program of this magnitude suggest that step-wise evolution of the SSF LSS is the most prudent approach to achieve success, rather than embarking on a totally new regenerative LSS design. Earlier emphasis should be placed on SSF regenerative LSS implementation to provide experience for reducing logistics, improving low-gravity hardware operation, extending component life, understanding system integration issues and improving system efficiencies. The resulting LSS can also be key to evolving to a Closed Ecological Life Support System (CELSS) as CELSS technology is phased into LSS to achieve more efficient closure. Additionally, regenerative LSS evolution is synergistic with SSF LSS logistics reduction objectives. Analytical techniques, such as those being developed for defining SSF logistics requirements, should be employed to assist in identifying LSS hardware technology development needs.

Meeting the goals of reduced logistics and high reliability requires a broad base technology development program for improving the current technology. For example, the selected SSF regenerative WRM system low technology risk design is partially offset by a significant logistics penalty for filter and sorbent bed replacement. Technology advancements can be made to reduce or eliminate expendables by developing technology for regenerative sorbent beds and filters. Additionally, regenerative hardware evolution should include waste recovery technology development that provides more complete closure by recovering metabolic and processor byproducts and food wastes. Regenerative AR technologies are generally considered more mature than water recovery technologies. Therefore, consideration should be given to beginning closure during PMC, rather than EMCC, in a building block fashion. This may be achieved by adding water electrolysis and later CO_2 reduction assemblies to the existing SSF AR system. AR technology evolution should concentrate in the areas of nitrogen supply, expendables reduction and closure efficiency improvement. This presentation will define the SSF regenerative LSS closure loop development history and current configuration, and a recommended LSS evolution plan for a Mars mission.

AAS 93-853
FIELD INVESTIGATIONS IN ANTARCTICA: A REGIMEN FOR MANNED MARS MISSION ASTRONAUT TRAINING ACTIVITIES
James W. Rice, Jr., Arizona State University

This work is based upon my experiences of performing geological investigations in the Bunger Hills region of Eastern Antarctica (Rice, 1992). Since the ice free regions of Antarctica are considered to be the best analog on Earth to the Martian surface (McKay, 1984), it then seems only logical to train future Mars Mission crews in these regions. The harsh conditions and isolation would also approximate Martian surface operations.

It is proposed that the crew, equipment, and techniques be field tested in the Antarctic as part of the mission training regimen. The idea of using the Moon as a testbed for Mars is both dubious and costly. First of all, the physical geomorphic processes operating on the Moon are not analogous to Mars (fluvial, lacustrine, aeolian, and periglacial processes have never operated on the lunar surface). The ice free regions of Antarctica have undergone the above mentioned processes, volcanism can also be added to the list. Perhaps more importantly, in these times of limited budgets for large space endeavors, the return to the Moon will only delay at best the more important objective: the manned exploration of Mars. The lunar outpost/base would siphon off funding that could be used to support Mars Missions. Unfortunately, one can't have both. As a scientist I do believe that the Moon requires further study, but not at the expense of Manned Mars Missions.

One possible scenario envisions a Mars crew living and training aboard the space station to simulate the experience of the outbound journey to Mars. After the crew has spent the designated time aboard the station they would then return to Earth and be transported to the selected site in Antarctica to begin their simulated Mars surface operations. When the field investigations are completed the crew would either return to JSC or possibly back to the space station to simulate their return voyage to Earth. The duration of this simulated mission would be performed at a fraction of the time required for the real mission.

References:

McKay, C.P., "Antarctica: Lessons for a Mars exploration program," AAS 84-156, in *The Case for Mars II*, C.P. McKay, ed., AAS *Science and Technology Series*, Vol. 62, pp. 79-87, 1984.

Rice, J.W., Jr., "Antarctic lakes (above and beneath the ice sheet): Analogues for Mars," MSATT Workshop on the polar regions of Mars: geology, glaciology, and climate history, LPI Technical Report Number 92-08, Part 1, 23-24, 1992.

AAS 93-854
MARS MISSION GRAVITY PROFILE SIMULATION

Lawrence H. Kuznetz, NASA Ames Research Center

An international cooperative mission is proposed to address the issue of whether artificial gravity is required for a manned mission to Mars. Current mission architectures use complex, heavy, and costly designs using rotating or tethered spacecraft to provide artificial gravity. This proposal presents a Space Shuttle mission to the Soviet MIR Space Station combined with an earth-based study as a means of answering the question of, "Do we need artificial gravity?" The mission is divided into 8 phases, each designed to simulate a specific phase of the manned Mars mission. Launch from the Earth to the MIR station is initially provided by the U.S.S.R., while transfer of the crew and return to Earth, reboost back to MIR and final return to Earth are provided by the U.S. Space Shuttle. Zero gravity simulating the outbound leg to Mars and the return leg to Earth is provided on the MIR Space Station. Simulated Mars gravity is provided by several different methods at NASA Ames Research Center. Medical and physiological monitoring and mission activity simulation is provided continuously. A follow-on program is also proposed providing for construction of an international variable gravity research facility linked to MIR by a tether. This would be the second step of an evolutionary process leading to a joint international manned mission to Mars.

14. BIOSPHERES

AAS 93-855
COVERED CRATER BIOME - A MARTIAN LIFE SUPPORT SYSTEM

R. P. Van Zandt

Until such time as the possibility of creating a livable atmosphere on Mars by terraforming techniques has been evaluated and developed, a much more practical means of trapping an atmosphere and solar heat in a roomy environment is possible with present technology. This paper proposes a means by which impact craters, abundant on Mars, could be converted into living biomes by providing them with a flattened dome-like cover consisting of an underlying plastic seal and loaded down with about 17 feet of Martian soil to balance the atmospheric pressure trapped within. A slight over-pressure would be desirable, balanced by a system of hold-down cables to provide stability. The interior could be flooded with sunlight by a system of guided mirrors on top of the cover which would concentrate sunlight to a central tower, like presently existing solar collectors on Earth. Instead of converting this into steam for power generation as done on Earth, this light could be directed down the tower and scattered uniformly throughout the interior by a suitable reflecting system (perhaps of spherical shape) at the center of the dome.

The initial construction could start with an air-tight plastic film laid down on the crater floor (also suitably sealed) and gradually inflated with Martian atmosphere and pressurized as Martian soil is flung over the cover until the desired atmospheric pressure is achieved. Following this, the mostly carbon dioxide atmosphere could be converted into a livable oxygen atmosphere by suitable biochemical means. The already present nitrogen could be enriched from the outside atmosphere. Martian water could be stored on the crater floor and used to supply the bio system and recondensed by outside cooling to maintain the proper humidity. Once a living environment is established, a 1 km diameter crater could provide life support for two to three hundred humans. The findings of the Biosphere 2 experiment in Arizona will be helpful in determining essential living requirements.

A preliminary study shows a more than ample supply of solar energy brought in by the light transfer system to maintain a suitable living temperature inside and to serve as a much needed heat source for other purposes.

This paper will consist of a study of the mechanical, Earth material and energy requirements in the construction of the biome and a more complete thermal study of the amount of solar energy that can be trapped and subsequently utilized. A small scale model involving the construction sequence of covering a dished out "crater" on Earth will be described as a proposal to determine the stability requirements during and after completion.

It is contemplated that the first covered crater built on Mars would serve as a prototype for additional craters and to determine what the eventual requirements would be for creating the necessary plastics and reflective coatings from indigenous sources on Mars.

AAS 93-856
OPERATION OF A REAL-TIME BIOLOGICAL LIFE SUPPORT SYSTEM LESSONS FROM BIOSPHERE 2
M. Nelson, W. F. Dempster, N. Alvarez-Romo, G. Hudman, T. MacCallum, S. Silverstone and M. Van Thillo, Space Biospheres Ventures
Since 26 September 1991, the Biosphere 2 facility has operated with a crew of eight people. The system is nearly airtight, with a current air exchange rate of less than ten percent per year. All water, human and domestic animal wastes are recycled within the structure. An agricultural area of approximately twenty thousand square feet has produced about 90 percent of food during the first year of closure (the rest was storage previously grown inside the Biosphere 2). Ecosystems ranging from rainforest to coral reef are contained within the facility and participate in the nutrient cycles and harnessing of solar energy for biomass production. A significant development was the decline of atmospheric oxygen concentration since closure. Additional oxygen was injected in January 1993 to boost atmospheric oxygen levels from 14 to 19 percent.

This paper will highlight some of the initial results from this experiment which is the longest human closure in a biological life support system. These include: strategies for modulating fluxes of critical environmental parameters such as atmospheric carbon dioxide, labor requirements for major tasks; real-time monitoring, control and communications of critical variables; the crop diversity required to overcome unforeseen pest and disease infestations; supply of a nutritionally adequate and satisfying cuisine; acceleration and seasonality of atmospheric fluxes; performance of analytic systems inside a closed system; and discussion of options on performing analyses inside vs. exporting from the closed system.

15. LIFE SUPPORT SYSTEMS

AAS 93-857
BENEFITS FROM NASA LIFE SUPPORT TECHNOLOGIES
D. R. Roberts and M. Averner, Lockheed Engineering and Sciences Company

As human exploration missions in space grow longer up to 3 years for a mission to and from Mars and occur at greater distances from Earth, life support and environmental monitoring systems, with long-term efficiency, will be critical to the health and well-being of the human crewmembers. Closed-loop life support systems will be required to eliminate the need for costly and logistically complex resupply. Researchers in the Life and Biomedical Sciences Divisions Controlled Ecological Life Support System (CELSS) Program are developing a bioregenerative life support system based upon photosynthesis that will provide food, potable water, and air by recycling waste in the closed environment of the spacecraft or planetary habitat. These advanced life support systems will not only benefit NASA, but provide valuable information as a model of another closed environmental life support system - the Earth. New technologies developed by the CELSS Program for space applications will also be valuable on Earth in our efforts to protect the environment. To reduce the build-up of toxic waste and pollution, more efficient means of waste management and disposal are required. CELSS, through recycling, requires fewer quantities of inputs to the system such as food and water, while reducing the output of wastes. NASA's technology development efforts will be used to stimulate American competitiveness through technology transfer programs, such as the proposed Mission to America. For example, NASA research has already provided a boost to the fledgling controlled environmental agriculture industry. Work by CELSS researchers has led to the refinement of techniques for growing crops in controlled environments. More nutritious and higher-quality crops are being grown hydroponically without the use of pesticides or other organic chemicals. The CELSS Program, while furthering our efforts to colonize Mars, is bringing the U.S. scientific, technological, and environmental benefits.

AAS 93-858
THE NATURAL CYCLES:
AN OVERVIEW OF A CONTROLLED ECOLOGICAL LIFE SUPPORT SYSTEM
D. J. Chamberlain, Colorado State University

A Controlled Ecological Life Support System (CELSS) also known as Bioregenerative life support, is an overview of the intricate webs of an ecosystem. The whole system can be represented by a simple idea. Primary producers (plant) photosynthesize carbohydrates, give off oxygen, and purify water by transpiration. Animals consume the carbohydrates, breath the oxygen (in turn giving off carbon dioxide for the plants) and drink the purified water. The waste produced by the animals can be transformed by decomposers into usable forms of essential elements for plants along with methane (CH_4) which can be used as a fuel source. Plants, in turn, will assimilate the nitrogen, phosphorous, potassium, and other trace essential elements produced by the decomposition.

A major boost for the whole concept is the surprisingly good crop yields produced in a controlled environment. To sustain 100 people indefinitely a crop growth area of around 5000 m^2 (the size of a US football field) and an area reserved for crew living space and resource recovery is required. It is conceivable to have a prototype CELSS small enough to sustain a small crew. As this system grows the capacity to buffer undesired chemicals increases. Therefore, this system could be used on a very large scale. This paper is a general look at an *entire* CELSS, mapping how each subsystem could be interconnected to other subsystems. A knowledge of a map of the systems is vital to the engineering and research perspectives working on a CELSS.

AAS 93-859
A LIFE SUPPORT SYSTEM ARCHITECTURE FOR THE MARS SURFACE MISSION
Andrew A. Gonzales, Lynn D. Harper, Elizabeth Dunsky, NASA Ames Research Center
Barney B. Roberts, NASA Johnson Space Center

This paper will propose a Life Support System (LSS) architecture for the initial crewed Mars exploration missions. The Mars LSS architecture is designed to meet habitation and safety mandates and enhance mission capability, while significantly minimizing the logistics burden and mission cost. Embedded in the architecture's supporting strategy is a key experiment to demonstrate the potential for extending human habitation to other worlds. Three major elements of a Mars LSS are discussed: (1) In situ resource utilization for oxygen and water generation; (2) regenerative systems for air and water recycling, and food production; and (3) caching techniques for life support consumables to provide redundancy and enable Mars base expansion with minimum additional resources from Earth. A combination of Bio-regenerative and Physical/Chemical technologies is proposed in order to take advantage of the unique capabilities of each type of system and to allow sharing of demands and resources. A technology assessment of today's state of the art is summarized to evaluate the feasibility of successfully implementing this strategy.

AAS 93-860
A NOVEL APPROACH TO WASTE REMEDIATION IN CONTROLLED ENVIRONMENTAL SYSTEMS UTILIZING PHOTOTROPHIC BACTERIA
Angela Asher and Dennis Ray Schneider, Micro-Bac International, Inc.
A new waste treatment system based upon the photoassimilation of organic compounds has been developed. Various strains of phototrophic bacteria were selected and adapted for optimal waste degradation parameters. These strains were combined and grown in a prototype waste treatment cell. This prototype could be operated for over 60 days and was capable of achieving a greater than 95% reduction of wastewater parameters. The system was also shown to degrade a variety of hydrocarbon compounds associated with hazardous waste parameters. Generation of potentially toxic gases was minimal or non-existent and there was little or no oxygen requirement for operation. Biomass from the system was shown to support germination and growth of a variety of food plants. Energy requirements for waste processing were considerably lower than conventional or physico-chemical methods. The system developed represents a new concept in waste treatment with significant advantages over existing anaerobic and aerobic systems. This work illustrated the need for more innovative approaches to ecosystem management and greater understanding of biosystem recycling mechanisms to accommodate long duration space missions and to establish self-sufficient settlement operations on Mars and other extraterrestrial bodies. The understanding of such mechanisms will be essential to establishing planetary scale biospheres.

AAS 93-861
SYSTEMS FOR MARS EMU DESIGN
George W. Herbert, Retro Aerospace
A research project at UC Berkeley, in conjunction with NASA Ames RC and Stanford University, analyzed systems requirements for Mars EMU (Spacesuit) design. This paper will report some of the results of that work, including overall suit design considerations, interaction with the Mars environment, passive thermal control options, system design options, and other work.

AAS 93-862
SPACE SUITS AND LIFE SUPPORT SYSTEMS FOR THE EXPLORATION OF MARS
Lawrence H. Kuznetz, NASA Ames Research Center
The manned exploration of Mars looms as a distinct probability within our lifetimes. When mankind finally plants his foot on the first true planet in the solar system other than our own, it will be inside a space suit entirely different from any that has preceded it. The gravity field, environment and conditions on Mars render all previous designs useless. This paper explores the requirements and technologies needed and proposes alternative concepts for space suits and portable life support systems for the exploration of Mars.

STATEMENT OF PROBLEM: The Martian surface is substantially different from any that man has visited. It is instructive to compare it with previous and current manned space environments. Such a comparison is shown in Table 1. From an examination of these data, it is clear that Mars presents a unique challenge. It has an atmosphere, essentially CO_2; it has substantial pressure at the surface, yet not enough to obviate a space suit; it has a temperature range that varies from sub-Arctic to near tropic; thermal and cosmic radiation hazards exist; winds, seasons and storms prevail; the

Mars day is just over 24 hours; the topography is diverse and demands high mobility and so on. Contemporary space suits and their life support systems for the Space Shuttle and *Freedom* Station Programs have not been designed to operate in the diverse Martian environment. A fresh design approach is required, specific to Mars, and enabling technologies must be identified.

Table 1
ENVIRONMENTAL COMPARISONS

Item	Mars	LEO	Lunar	Earth
Temp Range (degC)	-140 to 30	-160 to 150	-172 to 150	-50 to 40
Radiation Flux (solar constant W/atm	0.0615	0.1387	0.1387	0.1387
Top Winds (m/s)	15	0	0	10
Atm.Composition	95% CO_2	none	none	80% N_2
Gravity (G's)	0.38	0	0.166	1
Terrain Variance	High	none	Low	Med-High
Needed Leg Mobility	High	none	Med	High
Micrometeorite Danger	Low	Med	Med	Low
Geologic Activity	Some	none	none	Some
GCR Shielding (g/cm_2 AL)	27	2	4	
Weather	Storms, Seasons	none	none	Rain Seasons
Day/Year	24.5 hr, 656 Sols	-	-	24 hr, 365 Sols
Atm. Pressure (atm)	1/100	0	0	1

AAS 93-863
PASSIVE COOLING OF MARS EMU
Frank Crary, University of Colorado

Earlier studies have shown the need for extremely light EMU (spacesuits) on Mars and that the mass requirements are possible if passive cooling were possible on Mars (as opposed to the Moon or Earth orbit where highly variable environments preclude this). Using a thermal model similar to the "41-node" model, the present paper shows that passively cooled EMU are feasible on Mars given and demonstrates this with examples of thermal designs tailored to the Viking 1 and 2 lander sites.

16. NEW TECHNOLOGY

AAS 93-864
MARTIAN IONOSPHERIC DENSITY ANALYSIS SYSTEM (MIDAS)
Robert J. Yowell, NASA Johnson Space Center
C.D. Fry, United States Air Force

Future human and robotic precursor missions to Mars will desire the greatest mobility for extracting data over large areas of the Martian surface. During the Apollo lunar missions, an astronaut's mobility was not only limited by the amount of oxygen contained in a life support system, but also by the range of a UHF radio transceiver. The moon, having no atmosphere, required all surface communication to be line-of-sight, greatly restricting the distance an astronaut could venture. On Mars, its atmosphere, specifically its ionosphere, provides a natural "radio mirror" much like Earth's. Future networks of unmanned surface landers as well as multiple groups of human explorers can take advantage of the Martian ionosphere to provide a relay for data and voice over very long distances, without having to wait for a network of "areostationary" satellites to be assembled.

The existence of a Martian ionosphere has been known since the first radio occultation experiments conducted on Mariners 4, 6 and 7. These spacecraft measured electron density profiles at various altitudes of the Martian atmosphere, showing a definite layer of ionization, varying in density between day and night. These profiles matched closely with those of the terrestrial F1 layer, and were also seen in subsequent American and Russian missions. What has not been currently determined however, are the exact radio propagation characteristics of the Martian ionosphere. The only manner in which one could conduct a bottom-side survey of the Martian ionosphere, would be to deploy several stationary radio transceivers on the surface, located thousands of kilometers apart, at various latitudinal locations.

The MIDAS mission would deploy ten landers on the Martian surface, each equipped with a 50-100 Watt HF radio transceiver, capable of transmitting and receiving frequencies between 3 and 10 MHz. Each lander would have its own onboard telemetry processor and S-band transmitter, as well as a 20 meter tall deployable dipole antenna. Each lander would be powered by a small RTG.

Individual unique frequencies will be assigned to each pair of landers. These prestored frequencies will be used by the onboard transceiver to operate in alternating transmit and listen modes. The protocol will be for the transmitting lander to send a 10 Hz tone to the receiving lander. The receiving lander will digitally record the signal, time-tag it, and assign a five digit SINPO code to the signal. This SINPO code (Signal strength, Interference, Noise, Propagation, and Overall), will provide an easy way to analyze the telemetry data on Earth. The transmit and listen function would occur at least once an hour. The recording of each signal, the time, and SINPO code would be relayed to Earth via the S-band transmitter in a combined voice and telemetry Bi-phase-L packet of 2 kbps.

By continuing these broadcasts through Martian seasonal changes, as well as during varying periods of solar activity, the MIDAS experiment should provide a comprehensive database of Martian radio propagation conditions for future human exploration.

AAS 93-865

SOLVING THE DATA DILEMMA FOR SMALL MISSIONS TO MARS

Alan J. Willoughby, Analex Corporation
Denise S. Ponchak and John E. Zuzek, NASA Lewis Research Center

The value of any planetary mission is proportionate to the information which it returns to earth. In our desire to lower the cost of robotic exploration, we have been designing much smaller spacecraft and eliminating communications orbiters from our plans. Such small landers cannot readily support large antennas and are often power limited. Thus, over the challenging distances from Mars to Earth, only a trickle of data would be returned from such systems. Certainly, there could be a paucity of images to excite the public. And certainly also, there could be a great sacrifice in coverage, resolution and science value to the planetary science community.

It is not enough to deliver payloads and instruments, we must also send back answers! To enable data rich Mars missions within the confines of "smaller" and "cheaper," we are investigating the innovative combination of modern communications technology with kilowatt-class solar electric propulsion (SEP). The data dilemma is solved through three advantages of SEP: the excellent payload fraction delivered, the dual use of our vehicle as a transporter and orbiter, and the surplus power after arrival at destination.

For orbital science, the trans-Mars SEP stage, the science platform and the communications relay can be a single, power-rich, vehicle. For the other case of science from Mars' surface, the SEP stage can be the foundation of a communications orbiter after its delivery duties are complete. In either case, the orbiter becomes a robust communications link once a larger antenna and more robust communications subsystems are added. Cost savings are realized because it is not necessary to build and transport a separate complete communications relay vehicle, and because the SEP stage has excellent payload fraction. Because of it's maneuverability and excess power, our multi-function orbiter has capabilities beyond the ordinary communications orbiter. Anticipated long orbiter life also opens the opportunity for the vehicle to support subsequent and multiple landers and rovers.

This presentation will cover work in progress on some specific, nonoptimized cases investigated thus far.

AAS 93-866
ADAPTABLE SOFTWARE NEEDED FOR MARS MISSIONS
Ned Chapin, InfoSci Inc.

Mars missions will require more software adaptation to be done than has been true of space missions generally. Some of this adaptation will arise from the greater role of software in the missions, and some from the need personnel will experience to change mission tasks, functions, and resources on short notice. Software adaptation often will be on the critical path for Mars missions. Hence, in these times of lean funding and high demand for performance, Mars missions will stand to benefit from adaptable software. Drawing from accumulated experience, a fresh approach points to nine ways for use in Mars missions to improve software adaptation: 1—organize the software-hardware components as a network of nodes; 2—provide uniform human interfaces in limited variety; 3—transparently and routinely store data at two or more nodes; 4—allow transparent message response; 5—link command functions to personnel rather than to nodes; 6—provide common software for common tasks and functions; 7—use a uniform on-line help and documentation procedure; 8—actively manage the software; and 9—prefer adapting existing software to writing new software. While these can be effective for Mars missions when taken in combination, six barriers rooted in tradition stand as hurdles: 1—no suitable software exists; 2—the software is too difficult to use; 3—the hardware is different; 4—the software does not fit the task, function, or environment; 5—it is not good for the organization; and 6—collaborative work demands better communication. Success in improving software adaptation will be critical in the success of Mars missions. Being able to change mission tasks and functions at low cost, quickly and accurately, can enhance effectiveness and accomplishment for Mars missions.

17. MISSION STRATEGY

AAS 93-867
PROJECT EMA: A NEW PLAN FOR THE EXPLORATION OF MARS
Philip Turek, Orange County Space Society

Project EMA is based on the premise that Apollo and SEI failed because their mission planners overlooked the importance of building and sustaining the direct involvement of key elements of society in the exploration effort. A new Mars mission model has been developed, one that incorporates the participation of large segments of an earthbound population. Specific elements targeted are: the US Congress, environmentalists, and students from K through college. An indispensable feature of the EMA Mars exploration mission is the inclusion of the formal exploration of Earth, both before and after the manned time spent on the surface of Mars. This mechanism allows the public to make a direct connection between their everyday lives and the otherwise remote subject of space exploration.

AAS 93-868
EPIC - AN APPROACH TO AN INTEGRATED MANNED SPACE EXPLORATION PROGRAM
Marcus S. Murbach, NASA Ames Research Center

An integrated space program is suggested which attempts to optimally link the selection of a heavy lift launch vehicle, a large space station, and finally, Lunar and Mars missions. The first step is shown to be the selection of the in-line Shuttle Derived Vehicle (SDV) which can provide the necessary and inexpensive near-term heavy lift capability, This is used to launch large sections of a space station, the major section of which is a Supermodule which incorporates the significant pressurized volume element. The utility of the in-line SDV is further illustrated by the development of a propellant transfer technique which facilitates the integration of multi-launch Lunar and Mars missions. This is shown to allow relatively simple on-orbit integration without having to resort to the development of a "magnum class" heavy lift launch vehicle. The resulting mission vehicles are monolithic structures which incorporate Supermodule derivatives for surface and transit pressurized volume elements. The approach is intended to minimize system, and therefore organizational, interfaces with large potential savings in overall program cost. Finally, the resulting mission architectures are shown to be amenable to international cooperation.

AAS 93-869
SPACESHIP PROPULSION BY MOMENTUM TRANSFER
Christopher Willis

For years space scientists the world over have sought a propulsion system which could significantly reduce the time needed for interplanetary travel. My colleague, Albert Wu, and I have discovered such a propulsion system. This scheme works by transferring momentum from a series of "smart" projectiles launched from an electromagnetic projectile launcher located at the North pole of the Moon, to the spaceship by catching the projectiles with another electromagnetic projectile launcher on board the spaceship. A spaceship with a crew of 1000 can now travel to Mars in under two months and to Jupiter in under a year.

The prime mover is nuclear power which converts mass to energy according to Einstein's equation and drives the EMPL's which launch the projectiles. As these fast moving projectiles are caught by the spaceship's EMPL, their momentum will be transferred to the spaceship according to Newton's third law of motion. This reaction will accelerate the spaceship in the desired direction by an amount proportional to the ratio of the mass of the projectile to the mass of the spaceship. As each projectile is caught, the combined mass will be increased by the mass of one projectile. This means that some of the momentum of subsequent projectiles will be used to accelerate projectiles that have already been caught. That energy is wasted. For that reason, the projectiles will be launched toward the spaceship in groups - each projectile separated by the time and distance required by the EMPL on board the spaceship to charge up its capacitors to catch the next projectile. Once the spaceship has caught all of the projectiles in one group, it will proceed to launch them again in the direction opposite to the desired trajectory of the spaceship. In this manner each projectile will contribute two small acceleration pulses to the spaceship.

In order to accelerate the spaceship to cruising velocity it will be necessary to throw several thousand projectiles from the Moon to the spaceship (assuming none is reused). Since each individual projectile will be separated by about two minutes and each group by several hours, this process will take several days. Thus the Moon will rotate significantly during the launching of the projectiles - perhaps as much as 45 degrees. The tilt of the orbital axis of the Moon relative to a normal to the ecliptic plane may prevent launching of projectiles during part of the Moon's orbit, but this will be only a small percentage of the time and will not interfere with our plans.

The spaceship itself will have three major components: an electromagnetic projectile launcher/catcher, a nuclear power system to provide the electricity to drive the EMPL (and everything else), and the crew's quarters. The entire spaceship will be rotated axially about the long axis of the EMPL to provide gyroscopic stabilization and artificial gravity for the crew. In order to make the artificial gravity relatively uniform, the crew's quarters will be constructed in the shape of a ring. The anticipated mass of the spaceship will be 3000 metric tons. This includes 1900 tons for the EMPL (317 kg/meter), 100 metric tons for the nuclear power system (0.2 tons/Mw), and 1 metric ton per crew member for each of the 1000 crew members. Due to the artificial gravity, the crew will be able to move about relatively normally - to eat, to sleep, and to accomplish simple bodily functions normally. In addition, human muscles should not atrophy and extensive daily exercise programs will not be necessary. Perhaps most importantly, the artificial gravity should be sufficient to inhibit the bones of the crew from decalcifing - a known and very dangerous consequence of weightlessness.

The first expedition to Mars can be sketched out in the following paragraphs. Using materials derived from the Moon, a 10 kilometer EMPL will be built at the North pole of the Moon. From there it will be possible to launch objects into the plane of the orbit of the Moon and to either of the Lagrangian points, L4 or L5. Again using materials derived from the Moon, a spaceship will be assembled in orbit around the Earth at L4 or L5. This will be an unmanned spaceship consisting primarily of a 6 kilometer EMPL and a 500 Mw nuclear power system to drive it. The fissionable material for the nuclear power system will be supplied from Earth. This spaceship will carry a cargo of several thousand "smart" projectiles, which will be produced on the Moon and lifted to the spaceship. The ship will break away from the Earth's gravity by launching a few hundred of the projectiles in the opposite direction. Next, the Lunar EMPL will launch more projectiles toward the spaceship. These will be caught and saved by the spaceship, but they will accelerate the ship on its way to Mars. After a leisurely trip it will approach Mars. By launching several hundred of its projectiles, the ship will enter a circular orbit about Mars. There it will locate and match orbits with Phobos, the larger satellite of Mars. It will then attach itself to Phobos and await the arrival of the next spaceship.

The second spaceship will be assembled in a similar manner at L4 or L5. But this ship will have a crew of 1000, who will be lifted from Earth by one of the various ground to space planes currently being developed such as Hotol or Spacebus or Space Van or the German Sänger project. They will be lifted from the *Freedom* space station to L4 or L5 with the same space plane.

A few hundred projectiles will also be carried on board the spaceship. They will be used to break out of Earth orbit as the first ship did. Once on its way, the Lunar EMPL will begin launching groups of projectiles toward the spaceship. It will require about 160 projectiles to increase the velocity of the ship by one kilometer per second. After the ship catches and relaunches 20 groups, the ship's velocity will be about 19 kilometers per second. Cruising time to Mars will be about 41 days. When the second ship is still several days from Mars, the first ship will launch projectiles to slow down the second ship. It will then be maneuvered into orbit around Mars. Actual landing on Mars and return to Martian orbit will be accomplished with the usual LOX-LH2 produced from water present on Phobos.

The return trip will be similar, beginning with breakaway from Martian orbit and the launching of more groups of projectiles from the EMPL now on Phobos. As the returning ship approaches the Moon, projectiles will be launched from the Lunar EMPL to slow down the ship and help maneuver it into high Earth orbit again.

Note that Phobos has a very low gravitational field so that the first ship will be able to attach itself to Phobos without damage. Conversely, Phobos is large enough that the launching of several thousand (or even millions) projectiles will not disturb its orbit.

AAS 93-870
MARS MISSION DESIGNS: COMPARING THE NEAR TERM OPTIONS
Malcolm A. LeCompte, Aerodyne Research, Inc.
Julie P. Stets
The benefits to be derived from manufacturing chemical rocket propellants from native resources on the Martian surface are so great that it will almost certainly be a component of any future manned Mars exploration program. Different Mars mission designs recently proposed are examined to illustrate their salient features. A simple analytic tool has been developed to compare different mission approaches including the impact of in-situ propellant production. Quantitative criteria to evaluate competing approaches to achieve global Mars exploration are proposed. The immediate advantages of each design option, and its impact for long term programs will be compared. The prospects for, and benefits to be derived from, an evolutionary program of propellant manufacture will be discussed.

AAS 93-871
GETTING TO MARS WITHOUT A SATURN X
Benton C. Clark, Planetary Sciences Laboratory, Martin Marietta Astronautics
Human missions to Mars have continually faced the stumbling block of the extremely high masses involved. A roundtrip for sending astronauts to the moon requires less than 150 tonnes of hardware and propellants launched into space. This was accomplished by a single launch of a Saturn V. Missions to Mars invoke starting masses of between 250 and 1500 tonnes just to transport and support its 0.5 tonnes of human occupants. The exact number depends in complex ways upon the mission architecture, scenario, and technology content. Critical factors include number of crew members, trajectory type, trip times, staging strategy for cargo and crew, nuclear propulsion vs. the more conventional chemical propulsion, and the option of manufacturing chemical propellant at the destination. Longer trips require more mass because of the increased life support requirements. Shortening

the trip also increases mass because the propulsion must be stretched to accommodate the energetics of the trajectory.

Although a fully loaded Shuttle has a mass of 104 tonnes, the lift capability for cargo is currently approximately 20 tonnes whether it is mounted in the Shuttle cargo bay or on a Titan IV. Thus from one to six dozen Shuttle launches would be required a single mission to Mars, with an attendant launch cost of $5 to 30 B. The obvious solution has seemed to be to commence development of the long-awaited Heavy Lift Launch Vehicle (HLLV). However, it has finally become manifestly apparent that because such a vehicle has very little use other than for Mars missions an HLLV of the 250 t class is simply an unreal objective for the next several decades, if ever.

The obvious answer is assembly of the Mars mission vehicle from several payloads launched into orbit. This solution has been rejected numerous times in favor of single "all-up" launches, in the manner of Apollo. It will be argued here that from the standpoints of reliability and mission safety it is imperative that the Mars mission vehicle be operated and tested extensively while still in orbit around Earth. It is also pointed out that assembling components in orbit is in many respects superior to launching a single stack on a very large rocket.

It has already been two decades since the Saturn V was abandoned. Shuttle-derived cargo vehicle would lift only about 68 tonnes of useful payload and the recurring costs for such vehicles would be extremely high. The only HLLV which currently exists is the Energiya, with a launch payload of 100 t and growth paths to 150 t or so of life capability. If its production capability can be preserved, the rocket roadblock to the human exploration of Mars can be avoided.

AAS 93-872

A NEW DECADE OF PLANETARY EXPLORATION, A PEACE DIVIDEND

Larry Colin, Stanford University

Lew Franklin, Center for International Arms Control, Stanford University

As part of the START II treaty the U.S. with Russia will decommission over 2000 ICBM's. Joint studies with U.S. and Russian military have defined the cost and capability to use many of these vehicles for peaceful exploration of the solar system. Generic spacecraft, able to be manufactured to carry a variety of scientific payloads, are available from several U.S. firms, at costs ranging from $20 Million to $40 Million each.

Improvements in the size, cost, and power consumption of scientific instruments make the science packages for multi-launch projects feasible and affordable within current budget allocation for space. Laser transmitters and relatively low cost receivers (small telescopes) that can be placed at scientific institutions allow receiving and analyzing data from multiple space probes.

A proposal is presented to initiate a decade of planetary exploration as a part of the peace conversion process. A set of missions to the inner and outer planets, solar and galactic observatories and a network of gravity probes are described and priced. The set includes precursor Mars missions including MESUR probes and surface robots as logical extensions of the U.S. Mars Observer and Russian Mars '94 '96 missions.

The budgets are similar to the current unmanned space science expenditures but cover approximately 20 times as many missions through the use of generic spacecraft *and* decommissioned military rockets. Through international participation, partial solutions to the balance of trade for Russia and the U.S. is accomplished and stability is brought to the aerospace industries, reducing the danger of arms dissemination.

18. SPACECRAFT DESIGN

AAS 93-873

MICROMARS: A LIGHT UPPER STAGE

George W. Herbert, Retro Aerospace

As the scientific utility of microsat-class explorers becomes more evident, a wide spectrum of launch options for small Mars exploration vehicles should be considered. One such option is a single launch on a Pegasus-class vehicle with a custom upper stage for Mars transfer orbit injection. A preliminary design for such a transfer stage is presented, along with some proposed miniature Mars probe vehicle designs.

AAS 93-874
MARS EXPLORATION VEHICLE
Sergi Stoiko and Steve Merrihew, Stanford University

The Mars Exploration vehicles to send a crew to Mars as early as 2009 have been defined by a U.S.-Russian team working at Stanford University. The Mars Transfer Vehicle, to carry the explorers on the 8 to 9 month journey to Mars and return them a year and a half later, uses existing technologies available from the world's space powers. Two Energia heavy launch vehicles place the vehicle in Earth staging orbit where it refuels and then starts on its mission.

The Mars Exploration Vehicle is a biconic aerodynamic entry vehicle which will use Mars' atmosphere to decelerate on arrival. Seven meters in diameter at the base and 18 meters high, the ship has four working decks, a water-shielded radiation shelter and a small control deck and escape-return module. It provides artificial gravity at 3/8 of Earth gravity by spinning at 3 1/2 revolutions per minute about a tethered 15 ton counterweight. The Mars mission is supported by Cargo Vehicles, similar in size to the Mars Transfer Vehicle, which deliver supplies to the surface base, and return fuel and Mars landing and Ascent Vehicle to Mars Orbit. The supply vehicles are sent to Mars in 2007 before the crew flight.

The first flight of the large crew vehicle can occur in 2003, launched to low earth orbit by a single Energia rocket. This first system flight can provide all the mission goals of an international space station with crew transport and resupply being provided by both the Russian Voshod and Proton Rockets and the American Space Shuttle. Because of simplicity in design made possible by the Energia launch vehicle, the space station cost launched is about one half the cost of the U.S. Space Station. The development of the Mars Transfer Vehicle and Station flight represents about half of the full Mars mission.

19. PROPULSION AND TRANSPORTATION SYSTEMS

AAS 93-875
MAGNETOBRAKING FOR MARS RETURN VEHICLES
Geoffrey A. Landis, Sverdrup Technology, NASA Lewis Research Center

It is often proposed that a vehicle returning from Mars will use aerobraking in the Earth's atmosphere to dissipate hyperbolic excess velocity to capture into Earth orbit. Here it is proposed that a different scheme, magnetobraking, be used. In magnetobraking, an electrodynamic tether is deployed from the spacecraft. The Earth's magnetic field produces a force on electrical current in the tether. If the tether is oriented perpendicularly to the Earth's magnetic field and to the direction of motion of the spacecraft, force produced by the Earth's magnetic field can be used to either brake or accelerate the spacecraft without expenditure of reaction mass. The basic equations for magnetobraking are analyzed. The peak acceleration on the Mars return is 1.5 m/sec^2, and it is shown that the amount of braking possible is dependent on the density and current-carrying capacity of the tether, but is independent of length. Superconducting and non-superconducting tethers are analyzed.

Since energy is produced as the spacecraft velocity decreases, no on-board power source is required. A difficulty in magnetobraking is dissipating the energy produced. As the spacecraft approaches the Earth, the magnetic field increases and the power produced by the tether increases, reaching a maximum of about 150 kW per kilogram of spacecraft mass at closest approach. If the tether current is deceased during closest approach, this peak can be reduced, at the cost of reducing the braking effectiveness.

The energy produced as the spacecraft crosses the Earth's magnetic field can be used, for example, to drive an electric-propulsion engine such as an ion engine or a pulsed-inductive thruster. The power densities produced by an electrodynamic tether in the Earth's magnetic field can be orders of magnitude higher than power densities produced by other means, and hence extremely effective electric propulsion is possible. Used together, an electrodynamic tether and electric propulsion could be the most effective solution for braking a returning Mars vehicle.

AAS 93-876
EFFECT OF INERT PROPELLANT INJECTION ON MARS ASCENT VEHICLE PERFORMANCE
James E. Colvin and Geoffrey A. Landis, NASA Lewis Research Center

A Mars ascent vehicle is limited in performance by the propellant which can be brought from Earth. In some cases the vehicle performance can be improved by injecting non-reactive (inert) gas into the engine, if the inert gas is available as an in-situ resource and does not have to be brought from Earth. The Martian atmosphere contains abundant carbon dioxide, nitrogen, and argon, which could be obtained by compressing the atmosphere, without any chemical processing step. The effect of inert gas injection on rocket engine performance was analyzed with a numerical combustion code that calculates specific impulse for engines of varying combustion chamber pressure, expansion ratio, oxidizer/fuel ratio, and inert injection fraction. Results of this analysis were applied to several Mars ascent scenarios. For the missions studied, the presence of inert impurities in the propellant does not significantly decrease performance, and for some missions, the required mass of return propellant brought to Mars can be decreased by up to 13\% using inert propellant injection.

AAS 93-877
NUCLEAR THERMAL ROCKET/STAGE TECHNOLOGY OPTIONS FOR NASA'S FUTURE HUMAN EXPLORATION MISSIONS TO MARS
Stanley K. Borowski, Nuclear Propulsion Office, NASA Lewis Research Center

The nuclear thermal rocket (NTR) provides a unique propulsion capability to planners/designers of future piloted missions to Mars. In addition to its high specific impulse (850-1000 sec) and engine thrust to weight ratio (2-10), the NTR can also be configured as a "dual mode" system generating electrical power for spacecraft environmental systems, communications, and stage operations, such as refrigeration of liquid hydrogen propellant. At present NASA is considering a range of Mars mission applications for the NTR extending from a "single burn" trans-Mars injection maneuver to a "4-burn" reusable spacecraft. Each application results in a particular set of requirements in areas such as engine restart, fuel operating temperature and lifetime, cryo-fluid storage, and stage size. A variety of solid care NTR concepts are also under study at NASA's Nuclear Propulsion Office. Included among the concept options are engine designs based on (1) NERVA-derivative reactor (NDR) technology, (2) CERMET fast reactor technology, (3) particle bed reactor (PBR) technology and (4) a ternary carbide "twisted ribbon" fuel form developed by the Commonwealth of Independent States (CIS). Representative NTR concepts and vehicle designs satisfying the requirements/constraints of NASA's various Mars mission scenarios will be presented. Attractive performance characteristics and important "high leverage" technologies associated with both the engine and stage will be identified and parametric sensitivity data will be generated. The potential for "commonality" of engine and stage components with piloted lunar and "robotic" planetary science missions will also be shown.

AAS 93-878
A STUDY OF POWERED ABORT-TO-EARTH ORBITS FROM EARTH-TO-MARS TRANSFER ORBITS
Steve Merrihew, Stanford University

As a significant fraction of the duration of a human exploration mission to Mars will be devoted to the interplanetary transfer orbit, in-flight failures leading to mission degradation or termination may occur. Additionally, the failure of other mission elements such as pre-supplied fuel caches or surface habitats may preclude landing on Mars. Therefore, emergency mission operations will depend upon the failure type and the mission phase. This paper investigates powered abort to earth orbits as a mission abort mode for failures not affecting the vehicle's propulsion system. These powered abort to earth orbits are analyzed for both high and low energy nominal earth to Mars transfer orbits. This analysis includes contour plots of both the required abort orbit change in velocity (delta-V) and atmospheric entry velocity as a function of the time of abort orbit initiation and duration. These plots indicate regions of powered abort to earth orbit opportunities for any given propulsive capability. Several proposed Mars exploration mission architectures are evaluated to determine their ability to perform such maneuvers. Finally, design guidelines are presented to enable abort to earth capabilities in future Mars exploration mission architectures.

20. RESOURCE UTILIZATION

AAS 93-879
RELATIVE BENEFITS OF IN-SITU MINING ON MARS
Paulo R. Pereira and Russel J. Miller, Colorado School of Mines
Gary S. Brierley, Haley and Aldrich, Inc.
Inherent advantages of in-situ mining are identified, and include: (1) Less launch weight, (2) Less operating personnel, (3) Less operating cost, (4) Dramatic reduction in the environmental/surface impact caused by mining, (5) Less maintenance, (6) Less unit operations, and (7) No handling of raw materials on the surface. Also, since in-situ mining is generally a much simpler operation than other mining methods, extraterrestrial application will require less EVA time, be more amenable to automation, need less support form Earth, and minimize the necessary infrastructure. Surface and in-situ mining are briefly compared, unit operations identified, and a cost comparison provided. The distinct cost and operational advantages of in-situ recovery of local resources make a compelling case for considering in- situ mining on Mars.

AAS 93-880
METHANOL: A FUEL FOR EARTH AND MARS
Kelly R. McMillen, Thomas R. Meyer, Boulder Center for Science and Policy
Benton C. Clark, Martin Marietta Astronautics
To sustain a continued human presence on Earth as well as to investigate the human exploration and habitation of Mars, development of renewable fuel technologies for use on the Earth and on Mars is necessary. Methanol is an outstanding candidate for the fuel because of its storability as a low-pressure liquid, its use in fuel cells as well as internal combustion engines, and its potential for on-site manufacturability from local resources such as natural gas or atmospheric carbon dioxide. Development of advanced methanol vehicle propulsion for use in Mars exploration would furthermore, help speed the future transition to this extremely versatile alternative fuel for terrestrial automotive transportation. In this way, the benefits of on-going research and development of advanced technologies utilizing this alternative fuel will provide feedback of the space-program developments to help solve pressing human needs on Earth by developing power plants not dependent upon petroleum and with less pollutant emissions (B.C. Clark 1991). This paper investigates the progress of the production and use of methanol as a fuel.

AAS 93-881
MARS MISSIONS WITH ISRU:
AVAILABLE TECHNOLOGIES AT THE ARIZONA NASA USERC
Kumar Ramohalli, NASA Space Engineering Research Center, University of Arizona
This paper outlines the basic problems associated with Mars missions in the specific light of cost/mass reductions that are necessary for a viable program. In Situ Resource Utilization (IRSU) is presented as the natural class of technologies that can introduce this economy. The paper presents data from full-system working hardware at the University of Arizona NASA Space Engineering research Center. The importance of hardware data in making a strong case for ISRU is stressed. Fundamental component processes are discussed within a theoretical framework. The advantages of ISRU are briefly reviewed quantitatively through a Figure-of Merit for the overall missions.

Specific data to be presented include, oxygen production from a single-cell testbed that has yielded a factor of four increase after many of the details were improved, such as the tube preparation, electrode application, and catalyst purity. In addition, the reduction in size, weight, and power requirements of a disk system is also mentioned. Non-mechanical pumping of oxygen to 400 psia is reported. The spent-stream use in producing methane and higher hydrocarbons is reported. Other recent data include Martian permafrost modeling that uses fundamental constitutive relationships to predict the larger scale mechanical behavior. Experiments are conducted to characterize mixtures of soil and ice (water) at realistic Martian temperatures.

It is the purpose of this paper to present the rapid advances made in the Technology Readiness Level (TRL) of these technologies that should inspire confidence in mission planners, to include ISRU in the mainstream of future missions, so that the extraordinary promise of low cost can be realized.

AAS 93-882
COLD PRESS METHOD TO MANUFACTURE CONSTRUCTION MATERIALS ON MARS
Yoji Ishikawa, Tetsuo Sasaki, and Tetsumi Higasayama, Bio-Environmental Engineering Department, Technical Research Institute, Obayashi Corporation

Local Material utilization for construction of a Mars Base is essential in the era of Martian settlement. With water and clay minerals available on its surface, Mars offers a special opportunity to the builder for material production. The cold press method, using mechanical pressure without heating, is one of the promising material manufacturing methods on Mars. In this paper, we report the result of the cold press experiment using simulated Martian soil. The compressive strength of the final product reaches 7,000 kPa.

21. MARS BASE DESIGN AND TECHNOLOGY

AAS 93-883
MARS SURFACE MISSION DESIGN AND SAFETY PHILOSOPHY
Marc M. Cohen, Advanced Space Technology Office, NASA Ames Research Center

This paper presents an overview of the objectives, mission description, design philosophy and safety philosophy for Mars surface exploration developed by the 1992 NASA Mars study at Ames Research Center. The Mars Habitat and Architecture Working Group refined this particular philosophy to optimize the approach for surface habitat design. The three mission objectives are Science, Habitation and Resource Utilization. The Science Objectives include the search for life, search for planetary origins and the search for resources. The Habitation Objectives focus on demonstrating the capability for humans to live on the surface of Mars such that the risk level is no greater than expected from comparable Earth-based hazardous activities. The Resource Utilization Objective seeks to develop the capability to use Mars in situ resources to reduce the dependence upon Earth supply of atmosphere, water and other consumables.

The mission design philosophy reflects the assumptions that a crew of six will visit the surface of Mars for 600 days, with the first departure date in 2008. The first launch is a series of cargo landers, from which robots will unload and assemble the First Mars Outpost prior to the crew's arrival. The crew will follow on the next launch opportunity, 26 months later.

The design problem definition decomposes the mission into two major segments: getting there and being there. Mars 2008 addresses the surface portion of the Mars mission exclusively. The Mars 2008 design philosophy adopts an innovative approach to system engineering that looks at the total resource allocation, particularly crew time on Mars. The approach to reliability emphasizes the availability and maintainability of systems. By definition, reliability is the control of failures and the definition of safety is to control the loss resulting from failures.

Given this problem decomposition the safety philosophy approaches the Mars surface facilities in a manner more like all earth-based facility than a spacecraft. It recognizes that the level of risk and required reliability vary with the activities that the crew members undertake and the assured capabilities necessary to support those activities. The three levels of safety criticality and risk are Life-Critical, Mission-Critical and Mission-Discretionary. The overview includes an analysis of the detailed mission objectives in relation to the safety criticality levels.

AAS 93-884
EVA SURFACE MISSION STRATEGIES FOR MARS EXPLORATION
Theresa M. Buckley, NASA Ames Research Center
An agency-wide effort, headed by the EXPO office, was begun in late 1992 to establish and study various Mars surface mission strategies and requirements for a NASA manned mission to Mars. An advanced surface Extravehicular Activity (EVA) system design must consider EVA task scenario requirements, Mars base construction and maintenance needs and interface resources such as those available at the lander, the base or transportation aids such as rovers. Conversely, EVA surface mission strategies, or the means of interfacing man with the Mars surface for routine and productive work, can be chosen to minimize unfavorable design impacts on a surface suit. Hence, the design of a surface EVA system is driven by many factors unique to the mission and planet explored and an optimal design must consider all of these factors. This paper will illuminate the unique challenges for Mars EVA systems and, from a systems level or global approach, will describe and examine three strategies for approaching EVA on Mars. It is the intent of this paper, firstly, to illustratively explore how one might integrally optimize an EVA system with the broader based Mars mission goals. Secondly, it is intended to offer new thinking on approaches to EVA systems for Mars or other planetary surfaces.

AAS 93-885
PROPOSAL FOR A FIRST MARTIAN OUTPOST AND PERMANENT MARTIAN BASE
Gary T. Moore, Janis Huebner-Moths, Kerry L Paruleski, and Patrick J. Rebholz,
University of Wisconsin-Milwaukee
This paper will discuss human factors/environment-behavior issues and design solutions regarding the design of the first outpost and permanent base on the surface of Mars. The work was done in support of the 1991 Stafford et al. *Synthesis Report*, and in particular two of its scenarios—Architecture 1 and Architecture 4—and the NASA-JSC Weaver ExPO report on near-term Martian extraterrestrial explorations [1]. While the project investigated the implications of different mission scenarios, the Martian environment, supporting technologies, and other issues, attention in this paper will be focused on human factors and environment-behavior considerations for the design of the first Martian outpost and permanent base. This paper will present the results of that investigation. It will summarize habitability design requirement based environment-behavior research. It will also illustrate a full concept design and design development for the outpost and base. The design shows an evolutionary scheme that begins with a First Martian Outpost (FMO) and subsequent expansion into a First Martian Base for 18 crew members including all laboratory, mission control, and crew support spaces [2].
References:
[1]. T.P. Stafford (1991), *America at the Threshold: Report of the Synthesis Group on America's Space Exploration Initiative*, Washington, DC.
[2]. J. Huebner-Moths, P.J. Rebholz, J.P. Fieber, and K.L. Paruleski (1992), *Pax Permanent Base for Mars: Space Architecture for the First Human Habitation on Mars*, University of Wisconsin-Milwaukee.

AAS 93-886
CONSTRUCTION METHOD AND PROCEDURE FOR MARS BASE
Yoji Ishikawa, Owen Gwynne, Yukinobu Yamamoto, Hisashi Ueda, and Thomas Bongiovi,
Bio-Environmental Engineering Department, Technical Research Institute, Obayashi Corporation
One of the major tasks the travelers to Mars confront is habitat construction on Mars. Both high cost and weight limitation of the interplanetary travel require careful planning in advance and ceaseless support of a series of manned Mars missions. In this paper, we determine a construction strategy and sequences, essential on-surface components, and construction machines needed to prepare such infrastructure. Considering the estimated time requirement, we found that a majority of the work will be completed within a reasonable time scale. However, the soil covering process requires considerably longer time, assuming the use of conventional types of construction machines. Alternative types of machines are required for this type of work.

AAS 93-887
DESIGN AND CONSTRUCTION FOR A PERMANENT PRESENCE AT MARS
Ethan W. Cliffton

Draglines, batch plants, concrete pumps, tilt-up and lift-slab technology will be expanded to utilize flexible formwork in the production of safe, repeatable, and permanent buildings for the expanding scientific community at Mars. A mature base will begin to test locally available erz and sulfur cements as candidates for use in this adaptive concrete construction. The incremental addition of flexible formwork to the lexicon of concrete construction technology will reduce imports and lead to more elaborate structures over time. Validating this technology in small scale habitats is essential to perfecting the framework necessary for large buildings intended to integrate habitation, research, and biotic environments. Permanent buildings will represent the third phase in the evolution of Mars from a scientific outpost to a mature research station.

AAS 93-888
INFLATABLE STRUCTURES FOR A MARTIAN BASE
Willy Z. Sadeh and Marvin E. Criswell,
Center for Engineering Infrastructure and Science in Space (CEISS)

The first steps in the human expansion into space, which inevitably will occur during the next century, consist of the establishment of a human-tended base on the Moon followed by a human mission to and a human-tended base on Mars. Structures for both lunar and Martian bases face, to a large extent, similar requirements and design criteria despite the differences between the corresponding environmental conditions. Consequently, a unified approach to both lunar and Martian structures, with the Moon as the testbed, can result in higher efficiency and significant economy in design and construction. In both cases, the internal pressure is the major load applied since the external pressure is essentially absolute vacuum on the Moon and negligible on Mars. The reduced gravity on the Moon and Mars induces dead loads from the material weight and from a protective regolith layer for radiation shielding and micro meteoroid impact of about the same order of magnitude on both bodies. An inflatable structure made of a thin membrane is highly efficient in resisting the dominant internal pressure loading and dead loads, and can be designed to meet all the functional requirements. A generic modular inflatable structure that can be used on both the Moon and Mars is proposed and described.

AAS 93-889
LIVING ON MARS
Michael Scott and Lee Weaver, Stanford University

When a crew flies to Mars in a vehicle with artificial gravity and when they land on Mars they will encounter living at 3/8 of Earth's gravity. Everyday human activity will be affected by the lower gravity. The stride explorers use in moving on the surface is different, the handling of materials in constructing habitats, and the preparation and eating of foods will be different. The crew will have to adjust to the new conditions. As members of the Stanford Mars team, Michael Scott and Lee Weaver have conducted a number of 3/8 gravity flights in the aircraft of Weaver Aerospace to test conditions to be found on Mars. The aircraft fly an arc that gives Mars gravity conditions for approximately 30 seconds. Thirty such flights have been made testing pouring of liquids, and drinking of liquids. The video taped results will be presented.

The results show that those simple functions are often quite different in the reduced gravity but can be adapted to very quickly by the subject. The heating and boiling of water, dependent on density differences to establish convection, also changes significantly but not in a way that would be hard to adjust to. Walking in Mars gravity is also under test at NASA Ames in their neutral buoyancy facility. Some tapes of the adjusted gaits will be shown. Work is continuing to determine metabolic work and to guide the design of flexible space suits for Mars.

AAS 93-890

MARS BASE PLACEMENT:
GEOLOGICAL CONSIDERATIONS OF CANDIDATE LOCATIONS

Shawn Laatsch, Bishop Planetarium

Mars is a planet that contains a number of striking geological features from the largest volcanic mountain in the solar system, to vast canyons, to dry river beds. The placement of a future manned outpost on Mars is critical to the further colonization of the planet. The base must be placed in areas which provide the best of Martian resources and optimal protection from the severe weather on Mars. This paper will examine several candidates for a future Mars base and discuss the geological benefits of each candidate. Some of the candidates include Noctis Labirinthus, Memnonia Fossae, Capri Chasma, and Mare Cinmerium. It will also suggest the exploration of these areas in detail by precursor missions in order to determine the levels of resources available at the candidate sites. Emphasis will be placed on soil fertility, availability of water and possible mineral resource deposits and how they affect a Mars base placement.

APPENDIX B

PUBLICATIONS OF THE AMERICAN ASTRONAUTICAL SOCIETY

Following are the principal publications of the American Astronautical Society:

JOURNAL OF THE ASTRONAUTICAL SCIENCES (1954 -)
Published quarterly and distributed by AAS Business Office, 6352 Rolling Mill Place, Suite #102, Springfield, Virginia 22152. Back issues available from Univelt, Inc., P.O. Box 28130, San Diego, California 92198.

SPACE TIMES (1986 -)
Published bi-monthly and distributed by AAS Business Office, 6352 Rolling Mill Place, Suite #102, Springfield, Virginia 22152.

AAS NEWSLETTER (1962 - 1985)
Incorporated in *Space Times*. Back issues available from AAS Business Office, 6352 Rolling Mill Place, Suite #102, Springfield, Virginia 22152.

ASTRONAUTICAL SCIENCES REVIEW (1959 -1962)
Incorporated in *Space Times*. Back issues still available from Univelt, Inc., P.O. Box 28130, San Diego, California 92198.

ADVANCES IN THE ASTRONAUTICAL SCIENCES (1957 -)
Proceedings of major AAS technical meetings. Published and distributed for the American Astronautical Society by Univelt, Inc., P.O. Box 28130, San Diego, California 92198.

SCIENCE AND TECHNOLOGY SERIES (1964 -)
Supplement to *Advances in the Astronautical Sciences*. Proceedings and monographs, most of them based on AAS technical meetings. Published and distributed for the American Astronautical Society by Univelt, Inc., P.O. Box 28130, San Diego, California 92198.

AAS HISTORY SERIES (1977 -)
Supplement to *Advances in the Astronautical Sciences*. Selected works in the field of aerospace history under the editorship of R. Cargill Hall. Published and distributed for the American Astronautical Society by Univelt, Inc., P.O. Box 28130, San Diego, California 92198.

AAS MICROFICHE SERIES (1968 -)
Supplement to *Advances in the Astronautical Sciences*. Consists principally of technical papers not included in the hard-copy volume. Published and distributed for the American Astronautical Society by Univelt, Inc., P.O. Box 28130, San Diego, California 92198.

Subscriptions to the *Journal of the Astronautical Sciences* and the *Space Times* should be ordered from the AAS Business Office. Back issues of the *Journal* and all books and microfiche should be ordered from Univelt, Incorporated.

ADVANCES IN THE ASTRONAUTICAL SCIENCES SERIES (1957-)

ISSN 0065-3438,

LIBRARY OF CONGRESS CARD NO. 57-43769

Proceedings of Major AAS Technical Meetings

Vol. 1 Third Annual AAS Meeting, Dec. 6-7, 1956, New York, NY, 1957, 184p., ed. Norman V. Petersen, Microfiche only, $20 (ISBN 0-87703-002-2)

Vol. 2 Fourth Annual AAS Meeting, Jan. 29-31, 1958, New York, NY, 1958, 440p., eds. Norman V. Petersen, Horace Jacobs, Microfiche only, $20 (ISBN 0-87703-003-0)

Vol. 3 First Western National AAS Meeting, Aug. 18-19, 1958 530p., eds. Norman V. Petersen, Horace Jacobs, Microfiche only, $20 (ISBN 0-87703-004-9)

Vol. 4 Fifth Annual AAS Meeting, Dec. 27-31, 1958, Washington, D.C., 1959, 462p., ed. Horace Jacobs, Microfiche only, $20 (ISBN 0-87703-005-7)

Vol. 5 Second Western National AAS Meeting, Aug. 4-5, 1959, Los Angeles, CA, 1960, 364p., ed. Horace Jacobs, Microfiche only, $20 (ISBN 0-87703-006-3)

Vol. 6 Sixth Annual AAS Meeting, Jan. 18-21, 1960, New York, NY, 1961, 968p., eds. Horace Jacobs and Eric Burgess, Hard Cover $45 (ISBN 0-87703-007-3)

Vol. 7 Third Western National AAS Meeting, Aug. 4-5, 1960, Seattle, WA, 1961, 464p., eds. Horace Jacobs and Eric Burgess, Microfiche only, $20 (ISBN 0-87703-008-1)

Vol. 8 Seventh Annual AAS Meeting, Jan. 16-18, 1961, Dallas, TX, 1963, 602p., ed. Horace Jacobs, Microfiche only, $20 (ISBN 0-87703-009-X)

Vol. 9 Fourth Western Regional AAS Meeting, Aug. 1-3, 1961, San Francisco, CA, 1963, 910p., ed. Eric Burgess, Hard Cover $45 (ISBN 0-87703-010-3)

Vol. 10 Manned Lunar Flight (AAS/AAAS Symposium) Dec. 19, 1961, Denver, CO, 1963, 310p., eds. George W. Morgenthaler and Horace Jacobs, Hard Cover $35 (ISBN 0-87703-011-1)

Vol. 11 Eighth Annual AAS Meeting, Jan. 16-18, 1962, Washington, D.C., 1963, 808p., ed. Horace Jacobs, Hard Cover $45 (ISBN 0-87703-012-X)

Vol. 12 Scientific Satellites - Mission and Design (AAS/AAAS Symposium), Dec. 27, 1962, Philadelphia, PA, 1963, 262p., ed. Irving E. Jeter, Hard Cover $25 (ISBN 0-87703-013-8)

Vol. 13 Interplanetary Missions, 9th Annual AAS Meeting, Jan. 15-17, 1963, Los Angeles, CA, 1963, 690p., ed. Eric Burgess, Hard Cover $45 (ISBN 0-87703-014-6)

Vol. 14 Second AAS Symposium on Physical and Biological Phenomena under Zero G Conditions, Jan. 18, 1963, Los Angeles, CA, 1963, 382p., eds. Elliot T. Benedikt and Robert W. Halliburton, Hard Cover $30 (ISBN 0-87703-015-4)

Vol. 15 Exploration of Mars Symposium, Jun. 6-7, 1963, Denver, CO, 1963, 634p., ed. George W. Morgenthaler, Hard Cover $45 (ISBN 0-87703-016-2)

Vol. 16 Space Rendezvous, Rescue, and Recovery Symposium, Sept. 10-12, 1963, Edwards, CA 1963, 1408p., ed. Norman V. Petersen, Hard Cover, Part 1, 1028p., $45 (ISBN 0-87703-017-0); Part 2, 380p., $30 (ISBN 0-87703-018-9)

Vol. 17 Bioastronautics - Fundamental and Practical Problems (AAS/AAAS Symposium), Dec. 30, 1963, Cleveland, OH, 1964, 128p., ed. William C. Kaufman, Microfiche only, $10 (ISBN 0-87703-019-7)

Vol. 18 Lunar Flight Programs, 10th Annual AAS Meeting, May 4-7, 1964, New York, NY, 1964, 630p., ed. Ross Fleisig, Hard Cover $45 (ISBN 0-87703-020-0)

Vol. 19 Unmanned Exploration of the Solar System Symposium, Feb. 8-10, 1965, Denver, CO, 1965, 1000p., eds. George W. Morgenthaler, Robert G. Morra, Hard Cover $45 (ISBN 0-87703-021-9)

Vol. 20 Post Apollo Exploration, 11th Annual AAS Meeting, May 3-6, 1965, Chicago, IL, 1966, 1220p., ed. Francis Narin, Microfiche only, Part I, 572p., $30 (ISBN 0-87703-022-7); Part 2, 648p., $35 (ISBN 0-87703-023-5)

Vol. 21 Practical Space Applications Symposium, Feb. 21-23, 1966, San Diego, CA, 1967, 508p., ed. Lawrence L. Kavanau, Microfiche Only $40 (ISBN 0-87703-024-3)

Vol. 22 The Search for Extraterrestrial Life, 12th Annual AAS Meeting, May 23-25, 1966, Anaheim, CA, 1967, 388p., ed. James S. Hanrahan, Microfiche only $30 (ISBN 0-87703-025-1); Microfiche Suppl. (Vol. 1 AAS Microfiche Series) $12 (ISBN 0-87703-132-0)

Vol. 23 Commercial Utilization of Space, 13th Annual AAS Meeting, May 1-3, 1967, Dallas, TX, 1968, 512p., eds. J. Ray Gilmer, Alfred M. Mayo, Ross C. Peavey, Hard Cover (ISBN 0-87703-026-X); plus Microfiche Suppl. (Vol. 3 AAS Microfiche Series) $60 (ISBN 0-87703-216-5)

Vol. 24 Exploitation of Space for Experimental Research, 14th Annual AAS Meeting, May 13-15, 1968, Dedham, MA, 1968, 363p., ed. Harry Zuckerberg, Hard Cover $30 (ISBN 0-87703-027-8)

Vol. 25 Advanced Space Experiments, Sept. 16-18, 1968, Ann Arbor, MI, 1969, 530p., eds. O. Lyle Tiffany and Eugene M. Zaitzeff, Hard Cover $40 (ISBN 0-87703-028-6)

Vol. 26 Planning Challenges of the 70"s in Space, 15th Annual AAS Meeting, Jun. 17-20, 1969, Denver, CO, 1970, 470p., eds. George W. Morgenthaler and Robert G. Morra, Hard Cover $35 (ISBN 0-87703-053-7); Microfiche Suppl. (Vol. 14 AAS Microfiche Series) $20 (ISBN 0-87703-130-4)

Vol. 27/28 Space Stations (v27) and Space Shuttles and Interplanetary Missions (v28), 16th Annual AAS Meeting, Jun. 8-10, 1970, Anaheim, CA, 1970, Vol. 27, eds. Lewis Larmore and Robert L. Gervais, 606p., Hard Cover $45 (ISBN 0-87703-054-5); Vol. 28, eds. Lewis Larmore and Robert L. Gervais, 488p., Hard Cover $35 (ISBN 0-87703-055-3)

Vol. 29 The Outer Solar System, 17th Annual AAS Meeting, Jun. 28-30, 1971, Seattle, WA, 1971, ed. Juris Vagners, Part 1 (on Microfiche Only), 618p., $40 (ISBN 0-87703-059-6); Part 2, Hard Cover, 740p., $45 (ISBN 0-87703-060-X)

Vol. 30 International Congress of Space Benefits, 19th Annual AAS Meeting, Jun. 19-21, 1973, Dallas, TX, 1974, 528p., ed. Francis S. Johnson, Hard Cover $40 (ISBN 0-87703-065-0)

Vol. 31 The Skylab Results, 20th Annual AAS Meeting, Aug. 20-22, 1974, Los Angeles, CA, 1975, 1174p., eds. William C. Schneider and Thomas E. Hanes, Microfiche only (ISBN 0-87703-072-3); Plus Microfiche Suppl. (Vol. 22 AAS Microfiche Series) $60 (ISBN 0-87703-143-6)

Vol. 32 Space Shuttle Missions of the 80's, 21st Annual AAS Meeting, Aug. 26-28, 1975, Denver, CO, 1977, 1364p., eds. William J. Bursnall, George W. Morgenthaler, Gerald E. Simonson, Hard Cover, Part 1, 598p., $40 (ISBN 0-87703-078-2); Hard Cover, Part 2, 766p., $55 (ISBN 0-87703-087-1); Microfiche Suppl. (Vol. 25 AAS Microfiche Series $65 (ISBN 0-87703-133-9)

Vol. 33 AAS/AIAA Astrodynamics Conference, July 28-30, 1975, Nassau, Bahamas, 1976, 390p., eds. William F. Powers, Herbert E. Rauch, Byron D. Tapley, Carmelo E. Velez, Hard Cover $35 (ISBN 0-87703-079-0); Microfiche Suppl. (Vol. 26 AAS Microfiche Series) $40 (ISBN 0-87703-142-8)

Vol. 34 Apollo Soyuz Mission Report, 1977, 336p., ed. Chester M. Lee, Hard Cover $35 (ISBN 0-87703-089-8)

Vol. 35 The Bicentennial Space Symposium - New Themes for Space: Mankind"s Future Needs and Aspirations, 22nd AAS Meeting, Oct. 6-8, 1976, Washington, D.C., 1977, 242p., ed. William C. Schneider, Hard Cover $25 (ISBN 0-87703-090-1)

Vol. 36 The Industrialization of Space, 23rd Annual AAS Meeting, Oct. 18-20, 1977, San Francisco, CA, 1978, 1160p., eds. Richard A. Van Patten, Paul Siegler, Edward V.B. Stearns, Hard Cover, Part 1, 610p., $55 (ISBN 0-87703-094-4); Hard Cover, Part 2, 550p., $45 (ISBN 0-87703-095-2); Microfiche Suppl. (Vol. 28 AAS Microfiche Series) $15 (ISBN 0-87703-121-5)

Vol. 37 Space Shuttle and Spacelab Utilization, What are the Near-Term and Long-Term Benefits for Mankind?, 16th Goddard Memorial Symposium, 24th Annual AAS Meeting, March 8-10, 1978, Washington, D.C., 1978, 865p., eds. George W. Morgenthaler and Manfred Hollstein, Hard Cover, Part 1, 400p., $40 (ISBN 0-87703-096-0); Hard Cover, Part 2, 465p., $45 (ISBN 0-87703-097-9)

Vol. 38 The Future U.S. Space Program, 25th Anniversary Conference, Oct. 20 - Nov. 2, 1978, Houston, TX, 1979, 880p., eds. Richard S. Johnston, Albert Naumann, Jr., Clay W. G. Fulcher, Hard Cover, Part 1, 444p., $45 (ISBN 0-87703-098-7); Hard Cover, Part 2, 436p., $40 (ISBN 0-87703-099-5); Microfiche Suppl. (Vol. 30 AAS Microfiche Series) $15 (ISBN 0-87703-129-0)

Vol. 39 Guidance and Control 1979, Feb. 24-28, 1979, Keystone, CO, 1979, 492p., ed. Robert D. Culp, Hard Cover $45 (ISBN 0-87703-100-2); Microfiche Suppl. (Vol. 31 AAS Microfiche Series) $10 (ISBN 0-87703-128-2)

Vol. 40 AAS/AIAA Astrodynamics Conference, Jun. 25-27, 1979, Provincetown, MA, 1980, 996p., eds. Paul A. Penzo, Bernard Kaufman, Louis Friedman, Richard Battin, Hard Cover, Part 1, 494p., $45 (ISBN 0-87703-107-X); Soft Cover $35 (ISBN 0-87703-108-8); Hard Cover, Part 2, 502p., $45 (ISBN 0-87703-109-6); Soft Cover $35 (ISBN 0-87703-110-X); Microfiche Suppl. (Vol. 32 AAS Microfiche Series) $20 (ISBN 0-87703-139-8)

Vol. 41 Space Shuttle: Dawn of an Era, 26th Annual AAS Meeting, Oct. 29-Nov. 1, 1979, Los Angeles, CA, 1980, 980p., eds. William F. Rector, III and Paul A. Penzo, Hard Cover, Part 1, 452p., $45 (ISBN 0-87703-111-8); Soft Cover $35 (ISBN 0-87703-112-6); Hard Cover, Part 2, 528p., $55 (ISBN 0-87703-113-4); Soft Cover $40 (ISBN 0-87703-114-2); Microfiche Suppl. (Vol. 33 AAS Microfiche Series) $10 (ISBN 0-87703-136-3)

Vol. 42 Guidance and Control 1980, Feb. 17-21, 1980, Keystone, CO, 1980, 738p., ed. Louis A. Morine, Hard Cover $60 (ISBN 0-87703-137-1); Soft Cover $45 (ISBN 0-87703-138-X)

Vol. 43 Shuttle/Spacelab - The New Transportation System and its Utilization, (3rd DGLR/AAS Symposium), Apr. 28-30, 1980, Hannover, Germany, 1981, 342p., eds. Dietrich E. Koelle and George V. Butler, Hard Cover $45 (ISBN 0-87703-144-4); Soft Cover $35 (ISBN 0-87703-146-0)

Vol. 44 Space-Enhancing Technological Leadership, 27th Annual AAS Meeting, Oct. 20-23, 1980, Boston, MA, 1981, 580p., ed. Lawrence P. Greene, Hard Cover $65 (ISBN 0-87703-147-9); Soft Cover $50 (ISBN 0-87703-148-7); Microfiche Suppl. (Vol. 35 AAS Microfiche Series) $10 (ISBN 0-87703-164-9)

Vol. 45 Guidance and Control 1981, Jan. 31-Feb. 4, 1981, Keystone, CO, 1981, 506p., ed. Edward J. Bauman, Hard Cover $60 (ISBN 0-87703-150-9); Soft Cover $50 (ISBN 0-87703-151-7); Microfiche Suppl. (Vol. 36 AAS Microfiche Series) $15 (ISBN 0-87703-156-8)

Vol. 46 AAS/AIAA Astrodynamics Conference, Aug. 3-5, 1981, North Lake Tahoe, NV, 1982, 1124p., eds. Alan L. Friedlander, Paul J. Cefola, Bernard Kaufman, Walt Williamson, G.T. Tseng, Hard Cover, Part 1, 552p., $55 (ISBN 0-87703-159-2); Soft Cover $45 (ISBN 0-87703-160-6); Hard Cover, Part 2, 572p., $55 (ISBN 0-87703-161-4); Soft Cover $45 (ISBN 0-87703-162-2); Microfiche Suppl. (Vol. 37 AAS Microfiche Series) $40 (ISBN 0-87703-163-0)

Vol. 47 Leadership in Space - For Benefits on Earth, 28th Annual AAS Meeting, Oct. 26-29, 1981, San Diego, CA, 1982, 310p., ed. William F. Rector, III, Hard Cover $45 (ISBN 0-87703-168-1); Soft Cover $35 (ISBN 0-87703-169-X)

Vol. 48 Guidance And Control 1982, Jan. 30 -Feb. 3, 1982, Keystone, CO, 1982, 558p., eds. Robert D. Culp, Edward J. Bauman, W. E. Dorroh, Jr., Hard Cover $65 (ISBN 0-87703-170-3); Soft Cover $50 (ISBN 0-87703-171-1); Microfiche Suppl. (Vol. 38 AAS Michrofiche Series) $10 (ISBN 0-87703-180-0)

Vol. 49 Spacelab, Space Platforms, and the Future, Fourth AAS/DGLR Symposium and 20th Goddard Memorial Symposium, Mar. 17-19, 1982, Greenbelt, MD, 1982, 502p., eds. Peter M. Bainum, Dietrich E. Koelle, Hard Cover $55 (ISBN 0-87703-174-6); Soft Cover $45 (ISBN 0-87703-175-4); Microfiche Suppl. (Vol. 42 AAS Microfiche Series) $15 (ISBN 0-87703-181-9)

Vol. 50 Proceedings on an International Symposium on Engineering Sciences and Mechanics, Dec. 29-31, Tainan, Taiwan, 1983, two parts, 1570p., eds. Han-Min Hsia, Richard W. Longman, You-Li Chou, Hard Cover $120 (ISBN 0-87703-176-2); Microfiche Suppl. (Vol. 43 AAS Microfiche Series) $10 (ISBN 0-87703-215-7)

Vol. 51 Guidance and Control 1983, Feb. 5-9, 1983, Keystone, CO, 1983, 494p., eds. Edward J. Bauman, Zubin W. Emsley, Hard Cover $60 (ISBN 0-87703-182-7); Soft Cover $50 (ISBN 0-87703-183-5); Microfiche Suppl. (Vol. 44 AAS Microfiche Series) $10 (ISBN 0-87703-214-9)

Vol. 52 Developing the Space Frontier, 29th Annual AAS Meeting, Oct. 25-27, 1982, Houston, TX, 1983, 436p., eds. Albert Naumann, Grover Alexander, Microfiche Only $45 (ISBN 0-87703-189-4)

Vol. 53 Space Manufacturing 1983, May 9-12, 1983, Princeton, NJ, 1983, 496p., eds. James D. Burke, April S. Whitt, On Microfiche Only $50 (ISBN 0-87703-188-6)

Vol. 54 AAS/AIAA Astrodynamics Conference, Aug. 22-25, 1983, Lake Placid, NY, 1984, two parts, 1370p., eds. G.T. Tseng, Paul J. Cefola, Peter M. Bainum, David A. Levinson, Hard Cover $120 (ISBN 0-87703-190-8); Soft Cover $90 (ISBN 0-87703-191-6); Microfiche Suppl. (Vol. 45 AAS Microfiche Series) $40 (ISBN 0-87703-192-4)

Vol. 55 Guidance and Control 1984, Feb. 4-8, 1984, Keystone, CO, 1984, 500p., eds. Robert D. Culp, Parker S. Stafford, Hard Cover $60 (ISBN 0-87703-199-1); Soft Cover $50 (ISBN 0-87703-200-9); Microfiche Suppl. (Vol. 48 AAS Microfiche Series $15 (ISBN 0-87703-201-7)

Vol. 56 From Spacelab to Space Station, Fifth DGLR/AAS Symposium, Oct. 3-5, 1984, Hamburg, Germany, 1985, 270p., eds. H. Stoewer, Peter M. Bainum, Microfiche Only $30 (ISBN 0-87703-209-2)

Vol. 57 Guidance and Control 1985, Feb. 2-6, 1985, Keystone, CO, 1985, 618p., eds. Robert D. Culp, Edward J. Bauman, Charles A. Cullian, Hard Cover $65 (ISBN 0-87703-211-4); Soft Cover $50 (ISBN 0-87703-212-2); Microfiche Suppl. (Vol. 50 AAS Microfiche Series) $15 (ISBN 0-87703-213-0)

Vol. 58 AAS/AIAA Astrodynamics Conference, Aug. 12-15, 1985, Vail, CO, 1986, two parts, 1556p., eds. Bernard Kaufman, Joseph J.F. Liu, Robert A. Calico, Felix R. Hoots, Hard Cover $140 (ISBN 0-87703-245-9); Soft Cover $110 (ISBN 0-87703-246-7); Microfiche Suppl. (Vol. 51 AAS Microfiche Series); $60 (ISBN 0-87703-247-5)

Vol. 59 Space Station Beyond IOC, 32nd Annual AAS Meeting, Nov 6-7, 1985, Los Angeles, CA, 1986, 188p., ed. M. Jack Friedenthal, Hard Cover $40 (ISBN 0-87703-252-1); Soft Cover $30 (ISBN 0-87703-253-X)

Vol. 60 Space Exploitation and Utilization, First AAS/JRS Symposium, Dec. 15-19, 1985, Honolulu, HI, 1986, 740p., eds. Gayle L. May, Peter M. Bainum, Kenji Ikeda, Tamiya Nomura, Tatsuo Yamanaka, Ryojiro Akiba, Hard Cover $70 (ISBN 0-87703-254-8); Soft Cover $55 (ISBn 0-87703-255-6); Microfiche Suppl. (Vol. 52 AAS Microfiche Series) $10 (ISBN 0-87703-256-4)

Vol. 61 Guidance and Control 1986, Feb. 1-5, 1986, Keystone, CO, 1986, 460p., eds. Robert D. Culp, John C. Durrett, Hard Cover $60 (ISBN 0-87703-257-2); Soft Cover $50 (ISBN 0-87703-258-0); Microfiche Suppl. (Vol. 53 AAS Microfiche Series) $10 (ISBN 0-87703-259-9)

Vol. 62 Tethers in Space, Proceedings of First International Conference on Tethers in Space (NASA & PSN Sponsors; AIAA, AAS, & AIDAA Co-Sponsors), Sept. 17-19, 1986, Arlington, VA, 1987, 784p., eds. Peter M. Bainum, Ivan Bekey, Luciano Guerriero, Paul A. Penzo, Hard Cover $80 (ISBN 0-87703-264-5); Soft Cover $70 (ISBN 0-87703-265-3)

Vol. 63 Guidance and Control 1987, Jan. 31 -Feb. 4, 1987, Keystone, CO, 1987, 638p., eds. Robert D. Culp, Terry J. Kelly, Hard Cover $75 (ISBN 0-87703-268-8); Soft Cover $60 (ISBN 0-87703-269-6)

Vol. 64 Aerospace Century XXI, 33rd AAS Annual Meeting, Oct. 26-29, 1986, Boulder, CO, 1987, all three parts, Hard Cover $225 (ISBN 0-87703-276-9); Soft Cover $180 (ISBN 0-87703-277-7); Part I, Space Missions and Policy, 686p., eds. George W. Morgenthaler, Gayle L. May, Hard Cover $75 (ISBN 0-87703-279-3); Soft Cover $60 (ISBN 0-87703-282-3); Part II, Space Flight Technologies, 608p., eds. George W. Morgenthaler, W. Kent Tobiska, Hard Cover $75 (ISBN 0-87703-280-7); Soft Cover $60 (ISBN 0-87703-283-1); Part III, Space Sciences, Applications, and Commercial Developments, 724p., eds. George W. Morgenthaler, Jean N. Koster, Hard Cover $75 (ISBN 0-87703-281-5); Soft Cover $60 (ISBN 0-87703-284-X); Microfiche Suppl. (Vol. 54 AAS Microfiche Series) $25 (ISBN 0-87703-278-5)

Vol. 65 AAS/AIAA Astrodynamics Conference, Aug. 10-13, 1987, Kalispell, MT, 1988, two parts, 1774p., eds. John K. Soldner, Arun K. Misra, Robert E. Lindberg, Walton Williamson, Hard Cover $180 (ISBN 0-87703-285-8); Soft Cover $150 (ISBN 0-87703-286-6); Microfiche Suppl. (Vol. 55 AAS Microfiche Series); $70 (ISBN 0-87703-287-4)

Vol. 66 Guidance and Control 1988, Jan. 30 -Feb. 3, 1988, Keystone, CO, 1988, 576p., eds. Robert D. Culp, Paul L. Shattuck, Hard Cover $75 (ISBN 0-87703-288-2); Soft Cover $60 (ISBN 0-87703-289-0); Microfiche Suppl. (Vol. 56 AAS Microfiche Series) $10 (ISBN 0-87703-290-4)

Vol. 67 Space - A New Community of Opportunity, 34th AAS Annual Meeting, Nov. 3-5, 1987, Houston, TX, 1989, 472p., eds. William G. Straight, Henry N. Bowes, Hard Cover $70 (ISBN 0-87703-297-1); Soft Cover $55 (ISBN 0-87703-298-X)

Vol. 68 Guidance and Control 1989, Feb. 4-8, 1989, Keystone, CO, 1989, 708p., eds. Robert D. Culp, Robert A. Lewis, Hard Cover $85 (ISBN 0-87703-299-8); Soft Cover $70 (ISBN 0-87703-300-5)

Vol. 69 Orbital Mechanics and Mission Design, Apr. 24-27, 1989, Greenbelt, MD, 1989, 862p., ed. Jerome Teles, Hard Cover $95 (ISBN 0-87703-311-0); Soft Cover $80 (ISBN 0-87703-312-9); Microfiche Suppl. (Vol. 57 AAS Microfiche Series) $10 (ISBN 0-87703-313-7)

Vol. 70 The 21st Century in Space, 35th AAS Annual Meeting, Oct. 24-26, 1988, St. Louis, MO, 1990, 446p., ed. George V. Butler, Hard Cover $90 (ISBN 0-87703-314-5); Soft Cover $75 (ISBN 0-87703-315-3); Microfiche Suppl. (Vol. 58 AAS Microfiche Series) $10 (ISBN 0-87703-316-1)

Vol. 71 AAS/AIAA Astrodynamics Conference, Aug. 7-10, 1989, Stowe, VT, 1990, two parts, 1472p., eds. Catherine L. Thornton, Ronald J. Proulx, John E. Prussing, Felix R. Hoots, Hard Cover $200 (ISBN 0-87703-317-X); Soft Cover $170 (ISBN 0-87703-318-8); Microfiche Suppl. (Vol. 59 AAS Microfiche Series) $50 (ISBN 0-87703-319-6)

Vol. 72 Guidance and Control 1990, Feb. 3-7, 1990, Keystone, CO, 1990, 676p., eds. Robert D. Culp, Arlo D. Gravseth, Hard Cover $95 (ISBN 0-87703-320-X); Soft Cover $80 (ISBN 0-87703-321-8)

Vol. 73 Space Utilization and Applications in the Pacific, Third (PISSTA) AAS/JRS/CSA Symposium, Nov. 6-8, 1989, Los Angeles, CA, 1990, 764p., eds. Peter M. Bainum, Gayle L. May, Tatsuo Yamanaka, Yang Jiachi, Hard Cover $95 (ISBN 0-87703-325-0); Soft Cover $80 (ISBN 0-87703-326-9)

Vol. 74 Guidance and Control 1991, Feb. 2-6, 1991, Keystone, CO, 1991, 730p., eds. Robert D. Culp, James P. McQuerry, Hard Cover $120 (ISBN 0-87703-334-X); Soft Cover $90 (ISBN 0-87703-335-8)

Vol. 75 AAS/AIAA Spaceflight Mechanics Meeting, Feb. 11-13, 1991, Houston, TX, 1991, two parts, 1354p., eds. John K. Soldner, Arun K. Misra, Lester L. Sackett, Richard Holdaway, Hard Cover $220 (ISBN 0-87703-338-2); Soft Cover $190 (ISBN 0-87703-339-0); Microfiche Suppl. (Vol. 62 AAS Microfiche Series) $50 (ISBN 0-87703-340-4)

Vol. 76 AAS/AIAA Astrodynamics Conference, Aug. 19-22, 1991, Durango, CO, 1992, three parts, 2590p., eds. Bernard Kaufman, Kyle T. Alfriend, Ronald L. Roehrich, Robert R. Dasenbrock, Hard Cover $390 (ISBN 0-87703-347-1); Microfiche Suppl. (Vol. 63 AAS Microfiche Series) $75 (ISBN 0-87703-348-X)

Vol. 77 International Space Year (ISY) in the Pacific Basin, Fourth ISCOPS (formerly PISSTA) AAS/JRS/CSA Symposium, Nov. 17-20, 1991, Kyoto, Japan, 1992, 798p., eds. Peter M. Bainum, Gayle L. May, Makoto Nagatomo, Yoshiaki Ohkami, Yang Jiachi, Hard Cover $120 (ISBN 0-87703-351-X); Soft Cover $90 (ISBN 0-87703-352-8)

Vol. 78 Guidance and Control 1992, Feb. 2-6, 1992, Keystone, CO, 1992, 754p., eds. Robert D. Culp, Richard P. Zietz, Hard Cover $120 (ISBN 0-87703-353-6); Soft Cover $90 (ISBN 0-87703-354-4); Microfiche Suppl. (Vol. 64 AAS Microfiche Series) $20 (ISBN 0-87703-355-2)

Vol. 79 AAS/AIAA Spaceflight Mechanics Meeting, Feb. 24-26, 1992, Colorado Springs, CO, 1992, two parts, 1312p., eds. Roger E. Diehl, Ralph G. Schinnerer, Walton E. Williamson, Daryl G. Boden, Hard Cover $240 (ISBN 0-87703-358-7); Microfiche Suppl. (Vol. 65 AAS Microfiche Series) $40 (ISBN 0-87703-359-5)

Vol. 80 Space Business Opportunities, 37th and 38th AAS Annual Meetings, Nov. 5-7, 1990, Dec. 3-5, 1991, Los Angeles, CA, 1992, 380p., eds. Wayne J. Esser, Don K. Tomajan, Hard Cover $90 (ISBN 0-87703-360-9); Soft Cover $70 (ISBN 0-87703-361-7); Microfiche Suppl. (Vol. 61 AAS Microfiche Series) $50 (ISBN 0-87703-331-5); (Vol. 66 AAS Microfiche Series) $60 (ISBN 0-87703-362-5)

Vol. 81 Guidance and Control 1993, Feb. 6-10, 1993, Keystone, CO, 1993, 648p., eds. Robert D. Culp, George Bickley, Hard Cover $120 (ISBN 0-87703-365-X); Soft Cover $90 (ISBN 0-87703-366-8); Microfiche Suppl. (Vol. 67 AAS Microfiche Series) $20 (ISBN 0-87703-367-6)

Vol. 82 AAS/AIAA Spaceflight Mechanics Meeting, Feb. 22-24, 1993, Pasadena, CA, 1993, two parts, 1454p., eds. Robert G. Melton, Lincoln J. Wood, Roger C. Thompson, Stuart J. Kerridge, Hard Cover $240 (ISBN 0-87703-368-4); Microfiche Suppl. (Vol. 68 AAS Microfiche Series) $15 (ISBN 0-87703-369-2)

Vol. 83 Dynamics of Space Tether Systems (English Language Edition), 1993, 508p., by Vladimir V. Beletsky, Evgenii M. Levin, Hard Cover $120 (ISBN 0-87703-370-6); Soft Cover $90 (ISBN 0-87703-371-4)

Vol. 84 AAS/GSFC International Symposium on Spaceflight Dynamics, Apr. 26-30, 1993, Greenbelt, MD, 1993, two parts, 1450p., eds. Jerome Teles, Mina V. Samii, Hard Cover $240 (ISBN 0-87703-378-1); Microfiche Suppl. (Vol. 69 AAS Microfiche Series) $10 (ISBN 0-87703-379-X)

Vol. 85 AAS/AIAA Astrodynamics Conference, Aug. 16-19, 1993, Victoria, British Columbia, Canada, 1994, three parts, 2750p., eds. Arun K. Misra, Vinod J. Modi, Richard Holdaway, Peter M. Bainum, Hard Cover $390 (ISBN 0-87703-380-3); Microfiche Suppl. (Vol. 70 AAS Microfiche Series) $30 (ISBN 0-87703-381-1)

Vol. 86 Guidance and Control 1994, Feb. 2-6, 1994, Keystone, CO, 1994, 700p., eds. Robert D. Culp, Ronald D. Rausch, Hard Cover $120 (ISBN 0-87703-384-6); Soft Cover $90 (ISBN 0-87703-385-4)

Vol. 87 AAS/AIAA Spaceflight Mechanics Meeting, Feb. 14-16, 1994, Cocoa Beach, FL, 1994, two parts, 1272p., eds. John E. Cochran, Jr., Charles D. Edwards, Jr., Stephen J. Hoffman, Richard Holdaway, Hard Cover $240 (ISBN 0-87703-386-2)

Vol. 88 Guidance and Control 1995, Feb. 1-5, 1995, Keystone, CO, 1995, 600p. eds. Robert D. Culp, James D. Medbery, Hard Cover $120 (ISBN 0-87703-399-4); Soft Cover $90 (ISBN 0-87703-400-1)

Vol. 89 AAS/AIAA Spaceflight Mechanics Meeting, Feb. 13-16, 1995, Albuquerque, NM, 1995, two parts, 1774p., eds. Ronald J. Proulx, Joseph J. F. Liu, P. Kenneth Seidelmann, Salvatore Alfano, Hard Cover $280 (ISBN 0-87703-401-X); Microfiche Suppl. (Vol. 71 AAS Microfiche Series) $15 (ISBN 0-87703-402-8)

Vol. 90 AAS/AIAA Astrodynamics Conference, Aug. 14-17, 1995, Halifax, Nova Scotia, Canada, 1996, two parts, 2270p., eds. K. Terry Alfriend, I. Michael Ross, Arun K. Misra, C. Fred Peters, Hard Cover $290 (ISBN 0-87703-407-9); Microfiche Suppl. (Vol. 72 AAS Microfiche Series) $15 (ISBN 0-87703-408-7)

Vol. 91 Strengthening Cooperation in the 21st Century, Sixth ISCOPS (formerly PISSTA) AAS/JRS/CSA Symposium, Dec. 6-8, 1995, Marina Del Rey, CA, 1996, 1154p., eds. Peter M. Bainum, Gayle L. May, Yoshiaki Ohkami, Kuninori Uesugi, Qi Faren, Li Furong, Hard Cover $145 (ISBN 0-87703-409-5)

Vol. 92 Guidance and Control 1996, Feb. 7-11, 1996, Breckenridge, CO, 1996, 744p. Eds. Robert D. Culp, Marv Odefey, Hard Cover $120 (ISBN 0-87703-412-5); Soft Cover $90 (ISBN 0-87703-413-3)

Vol. 93 AAS/AIAA Spaceflight Mechanics Meeting, Feb. 12-15, 1996, Austin, TX, 1996, two parts, 1776p., eds. G. Edward Powell, Robert H. Bishop, John B. Lundberg, Robert H. Smith, Hard Cover $280 (ISBN 0-87703-414-1); Microfiche Suppl. (Vol. 73 AAS Microfiche Series) $10 (ISBN 0-87703-415-X)

Vol. 94 Guidance and Control 1997, Feb. 5-9, 1997, Breckenridge, CO, 1997, 458p. Eds. Robert D. Culp, Stuart B. Wiens, Hard Cover $120 (ISBN 0-87703-430-3); Soft Cover $90 (ISBN 0-87703-431-1)

Vol. 95 AAS/AIAA Spaceflight Mechanics Meeting, Feb. 10-12, 1997, Huntsville, AL, 1997, two parts, 1168p., eds. Kathleen C. Howell, David A. Cicci, John E. Cochran, Jr., Thomas S. Kelso, Hard Cover $240 (ISBN 0-87703-432-X)

Vol. 96 Space Cooperation into the 21st Century, Seventh ISCOPS (formerly PISSTA) JRS/AAS/CSA Symposium, Jul. 15-18, 1997, Nagasaki, Japan, 1997, 1089p., eds. Peter M. Bainum, Gayle L. May, Makoto Nagatomo, Kuninori T. Uesugi, Fu Bing-chen, Zhang Hui, Hard Cover $145.00 (ISBN 0-87703-438-9)

Vol. 97 AAS/AIAA Astrodynamics Conference, Aug. 4-7, 1997, Sun Valley, ID, 1998, two parts, 2184p., eds. Felix R. Hoots, Bernard Kaufman, Paul J. Cefola, David B. Spencer, Hard Cover $310 (ISBN 0-87703-441-9)

Vol. 98 Guidance and Control 1998, Feb. 4-8, 1998, Breckenridge, CO, 1998, 706p. Eds. Robert D. Culp, David Igli, Hard Cover $120 (ISBN 0-87703-448-6); Soft Cover $90 (ISBN 0-87703-449-4)

Vol. 99 AAS/AIAA Spaceflight Mechanics Meeting, Feb. 9-11, 1998, Monterey, CA, 1998, two parts, 1638p., eds. Jay W. Middour, Lester L. Sackett, Louis A. D"Amario, Dennis V. Byrnes, Hard Cover $280 (ISBN 0-87703-450-8); Microfiche Suppl. (Vol. 78 AAS Microfiche Series) $10 (ISBN 0-87703-452-4)

Vol. 100 AAS/GSFC International Symposium on Space Flight Dynamics, May 11-15, 1998, Greenbelt, MD, 1998, two parts, 1092p., ed. Thomas H. Stengle, Hard Cover $250 (ISBN 0-87703-453-2)

Vol. 101 Guidance and Control 1999, Feb. 3-7, 1999, Breckenridge, CO, 1999, 528p. Eds. Robert D. Culp, Douglas Wiemer, Hard Cover $120 (ISBN 0-87703-456-7); Soft Cover $90 (ISBN 0-87703-457-5)

Vol. 102 AAS/AIAA Spaceflight Mechanics Meeting, Feb. 7-10, 1999, Breckenridge, CO, 1999, two parts, 1600p., eds. Robert H. Bishop, Robert D. Culp, Donald L. Mackison and Maria Evans, Hard Cover $280 (ISBN 0-87703-458-3)

Order from Univelt, Inc., P.O. Box 28130, San Diego, California 92198
STANDING ORDERS ACCEPTED

SCIENCE AND TECHNOLOGY SERIES (1964-)

ISSN 0278-4017

A Supplement to *Advances in the Astronautical Sciences*. Proceedings and monographs, most of them based on AAS technical meetings.

Vol. 1 Manned Space Reliability Symposium, Jun. 9, 1964, Anaheim, CA, 1964, 112p., ed. Paul Horowitz, Hard Cover $20 (ISBN 0-87703-029-4)

Vol. 2 Towards Deeper Space Penetration (AAS/AAAS Symposium), Dec. 29, 1964, Montreal, Canada, 1964, 182p., ed. Edward R. Van Driest, Hard cover $20 (ISBN 0-87703-030-8)

Vol. 3 Orbital Hodograph Analysis, 1965, 150p., ed. Samuel P. Altman, Hard Cover $20 (ISBN 0-87703-031-6)

Vol. 4 Scientific Experiments for Manned Orbital Flight, 3rd Goddard Memorial Symposium, Mar. 18-19, 1965, Washington, D.C., 1965, 372p., ed. Peter C. Badgley, Hard Cover $30 (ISBN 0-87703-032-4)

Vol. 5 Physiological and Performance Determinants in Manned Space Systems (AAS/HFS Symposium, Apr. 14-15, 1965, Northridge, CA, 1965, 220p., ed. Paul Horowitz, Microfiche Only $20 (ISBN 0-87703-033-2)

Vol. 6 Space Electronics Symposium (AAS/AES Meeting), May 25-27, 1965, Los Angeles, CA, 1965, 404p., ed. Chung-Ming Wong, Hard Cover $30 (ISBN 0-87703-034-0)

Vol. 7 Theodore von Karman Memorial Seminar, May 12, 1965, Los Angeles, CA, 1966, 140p., ed. Shirley Thomas, Hard Cover $30 (ISBN 0-87703-035-9)

Vol. 8 Impact of Space Exploration on Society, Aug. 18-20, 1965, San Francisco, CA, 1966, 382p., ed. William E. Frye, Hard Cover $30 (ISBN 0-87703-036-7)

Vol. 9 Recent Developments in Space Flight Mechanics, (AAS/AAAS Symposium), Dec. 29, 1965, Berkeley, CA, 1966, 280p., ed. Paul B. Richards, Hard Cover $25 (ISBN 0-87703-037-5)

Vol. 10 Space in the Fiscal Year 2001, 4th Goddard Memorial Symposium, Mar. 15-16, 1966, Washington, D.C., 1967, 458p., eds. Eugene B. Konecci, Maxwell W. Hunter, II, Robert F. Trapp, Hard Cover $35 (ISBN 0-87703-038-3)

Vol. 11 Space Flight Specialist Conference, Jul. 6-8, 1966, Denver, CO, 1967, 618p., ed. Maurice L. Anthony, Microfiche Only (ISBN 0-87703-039-1); Plus Microfiche Suppl. (Vol. 2 AAS Microfiche Series) $60 (ISBN 0-87703-221-1)

Vol. 12 Management of Aerospace Programs Conference, Nov. 16-18, 1966, Columbia, MO, 1967, 392p., ed. Walter K. Johnson, Hard Cover $30 (ISBN 0-87703-040-5)

Vol. 13 Physics of the Moon (AAS/AAAS Symposium), Dec. 29, 1966, Washington, D.C., 1967, 260p., ed. S. Fred Singer, Hard Cover $25 (ISBN 0-87703-041-3)

Vol. 14 Interpretation of Lunar Probe Data, Sept. 17, 1966, Huntington Beach, CA, 1967, 270p., ed. Jack Green, Hard Cover $25 (ISBN 0-87703-042-1)

Vol. 15 Future Space Program and Impact on Range and Network Development Symposium, Mar. 22-24, 1967, Las Cruces, NM, 1967, 588p., ed. George W. Morgenthaler, Hard Cover $40 (ISBN 0-87703-043-X)

Vol. 16 Voyage to the Planets, 5th Goddard Memorial Symposium, Mar. 14-15, 1967, Washington, D.C., 1968, 184p., ed. S. Fred Singer, Hard Cover $20 (0-87703-044-8)

Vol. 17 Use of Space Systems for Planetary Geology and Geophysics Symposium, May 25-27, 1967, Boston, MA, 1968, 623p., ed. Robert D. Enzmann, Hard Cover $45 (ISBN 0-87703-045-6); Microfiche Suppl. (Vol. 5 AAS Microfiche Series) $15 (ISBN 0-87703-135-5)

Vol. 18 Technology and Social Progress, 6th Goddard Memorial Symposium, Mar. 12-13, 1968, Washington, D.C., 1969, 170p., ed. Philip K. Eckman, Hard Cover $20 (ISBN 0-87703-046-4)

Vol. 19 Exobiology - The Search for Extraterrestrial Life (AAS/AAAS Symposium) Dec. 30, 1967, New York, NY, 1969, 184p., eds. Martin M. Freundlich, Bernard W. Wagner, Hard Cover $20 (ISBN 0-87703-047-2)

Vol. 20 Bioengineering and Cabin Ecology (AAS/AAAS Symposium) Dec. 30, 1968, Dallas, TX, 1969, 162p., ed. William Cassidy, Hard Cover $20 (ISBN 0-87703-048-0)

Vol. 21 Reducing the Cost of Space Transportation, 7th Goddard Memorial Symposium, Mar. 4-5, 1969, Washington, D.C., 1969, 264p., ed. George K. Chacko, Microfiche only $25 (ISBN 0-87703-049-9)

Vol. 22 Planning Challenges of the 70"s in the Public Domain, 15th Annual AAS Meeting, Jun. 17-20, 1969, Denver, CO, 1970, 504p., eds. William J. Burnsnall, George K. Chacko, George W. Morgenthaler, Hard Cover $40 (ISBN 0-87703-050-2); Microfiche Suppl. (Vol. 13 AAS Microfiche Series) $20 (ISBN 0-87703-131-2); See also Vols. 15-17, AAS Microfiche Series

Vol. 23 Space Technology and Earth Problems Symposium, Oct. 23-25, 1969, Las Cruces, NM, 1970, 418p., ed. C. Quentin Ford, Hard Cover $35 (ISBN 0-87703-051-0); Microfiche Suppl. (Vol. 12 AAS Microfiche Series) $20 (ISBN 0-87703-134-7)

Vol. 24 Aerospace Research and Development, Jul. 14, 1966, Holloman AFB, NM, 1970, 500p., ed. Ernst A. Steinhoff, Microfiche Only $40 (ISBN 0-87703-052-9)

Vol. 25 Geological Problems in Lunar and Planetary Research, Feb. 17-18, 1969, Huntington Beach, CA, 1971, 750p., ed. Jack Green, Hard Cover $45 (ISBN 0-87703-056-1)

Vol. 26 Technology Utilization Ideas for the 70s and Beyond, Oct. 30, 1970, Winrock, AR, 1971, 312p., eds. Fred W. Forbes, Paul Dergarabedian, Microfiche only $30 (ISBN 0-87703-057-X)

Vol. 27 International Cooperation in Space Operations and Exploration, 9th Goddard Memorial Symposium, Mar. 11, 1971, Washington, D.C. 1971, 194p., ed. Michael Cutler, Hard Cover $20 (ISBN 0-88703-058-8)

Vol. 28 Astronomy from a Space Platform (AAS/AAAS Symposium) Dec. 27-28, 1971, Philadelphia, PA, 1972, 416p., eds. George W. Morgenthaler, Howard D. Greyber, Hard Cover $35 (ISBN 0-87703-061-8)

Vol. 29 Space Technology Transfer to Community and Industry, 10th Goddard Memorial Symposium, 18th Annual AAS Meeting, Mar. 13-14, 1972, Washington, D.C., 1972, 196p., eds. Ralph H. Tripp, John K. Stotz, Jr., Hard Cover $20 (ISBN 0-87703-062-6); on Microfiche $15

Vol. 30 Space Shuttle Payloads (AAS/AAAS Symposium) Dec. 27-28, 1972, Washington, D.C., 1973, 532p., eds. George W. Morgenthaler, William J. Bursnall, Hard Cover $40 (ISBN 0-87703-063-4)

Vol. 31 The Second Fifteen Years in Space, 11th Goddard Memorial Symposium, Mar. 8-9, 1973, Washington, D.C., 1973, 212p., ed. Saul Ferdman, Hard Cover $25 (ISBN 0-87703-064-2)

Vol. 32 Health Care Systems Conference, Nov. 21-22, 1972, Dallas, TX, 1974, 265p., ed. Eugene B. Konecci, Hard Cover $25 (ISBN 0-87703-067-7)

Vol. 33 Orbital International Laboratory, 3rd and 4th IAF/OIL Symposia, Oct. 5-6, 1970, Constance, Germany, Sept. 24-25, 1971, Brussels, Belgium, 1974, 322p., ed. Ernst A. Steinhoff, Hard Cover $30 (ISBN 0-87703-068-5)

Vol. 34 Management and Design of Long-Life Systems, Apr. 24-26, 1973, Denver, CO, 1974, 198p., ed. Harris M. Schurmeier, Hard Cover $20 (ISBN 0-87703-069-3)

Vol. 35 Energy Delta, Supply vs. Demand, (AAS/AAAS Symposium) Feb. 25-27, 1974, San Francisco, CA, 1975, 2nd Printing 1976, 604p., eds. George W. Morgenthaler, Aaron N. Silver, Hard Cover $35 (ISBN 0-87703-070-7); Soft Cover $25 (ISBN 0-87703-082-0); on Microfiche $20

Vol. 36 Skylab and Pioneer Report, 12th Goddard Memorial Symposium, Mar. 8, 1974, Washington, D.C., 1975, 160p., eds. Philip H. Bolger, Paul B. Richards, Hard Cover $20 (ISBN 0-87703-071-5)

Vol. 37 Space Rescue and Safety 1974, 7th International IAA Symposium, Sept. 30 - Oct. 5, 1974, Amsterdam, Netherlands, 1975, 294p., ed. Philip H. Bolger, Microfiche Only $25 (ISBN 0-87703-073-1)

Vol. 38 Skylab Science Experiments, (AAS/AAAS Symposium) Feb. 28, 1974, San Francisco, CA, 1976, 274p., eds. George W. Morgenthaler, Gerald E. Simonson, Microfiche only $20 (ISBN 0-87703-074-X)

Vol. 39 Environmental Control and Agri-Technology, 1976, 346p., ed. Eugene B. Konecci, Microfiche only $20 (ISBN 0-87703-075-8)

Vol. 40 Future Space Activities, 13th Goddard Memorial Symposium, Apr. 11, 1975, Washington, D.C., 1976, 182p., ed. Carl H. Tross, Microfiche only $20 (ISBN 0-87703-076-6)

Vol. 41 Space Rescue and Safety 1975, 8th International IAA Symposium, Sept. 21-27, 1975, Lisbon, Portugal, 1976, 230p., ed. Philip H. Bolger, Hard Cover $25 (ISBN 0-87703-077-4)

Vol. 42 The End of an Era in Space Exploration, From International Rivalry to International Cooperation, 1976, 216p., by J.C.D. Blaine, Hard Cover $25 (ISBN 0-87703-084-7); without volume number (ISBN 0-87703-080-4)

Vol. 43 The Eagle Has Returned, Part I, International Space Hall of Fame Dedication Conference, Oct. 5-9, 1976, Alamogordo, NM, 1976, 370p., ed. Ernst. A. Steinhoff, Hard Cover $30 (ISBN 0-87703-086-3)

Vol. 44 Satellite Communications in the Next Decade, 14th Goddard Memorial Symposium, Mar. 12, 1976, Washington, D.C., 1977, 188p., ed. Leonard Jaffe, Hard Cover $20 (ISBN 0-87703-088-X)

Vol. 45 The Eagle Has Returned, Part 2, International Space Hall of Fame Dedication Conference, Oct. 5-9, 1976, Alamogordo, NM, 1977, 454p., ed. Ernst A. Steinhoff, Hard Cover $35 (ISBN 0-87703-092-8)

Vol. 46 Export of Aerospace Technology, 15th Goddard Memorial Symposium, Mar. 31 - Apr. 1, 1977, Washington, D.C., 1978, 174p., ed. Carl H. Tross, Hard Cover $20 (ISBN 0-87703-093-6)

Vol. 47 Handbook of Soviet Lunar and Planetary Exploration, 1979, 276p., by Nicholas L. Johnson, Microfiche Only $25 (ISBN 0-87703-105-3)

Vol. 48 Handbook of Soviet Manned Space Flight, 2nd Edition, 1988, 474p., by Nicholas L. Johnson, Hard Cover $60 (ISBN 0-87703-115-0); Soft Cover $45 (ISBN 0-87703-116-9)

Vol. 49 Space - New Opportunities for International Ventures, 17th Goddard Memorial Symposium, Mar. 28-30, 1979, Washington, D.C., 1980, 300p., ed. William C. Hayes, Jr., Hard Cover $35 (ISBN 0-87703-124-X); Soft Cover $25 (ISBN 0-87703-125-8); see also Vol. 2 AAS History Series

Vol. 50 Remember the Future - The Apollo Legacy, Jul. 20-21, 1979, San Francisco, CA, 1980, 218p., ed. Stan Kent, Hard Cover $25 (ISBN 0-87703-126-6); Soft Cover $15 (ISBN 0-87703-127-4)

Vol. 51 Commercial Operations in Space 1980-2000, 18th Goddard Memorial Symposium, Mar. 27-28, 1980, Washington, D.C., 1981, 214p., eds. John L. McLucas, Charles Sheffield, Hard Cover $30 (ISBN 0-87703-140-1); Soft Cover $20 (ISBN 0-87703-141-X); Microfiche Suppl. (Vol. 34 AAS Microfiche Series) $10 (ISBN 0-87703-165-7); see also Vols. 2 and 3, AAS History Series

Vol. 52 International Space Technical Applications, 19th Goddard Memorial Symposium, Mar. 26-27, 1981, Washington, D.C., 1981, 186p., eds. Andrew Adelman, Peter M. Bainum, Hard Cover $30 (ISBN 0-87703-152-5); Soft Cover $20 (ISBN 0-87703-153-3); see also Vol. 5, AAS History Series

Vol. 53 Space in the 1980"s and Beyond, 17th European Space Symposium, Jun. 4-6, 1980, London, England, 1981, 302p., ed. Peter M. Bainum, Hard Cover $40 (ISBN 0-87703-154-1); Soft Cover $30 (ISBN 0-87703-155-X)

Vol. 54 Space Safety and Rescue 1979-1981 (with abstracts 1976-1978), Proceedings of symposia of the International Academy of Astronautics held in conjunction with the 30th, 31st, and 32nd International Astronautical Federation Congresses, Munich, Germany, 1979, Tokyo, Japan, 1980, and Rome, Italy, 1981, 1983, 456p., ed. Jeri W. Brown, Hard Cover $45 (ISBN 0-87703-177-0); Soft Cover $35 (ISBN 0-87703-178-9); Microfiche Suppl. (Vols. 39-41 AAS Microfiche Series) $39 (ISBN 0-87703-222-X); (ISBN 0-87703-223-8); (ISBN 0-87703-224-6)

Vol. 55 Space Applications at the Crossroads, 21st Goddard Memorial Symposium, Mar. 24-25, 1983, Greenbelt, MD, 1983, 308p., eds. John H. McElroy, E. Larry Heacock, Hard Cover $45 (ISBN 0-87703-186-X); Soft Cover $35 (ISBN 0-87703-187-8)

Vol. 56 Space: A Developing Role for Europe, 18th European Space Symposium, Jun. 6-9, 1983, London, England, 1984, 278p., eds. Len J. Carter, Peter M. Bainum, Hard Cover $45 (ISBN 0-87703-193-2); Soft Cover $35 (ISBN 0-87703-194-0); Microfiche Suppl. (Vol. 46 AAS Microfiche Series) $15 (ISBN 0-87703-195-9)

Vol. 57 The Case for Mars, Apr. 29 - May 2, 1981, Boulder, CO, 1984, Second Printing 1987, 348p., ed. Penelope J. Boston, on Microfiche only $25 (ISBN 0-87703-198-3)

Vol. 58 Space Safety and Rescue 1982-1983, Proceedings of the International Academy of Astronautics held in conjunction with the 33rd and 34th International Astronautical Congresses, Paris, France, Sept. 27 - Oct. 2, 1982, and Budapest, Hungary, Oct. 10-15, 1983, 1984, 378p., ed. Gloria W. Heath, Hard Cover $50 (ISBN 0-87703-202-5); Soft Cover $40 (ISBN 0-87703-203-3)

Vol. 59 Space and Society - Challenges and Choices, April 14-16, 1982, University of Texas at Austin, 1984, 442p., eds. Paul Anaejionu, Nathan C. Goldman, Philip J. Meeks, Hard Cover $55 (ISBN 0-87703-204-1); Soft Cover $35 (ISBN 0-87703-205-X)

Vol. 60 Permanent Presence - Making It Work, 22nd Goddard Memorial Symposium, Mar. 15-16, 1984, Greenbelt, MD, 1985, 190p., ed. Ivan Bekey, Hard Cover $40 (ISBN 0-87703-207-6); Soft Cover $30 (ISBN 0-87703-208-4)

Vol. 61 Europe/United States Space Activities - With a Space Propulsion Supplement, 23rd Goddard Memorial Symposium/19th European Space Symposium, Mar. 27-29, 1985, Greenbelt, MD, 31st Annual AAS Meeting, Oct. 22-24, 1984, Palo Alto, CA, 1985, 442p., eds. Peter M. Bainum, Friedrich von Bun, Hard Cover $55 (ISBN 0-87703-217-3); Soft Cover $45 (ISBN 0-87703-218-1)

Vol. 62 The Case for Mars II, July 10-14, 1984, Boulder, CO, 1985, 730p., ed. Christopher P. McKay, Hard Cover $60 (ISBN 0-87703-219-1); Soft Cover $40 (ISBN 0-87703-220-3)

Vol. 63 Proceedings of 4th International Conference on Applied Numerical Modeling, Dec. 27-29, 1984, Tainan, Taiwan, 1986, 800p., ed. Han-Min Hsia, You-Li Chou, Shu-Yi Wang, Sheng-Jii Hsieh, Hard Cover $70 (ISBN 0-87703-242-4)

Vol. 64 Space Safety and Rescue 1984-1985, Proceedings of the International Academy of Astronautics held in conjunction with the 35th and 36th International Astronautical Congresses, Lausanne, Switzerland, Oct. 7-13, 1984, and Stockholm, Sweden, Oct. 7-12, 1985, 1986, 400p., ed. Gloria W. Heath, Hard Cover $55 (ISBN 0-87703-248-3); Soft Cover $45 (ISBN 0-87703-249-1)

Vol. 65 The Human Quest in Space, 24th Goddard Memorial Symposium, Mar. 20-21, 1986, Greenbelt, MD, 1987, 312p., ed. Gerald L. Burdett, Gerald A. Soffen, Hard Cover $55 (ISBN 0-87703-262-9); Soft Cover $45 (ISBN 0-87703-263-7)

Vol. 66 Soviet Space Programs 1980-1985, 1987, 298p., by Nicholas L. Johnson, Microfiche Only $45 (ISBN 0-87703-266-1)

Vol. 67 Low-Gravity Sciences, Seminar Series 1986, University of Colorado at Boulder, 290p., ed. Jean N. Koster, Hard Cover $55 (ISBN 0-87703-270-X); Soft Cover $45 (ISBN 0-87703-271-8)

Vol. 68 Proceedings of the Fourth Annual L5 Space Development Conference, Apr. 25-28, 1985, Washington, D.C., 1987, 268p., ed. Frank Hecker, Hard Cover $50 (ISBN 0-87703-272-6); Soft Cover $35 (ISBN 0-87703-273-4)

Vol. 69 Visions of Tomorrow: A Focus on National Space Transportation Issues, 25th Goddard Memorial Symposium, Mar. 18-20, 1987, Greenbelt, MD, 1987, 338p., ed. Gerald A. Soffen, Hard Cover $55 (ISBN 0-87703-274-2); Soft Cover $45 (ISBN 0-87703-275-0)

Vol. 70 Space Safety and Rescue 1986-1987, Proceedings of the International Academy of Astronautics held in conjunction with the 37th and 38th International Astronautical Congresses, Innsbruck, Austria, Oct. 4-11, 1986, and Brighton, England, Oct. 11-16, 1987, 1988, 360p., ed. Gloria W. Heath, Hard Cover $55 (ISBN 0-87703-291-2); Soft Cover $45 (ISBN 0-87703-292-0)

Vol. 71 The NASA Mars Conference, Jul. 21-23, 1986, Washington, D.C., 1988, 570p., ed. Duke B. Reiber, Hard Cover $50 (ISBN 0-87703-293-9); Soft Cover $30 (ISBN 0-87703-294-7)

Vol. 72 Working in Orbit and Beyond: The Challenges for Space Medicine, Jun. 20-21, 1987, Washington, D.C., 1989, 188p., ed. David Lorr, Victoria Garshnek, Hard Cover $45 (ISBN 0-87703-295-5); Soft Cover $35 (ISBN 0-87703-296-3)

Vol. 73 Technology and the Civil Future in Space, 26th Goddard Memorial Symposium, Mar. 16-18, 1988, Greenbelt, MD, 1989, 246p., ed. Leonard A. Harris, Hard Cover $50 (ISBN 0-87703-301-3); Soft Cover $35 (ISBN 0-87703-302-1)

Vol. 74 The Case for Mars III: Strategies for Exploration - General Interest and Overview, July 18-22, 1987, Boulder, CO, 1989, 744p., ed. Carol Stoker, Hard Cover $75 (ISBN 0-87703-303-X); Soft Cover $55 (0-87703-304-8)

Vol. 75 The Case for Mars III: Strategies for Exploration - Technical, July 18-22, 1987, Boulder, CO, 1989, 646p., ed. Carol Stoker, Hard Cover $70 (ISBN 0-87703-305-6); Soft Cover $50 (ISBN 0-87703-306-4)

Vol. 76 Global Environmental Change: The Role of Space in Understanding Earth, 27th Goddard Memorial Symposium, Mar. 8-10, 1989, Washington, D.C., 1990, 178p., ed. Richard G. Johnson, Hard Cover $50 (ISBN 0-87703-322-6); Soft Cover $40 (ISBN 0-87703-323-4); Microfiche Suppl. (Vol. 60 AAS Microfiche Series) $10 (ISBN 0-87703-324-2)

Vol. 77 Space Safety and Rescue 1988 - 1989, Proceedings of the International Academy of Astronautics held in conjunction with the 39th and 40th International Astronautical Congresses, Bangalore, India, Oct. 8-15, 1988, and Málaga, Spain, Oct. 7-12, 1989, 1990, 500p., ed. Gloria W. Heath, Hard Cover $70 (ISBN 0-87703-327-7); Soft Cover $55 (ISBN 0-87703-328-5)

Vol. 78 Leaving the Cradle: Human Exploration of Space in the 21st Century, 28th Goddard Memorial Symposium, Mar. 14-16, 1990, Washington, D.C., 1991, 348p., ed. Thomas O. Paine, Hard Cover $70 (ISBN 0-87703-336-6); Soft Cover $55 (ISBN 0-87703-337-4)

Vol. 79 Space Safety and Rescue 1990, Proceedings of the International Academy of Astronautics held in conjunction with the 41st International Astronautical Congress, Dresden, Germany, Oct. 6-12, 1990, 1991, 232p., ed. Gloria W. Heath, Hard Cover $65 (ISBN 0-87703-341-2); Soft Cover $50 (ISBN 0-87703-342-0)

Vol. 80 Prospects for Interstellar Travel, 1992, 390p., by John H. Mauldin, Hard Cover $50 (ISBN 0-87703-344-7); Soft Cover $27 (ISBN 0-87703-345-5)

Vol. 81 Humans and Machines in Space: The Vision, The Challenge, The Payoff, 29th Goddard Memorial Symposium, Mar. 14-15, 1991, Washington, D.C., 1992, 204p., ed. Bradley Johnson, Gayle L. May, Paula Korn, Hard Cover $50 (ISBN 0-87703-356-0); Soft Cover $35 (ISBN 0-87703-357-9)

Vol. 82 Space Safety and Rescue 1991, Proceedings of the International Academy of Astronautics held in conjunction with the 42nd International Astronautical Congress, Montreal, Canada, Oct. 5-11, 1991, 1993, 270p., ed. Gloria W. Heath, Hard Cover $65 (ISBN 0-87703-372-2); Soft Cover $50 (ISBN 0-87703-373-0)

Vol. 83 Space: A Vital Stimulus to Our National Well-Being, 31st Goddard Memorial Symposium, March 9-10, 1993, Arlington, Virginia, and World Space Programs and Fiscal Reality, 30th Goddard Memorial Symposium, April 9-10, 1992, Alexandria, Virginia, 1994, 334p., ed. Gayle L. May, Saunders B. Kramer, Paula Korn, Leonard David, Barbara Sprungman, Hard Cover $70 (ISBN 0-87703-389-7); Soft Cover $50 (ISBN 0-87703-390-0)

Vol. 84 Space Safety and Rescue 1992, Proceedings of the International Academy of Astronautics held in conjunction with the World Space Congress, Washington, D.C., Aug. 28 to Sept. 5, 1992, 1994, 372p., ed. Gloria W. Heath, Hard Cover $70 (ISBN 0-87703-391-9); Soft Cover $55 (ISBN 0-87703-392-7)

Vol. 85 Civil Space in the Clinton Era, 32nd Goddard Memorial Symposium, March 1-2, 1994, Crystal City, Virginia, and Partners in Space . . . 2001, 41st Annual Meeting, November 14-16, 1994, Crystal City, Virginia, 1995, 292p., ed. Donald R. McConathy, Paula Korn, Hard Cover $70 (ISBN 0-87703-397-8); Soft Cover $50 (ISBN 0-87703-398-6)

Vol. 86 Strategies for Mars: A Guide to Human Exploration, 1996, 644p, ed. Carol R. Stoker, Carter Emmart, Hard Cover $70 (ISBN 0-87703-405-2); Soft Cover $45 (ISBN 0-87703-406-0)

Vol. 87 Space Safety and Rescue 1993, Proceedings of the International Academy of Astronautics held in conjunction with the 44th International Astronautical Congress, Graz, Austria, Oct. 16-22, 1993, 1996, 344p., ed. Gloria W. Heath, Hard Cover $70 (ISBN 0-87703-410-9); Soft Cover $50 (ISBN 0-87703-411-7)

Vol. 88 Space Safety and Rescue 1994, Proceedings of the International Academy of Astronautics held in conjunction with the 45th International Astronautical Congress, Jerusalem, Israel, Oct. 9-14, 1994, 1996, 326p., ed. Gloria W. Heath, Hard Cover $70 (ISBN0-87703-416-8); Soft Cover $50 (ISBN 0-87703-417-6)

Vol. 89 The Case for Mars IV: The International Exploration of Mars"Mission Strategy and Architectures, June 4-8, 1990, Boulder, CO, 1997, 790p., ed. Thomas R. Meyer, Hard Cover $80 (ISBN 0-87703-418-4); Soft Cover $55 (ISBN 0-87703-419-2)

Vol. 90 The Case for Mars IV: The International Exploration of Mars"Considerations for Sending Humans, June 4-8, 1990, Boulder, CO, 1997, 502p., ed. Thomas R. Meyer, Hard Cover $70 (ISBN 0-87703-420-6); Soft Cover $55 (ISBN 0-87703-421-4)

Vol. 91 From Imagination to Reality: Mars Exploration Studies of the Journal of the British Interplanetary Society (Part I, Precursors and Early Piloted Missions), 1997, 388p., ed. Robert M. Zubrin, Hard Cover $70 (ISBN 0-87703-426-5); Soft Cover $45 (ISBN 0-87703-427-3)

Vol. 92 From Imagination to Reality: Mars Exploration Studies of the Journal of the British Interplanetary Society (Part II, Base Building, Colonization and Terraformation), 1997, 376p., ed. Robert M. Zubrin, Hard Cover $70 (ISBN 0-87703-428-1); Soft Cover $45 (ISBN 0-87703-429-X)

Vol. 93 Space Safety and Rescue 1995, Proceedings of the International Academy of Astronautics held in conjunction with the 46th International Astronautical Congress, Jerusalem, Israel, Oct. 2-6, 1995, 1997, 482p., ed. Gloria W. Heath, Hard Cover $80 (ISBN 0-87703-416-8); Soft Cover $55 (ISBN 0-87703-417-6)

Vol. 94 Fluid and Electrolyte Regulation in Spaceflight, 1998, 238p., Carolyn S. Leach Huntoon, Anatoliy I. Grigoriev, Yuri V. Natochin, Hard Cover $60 (ISBN 0-87703-442-7); Soft Cover $40 (ISBN 0-87703-443-5)

Vol. 95 Space Safety and Rescue 1996, Proceedings of the International Academy of Astronautics held in conjunction with the 47th International Astronautical Congress, Beijing, China, Oct. 7-11, 1996, 1998, 362p., ed. Gloria W. Heath, Hard Cover $75 (ISBN 0-87703-446-X); Soft Cover $50 (ISBN 0-87703-447-8)

Vol. 96 Space Safety and Rescue 1997, Proceedings of the International Academy of Astronautics held in conjunction with the 48th International Astronautical Congress, Turin, Italy, Oct. 6-10, 1997, 1999, 412p., ed. Gloria W. Heath, Hard Cover $80 (ISBN 0-87703-454-0); Soft Cover $55 (ISBN 0-87703-455-9)

Vol. 97 The Case for Mars V, May 26-29, 1993, Boulder, CO, 2000, 564p., ed. Penelope J. Boston, Hard Cover $80 (ISBN 0-87703-459-1); Soft Cover $55 (ISBN 0-87703-460-5)

Vol. 98 The Case for Mars VI: Making Mars an Affordable Destination, July 17-20, 1996, Boulder, CO, 2000, 578p., ed. Kelly R. McMillen, Hard Cover $80 (ISBN 0-87703-461-3); Soft Cover $55 (ISBN 0-87703-462-1)

Vol. 99 Space Safety and Rescue 1998, Proceedings of the International Academy of Astronautics held in conjunction with the 49th International Astronautical Congress, Melbourne, Australia, Sept. 28 - Oct. 2, 1998, 2000, 410p., ed. Macgregor S. Reid and Walter Flury, Hard Cover $80 (ISBN0-87703-463-X); Soft Cover $55 (ISBN 0-87703-464-8)

Order from Univelt, Inc., P.O. Box 28130, San Diego, California 92198
STANDING ORDERS ACCEPTED

AAS HISTORY SERIES

Vol. 1 Two Hundred Years of Flight in America: A Bicentennial Survey, Edited by Eugene M. Emme, 1977, 326p, Third Printing 1981, Hard Cover $35 (ISBN 0-87703-091-X); Soft Cover $25 (ISBN 0-87703-101-0); special price for classroom text or bulk purchase.

Vol. 2 Twenty-Five Years of the American Astronautical Society: Historical Reflections and Projections, 1954-1979, Edited by Eugene M. Emme, 1980, 248p, Hard Cover $25 (ISBN 0-87703-117-7); Soft Cover $15 (ISBN 0-87703-118-5).

Vol. 3 Between Sputnik and the Shuttle: New Perspectives on American Astronautics, 1957-1980, Edited by Frederick C. Durant, III, 1981, 350p, Hard Cover $40 (ISBN 0-87703-145-2); Soft Cover $30 (ISBN 0-87703-149-9).

Vol. 4 The Endless Space Frontier: A History of the House Committee on Science and Astronautics, By Ken Hechler, Abridged and edited by Albert E. Eastman, 1982, 460p, Hard Cover $45 (ISBN 0-87703-157-6); Soft Cover $35 (ISBN 0-87703-158-4).

Vol. 5 Science Fiction and Space Futures: Past and Present, Edited by Eugene M. Emme, 1982, 278p, Hard Cover $35 (ISBN 0-87703-172-X); Soft Cover $25 (ISBN 0-87703-173-8).

Vol. 6 First Steps Toward Space, Edited by Frederick C. Durant, III and George S. James, 1986, 318p, Hard Cover $45 (ISBN 0-87703-243-2); Soft Cover $35 (ISBN 0-87703-244-0).

Vol. 7 History of Rocketry and Astronautics, Edited by R. Cargill Hall, 1986, Part I, 250p, Part II, 502p, sold as a set, Hard Cover $100 (ISBN 0-87703-260-2); Soft Cover $80 (ISBN 0-87703-261-0).

Vol. 8 History of Rocketry and Astronautics, Edited by Kristan R. Lattu, 1989, 368p, Hard Cover $50 (ISBN 0-87703-307-2); Soft Cover $35 (ISBN 0-87703-308-0).

Vol. 9 History of Rocketry and Astronautics, Edited by Frederick I. Ordway, III, 1989, 330p, Hard Cover $50 (ISBN 0-87703-309-9); Soft Cover $35 (ISBN 0-87703-310-2).

Vol. 10 History of Rocketry and Astronautics, Edited by Å. Ingemar Skoog, 1990, 330p, Hard Cover $60 (ISBN 0-87703-329-3); Soft Cover $40 (ISBN 0-87703-330-7)

Vol. 11 History of Rocketry and Astronautics, Edited by Roger D. Launius, 1994, 236p, Hard Cover $60 (ISBN 0-87703-382-X); Soft Cover $40 (ISBN 0-87703-383-8).

Vol. 12 History of Rocketry and Astronautics, Edited by John L. Sloop, 1991, 252p, Hard Cover $60 (ISBN 0-87703-332-3); Soft Cover $40 (ISBN 0-87703-333-1).

Vol. 13 History of Liquid Rocket Engine Development in the United States 1955-1980, Edited by Stephen E. Doyle, 1992, 176p, Hard Cover $50 (ISBN 0-87703-349-8); Soft Cover $35 (ISBN 0-87703-350-1).

Vol. 14 History of Rocketry and Astronautics, Edited by Tom D. Crouch, Alex M. Spencer, 1993, 222p, Hard Cover $50 (ISBN 0-87703-374-9); Soft Cover $35 (ISBN 0-87703-375-7).

Vol. 15 History of Rocketry and Astronautics, Edited by Lloyd H. Cornett, Jr., 1993, 452p, Hard Cover $60 (ISBN 0-87703-376-5); Soft Cover $40 (ISBN 0-87703-377-3).

Vol. 16 Out From Behind the Eight-Ball: A History of Project Echo, by Donald C. Elder, 1995, 176p, Hard Cover $50 (ISBN 0-87703-387-0); Soft Cover $30 (ISBN 0-87703-388-9).

Vol. 17 History of Rocketry and Astronautics, Edited by John Becklake, 1995, 480p, Hard Cover $60 (ISBN 0-87703-395-1); Soft Cover $40 (ISBN 0-87703-396-X).

Vol. 18 Organizing for the Use of Space: Historical Perspectives on a Persistent Issue, Edited by Roger D. Launius, 1995, 234p, Hard Cover $60; Soft Cover $40

Vol. 19 History of Rocketry and Astronautics, Edited by J. D. Hunley, 1997, 318p, Hard Cover $60 (ISBN 0-87703-422-2); Soft Cover $40 (ISBN 0-87703-423-0).

Vol. 20 History of Rocketry and Astronautics, Edited by J. D. Hunley, 1997, 344p, Hard Cover $60 (ISBN 0-87703-424-9); Soft Cover $40 (ISBN 0-87703-425-7).

Vol. 21 History of Rocketry and Astronautics, Edited by Philippe Jung, 1997, 368p, Hard Cover $60 (ISBN 0-87703-439-7); Soft Cover $40 (ISBN 0-87703-440-0).

Vol. 22 History of Rocketry and Astronautics, Edited by Philippe Jung, 1998, 418p, Hard Cover $60 (ISBN 0-87703-444-3); Soft Cover $40 (ISBN 0-87703-445-1).

INDEX

NUMERICAL INDEX

VOLUME 97	**SCIENCE AND TECHNOLOGY SERIES**, *THE CASE FOR MARS V* (1999)
	(Conference held May 26-29, 1993, University of Colorado, Boulder, Colorado)
AAS 93-801	Status Report On Strategic Planning at NASA, Charles Pellerin (Abstract)
AAS 93-802	The Long Goodbye: A Writer's Perspective on the Psychological Imperatives of Space Exploration, Thomas Simmons (Abstract)
AAS 93-803	The Third Era of Space Exploration, Dwayne A. Day
AAS 93-804	Earth And Mars in Historical Perspective, Tom Meyer (Abstract)
AAS 93-805	Is Mars a Solution to the Problem of Exponential Growth on Earth?, Albert A. Bartlett (Abstract)
AAS 93-806	Marketing Mars: A Preliminary Survey of Public Attitudes, David T. Cadden
AAS 93-807	Footsteps To Mars: An Incremental Approach To Mars Exploration, Geoffrey A. Landis (Abstract)
AAS 93-808	Is There a Short Term Economic and Social Justification for Manned Solar System Exploration?, Robert E. Becker
AAS 93-809	The Need for Policy Resiliency in Mars Exploration (or Why Haven't People Landed on Mars Yet), Matthew Fisk Marshall (Abstract)
AAS 93-810	A Commericial Basis for Mars Development, Mike H. Ryan (Abstract)
AAS 93-811	Case For Mars V — Education Outreach and Working Group, Coordinator: Barbara Sprungman (Abstract)
AAS 93-812	Mars Base Placement: Geological Considerations of Candidate Locations, Shawn Laatsch (Abstract)
AAS 93-813	Education Strategies for Management of a Global Biosphere, Dana Richard (Abstract)
AAS 93-814	The Mars Environmental Survey (MESUR) Network and Pathfinder Missions, Richard A. Cook and John B. McNamee (Abstract)
AAS 93-815	MARSNET: A European Contribution to Mars Exploration, A. F. Chicarro (Abstract)
AAS 93-816	MESUR-Based Mars Sample Return Study, Philip C. Knocke, R. Terry Gamber and Benton Clark (Abstract)
AAS 93-817	MESUR Pathfinder Microrover Flight Experiment - A Status Report, Donna Shirley

AAS 93-818 The Mars Aerial Platform Mission: Utilizing Super-Pressure Balloons to Carry Out a Global Reconnaissance of the Red Planet, Robert M. Zubrin, Ben Clark, Steve Price, Terry Gamber and Jim Cantrell

AAS 93-819 AEOLUS - An Unmanned Human Precursor to Mars Using High L/D Atmospheric Entry Systems, Marcus S. Murbach (Abstract)

AAS 93-820 The Nomad Explorer Assembly Assist Vehicle: An Architecture for Rapid Global Extraterrestrial Base Infrastructure Establishment, Madhu Thangavelu

AAS 93-821 Mars Sample Return and Biohazards: A Source of Public Concern and Controversy, Margaret S. Race (Abstract)

AAS 93-822 Relevance and Impact of Planetary Environment Protection on Future Mars Missions, François Spiero

AAS 93-823 Mars Direct: Humans To The Red Planet Within A Decade, Robert M. Zubrin (Abstract)

AAS 93-824 Mars Exploration and Settlement, Bruce Lusignan (Abstract)

AAS 93-825 Mars Exploration Strategies: A Reference Program and Comparisons of Alternate Architectures, Michael B. Duke and David B. Weaver (Abstract)

AAS 93-826 The IAA Mars Cosmic Study, George E. Mueller (Abstract)

AAS 93-827 Being There: The Mission on the Martian Surface from the 1992 NASA Mars Exploration Mission Study, L. Lemke (Abstract)

AAS 93-828 Summary of the Workshop on International Coordination of the Exploration of Mars, Roger D. Bourke (Abstract)

AAS 93-829 Let's Make Mars Exploration Truly International, Louis D. Friedman and John M. Logadon (Abstract)

AAS 93-830 Thirty Years Together: A Chronology of U.S.-Soviet Space Cooperation, David S. F. Portree (Abstract)

AAS 93-831 Power Systems for the First Mars Outpost, Paul S. Gill (Abstract)

AAS 93-832 Power Systems for Mars Exploration, Robert L. Cataldo (Abstract)

AAS 93-833 The Viability of Photovoltaics on the Martian Surface, James R. Gaier and Marla E. Perez-Davis

AAS 93-834 Dust Accumulation on Solar Arrays on Mars: Theory and Instrumentation, Geoffrey A. Landis (Abstract)

AAS 93-835 Effects of Dust on Radiator Surfaces on the Martian Surface, James R. Gaier, Marla E. Perez-Davis and Sharon K. Rutledge (Abstract)

AAS 93-836 Solar Array Corrections for Mars Surface, Dale R. Burger

AAS 93-837 Genetic Reinforcement Learning in a Robotic System for Space Exploration, Robert E. Smith and Lia Noemi Tusanotte Cisnero

AAS 93-838 Russia's Mars Rover, Daphne Burleson (Abstract)

AAS 93-839 Minimum Energy Paths on a Planetary Surface, George W. Morgenthaler and Kevin K. Gifford (Abstract)

AAS 93-840 A "Jumper" for the Mars Surface Exploration, G. Savu

AAS 93-841 Life on Mars: Did Viking Close the Book?, Benton C. Clark (Abstract)

AAS 93-842 Exopaleontology and the Search for a Fossil Record on Mars, Jack D. Farmer and David J. Des Marais (Abstract)

AAS 93-843 The Temporal and Spatial Distribution of Martian Paleolakes, James W. Rice, Jr. (Abstract)

AAS 93-844 Mars Exobiology Sites for Future Exploration, Ragnhild Landheim, Ronald Greeley, Jack D. Farmer, David J. Des Marais and Harold P. Klein

AAS 93-845 The Long-Term Habitation of Mars, Martyn J. Fogg

AAS 93-846 Technological Requirements for Terraforming Mars, Robert M. Zubrin and Christopher P. McKay (Abstract)

AAS 93-847 Paraterraforming: Construction, Energy, Environment and Hazard Control Strategies for a Quasi-Global Martian Worldhouse, Richard L. S. Taylor

AAS 93-848 Overview of Life Science Challenges for Human Exploration, James E. Lynne (Abstract)

AAS 93-849 Radiation Concerns on Long Term Deep Space Missions, Karl A. Pfitzer (Abstract)

AAS 93-850 Is Bone Loss in Space a Problem?, Christopher E. Cann

AAS 93-851 Artificial Gravity as a Countemeasure to the Effects of Spaceflight on Living Systems, Carl Rudiger, Jr. and Phillip A. Wagner (Abstract)

AAS 93-852 Space Station Freedom: A Key Step for a Mars Mission Life Support System, Richard P. Reysa (Abstract)

AAS 93-853 Field Investigations in Antarctica: A Regimen for Manned Mars Mission Astronaut Training Activities, James W. Rice, Jr. (Abstract)

AAS 93-854 Mars Mission Gravity Profile Simulation, Lawrence H. Kuznetz (Abstract)

AAS 93-855 Covered Crater Biome - A Martian Life Support System, Rollin P. Van Zandt

AAS 93-856 Operation of a Real-Time Biological Life Support System Lessons from Biosphere 2, M. Nelson, W. F. Dempster, N. Alvarez-Romo, G. Hudman, T. MacCallum, S. Silverstone and M. Van Thillo (Abstract)

AAS 93-857 Benefits from NASA Life Support Technologies, D. R. Roberts and M. Averner (Abstract)

AAS 93-858 The Natural Cycles: An Overview of a Controlled Ecological Life Support System, D. J. Chamberlain (Abstract)

AAS 93-859 A Life Support System Architecture for the Mars Surface Mission, Andrew A. Gonzales, Lynn D. Harper, Elizabeth Dunsky and Barney B. Roberts (Abstract)

AAS 93-860 A Novel Approach to Waste Remediation in Controlled Environmental Systems Utilizing Phototrophic Bacteria, Angela Asher and Dennis Ray Schneider (Abstract)

AAS 93-861 Systems for Mars EMU Design, George W. Herbert

AAS 93-862 Space Suits and Life Support Systems for the Exploration of Mars, Lawrence H. Kuznetz (Abstract)

AAS 93-863 Passive Cooling of Mars EMU, Frank Crary (Abstract)

AAS 93-864 Martian Ionospheric Density Analysis System (MIDAS), Robert J. Yowell and C. D. Fry (Abstract)

AAS 93-865 Solving the Data Dilemma for Small Missions to Mars, Alan J. Willoughby, Denise S. Ponchak and John E. Zuzek (Abstract)

AAS 93-866 Adaptable Software Needed for Mars Missions, Ned Chapin

AAS 93-867 Project EMA: A Plan for the Exploration of Mars, Philip A. Turek

AAS 93-868 EPIC - An Approach to an Integrated Manned Space Exploration Program, Marcus S. Murbach (Abstract)

AAS 93-869 Spaceship Propulsion By Momentum Transfer, Christopher Willis (Abstract)

AAS 93-870 Mars Mission Designs: Comparing The Near Term Options, Malcolm A. LeCompte and Julie P. Stets (Abstract)

AAS 93-871 Getting to Mars Without a Saturn X, Benton C. Clark (Abstract)

AAS 93-872 A New Decade of Planetary Exploration, A Peace Dividend, Larry Colin and Lew Franklin (Abstract)

AAS 93-873 Zephyr: A Microsat Transfer Orbit Stage, George W. Herbert

AAS 93-874 Mars Exploration Vehicle, Sergi Stoiko and Steve Merrihew (Abstract)

AAS 93-875 Magnetobraking for Mars Return Vehicles, Geoffrey A. Landis

AAS 93-876 Effect of Inert Propellant Injection on Mars Ascent Vehicle Performance, James E. Colvin and Geoffrey A. Landis

AAS 93-877 Nuclear Thermal Rocket/Stage Technology Options for NASA's Future Human Exploration Missions to Mars, Stanley K. Borowski (Abstract)

AAS 93-878 A Study of Powered Abort to Earth Orbits from Earth to Mars Transfer Orbits, Stephen C. Merrihew

AAS 93-879 Relative Benefits of In-Situ Mining on Mars, Paulo R. Pereira, Russel J. Miller and Gary S. Brierley (Abstract)

AAS 93-880 Methanol: A Fuel for Earth and Mars, Kelly R. McMillen, Thomas R. Meyer and Benton C. Clark

AAS 93-881 Mars Missions with ISRU: Available Technologies at the Arizona NASA USERC, Kumar Ramohalli (Abstract)

AAS 93-882 Cold Press Method to Manufacture Construction Materials on Mars, Yoji Ishikawa, Tetsuo Sasaki and Tetsumi Higasayama (Abstract)

AAS 93-883A Mars 2008 Surface Habitation Study: Part III - Safety Philosophy, Marc M. Cohen (Abstract)

AAS 93-883B Mars 2008 Surface Habitation Study: Part V - Mission Design Evaluation Process and Criteria, Marc M. Cohen

AAS 93-884 EVA Surface Mission Strategies for Mars Exploration, Theresa M. Buckley (Abstract)

AAS 93-885 Proposal for a First Martian Outpost and Permanent Martian Base, Gary T. Moore, Janis Huebner-Moths, Kerry L Paruleski and Patrick J. Rebholz (Abstract)

AAS 93-886 Construction Method and Procedure for Early-Phase Mars Base, Yoji Ishikawa, Owen Gwynne, Yukinobu Yamamoto and Hisashi Uyeda

AAS 93-887 Design and Construction for a Permanent Presence on Mars, Ethan W. Cliffton

AAS 93-888 Inflatable Structures for a Martian Base, Willy Z. Sadeh and Marvin E. Criswell (Abstract)

AAS 93-889 Living on Mars, Michael Scott and Lee Weaver (Abstract)

AAS 93-890 Mars Base Placement: Geological Considerations of Candidate Locations, Shawn Laatsch (Abstract)

AAS 93-891 Over-the-Horizon Communications on Mars via HF Radio Propagation, C. D. Fry and R. Yowell

AUTHOR INDEX*

* For each author the paper number is given. The page numbers refer to Volume 97, AAS *Science and Technology Series*.